中国植物病害化学防治研究

（第十卷）

周明国　主　编

刘西莉　刘　勇　陈长军　副主编

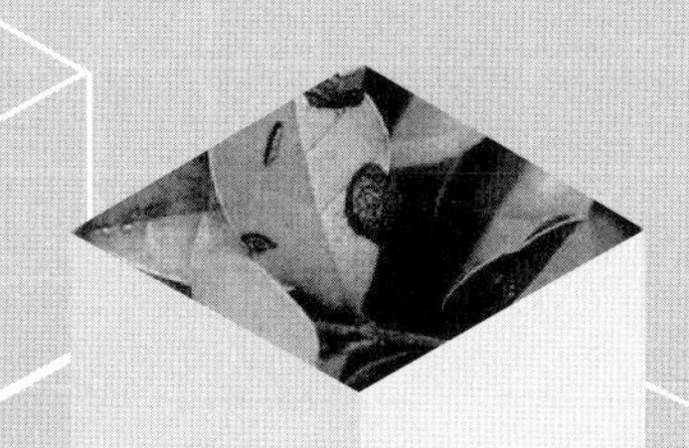

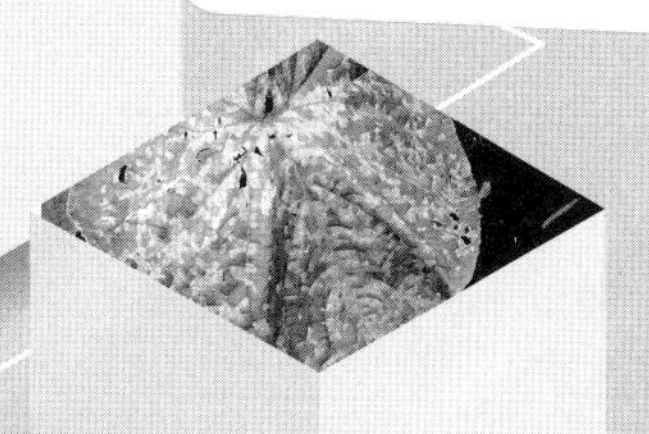

中国农业科学技术出版社

图书在版编目（CIP）数据

中国植物病害化学防治研究．第十卷／周明国主编．—北京：中国农业科学技术出版社，2016.10
ISBN 978－7－5116－2770－4

Ⅰ.①中… Ⅱ.①周… Ⅲ.①病害－农药防治－研究－中国
Ⅳ.①S432

中国版本图书馆 CIP 数据核字(2016)第 237499 号

责任编辑 姚 欢
责任校对 贾海霞

出 版 者 中国农业科学技术出版社
北京市中关村南大街 12 号 邮编：100081
电 话 (010) 82106636 (编辑室) (010) 82109702 (发行部)
(010) 82109709 (读者服务部)
传 真 (010) 82106631
网 址 http：//www.castp.cn
经 销 者 各地新华书店
印 刷 者 北京富泰印刷有限责任公司
开 本 787mm×1 092mm 1/16
印 张 26.5
字 数 600 千字
版 次 2016 年 10 月第 1 版 2016 年 10 月第 1 次印刷
定 价 70.00 元

《中国植物病害化学防治研究》

编　委　会

主　编　周明国

副主编　刘西莉　刘　勇　陈长军

编　委　(以姓氏拼音为序)

陈福如　陈绵才　高同春　郭井泉

李　明　李明立　梁桂梅　刘君丽

陆　凡　陆悦健　马忠华　区越富

时春喜　司乃国　宋玉立　王文桥

吴新平　徐大高　赵廷昌

中国植物病理学会化学防治专业委员会
第三届委员会组成名单

前　言

《中国植物病害化学防治研究》（第十卷）与以前出版的九卷共同组成了中国植物病理学会化学防治专业委员会编辑出版的系列论文集。该专业委员会自1998年成立以来，已经成功举办了十次全国性学术研讨会和多次小型学术活动，还开展了有关的科普宣传、科学考察和咨询服务；为我国广大植物病害防治科技工作者，特别是常年奋战在基层的科技工作者提供了学术交流、展示和了解国、内外最新科研成果的舞台，为推动和促进我国植物病害可持续防控和科技进步发挥了积极作用。本书共汇编了参加第十届中国植物病害化学防治学术研讨会84篇论文，充分反映了近两年来我国农药化工和植物保护科技工作者最新的科研成果和国际上植病防控的最新科技成果。

当今科技日新月异，人民生活水平稳步提高，现代农业生产正从追求粮食和食品生产数量放在首位的传统观念向追求“无公害食品”、“绿色食品”和“有机食品”改变。但是，应该清醒地认识到现代农用化学品不仅在保证生产足够数量的粮食和食品，满足不断增加的人口对食品的需求方面发挥了不可替代的巨大作用，而且科学使用现代农用化学品可以改善和提高农产品的质量。众所周知，罹病的农产品往往伴随品质下降，有的还因为病原微生物产生毒素而导致食用后的中毒事故，如小麦赤霉病菌产生的DON类等毒素。大量研究已经证明现代杀菌剂具有高效、低毒、低残留的特点，不仅能够有效防治多种植物病害、减少产量损失，而且能够调节植物生长、延缓植物衰老、增强光合作用、提高农产品的品质。毫无疑问，如果滥用现代杀菌剂，不仅会导致高残留，引发粮食和食品质量安全问题，而且抑制作物生长，甚至破坏农业可持续发展的生态环境，最终导致农产品的产量下降。因此，杀菌剂减施增效技术研发是植物病害可持续防控的关键！

中国植物病理学会在本次会议的筹备过程中给予了多方面的支持和指导，中国植物病理学会化学防治专业委员会第十届学术研讨会由南京农业大学植物保护学院承办，尤其是南京农业大学农药系陈长军教授、侯毅平副教授和匡静女士在筹备和承办这次会议中为会议的成功召开付出了辛勤劳动，在此一并致谢。

本书的编者和审稿人员仔细阅读了全部来稿，并对部分论文进行了删减和修改，部分论文由于内容不符合本次会议要求或其他原因未能录用，敬请谅解。由于时间仓促，书中仍然存在不少疏漏和错误，望读者和作者批评指正。

周明国

二零一六年九月

目 录

协同防控、减量用药，提高中国植物病害化学防治科技水平

周明国
（中国植物病理学会化学防治专业委员会，南京 210095）

18年前，在卓有远见的中国植物病理学家、前中国植物病理学会理事长曾士迈院士和裘维蕃院士、前秘书长唐文华教授，以及长期从事植物病害防治研究的叶钟音教授、刘国镕、季良、李明远研究员等老一辈科学家的呼吁、关心和支持下，在许多中青年学者勤奋工作和积极努力下，1998年12月中国植物病理学会化学防治专业委员会（以下简称化学防治专业委员会）在南京隆重成立，成为中国植物病理学会第11个专业委员会，挂靠南京农业大学植物保护学院。并于2000年得到中华人民共和国民政部注册登记，成为全国性群众学术组织，可以在全国范围内开展学术交流、理论研究、宣传普及及科学考察、国际合作、书刊编辑和咨询服务等。

化学防治专业委员会是适应时代发展而诞生的学术团体。改革开放以后，我国经济和科学技术得到快速发展，农业向着机械化、集约化、商品化和高效化方向变革，化学农药在科学技术驱动、市场经济推动、生产需求拉动和新型杀菌剂发展的带动下，快速发展成为我国植物保护不可或缺的重要武器之一。同时，植物病害化学防治在我国长期没有能够得到足够重视，科研立项和研究队伍人员少、经费短缺，几乎没有相应的学术活动，缺少植物病害化学防治的科普和培训项目，广大民众缺乏植物病害化学防治的基本知识。为了尽快改变中国植物病害化学防治科技落实的面貌，化学防治专业委员会从成立开始就在中国植物病理学会的领导下，针对国家和社会需求，开展理论与实际相结合的科学研究，开展学会赋予的各种学术活动。各位委员深入基层调查研究，从生产实践中发现问题、研究问题和解决问题，加强与有关部门的联系，积极争取科研立项，踏踏实实地辛勤工作，不仅发表了大量能够与国际同行交流的SCI论文，而且还通过各种方式展示了许多能让广大民众看得懂、可转化的研究成果，提供了大量用得着、解决生产问题的先进技术。18年来，化学防治专业委员会在学术交流、科学研究、人才培养、科学知识宣传普及与咨询服务和国际合作等方面开展了大量工作，解决了许多生产实际问题。

1 18年的主要工作回顾

1.1 学术交流

自化学防治专业委员会成立并于1998年12月在南京举办“第一届中国植物病害化学防治学术研讨会”以来，按每两年一届至今已经先后在安徽黄山、湖南张家界、贵州贵阳、内蒙古呼和浩特、辽宁沈阳、海南海口、陕西西安、河南郑州成功举办了第二至第九届全国性学术研讨会，现在又如期在江苏南京举办第十届中国植物病害化学防治学术研讨会。其中2008年在沈阳召开的第六届学术研讨会上，还及时总结和回顾了化学防治专业委员会10年的发展历程。“中国植物病害化学防治学术研讨会”已经成为定期举行的全国性植物病理学专业学术会议，成为中国植物病理学会学术活动的重要组成部分。

化学防治专业委员会积极鼓励和吸引了与植物病害化学防治有关的各个领域的代表踊跃

投稿参加学术交流，尤其是来自生产第一线的植保技术推广和应用部门、农药科研和生产及流通领域的代表参加学术交流，畅谈他们的研究成果和遇到的问题，牵手产、学、研、政、企合作。每届参加会议的学者均在130名以上，累计达到1 800多人次。

除了举办定期的全国性学术研讨会以外，化学防治专业委员会还不定期举办了许多专题学术研讨会。例如与全国农业技术推广服务中心及有关省市植保站合作，先后多次举办了植物病害化学防治高级论坛、杀菌剂安全使用、蔬菜病害化学防治、水稻病害化学防治、小麦病害化学防治、小麦赤霉病菌抗药性治理等专题研讨会。此外，还先后与先正达公司合作，在广州、三亚和南京举办了阿米西达新型杀菌剂学术研讨会；与巴斯夫合作在南京举办了中－德农业有害生物抗药性学术研讨会；与浙江新农化工股份有限公司合作在南京举办了植物细菌病害化学防治研讨会；与河北冠龙农化有限公司合作举办了植物病害化学防治论坛等学术活动。

为了表彰广大学者积极参加植物病害化学防治学术活动，中国植物病理学会化学防治专业委员会与BASF（中国）股份有限公司合作，自第六届学术研讨会开始设立了“BSAF-中国植物病害化学防治学术活动积极分子奖”，先后已经为41名代表颁发了纪念奖品。与先正达（中国）股份有限公司合作从第三届学术研讨会开始，对参加会议交流的论文设立了“中国植物病理学会化学防治专业委员会/先正达优秀论文奖”，先后有282名代表获得了一、二、三等奖，颁发了获奖证书及奖金。显著提高了会议交流论文的学术水平，许多基层学者还因在中国植物病害化学防治学术研讨会上进行学术论文交流或获奖晋级升职，培养了一批植物病害化学防治人才。与中国农业科学技术出版社合作，将参加每届中国植物病害化学防治学术研讨会交流的论文出版了论文集《中国植物病害化学防治研究》第1～10卷。

1.2　科学研究与人才培养

化学防治专业委员会成立以后，加强了与各部门的联系，积极参与有关科研项目立项指南的编写，例如参与教育部、科技部、中国科学院、国家自然科学基金委组织的10 000个科学难题的编写，参与中国科协和中国植物保护学会组织的《植物保护学科发展报告》中的“农药学科发展报告”编写，参与有关杀菌剂田间药效试验准则、杀菌剂抗性风险评估及植物病原菌抗药性监测等行业标准制定和审定等。植物病害化学防治及杀菌剂抗性研究先后分别在国家农业科技攻关/支撑计划、863计划、973计划、自然科学基金、行业（农业）科研专项、重点研发计划等项目中立项，壮大了植物病害化学防治研究队伍，培养了大量研究生和高水平人才，许多科技人员成为所在单位的科技骨干和领军人才，获得各级政府的荣誉称号，例如副主任委员刘勇研究员入选湖南省“121”第一层次人才，农业部杰出人才和产业体系岗位科学家，科技部创新团队负责人和中组部万人计划科技创新人才；副主任委员马忠华教授入选农业部杰出人才和岗位科学家，中组部万人计划科技创新领军人才，获得国家杰出青年基金；刘西莉教授入选教育部新世纪人才计划，农业部杰出人才，国家中青年科技创新领军人才，中组部万人计划科技创新领军人才等。

1.3　国际合作研究与交流

为尽快提高我国植物病害化学防治的学术水平，化学防治专业委员会十分注重国际科研合作和国际学术交流。与先正达（中国）股份有限公司合作，2002年设立了“植物病害化学防治创新技术”研究基金，从全国39份申请书中，评选出12项课题开展研究，这些研究课题包括了新型杀菌剂阿米西达在蔬菜、水果、烟草、马铃薯等作物上的使用技术，病原菌抗药性风险评估，杀菌剂对作物的安全性，杀菌剂对农产品产量和质量的影响等内容，取得了一批在生产上可以直接使用的新技术和对杀菌剂推广及使用具有指导意义的新成果。其中阿米西达对嘎啦苹果和云烟的植物毒性研究成果，对后来 QoIs 类杀菌剂的市场开发和科学应用发挥了重要指导性作用。2004—2007 年与巴斯夫合作设立了羧酸酰胺类（CAAs）杀菌剂抗性合作研究项目，研究了中国主要植物病原卵菌对 CAA 类杀菌剂的抗药性风险，为这类杀菌剂的推广应用提供了科学依据。

自第三届中国植物病害化学防治学术研讨会以来，还邀请了外国学者参加全国性学术研讨会和专题研讨会进行学术交流 30 多人次。化学防治专业委员会组织了 2004 年在北京召开的第 15 届国际植保大会上的杀菌剂抗性专题研讨会，2013 年在北京召开的第 10 届国际植物病理学大会上的植物病害化学防治专题研讨会。2006 年 10 月在南京举办了中 – 德农业有害生物抗药性学术研讨会，邀请了 4 位德国和英国专家和 4 位中国专家作报告，与会代表达到 160 多人。2013 年与美国植病学会合作，在奥斯汀美国植病学会和真菌学会年会上组织了中国专题学术交流会，报告了中国小麦赤霉病菌抗药性研究进展。与先正达合作，2005 年评选了 12 名中国植物病理学会化学防治专业委员会科技先进工作者，组织了两个访问代表团共 20 人，参观访问了先正达在亚太地区和欧洲地区的研究机构，并相互进行了学术报告交流。此外，化学防治专业委员会主任委员周明国教授单独或带队应邀于 2003 年 3 月赴先正达在英国和瑞士的生物研究中心学术访问交流、应邀参加日本 2004 年第 14 届杀菌剂抗性学术年会学术交流，访问了日本农业环境资源研究所和全营农实验站杀菌剂研究室，2006 年应邀在美国 232 届国际农药化学学术大会和 2008 年都灵第九届国际植病大会上做特邀学术报告，并顺访德国 BASF 公司。2007 年应邀访问了加拿大植病学会，2007 年和 2014 年应邀访问了美国斯坦福大学、哈佛大学、纽约大学、麻省理工学院、弗吉尼亚理工学院和密歇根州立大学等，2016 年应邀访问德国马普微生物研究所和参加了在图林根召开的第 18 届国际杀菌剂学术研讨会，并分别做了杀菌剂选择性分子靶标微管蛋白和肌球蛋白研究进展的主题报告。还有许多化学防治专业委员会委员先后应邀赴国外访问交流、进修、合作研究和参加国际会议，并进行学术报告。此外，化学防治专业委员会也邀请了许多外国同行专家来华访问交流，如 1998 年和 2008 年邀请了著名的英国杀菌剂抗性专家 Hollomon D 博士、2013 年邀请意大利真菌抗药性遗传学专家 Faretra F 教授、美国 Trail F 教授、2015 年邀请美国科学院院士 He S. Y 教授等来华访问交流和指导研究生的科研活动。先后选派多名优秀博士研究生赴国外高水平实验室，联合培养了一批从事植物病害化学防治和抗药性研究方向的人才。通过各种形式的国际交流，建立了广泛的合作渠道，推动了我国植物病害化学防治科技水平的不断进步。

1.4　学术调研与技术示范

化学防治专业委员会以了解生产需求、解决实际问题为己任，组织专家和研究生走进农户，深入田头，寻找和调查重大作物病害化学防治问题及解决问题的方法。例如通过调查和田间试验发现 2001—2002 年江苏一些多年使用咪鲜胺进行水稻种子处理的地区突然暴发水稻小穗、翘穗现象，是水稻干尖线虫病在新的品种和栽培条件下出现的新症状，通过推广使用高效广谱杀菌杀线虫剂“二硫氰基甲烷”代替“施宝克”进行种子处理，完全解决了生产问题。

针对油菜菌核病的毁灭性危害，2005 年起先后与江苏省通州市和姜堰市植保站合作，研究发现病害大面积流行原因是由于抗药性造成的，研发了一系列替代多菌灵的杀菌剂及其配套使用技术，取得了显著效果。

2008 年与黑龙江省农垦局植保站合作，对黑龙江省稻瘟病和褐变穗等水稻上重大病害发生状况进行了调研，提出了建设性解决方案。

针对小麦赤霉病的发展态势，与农业部种植业司植保植检处、全国农业技术推广服务中心药械处和有关省市植保站合作，连续监测了我国小麦赤霉病菌对多菌灵的抗药性群体发展态势，2006年以来系统研究并证明赤霉病菌抗药性会提高真菌致病力和大幅度增加毒素污染，2010年和2012年组织研究生对江苏省28个地区进行了用药情况及抗药性对小麦产量影响的系统调查，先后向江苏省委省政府、农业部、全国人大、全国政协、国务院和科技日报内参，反映了抗药性赤霉病对粮食安全和食品安全的严重威胁，得到了各级政府的高度重视。其间与江苏省农药研究所股份有限公司合作，成功研发和大面积推广应用了治理多菌灵抗性、高效防治小麦赤霉病的氰烯菌酯和氰戊复配制剂，得到了全社会的高度评价。研发的生物-化学协同高效综合防控小麦赤霉病、锈病和白粉病的一系列减量用药技术，经过连续4年多地试验和大面积示范，证明用药量之少、效果之高处于国际同类技术产品的最高水平，目前已经进入成果转化阶段。

1.5 科技宣传、普及与培训

化学防治专业委员会委员在各自单位工作任务繁重和没有任何专门经费和不计工作量考核的情况下，努力与各方面合作，积极开展相关的科技普及和培训工作。如在化学防治专业委员会成立早期，副主任委员顾宝根副所长和梁桂梅副处长，利用自己的工作特点和优势，不仅在各种场合开展了科技普及和科技培训工作，还与专业委员会合作，多次在召开的全国性农药大田药效试验工作会议和植保工作会议上，邀请化学防治专业委员会专家进行杀菌剂抗性、杀菌剂使用技术、杀菌剂安全性、蔬菜生产安全用药、植物病害化学防治研究进展等科学知识普及和培训，还在南京农业大学多次对全国植保技术推广系统的科技人员代表进行杀菌剂抗性监测技术培训。同时，化学防治专业委员会还与山东、江苏、河南、上海、黑龙江农垦等省、市、系统植保站、植病学会合作，对植保技术推广人员进行了类似内容的培训，与先正达公司、江苏农药研究所、西大华特、陕西上格、绵阳利尔、浙江新农、河北冠龙、江苏艾津等企业合作，在江苏、上海、安徽、河南、湖北、四川、陕西、广东、海南等地对种田大户和基层植保技术人员进行了杀菌剂生物测定技术、杀菌剂抗性检测技术、应用技术等培训。18年来，以各种方式进行技术普及和培训的人员超过1万人次。

在信息交流不够发达的化学防治专业委员会成立初期，专业委员会通过业余时间先后编印了7期20多万字的《中国植物病害化学防治快讯》内部资料，免费向全国各级植保技术人员寄送了近两万份快讯，受到了各界一致好评，发挥了巨大社会效益。最近又建立了中国植物病理学会化学防治专业委员会网站（www. ccccspp. org），可以通过该平台进行信息交流和回答及指导生产解决问题。

正是通过这些面向全国植保和农药系统的科技普及、宣传和培训，推广二硫氰基甲烷替代多菌灵，有效治理了我国水稻恶苗病菌抗药性问题；证明2002—2003年水稻生长后期使用烯唑醇是造成近千万亩水稻不抽穗的药害问题，停止了烯唑醇在水稻上的使用；证明了三唑酮防治黄瓜白粉病存在隐性药害问题，并通过宣传培训和试验示范停止了农民近20年在黄瓜上使用三唑酮的习惯，终止了使用多菌灵和洗衣粉防治黄瓜霜霉病及双氧水防治蔬菜病害的现象；解决了江苏水稻因干尖线虫病造成的大面积严重减产问题。

1.6 组织建设

18年来，通过有关部门、单位和专家推荐与协商，考虑便于工作和委员在各区域各部门的分布，及征求本人意见，确定委员候选人，并在候选人中进行投票选举，报请中国植物病理学会理事会批准的程序。1998年化学防治专业委员会成立以来，进行了三次专业委员会的改选。随着化学防治专业委员会大量工作的开展和植物病害化学防治科技队伍的迅速壮大，化学防治专业委员会从第一届的16人、第二届的21人，第三届的23人，增加到现在改选的第四届34人，并成立了6个工作组。经申请、审核和中国植物病理学会批准，第一、二届化学防治专业委员会还先后吸收了10家企业团体会员。提高了这些企业对植物病害化学防治和杀菌剂的学术和技术的重视程度。

自化学防治专业委员会成立以来，各位委员和专家学者团结一致，热心奉献，积极开展大量学会工作，为社会做出了有目共睹的贡献，受到了植物保护领域的广泛赞扬。化学防治专业委员会先后收到10多封来自全国人大代表、退休的植保专家和植保站在职人员的赞扬和鼓励信件，化学防治专业委员会被誉为中国植物病理学会中最有活力的专业委员会之一，本人于2006年还被评为中国科协先进工作者，这也是对化学防治专业委员会全体委员努力开展工作的充分肯定。

2　面临的挑战和任务

随着植物病害化学防治水平的提高，新型杀菌剂不断涌现和用量不断增加，使一些重大植物病害如小麦锈病和白粉病的流行得到有效控制，一些高产优质但比较感病的稻麦新品种得以大面积推广种植，为保障国家粮食安全作出了重大贡献。但是，我国幅员辽阔，气候多变，作物种类和品种多样，病害种类繁多，东西南北经济和科技发展水平及农药应用水平差距甚大。在农药应用总量零增长和建设绿色中国、生态农业、美丽家园的新形势下，化学防治专业委员会依然面临着许多新的挑战和任务。在此我提出以下4点建议。

（1）坚持“百花齐放、百家争鸣”的方针和“实事求是、理论联系实际”的科学发展观，倡导“学术民主、学术自由”的思想。希望大家团结在新一届化学防治专业委员会的周围，积极响应和开展形式多样的学术活动，推动学会的发展，提高化学防治专业委员会的影响力。

（2）发挥化学防治专业委员会的职能，履行开展学术交流、理论研究、宣传普及科学知识、科学考察、国际合作、书刊编辑和咨询服务的职责。尤其是开展学术交流、宣传普及科学知识和咨询服务，需要我们具有为学会发展和社会服务的奉献精神。我们不仅要把定期举办的中国植物病害化学防治学术研讨会和不定期举办的专题学术活动坚持下去，而且办得更好更有特色，力争成为中国植物病理学会的品牌学术活动之一。

（3）适应时代发展需求，发挥化学防治专业委员会联系产、学、研、企、政的桥梁作用，为三农服务，把化学防治专业委员会网站办成信息交流、科学知识普及和咨询服务的平台。群策群力，集思广益，为提高我国广大民众对杀菌剂及植物病害化学防治的科学认知，促进植物病害化学防治学科发展和产业发展，推动我国植物病害化学防治的科技进步，创新性地开展化学防治专业委员会的工作。

（4）瞄准科学发展前沿，探索超高效、特安全的新型杀菌剂创制理论和实用新技术。例如突破干扰能量形成（抑制孢子萌发）、生物合成（抑制生长）和诱导寄主抗性的杀菌剂研发理论与技术框架，采用基因编辑技术，创新性地研发干扰基因表达、RNA调控及致病力的新型杀菌剂或核酸杀菌剂及转基因杀菌剂；探索靶标组学理论，研发不易产生抗性的高效新型杀菌剂等。研发协同增效、减量用药、综合防控植物病害的新理论和新技术。在不断提高作物病害化学防治效果、扩大化学防治病害种类、提高欠发达地区的病害化学防治水平情况下，通过推广应用高效新型杀菌剂和科学用药，确保杀菌剂用量零增长。为建设美丽家园、生态农业和绿色中国贡献力量。

设施蔬菜病虫害防治问题和农药减施技术研究应用*

王文桥**
（河北省农林科学院植物保护研究所，保定　071000）

摘要：中国设施蔬菜病虫害发生呈现新的特点，化学农药被滥用、低效使用或过量施用的现象时有发生，政府已高度重视农药过度使用带来的问题，开始启动农药减量行动，2015年设立重点研发专项“化肥农药减施增效综合技术”，农业部2015年提出“到2020年化肥、农药使用量零增长”方案。2016年正值重点研发专项实施方案制订和初步实施阶段，很多研究思路尚不清晰。笔者于2016年在《中国蔬菜》杂志上对中国设施蔬菜农药减施增效提出了展望，本文针对中国设施蔬菜病虫害化学防治中存在问题，对比国内外农药减施技术应用研究现状，提出了设施蔬菜减施农药提高农药利用率的途径和农药减施技术与化肥减施技术和高产栽培技术集成应用对实现设施蔬菜生产增效必不可少的观点，对农药减施增效技术模式集成应用进行了再思考，供读者参考。

1　中国设施蔬菜病虫害防治现状及存在问题

中国设施蔬菜产业发展很快，对于保障中国四季蔬菜持续供应和调整蔬菜品种结构意义重大。北方蔬菜以日光温室栽培和塑料大棚栽培为主，南方蔬菜以中小拱棚栽培为主，2014年中国设施蔬菜种植面积已超过380.0hm^2（5 700万亩）。相对露地蔬菜，设施栽培可实现蔬菜周年生产；而棚室高湿，轮作倒茬困难，更容易发生病虫害，更加依赖化学防治，使农药投入量大幅增加，导致如“毒豇豆”“毒韭菜”“膨大剂西瓜”等事件的发生，西瓜、草莓、甜瓜被提前采收再靠植物生调节剂催熟而口感下降（不甜）的传闻已在社会上造成很大负面影响，消费者对产自棚室的蔬菜产品愈加忧心，担心农药过量施用引起中毒，人体健康受到不良影响。

1.1　设施蔬菜病虫害发生特点及防治现状

（1）黄瓜/莴苣/白菜霜霉病、黄瓜/番茄/辣椒/茄子/韭菜/莴苣灰霉病、瓜类作物/辣椒白粉病、番茄叶霉病、番茄晚疫病、番茄早疫病、黄瓜/辣椒疫病、黄瓜/辣椒炭疽病、菜豆锈病等气传病害频繁发生，主要依靠化学防治进行防控。但病原菌繁殖快，极易变异而对防治药剂产生抗性，并导致作物品种抗病性丧失。尽管已有一些抗病或耐病品种被培育并利用，借助棚室放风降湿和促根壮秧栽培可减轻气传病害的发生，但生产中仍很缺乏兼抗病虫的优良品种。因此，棚室在一个生长季节内仍需多次喷药防治，有些长季节蔬菜一个生长季节甚至要用近20次化学农药，阴雨天棚室不宜防风降湿，菜农往往用药熏蒸防病，显著提

* 基金项目：国家重点研发计划“设施蔬菜化肥农药减施增效技术集成研究与示范－环渤海暖温带区设施蔬菜化肥农药减施技术模式建立与示范”2016YFD0201006

** 作者简介：王文桥，研究员，从事杀菌剂应用技术、植物病原菌抗药性和植物病害化学防治与综合防治研究；E-mail：wenqiaow@163.com

高棚室中农药用量。

（2）由于连年种植，加之田间引入带菌带虫种苗和苗土，棚室土壤中病原菌长年累积，致使根系微生物环境恶化，蔬菜根结线虫病、黄瓜疫病、黄瓜枯萎病、茄子黄萎病、黄瓜/辣椒根腐病、茎基腐病、菌核病、立枯病、猝倒病等土传病害或苗期病害严重发生，而且往往根结线虫与镰刀菌、丝核菌、腐霉、疫霉等病原物并存，而黄萎病、枯萎病等土传病害也是系统侵染性病害。因病原物的构成复杂而引起的病害多样而很难防治，依靠施用一种杀菌剂往往难以奏效，还必须用杀虫剂防治线虫病，用五氯硝基苯、多菌灵、甲基硫菌灵、敌克松、甲霜灵混剂等土壤处理剂拌土、撒施或喷淋浇灌处理土壤或植物茎基部或根系处理及咯菌腈、多菌灵、精甲霜灵混剂等化学药剂进行拌种、蘸秧及种苗处理均难以获得很好的防治效果，而且用药量很大；而通过种植抗病品种、合理轮作倒茬、施用生物有机肥或生防菌剂改善土壤环境或修复土壤，再结合化学药剂处理土壤可提高土传病害或苗期根部病害的防治效果，同时减少化学药剂的投入量。

（3）种子远距离调拨及消毒检疫不严导致黄瓜黑星病、黄瓜蔓枯病、辣椒疮痂病、黄瓜细菌性角斑病、黄瓜靶斑病、番茄细菌性溃疡病、茄科蔬菜青枯病等种传病害蔓延。加强种子检疫，防止带病种子转运至非疫区，播种包衣种子或种子用药剂处理后再播种，生长期用药喷施处理植株地上部分，是防治种传病害的有效途径。

（4）粉虱、蚜虫、蓟马、潜叶蝇等小型害虫，繁殖快，世代重叠，发生危害严重，还能传播诸如黄化曲叶病毒、斑萎病毒等危害极大的蔬菜病毒病，需在一个生长季节多次喷药防治或采用烟熏剂熏蒸处理才能控制。培育利用抗病品种，采用防虫网室培育无毒苗，在棚室喷施高效杀虫剂切断虫传毒途径是控制虫传病毒病发生的有效途径。但目前生产上大多蔬菜品种不抗病毒病，采用防虫网室育苗结合喷施杀虫剂控制传毒昆虫种群成为主要的控毒方法。

（5）黄瓜靶斑病、番茄灰叶斑病和番茄叶霉病等次要病害上升为主要病害，番茄黄化曲叶病毒和番茄斑萎病毒病等新病害不断出现，生产上缺乏很有效的防治药剂和防治措施，迫切需要解决，呼唤着减少对化学药剂的依赖提升作物收益的有效综合治理对策和技术的到来。

（6）生产中缺乏高效的施药器械及与之配套的施药技术和农药剂型，烟雾机、高效喷粉机等施药器械还未能在生产中大量使用，有利于将棚室湿度控制在较低范围内，避免像喷雾处理带来的增加棚室湿度的问题出现，因此有利于棚室病害的控制。采用背负式喷雾器喷雾施药仍是防治病虫害的主要防治手段，导致农药的大量流失，农药利用率低，防治效果大打折扣，大量农药渗进土壤，污染地下水，恶化土壤结构和土壤环境。

（7）尽管已有有害生物综合治理、绿色防控的原则和方法，但很多单项绿色防控技术未被很好地集成，而一些高效低毒的化学农药的要么被低估，要么被过度依赖，未被很好地组装到以减施化学农药和促进作物增产增效为目标的综合防治体系中，生产中缺乏标本兼治的病虫害防治技术和防治理念，仍然停留在为治病治虫、为治病治虫、见病治病、见虫治虫、单病单虫单治的分散、治标管理阶段，忽视作物的健身栽培和高效栽培管理。而生产中呼唤着以作物为中心，以增强作物对病虫害危害等逆境胁迫的抵抗力的健身栽培为前提，以减施化学农药和促进作物增产增效为目标、适应现代设施蔬菜水肥药一体化管理的节肥减药高效栽培模式，病虫害的治理不能以牺牲作物的收益增加和环境保护为代价。

1.2 化学防治的作用及存在的问题

化学防治仍然是控制设施蔬菜作物病虫害的有效手段，是其他防治方法难于替代的，而实际中生态调控、生物防治、物理防治及农业防治往往被轻视甚至忽视。农药对保障蔬菜丰收和提质增效必不可少，化学农药具有使用见效快、药效发挥稳定等特点，可应急使用控制重要病虫害暴发成灾。为了获得较好或较快的防治效果，农民往往过量使用化学农药（杀菌剂和杀虫剂）作喷雾、熏蒸、蘸花处理蔬菜地上部分或土壤消毒防治蔬菜地上部病虫害、土传病害和根结线虫。此外，为了让瓜果蔬菜早上市，增加产量，大量的植物生长调节剂（激素）用作蘸花催熟，但往往带来的是蔬菜品质下降和农药的过度使用。

化学农药被过度依赖而滥用，低效使用或过量施用，导致农产品及生产环境中农药残留超标而影响农产品安全及环境受污染，有害生物对很多常用药剂产生抗性，蔬菜品质下降，使人畜健康受损害，消费者缺乏安全感；农药过量施用还导致病虫害再猖獗，土壤中有益微生物群落下降及有害微生物积累泛滥。而抗药性普遍发生又会引起药剂使用寿命缩短，药效变差，增加施药次数和用药剂量，提高用药成本及农药企业对化学药剂的开发风险和成本。不同靶标菌或害虫对单一药剂的抗性产生及同一病原菌或害虫同时对多个防治药剂出现抗性（多抗）的产生对化学防治提出更大的挑战，农药的应用迫切需要抗药性治理技术的实施和寻求更为有效的非化学防治措施，而不能过度地依靠化学防治，不能以牺牲环境为代价来使用化学农药。

生物防治、生态调控、物理防治等非化学防治在设施蔬菜病虫害防治中变得越来越重要，但也存在着一些问题难以克服，难以完全取代化学农药的使用。例如，枯草芽孢杆菌、木霉、BT、白僵菌、绿僵菌等含有活菌的生物农药药效发挥较慢，受环境条件的影响较大，药效不稳定。为了追求高产，农民往往乐于采用大水大肥的栽培措施提高产量但利于病害发生，环境调控的控病作用发挥得较慢。通过设立防虫网减轻虫传病毒发生，采用色板、性诱剂和黑光灯诱杀防虫往往药效较差，防治对象较少。化学防治中还存在着过于追求防治效果和产量、轻视投入产出比及生态效益评估、忽视农药过量施用带来的环境压力、重化学防治轻综合防控技术的集成应用、缺少农药减施增效技术模式等问题，还普遍存在着重视化学药剂开发利用、轻视生物农药及能提高农药利用率的高效施药器械的研发利用、忽视药械利用及病虫预测预报在精准用药及药剂减施技术中应发挥作用等现象。

随着人们环境保护意识的增强及对农产品质量安全越来越重视，对作物有害生物实行绿色防控，对农药和化肥实行减施增效成为紧要任务。如何将生产中化学药剂的用量降下来，成为当前的重要课题，进行设施蔬菜农药减施增效研究已变得必不可少。

2 国内外农药减施技术应用研究现状

如何正确理解农药减施增效很重要。狭义地讲农药减施增效就是在确保防治效果不下降或者增加的前提下少用药。广义地讲农药减施增效就是既要减少用药又要达到作物增产、农业效益增加的目的，还要确保产生良好的生态效益和社会效益。

2.1 国外农药减施技术应用现状

欧美、日本、以色列等发达国家有机或绿色蔬菜和果品受到消费者的青睐，蔬菜或园艺作物产业发展得很快，蔬菜行业协会发达，提供给农民专业化防治知识和技术培训指导，对农产品质量实行跟踪管理，重视蔬菜或园艺作物农药减施增效技术的研发应用、农产品安全性及农药对农产品安全性及环境的影响、环境友好型或绿色农药的研发及精准减量使用、生物农药与化学农药的协调使用、病虫害抗药性监测与治理。对蔬菜病虫害主要采取以生物防治和生态调控为主的绿色综合防控模式，减少高达50%的农药用量，使生产出的蔬菜产品达到绿色等级。采用一些适合棚室使用的剂型配合烟雾机、多喷头喷雾机等高效施药器械施药，棚室采取自动通风降湿控光、膜下滴灌降湿、高透光杀菌覆膜降湿增光，使用环境友好型生物农药，应用水肥药一体化控病防虫等先进施药技术。通过实施蔬菜或园艺作物的流水式、自动化程度较高的标准化健身栽培管理、有效调控设施内温度湿度光照等与病虫害发生直接相关的环境因素、限制某些化学农药的使用或禁止其超标使用、设定农药限量标准、提倡施用生物农药或释放昆虫天敌、突出生物防治、生态调控、物理防治和农业防治等非化学控制技术的作用、注重病虫害的预防等途径而对作物病虫害进行高效综合防控，减少化学药剂的投入，增加蔬菜园艺作物的收益。

美国、加拿大、荷兰、丹麦、瑞典等欧美国家和澳大利亚等发达国家农业普遍实施规模化和标准化种植管理模式，加强对病虫害的预测预报，在预测预报的基础上，采用一些高效低风险农药产品或生防产品，对小麦、马铃薯等大田作物、苹果、葡萄等果树作物病虫害普遍实施高效精准化学防治、专业化防治和统防统治。

国际上主要农化工业组织设有“杀菌剂或杀虫剂抗性专业委员会”，开展病原菌和害虫抗药性监测与治理研发项目，定期发布重要病原菌和害虫对重要类型药剂的抗性发展动态的信息，提出抗性治理对策，暂停使用已产生抗药性问题、防效很差的药剂，推荐采用不同作用机理的高效低风险的药剂替代，提倡对病虫害进行综合治理，减少对化学农药过度依赖，降低化学农药投入，提高化学农药利用率，符合绿色植保的理念。

2.2　中国农药减施技术应用研究现状及存在的问题

中国在蔬菜病虫害综合治理、环境相容性农药的安全、高效和精准使用、无公害或绿色蔬菜病虫害防治操作规范等方面取得较大进展，开展重大病虫害快速检测、抗药性早期检测与预警、抗药性风险评估，研发多功能生防制剂，开展生防与化学防治协调使用技术研究，构建了一系列的无公害或绿色蔬菜病虫害防治规范、无公害或绿色蔬菜生产规程，提倡节水、省药、提质、增效，提高蔬菜病虫害管理和安全生产水平，促进蔬菜产业发展。根据中国蔬菜生产的特点，将绿色植保等新的植保理念注入蔬菜生产中，研究蔬菜病虫害频发致害的成因，推广安全防控技术，以实现蔬菜生产的高产、优质、高效及生态安全，但实际应用与发达国家还有很大差距，一些无公害或绿色蔬菜病虫害防治操作规范技术复杂、使用成本过高，农民难以接受，导致推广起来很难 。

在农药研发和应用方面取得长足进步，有效成分趋于高效广谱低毒化；生物农药增长快、潜力大；剂型趋向环境友好、省工省力；使用方法趋于简便、高效、持久，每年农药防治面积达到70亿~80亿亩次，防治贡献率达到70%。

但中国开展农药减量增效研究起步较晚，尽管积累不少减少化学农药使用并提高化学农药利用率的单项技术，但未能进行充分集成验证，缺少切实可行的不同气候带不同作物不同种植制度下农药减施增效技术模式，农药利用率低。原药创新、制剂水平及施药器械和施药技术总体上仍落后，原药生产以仿制为主，缺少足够的自主创新产品，制剂落后，可湿性粉剂和乳油等落后老剂型仍在普遍使用，农药成品加工粗糙，粒径较大，菜农安全、高效用药意识较差，仍主要采用背负式喷雾器施药，喷头孔径大，造成施药过程中农药大量的浪费，农药的减施增效潜力很大。农药产品对水后悬浮性较差，展着性差，影响药效发挥，货架期较短，对作物安全性较差。

中国农药利用率低的原因在于：①对环境保护、农产品的安全性、农药化肥高效、精准、减量科学使用和对作物有害生物综合防控的先进理念、知识和技术的普及宣传不够，对农药科学使用的益处及非科学用药的危害性认识不足；②作物栽培和病虫害管理水平较差，未重视生态调控、预防和精准施药技术及病虫抗药性管理，过度依赖用药，打保险药。轻预防，重害后治理，重化学防治，对非化学防治技术研发不足，病虫害早期监测与抗药性预警技术较弱；③种植制度的差异。土地等资源利用强度很大，普遍存在的设施蔬菜种植模式一年多茬多熟，轮作倒茬较难，病虫害发生严重，依靠化学防治来保产，不像欧美国家多为一年一熟，有的还实行休耕和轮作，病虫害发生较轻，投入农药较少；④作物产量水平差异。中国人多地少，依靠增加化肥和农药投入来提高单产水平；⑤施药方式的差异。欧美等国家通过实施统防统治、专业化防治、机械化施药，而中国农户大多仍采用背负式喷雾器喷雾的落后施药方式和施药器械。

农药减施已成为制约中国农产品安全、缓解对环境保护的压力、保障粮食、蔬菜、果品等农产品可持续发展的紧迫任务。政府高度重视农药的过度使用带来的问题，开始启动农药减量行动，将风险评估技术引入农药管理，构建残留标准体系快速，加速淘汰高毒高风险农药，鼓励环境友好型农药的创制及高效绿色环保技术的研发利用。农业部 2015 年提出到 2020 年化肥、农药使用量零增长方案，引导农业绿色发展，旨在加快转变施肥用药方式，根据不同作物和病虫害推广应用农业防治、生物防治、物理防治等绿色防控技术，推进统防统治与绿色防控融合；加快推广新型高效农药，重点研发高效低毒低残留农药、生物农药等新型产品，推广先进施药机械；加快发展能扶持提供病虫统防统治的社会化服务组织，鼓励其参与农药使用量零增长行动。

3 实现设施蔬菜农药减施增效的途径

农药减施就是要提高农药利用率，减少农药的使用量。据估算，农药利用率提高 1.6%，农民就减少生产投入约 8 亿元，利于减少农药残留，保障农产品质量安全，保护土壤和水体环境。提高农药利用率，以确保农药减量施用，可通过以下途径来实现。

（1）创制新型高效环境友好型农药，使老的农药品种更新换代。

（2）在抗药性发展早期实施治理，加强抗药性监测，停用或淘汰不能或难于恢复靶标敏感度的农药品种，目前含甲霜灵或精甲霜灵的杀菌剂在中国防治黄瓜霜霉病、黄瓜白粉病、马铃薯晚疫病已经因靶标菌对甲霜灵或精甲霜灵普遍产生抗性而致防治效果大为降低，而且噁霜灵与甲霜灵或精甲霜灵之间具有正交抗关系，生产中应停止使用含甲霜灵、噁霜灵或精甲霜灵的杀菌剂防治黄瓜霜霉病、黄瓜白粉病、马铃薯晚疫病。黄瓜霜霉病菌、黄瓜白粉病菌对嘧菌酯普遍产生抗性，而且甲氧基丙烯酸酯类药剂之间具有正交互抗性关系，生产中应停止用嘧菌酯、醚菌酯、吡唑醚菌酯单剂、甲氧基丙烯酸酯类杀菌剂与不同作用机理的杀菌剂的混剂防治黄瓜霜霉病和黄瓜白粉病。番茄灰霉病菌对多菌灵、甲基硫菌灵、乙霉威、腐霉利、异菌脲、嘧霉胺普遍产生，生产中应停止使用多菌灵、甲基硫菌灵单剂及其与其他药剂的混剂、嘧霉胺单剂及其混剂、乙霉威单剂及其混剂防治番茄灰霉病。由于番茄灰霉病菌对腐霉利和异菌脲的抗性水平很低，生产中可将这两种药剂的单剂及其与除多菌灵、甲基硫菌灵、嘧霉胺外的药剂（咯菌腈、啶酰菌胺等）的混剂可用于预防番茄灰霉病。

（3）正确进行药剂轮用和混用，将不同作用机理的药剂进行合理混用或轮换使用，原则上保护性药剂用于病虫害预防，而具有内吸传导型药剂用于病虫害治疗，将无拮抗作用或作用方式与内吸传导性互补的药剂混配组合用于混用。

（4）建立与药剂、作物系统、病虫害相匹配的施药技术，以作物为中心，按照产前、产中及产后几个阶段进行管理，根据不同生长时期病虫害发生规律，制定合理的施药规程和健身栽培管理规程，在有效控制病虫害和作物增产的基础上少用化学药剂。

（5）提高栽培管理和病虫害防治水平，改进施药方式，强化技术培训和指导服务，大力推广先进适用技术（包括田园清洁技术、种子无毒无病虫处理技术、种子包衣技术、无病土或基质工厂化育苗、壮秧壮苗栽培技术、嫁接技术、土壤消毒技术、适期施药技术、对靶施药技术、利用新型药械高效施药技术、棚室内天敌释放技术、色板诱杀与防虫网结合、生态调控降湿控温技术、产后残体无害化处理技术等），推进机械施药（超声波布药器，常温烟雾机、热力烟雾机、轨道风送式喷雾机、自走式风送喷雾机），研发水肥药一体化施用技术。

（6）推进新农药新技术的应用，推广缓释性农药、生物农药、种衣剂、土壤调理剂。

（7）推进生物农药及有益微生物资源利用，加快生物菌肥的研发和推广应用进度。

（8）利用水溶性肥料，重视水肥药一体化节肥省药高效栽培技术的研发利用。

中国 2015 年设立重点研发专项“化肥农药减施增效综合技术”，“设施蔬菜化肥农药减施增效技术集成研究与示范”是首批启动的重大专项之一，即基于蔬菜病虫害防治指标与化学农药限量标准，集成配套与区域生产相适应的高效安全农药新产品、土壤消毒、种苗处理、设施农业智能高效施药等技术，优化物理防控、生物防治等绿色防控技术，形成不同生态区的设施蔬菜农药减施技术模式，建立相应技术规程。通过示范、培训及推广应用，将农药利用率提高 10 个百分点，减量 30%。如何实施好国家重大专项“设施蔬菜化肥农药减施增效技术集成研究与示范”直接关系到能否实现国家“到 2020 年化肥、农药使用量零增长方案”的目标。

4 设施蔬菜农药减施增效技术集成应用思路

农药减施技术与化肥减施技术和高产栽培技术集成实施才能实现设施蔬菜的增效，缺一不可。以设施番茄、黄瓜、辣椒、茄子、韭菜、芹菜等果菜类和叶菜类蔬菜为研究对象，调查清楚我国不同生态区域（例如黄淮海区、环渤海暖温带区、长江中下游亚热带区、华南亚热带区或热带区、西北高原温带区、东北温带区等）不同设施类型（日光温室、大棚、中小拱棚等）主要蔬菜病虫害发生及防治状况与农药使用状况合存在的主要问题、各种施药器械的使用情况、化肥、生物有机肥、农药等农资投入与产出情况、蔬菜生产效益情况，抓住关键问题和关键技术，确定病虫害防治指标与农药限量标准，做好本底研究和项目设计，制订好实施方案，开展有关农药减施增效关键共性技术研发。

（1）高效低风险农药的筛选及其高效减量用药技术的研发，监测抗药性动态，评估重要病原菌和害虫对主要防治药剂的抗性风险，制定药剂精准高效减量施用技术及重要病原菌抗药性治理技术。

（2）植物源农药及枯草芽孢杆菌、木霉、白僵菌等生防菌剂及天敌昆虫资源的筛选及利用，研发生物农药与化学药剂兼容协调使用技术，探查生防菌剂、天地昆虫与化学药剂交替使用对霜霉病、白粉病、灰霉病、粉虱等重要病虫害的防治效果。

（3）研发降低化学农药抗药性风险和精准减量施用技术。

（4）通过抽样检测蔬菜产品中及土壤中的农药残留动态、田间现场检测产量和防治效果评测农药减施增效，利用雾滴卡检测农药利用效率。

（5）利用好国内外已经成熟的各种设施蔬菜节肥减药关键技术，与蔬菜作物高效栽培及化肥减施增效技术研究者协同攻关，分工明确，齐头并进，研发各种化肥农药减施增效技术及高效栽培技术［包括抗病抗虫品种利用；基质工厂化嫁接育苗；植物生长调节剂处理种苗壮苗；高效栽培壮秧；防虫网室育苗隔离防治粉虱等虫害；滴灌、无滴膜、地膜覆盖等生态调控降湿；高效低风险农药精准施用及合理混用与轮换使用；生物农药与化学农药协调使用；太阳能消毒；石灰氮、水溶性生物菌肥修复土壤；设施农业智能高效施药（烟雾机、喷粉机）；水肥药一体化管理等］，边引进，边研发，边集成示范，构建设施蔬菜化肥农药减施增效综合技术模式，通过设立不同的课题、子课题，各个参加单位在蔬菜主产区建立示范基地，与遍布全国的各级植保、土肥、蔬菜管理及农技推广部门形成的网络及各种蔬菜标准园区、各种蔬菜生产专业合作社和种植大户结合，抓好科技示范户，进行各种蔬菜化肥农药减施增效技术模式示范及推广，不断优化，形成标准或规程，做到模式规模化应用，产生良好的经济、社会和生态效益。

试验示范过程中要与菜农常规施肥施药技术模式进行比较，通过召开培训会、借助现代媒体传播培训新型经营主体和现代职业农民，通过自评（项目执行者在试验示范过程中及时采集数据并进行比较分析整理评判）及第三方评价（例如召开现场会进行现场检测，给有资质的独立检测单位送样检测化验），明确项目集成研发的化肥农药减施增效技术模式的双减增效效果，实现水肥药一体化管理和高效栽培，解决设施蔬菜生产中化肥农药超量使用导致病虫害严重发生、蔬菜品质严重下降、畜禽粪便缺乏量化指标、有机无机配比不合理、畜禽粪便过量施用，致使土壤板结严重、土壤盐渍化、地下水污染等问题，提高化学农药利用率，明显减施农药，主要农药残留不超标，病虫害得到有效防控，满足防治增产及控制化肥农药使用的需求，蔬菜达到绿色产品标准，保证蔬菜增产和农民收益增加，提升农民种植蔬菜的水平和积极性，为各生态区绿色蔬菜产业的可持续发展及安全生产提供技术支撑，有力保护土壤和地下水资源环境。

参考文献

［1］ Yang N W, Zang L S, Wang S, et al. Biological pest management by predators and parasitoids in the greenhouse vegetables in China［J］. *Biological Control*, 2014, 68: 92－102.

［2］ 黄保宏，林桂坤，王学辉，等. 防虫网对设施蔬菜害虫控害作用研究［J］. 植物保护，2013，39（6）：164－169.

［3］ 姜军侠，白伟，何玲，等. 陕西蔬菜农药使用现状及问题思考［J］. 农药科学与管理，2016，37（5）：6－9.

［4］ 李姝，王甦，赵静，等. 释放异色瓢虫对北京温室甜椒和圆茄上桃蚜的控害效果［J］. 植物保护学报，2014，41（6）：699－704.

［5］ 刘云，梁玉芹，王灵敏，等. 河北省设施蔬菜突发性灾害防控技术探讨［J］. 河北农业科学，2015，19（1）：45－48.

［6］ 王文桥. 我国设施蔬菜农药减施增效展望［J］. 中国蔬菜，2016，5：1－3.

［7］ 王文桥，张小凤，韩秀英，等. 我国北方蔬菜病害抗药性问题及治理对策［J］. 中国蔬菜，2010，23：20－23.

［8］ 王青松. 无公害蔬菜病虫害综合防治规范操作模式［J］. 福建农业科技，2012，10：53－54.

［9］ 许立. 设施蔬菜栽培连作障碍分析及综合防治［J］. 中国园艺文摘，2015，12：184－185.

[10] 余朝阁，李颖，黄欣阳，等. 当前设施蔬菜病虫害防治中存在的问题及解决途径 [J]. 长江蔬菜，2013，8：58－61.

[11] 喻景权. “十一五”中国设施蔬菜生产和科技进展及其展望 [J]. 中国蔬菜，2011，2：11－23.

[12] 张博，高新昊，李长松，等. 山东省保护地蔬菜农药使用现状及建议 [J]. 农药学报，2012，2（19）：23－27.

[13] 张帆，张君明，罗晨，王甦. 蔬菜害虫的生物防治技术概述. 中国蔬菜，2011（1）：23－24.

[14] 张帆，李姝，肖达，等，中国设施害虫天地昆虫应用研究进展 [J]. 中国农业科学，2015，48（17）：3 463－3 476.

[15] 张君明，张帆，王兵. 以释放丽蚜小蜂为主的保护地番茄温室粉虱的控制技术 [J]. 蔬菜，2010，7：34－35.

[16] 张庆霞，高江霞. 利用栽培技术预防和控制设施蔬菜病虫害 [J]. 北方园艺，2010，8：178－179.

[17] 张臻，设施蔬菜病虫害发生规律及综合防治技术示范，2015，上海蔬菜，3：60－61.

[18] 张真和，陈青云，高丽红，等. 中国设施蔬菜产业发展对策研究（上）[J]. 蔬菜，2010（5）：1－3.

[19] 张真和，陈青云，高丽红，等. 中国设施蔬菜产业发展对策研究（下）. 蔬菜，2010（6）：1－3.

微管蛋白靶标类药物的离体筛选*

徐建强[1,2]**，陈长军[1]，周明国[1]***
（1. 南京农业大学植物保护学院，南京 210095；2. 河南科技大学林学院，洛阳 471003）

摘要：本文概述了生物体微管蛋白的制备方法、原核表达及纯化、三维结构及药物结合位点的模拟、体外聚合、同药剂的体外结合，分析了以微管蛋白为靶标进行药物离体筛选的可能性。
关键词：微管蛋白；靶标；离体筛选

In vitro Screening of Drugs Targeting on Tubulin*

Xu Jianqiang[1,2]**, Chen Changjun, Zhou Mingguo[1]***
(1. *College of Plant Protection*, *Nanjing Agricultural University*, *Nanjing* 210095;
2. *College of Forestry*, *Henan University of Science and Technology*, *Luoyang* 471003)

Abstract: The paper gives a brief introduction on the preparation method of tubulin, and some improvements on tubulin prokaryotic expression and purification, tubulin homology modeling and the binding sites of drugs, *in vitro* polymerization between α-and β-tubulin, and *in vitro* binding kinetics of tubulin with drugs. *In vitro* screening of drugs targeting on tubulin is also discussed in detail.
Key words: Tubulin; Target; *In vitro* screening

微管在生物体细胞的有丝分裂及胞内运输等各方面均具有重要作用，由 α-和 β-微管蛋白构成的微管一直处于聚合/解聚的动态平衡过程中。许多作用于微管蛋白的药物都是同靶标结合后抑制或促进了微管的动态平衡，进而影响细胞的有丝分裂而发挥作用的。尽管许多微管蛋白靶标类药物都产生了严重的抗性，但人们在寻找同已有的微管蛋白类药剂相近或更好活性的新的化合物方面进行了很多努力，微管蛋白依然是进行小分子药物发现的理想靶标[1]。从动物组织中提取的微管蛋白已在离体条件下被用作紫杉醇类抗肿瘤药物及抗寄生虫药物的筛选，而且已开始应用在药物的高通量筛选中[2-4]。在离体条件下以微管蛋白为靶进行药物筛选有非常广阔的发展前景。

1 微管蛋白的制备方法

离体条件下进行药物筛选首先面临的就是微管蛋白的大量制备。对于微管蛋白含量丰富的组织，如猪脑或牛脑等，可采用直接制备的方法[5-7]；而对于微管蛋白含量低的组织，则可以采用离子交换树脂的方法来进行[8-11]。

---

* 基金项目：国家自然科学基金（31401774）；农业部公益性行业专项（201303023）
** 第一作者：徐建强，男，博士，副教授，主要从事小麦及牡丹病害的综合防控研究；E-mail：xujqhust@126.com
*** 通讯作者：周明国，教授，主要从事植物病害化学防治研究

1.1 直接制备

从动物脑（牛脑、猪脑等）中直接制备微管蛋白，是利用微管高温聚合/低温解聚的原理（高温聚合收集沉淀舍弃上清，低温解聚收集上清舍弃沉淀）去除杂蛋白，经过多次（一般3~4次）循环，即可制备到纯度较高的微管蛋白。目前，此方法多用于从微管蛋白含量丰富的动物组织中提取。

但在实际应用中，此法也有许多缺点：步骤烦琐，需要经过多次的聚合/解聚过程，多次操作会造成微管蛋白含量降低及蛋白降解；在蛋白纯化过程中，其他蛋白，如MAPs、Mg^{2+}结合蛋白或其他因子会随着一同纯化，“污染源”较多[7]；不能保证所有蛋白均经过同样的翻译后修饰过程；更为重要的一点，对于微管蛋白含量低的生物，如丝状真菌及原生动物，此法并不适合。

1.2 离子交换树脂纯化

在原生动物及真菌中，由于微管蛋白含量较低，不能采用聚合/解聚的方法来进行微管蛋白的纯化，可采用离子交换树脂的方法。如使用DEAE Sephadex从样品中富集微管蛋白。

利用离子交换树脂（ionic exchange resins）的方法可以对寄生虫[8-9]、酵母[10]及丝状真菌[11]的微管蛋白进行直接提取。但对于丝状真菌，采用离子交换树脂的方法制备微管蛋白，在制备过程中需要构建过表达载体及进行遗传转化，对于那些尚未建立起遗传转化体系的真菌，显然是不可能的；步骤烦琐，纯化步骤较多，多次操作会造成微管蛋白量降低及蛋白降解；在蛋白纯化过程中，其他蛋白，如MAPs、Mg^{2+}结合蛋白或其他因子会随着一同纯化，且不能保证所有蛋白均经过同样的翻译后修饰过程；纯化过程中，为了能使微管蛋白热聚合充分，紫杉醇被用来促进聚合，这增加了蛋白纯化成本。所以，离子交换树脂制备真菌微管蛋白有一定局限性[12-13]。

植物中也有采用离子交换树脂进行微管蛋白纯化的报道。如Huang等[14]采用DEAE-Sepharose从百合花粉中制备了微管蛋白，Xu等[15]也取得了类似的结果。

从生物体内提取的微管蛋白，是α、β-微管蛋白的异源二聚体，很难将α-与β-微管蛋白分开，故不能对单个微管蛋白亚型进行研究，但却能反映出生物体细胞内微管蛋白的真实情况。

2 微管蛋白的原核表达

大肠杆菌表达系统在表达外源基因方面已经非常成熟，许多公司都可以提供进行原核表达的菌株及载体，所以利用大肠杆菌表达系统生产外源蛋白是目前应用最多、最广泛的。这不但是由于大肠杆菌表达系统具有生长快、遗传背景清晰、表达效率高及成本低等特点，还因为原核细胞没有微管相关蛋白等的污染，可以在短时间内得到大量的纯度很高的微管蛋白。目前，原生动物、植物及植物病原真菌的微管蛋白均有原核表达的报道。

2.1 寄生虫微管蛋白的原核表达

寄生虫体内微管蛋白含量较低，尽管前面提到了可采用离子交换树脂进行纯化，但大部分还是采用原核表达的方法进行。目前，微管蛋白已进行过原核表达的寄生虫包括：贾第虫（*Giardia duodenalis*）、微小隐孢子虫（*Cryptosporidium parvum*）及肠道脑胞内原虫（*Encephalitozoon intestinalis*）α-和β-微管蛋白[16-18]，*Reticulomyxa filosa* α-和β-微管蛋白[19]，细颈杯环线虫（*Cylicocyclus nassatus*）β-微管蛋白[20]，布氏锥虫（*Trypanosoma brucei*）α-和β-微管蛋白[21]，捻转血茅线虫（*Haemonchus contortus*）α-和β-微管蛋白[22]，*Trypanosoma danilewskyi* β-微管蛋白[23]，捻转血茅线虫（*Haemonchus contortus*）β-微管蛋白[24]。

2.2 植物微管蛋白的原核表达

植物微管蛋白除了上述的从花粉中直接制备的方法外，也有进行原核表达的报道。Jang 等[25]和 Koo 等[26]分别进行了甜椒（*Capsicum annuum*）和马铃薯（*Solanum tuberosum* L.）α-和 β-微管蛋白的原核表达，表达的蛋白都是以包涵体形式存在，通过复性可以对微管蛋白进行纯化。

2.3 丝状真菌微管蛋白的原核表达

相对于离子交换树脂制备真菌的微管蛋白，原核表达的方法不需要进行微管蛋白过表达载体的构建，且纯化步骤较少，不会受到微管相关蛋白或 Mg^{2+} 结合蛋白等其他因子的污染，纯化过程中也不需要紫杉醇，节约了试验成本。

目前，微管蛋白已进行过原核表达的真菌：粗糙脉孢菌（*Neurospora crassa*）β-微管蛋白[27]，辣椒疫霉（*Phytophthora capsici*）α-和 β-微管蛋白[13]，大麦云纹病菌（*Rhynchosporium secalis*）β-微管蛋白[12]，禾谷镰孢菌（*Fusarium graminearum*）β_1-和 β_2-微管蛋白[28]。

2.4 微管蛋白在其他系统里的表达

除了原核表达，微管蛋白也有在酵母、芽孢杆菌及昆虫细胞内进行表达的报道。如 Bell 等将引起人疟疾的疟原虫（*Plasmodium faiciparum*）的 β-微管蛋白在短芽孢杆菌（*Bacillus brevis*）中成功表达[29]；Vats-Mehta 和 Yarbrough 利用昆虫杆状病毒表达系统（Baculovirus expression system）在昆虫细胞内表达了小鸡及酿酒酵母（*S. cerevisiae*）的 β-微管蛋白[30]。

Linder 等在毕赤酵母（*Pichia pastoris*）中表达了 *Reticulomyxa filosa* 的 α-和 β-微管蛋白，并发现单独表达 *Reticulomyxa filosa* 一个微管蛋白，表达量都很少，几乎检测不到；而若同时表达 α-和 β-微管蛋白，微管蛋白的表达量增加非常明显，而此时毕赤酵母自身的微管蛋白表达量反而下降；进一步的研究发现，同时表达 α-和 β-微管蛋白，两者的相对量及毕赤酵母体内微管蛋白总量都处于一个平衡状态。（*Pichia pastoris*）/（*Reticulomyxa filosa*）系统在同时研究微管调节的各种机制上具有重要作用[31]。

采用原核表达方法制备的微管蛋白，利于对微管蛋白单个亚型进行分析，但却不能反映生物体细胞内微管蛋白的真实状态。

3 微管蛋白的同源模建及药剂结合位点

1998 年，Nogales 等采用电子晶体衍射的方法解析了猪脑 α/β-微管蛋白异源二聚体的结构[32]，这极大地促进了微管蛋白类药物在微管蛋白上结合位点的研究。结合计算模拟及蛋白同源模建的技术，多类药物在微管蛋白上的结合位点及作用方式都被模拟出来。

3.1 真菌微管蛋白的同源模建

Oakley 在 Swiss-Model Automated Comparative Protein Modeling Server（http://www.expasy.ch/swissmod/）上根据牛脑微管蛋白的结构（1JFF），模拟了构巢曲霉（*Aspergillus nidulans*）α-和 β-微管蛋白的结构，并利用 SwissPdb Viewer version 3.7（http://www.expasy.ch/spdbv/）观察。抗性、敏感菌株 β-微管蛋白存在着氨基酸序列差异，抗性蛋白序列上变化的那些位点构成的区域可能就是药剂分子在靶标蛋白上的结合位点。由此，Oakley 在 β-微管蛋白的结构图上标上了突变位点，发现这些位点尽管在一级结构上相距较远，但在空间结构上的距离却非常近，排列成一簇，由此可以推断苯并咪唑类药剂分子的结合区域应当由这些位点构成[33]。

Li 和 Yang 等模建了木霉 β-微管蛋白的空间结构，并将突变后可以对苯并咪唑类杀菌剂产生抗性的位点，包括 His6、Tyr50、Gln134、Ala165、Phe167、Glu198、Phe200、Leu240、Arg241 和 Met257 等在三维结构图上标出，发现药物结合位点位于微管蛋白的内部，而报道较多的同微管蛋白发生相互作用的蛋白多位于微管的外部[34]；Qiu 等模建了禾谷镰孢菌 β_2-微管蛋白的空间结构，结果表明 Phe167、Glu198 和 Phe200 三个氨基酸组成了一个袋状结构，苯并咪唑类药剂分子可能位于袋状结构内，而若 3 个位点的氨基酸突变后，影响了药剂分子的结合，结合力下降，从而表现出抗药性，而其他引起抗性的位点发生突变可能会使袋状结构内部的力减弱，对药剂分子的结合会产生一定影响[35]；Zou 等模拟了球孢白僵菌（*Beauveria bassiana*）β-微管蛋白的空间结构，发现对苯并咪唑类药剂有抗性的突变位点多位于 β-微管蛋白同 α-微管蛋白相互接触的一面[36]。

微管蛋白的同源模建在分析药物结合位点时尽管可以看到突变位点在靶标上的空间位置，大概推测药剂的结合域，但起的作用有限，借助于分子动力学等手段将药剂分子锚定在靶标蛋白内部，就更有助于对药剂结合域的分析。

3.2　二硝基苯胺类药剂在牛筋草 α-微管蛋白上的结合位点

Blume 等模建了牛筋草（*Eleusine indica*）α-微管蛋白，比较敏感和抗性的 α-微管蛋白电子密度图的差异，并根据范德华力及电势高低模拟了氟乐灵的空间结构，从蛋白及药剂分子电子密度图的角度分析两者的结合部位。研究发现，牛筋草对药剂产生抗性的突变位点 Thr239 并不是暴露于 α-微管蛋白的表面，而是临近药剂分子同靶标蛋白二者接触穴部位。在接触穴上，药剂分子上的 NO_2 具有重要的功能，可对微管的进一步聚合产生阻止作用；药剂分子骨架上的 CH 侧链围绕着 α-微管蛋白表面的 Gln133、Asn253、Gln256 的排列，这些氨基酸参与同下一个异源二聚体上 β-微管蛋白氨基酸的互作[37]。

狗尾草对二硝基苯胺类除草剂的抗性是由 α_2-微管蛋白 Leu136 和 Thr239 突变引起。Delye 等在模建的 α_2-微管蛋白空间结构上发现，两个突变位点邻近微管蛋白异源二聚体互作的区域；同对药剂表现敏感的杂草 α-微管蛋白空间结构比较后发现，253 位氨基酸及 251 位氨基酸侧链附近是决定杂草对药剂敏感性差异的关键部位；16、24、136、239、252 和 268 位也同药剂的敏感性有关[38]。

Morrissette 和 Sept 等采用分子力学的方法研究了二硝基苯胺类药剂对牛筋草（*Eleusine indica*）和刚地弓形虫（*Toxoplasma gondii*）的敏感性，模拟了药剂分子在 α-微管蛋白上的结合位点。α-微管蛋白存在着两个 loop 结构，分别为 M-loop 和 N-loop；α-微管蛋白有 8 个氨基酸插入到 β-微管蛋白内部，稳定 α-微管蛋白上的 M-loop；紫杉醇结合到 β-微管蛋白 M-loop 结构处，对微管的聚合起到促进作用；而二硝基苯胺类药剂分子是结合在 α-微管蛋白 N-loop（H1-S2）结构的下面。刚地弓形虫对二硝基苯胺类药剂产生抗性的突变位点都聚集在 N-loop 结构一个核心 domain 附近，此 domain 正是二硝基苯胺类药剂分子的结合区域[39]。

3.3　苯并咪唑类及二硝基苯胺类药剂在寄生虫微管蛋白上的结合位点

Morrissette 等模拟了氨磺乐灵（Oryzalin）在弓形虫 α-微管蛋白上的结合位点，包括 Arg2、Glu3、Val4、Trp21、Phe24、His28、Ile42、Asp47、Arg64、Cys65、Thr239、Arg243 和 Phe244，Phe24、His28、Thr239 和 Arg243 四个位点的突变能引起抗药性，有几个位点位于 N-loop 的下面，但大部分位点都是在 α-微管蛋白的内部，Oryzalin 上的 SO_2 基团同 Arg64 上的 NH 基团形成了一个氢键；oryzalin 结合到 α-微管蛋白的 N-loop 结构后，使 N-loop 向分子内部发生移动，影响了 N-loop 同相邻原纤丝上 M-loop 的互作，阻断了原纤丝同微管的结合，从而影响了微管的进一步生长[40]。

Prichard 在捻转血矛线虫同药剂分子互作时，根据 α-及 β-微管蛋白上抗药性突变位点，提出了一个药剂结合的简单模式。他认为：α-和 β-微管蛋白分子共同参与了药剂的结合，β-微管蛋白上 Phe167 和 Phe200 等也参与了此作用；药剂分子上的苯环位于 Phe167 和 Phe200 2 个苯环之间，3 个苯环形成一个堆积作用，这是药剂分子同靶标结合力之一；第二个结合力来自药剂分子上的氨基甲酸酯同 Cys201 的共价连接；Ser166、Phe167、Phe200 和 Cys201 四个氨基酸残基同药剂分子发生作用。在产生抗性的线虫个体内，Phe167 和 Phe200 突变为 Tyr 后，阻止了药剂分子锚定在 Phe167 或 Phe200 的结合位点上[41]。

但随后 Robinson 等对此模型提出了怀疑：第一，Phe167 和 Phe200 发生苯环堆积的距离是在 6.0 ~ 6.5Å，而在已发表的结构上，Phe167 和 Phe200 的距离在 5Å 以内，不足以形成堆积力，而若容纳药剂分子的苯环，则分子必须移动 5 ~ 6Å，超过了微管蛋白结构解析时的分辨率 3.9Å；第二，在已发表的结构中，β-微管蛋白上 200 ~ 206 位残基在构象上位于 200 位残基的相反的一方，不可能同药剂分子发生接触；第三，Ser166、Phe167、Phe200 和 Cys201 均埋在 β-微管蛋白的内部，不能同溶剂分子发生接触，药剂如何进入该区域、如何同特异性的残基结合及如何发挥它们的杀虫活性等都没有解释清楚[42]。

随后，Robinson 等提出了自己的看法[43]。他们在 Silicon Graphics O_2 工作站上，采用分子动力学等手段进行模拟，利用 Insight Ⅱ，Biopolymer and Discover 软件，将阿苯达唑亚砜锚定在捻转血矛线虫 *H. contortus* β-微管蛋白内部，对结合区域进行了分析。因为真菌上产生抗药性的位点均位于 β-微管蛋白分子内部，他们分析了药剂分子是如何进入到分子内部的。研究发现，产生抗药性的位点均位于 β-微管蛋白 N 端区域同中间区域发生互作的界面处。在微管运动的动力学过程中，β-微管蛋白 N 端区和中间区的转动有效地改变了微管构象，使原纤丝上直的微管蛋白异源二聚体变成弯曲的解聚的微管蛋白单体。在微管的生长端，为了下一个异源二聚体分子的加入，已经聚合的微管上 β-微管蛋白的 GTP 会发生水解释放能量，而这也导致了 β-微管蛋白 N 端区和中间区的转动，导致两者之间的缝隙扩大，使药剂分子有机会进入到微管蛋白内部；药剂分子进入后，蛋白构象受药剂诱导发生改变，药剂锚定在 β-微管蛋白单体构象上，导致微管发生空解聚（只解聚不再发生聚合作用）。药剂是在 α-，β-微管蛋白发生聚合形成微管之前同微管蛋白发生作用。

微管蛋白同药剂分子的共结晶将最终能揭示药剂分子在微管蛋白上结合部位。

4 α-、β-微管蛋白的体外聚合

无论是直接从动物组织中提取的微管蛋白还是通过原核表达方法制备的微管蛋白，在体外适当的条件下均可以发生聚合，形成类似于生物体内状的微管结构。

α-、β-微管蛋白的聚合条件：缓冲液体系包括 MES（4-吗啉乙磺酸）缓冲液或 Pipes［哌嗪-N，N′-双（2-乙磺酸）］缓冲液；除此之外，还应包括 Mg^{2+}，GTP，EGTA，以及一些蛋白酶抑制剂；体外聚合条件同微管蛋白提取时一样，可在 37℃ 聚合 30 ~ 90m；聚合时 α-、β-微管蛋白的浓度要达到一定程度，一般在 1mg/mL 以上[16,22]。

α-、β-微管蛋白体外聚合主要有以下几种检测方法：在分光光度计上检测聚合动力学：微管蛋白聚合会导致聚合液在 350nm 下的吸光度发生变化，并且吸光值变化会随着聚合进程而发生改变，通过在一定条件下对发生聚合的 α-、β-微管蛋白反应液进行定时扫描，可以判断聚合进程[13,25]；Western Blot 及在自然条件下进行聚丙烯酰胺凝胶电泳（Native

Page)：因为聚合后的微管分子量加大，采用 Western Blot 或 Native Page 可对聚合的二聚体的分子量进行判定，但不能采用 SDS-PAGE，因为变性条件下二聚体会解聚形成单体，无法判断；透射电镜：将 α-、β-微管蛋白聚合后的反应液进行高速离心，收集沉淀，负染后在透射电镜下进行观察，可以观察到微管类结构[22,25]。

从动植物组织中提取的微管蛋白，在低温下解聚成单体状态后，即可在高温下发生聚合反应；此时若在解聚液中加入一些微管蛋白聚合的抑制物质，就可以研究其对微管蛋白聚合的影响。

微管稳定试剂或微管“种子”对植物微管蛋白的体外聚合是必需的。Huang 等研究了从植物花粉中提取的微管蛋白在体外条件下的聚合，发现在无微管稳定试剂或微管“种子”的情况下微管蛋白可以发生体外聚合；在电镜下也观察到了典型的微管结构；微管蛋白聚合动力学呈现出典型的“抛物线”形状；紫杉醇可以显著地改变植物微管蛋白的聚合特征，形成了异常的微管结构，微管蛋白最低聚合浓度也从 3mg/mL 降到 0. 043mg/mL[14]。

第一例对原核表达的微管蛋白进行体外聚合研究的是 Oxberry 等[22]。他们在大肠杆菌中表达了捻转血矛线虫 *H. contortus* 的 α-、β-微管蛋白，并通过柱上复性对包涵体进行了纯化，然后研究两者在体外的聚合。研究发现：纯化的 α-、β-微管蛋白在没有微管相关蛋白（MAPs）的条件下，体外可以发生聚合；纯化的 β-微管蛋白亚型，β12-16，在没有 α-微管蛋白的条件下，单独也可以聚合形成微管状结构。下列条件有助于微管蛋白的聚合：微管蛋白浓度在 0. 25mg/mL 以上；温度在 20℃ 以上，GTP 浓度在 2mmol/L，有甘油、EGTA 及 Mg^{2+}。GTP 浓度超过 2mmol/L 及在阿苯达唑存在时均对聚合有抑制作用；Ca^{2+} 及 6 ~ 8. 5 的酸碱度对聚合无明显的影响；微管蛋白单个亚型聚合程度同 α-、β-微管蛋白时一样。作者最后指出，尽管分光光度计法可以检测聚合，但它却不能显示聚合形成的结构；电镜观察表明：α-、β-微管蛋白体外聚合除可以形成微管状结构外，还可以形成片状、折叠片及环状结构。

Yoon 课题组研究了辣椒、马铃薯和辣椒疫霉的微管蛋白体外聚合[13,25-26,44]。Jang 等在大肠杆菌里表达了辣椒的 α-、β-微管蛋白，在 BL21（DE3）中，主要以包涵体形式出现，包涵体复性后，两种蛋白在有 2mmol/L GTP 存在时可以在离体条件下发生聚合[25]。随后 Koo 等对辣椒 α-、β-微管蛋白的离体聚合进行了更深的研究，结果发现：α-、β-微管蛋白的离体聚合受多种因素的影响，1mmol/L $MgCl_2$ 可以使聚合降低 36%；5% 和 10% 的甘油浓度对聚合的影响不大；100μmol/L 的紫杉醇可以使聚合增加 23%，1mmol/L 的 $CaCl_2$，二甲基亚砜（Dimethyl sulfoxide，DMSO）浓度≥1%，低温条件使微管蛋白聚合降低；0. 089mg/mL 的浓度是微管蛋白聚合的最低限度[44]。Koo 等在大肠杆菌中表达了辣椒疫霉（*Phytophthora capsici*）的 α-、β-微管蛋白，在变性条件下进行了纯化，对复性的微管蛋白在体外进行了聚合，研究发现：0. 12mg/mL 是微管蛋白体外聚合的最低浓度，苯菌灵对聚合的抑制终浓度在（468 ±20）μmol/L[13]。Koo 等又表达了来源于马铃薯叶片的 α-、β-微管蛋白，采用 Western blot 证实了其体外可以发生聚合[26]。

5 微管蛋白与药剂的体外结合

通过原核表达的方法制备的微管蛋白，可以在离体条件下同药物分子进行结合，根据亲和力的大小来判断药物分子的活性，或研究来源于对药物不同表型的生物体靶标分子同药物结合力的强弱。

5.1 同位素标记法

Hollomon 等在大肠杆菌里表达了大麦云纹病菌 *Rhynchosporium secalis* 多菌灵抗性及敏感菌株的 β-微管蛋白，然后将蛋白提取液同^{14}C标记的多菌灵及^{14}C标记的乙霉威结合，并在4℃下孵育，再将蛋白提取液过凝胶柱，并用磷酸缓冲液分离未结合的多菌灵及乙霉威，将结合到柱上的蛋白洗脱后，在分光光度计上检测洗脱液每分钟的衰变值（disintegrations per minute，衰变/分）。野生敏感菌株（Glu_{198}）同抗性菌株（Gly_{198}）相比，放射性高很多，说明野生敏感菌株同多菌灵的亲和性大于抗性菌株；而同^{14}C标记的乙霉威的亲和力正好相反，抗性菌株大于野生敏感菌株，说明乙霉威同多菌灵存在着负交互抗性，可以在抗药性治理中应用[12]。

5.2 荧光淬灭法

由于放射性标记的药剂分子不容易获取而且对人体有害，所以应用并不普遍。随后人们采用荧光淬灭的方法在离体条件下研究微管蛋白同药剂分子的结合。所谓荧光淬灭指的是能产生荧光的物质在与溶剂或盐、药剂分子相互作用后，使荧光强度减弱的现象，引起荧光强度降低的物质称为猝灭剂。

分子结构和化学环境是影响物质发射荧光和荧光强度的重要因素：饱和的或只有一个双键的化合物不呈现显著的荧光；最简单的杂环化合物，如吡啶、呋喃、噻吩和吡咯等，不产生荧光；大多数无机盐类金属离子不产生荧光，某些情况下，金属螯合物却能产生很强的荧光；至少具有一个芳环或具有多个共轭双键的有机化合物容易产生荧光；稠环化合物也会产生荧光；具有刚性结构的分子容易产生荧光。另外，取代基的性质对荧光体的荧光特性和强度均有强烈影响：苯环上的取代基会引起最大吸收波长的位移及相应荧光峰的改变，通常给电子基团，如$-NH_2$、-OH、$-OCH_3$、$-NHCH_3$和$-N(CH_3)_2$等，使荧光增强；而苯环上的吸电子基团，如-Cl、-Br、-I、$-NHCOCH_3$、$-NO_2$和-COOH，使荧光减弱。

蛋白质分子中由于有色氨酸分子的存在，而色氨酸（侧链）含有苯环及共轭双键，所以蛋白质分子可以产生荧光。当向蛋白质溶液中加入药剂（猝灭剂）后，由于药剂同蛋白的结合，会导致荧光强度的降低，即发生荧光淬灭现象。根据蛋白同不同药剂反应后荧光降低的强度来判断药剂同蛋白亲和力的强弱，降低多，则两者结合强；降低少，则两者结合弱，从而进行蛋白同药剂的结合反应。

Giles 等将布氏锥虫 *Trypanosoma brucei* 的 α-、β-微管蛋白在大肠杆菌里进行了原核表达，在离体条件下采用荧光淬灭法研究了微管蛋白同几种新的氟乐灵类似物的结合力，并与哺乳动物微管蛋白的亲和力做比较。研究发现，不管氟乐灵类似物化学组成如何，这些化合物同布氏锥虫微管蛋白的亲和性均大于同哺乳动物微管蛋白的亲和性，对布氏锥虫 α-微管蛋白的亲和力远远大于同 β-微管蛋白的亲和力。由此看出，这些化合物均是选择性的结合到布氏锥虫 α-微管蛋白上，由于其同哺乳动物微管蛋白亲和力小，故可以被开发用作驱虫剂[21]。

MacDonald 等在大肠杆菌中表达了贾第虫（*Giardia duodenalis*）、微小隐孢子虫（*Cryptosporidium parvum*）及肠道脑胞内原虫（*Encephalitozoon intestinalis*）的 α-和 β-微管蛋白，并在离体条件下采用荧光淬灭法研究了同苯并咪唑类驱虫剂的亲和力。其中，贾第虫、肠道脑胞内原虫对苯并咪唑类药剂敏感，而微小隐孢子虫对苯并咪唑类药剂不敏感。在荧光淬灭实验中，要测定微管蛋白同药剂的结合速率 Kon，解离速率 Koff，两者的比值是亲和常数 Ka（Kon/Koff）。根据 Ka 值的高低即可判断药剂同微管蛋白的亲和性。研究发现：对苯并咪唑类药剂敏感的寄生虫微管蛋白同药剂的亲和力大于对药剂不敏感的寄生虫微管蛋白；来源于敏感寄生虫体内的微管蛋白同药剂表现高的结合速率和低的解离速率而不敏感虫体内的微管蛋白同此正好相反；结合药剂分子后的α、β-微管蛋白异源二聚体同没有结合药剂分子的相比，其聚合速率显著降低，药剂分子抑制了聚合；药剂与β-微管蛋白的亲和常数大于同α-微管蛋白的，说明苯并咪唑类驱虫剂的靶标是β-微管蛋白；而二硝基苯胺类药剂（如氟乐灵等）正好与此相反，其结合靶标是α-微管蛋白；α、β-微管蛋白体外二聚体同β-微管蛋白对苯并咪唑类驱虫剂有相同的亲和常数，说明药剂同二聚体的反应和单体一样；苯并咪唑类驱虫剂不能促使已经发生聚合的α、β-微管蛋白体外二聚体发生解聚。由此，作者推断：药剂可能在α、β-微管蛋白聚合之前结合到微管蛋白上，阻止两者的聚合，因为亲和力高的药剂可大大降低微管聚合的速率和数量[16]。

尽管荧光淬灭法得到了一定的应用，但它也有一定的不足：如微管蛋白上色氨酸残基不多；另外，色氨酸不一定位于药剂同微管蛋白的结合部位或邻近部位，有时荧光变化值不足以反映药剂同微管蛋白的结合强弱。

5.3 半胱氨酸巯基化法

5，5-二硫代双（2-硝基苯甲酸）［DTNB（5，5′-Dithiobis（2-nitrobenzoic acid））］可检测蛋白质或多肽上游离的巯基（-SH）（蛋白质上的半胱氨酸带有巯基）。在有蛋白质存在的情况下，无色的DTNB将被转变成黄色的5-巯基-2-硝基苯甲酸。由于5-巯基-2-硝基苯甲酸在412nm处具有最大吸收，且DTNB的吸收光谱并不干扰巯基的测定，故根据412nm处吸光度的大小就可判断产生的5-巯基-2-硝基苯甲酸的多少，进而判断蛋白质浓度。但当蛋白质同药剂结合后，就会影响到半胱氨酸中游离的巯基数量，进而影响到412nm处的光吸收，根据光吸收的降低值可判断药剂同蛋白亲和力的大小：光吸收值降低的多，说明药剂同蛋白亲和力强。

Koo等研究发现，原核表达产生的辣椒疫霉微管蛋白，在同DTNB结合后可判断α-、β-微管蛋白异源二聚体有（18.66 ±0.13）个半胱氨酸残基；而当有200μmol/L苯菌灵时，游离的半胱氨酸残基数变为（12.43 ±0.12）[13]。在另一项研究中，Koo等发现在有50μmol/L和100μmol/L苯菌灵时，该数值变为13.05和14.33，被保护的半胱氨酸残基数分别为4.77和3.49。被保护的半胱氨酸数即是由于苯菌灵结合而造成的，说明苯菌灵结合到微管蛋白上对微管蛋白构象造成一定影响。对半胱氨酸影响的大小顺序为苯菌灵 > 秋水仙碱 > GTP > 紫杉醇[44]。

随后，Koo等又研究原核表达的马铃薯α-/β-微管蛋白同苯菌灵的结合。在没有苯菌灵存在的情况下，每个α-、β-微管蛋白异源二聚体有（16.93 ±0.11）个半胱氨酸残基，但苯菌灵却可显著地降低半胱氨酸的游离巯基数量，50μmol/L和100μmol/L苯菌灵时半胱氨酸数量分别为（14.99 ±0.32）和（12.27 ±0.21），被保护而不参与DTNB成色反应的半胱氨酸分别为1.94和4.66；GTP和秋水仙碱存在时半胱氨酸的数量为（14.55 ±0.1）和（14.79 ±0.12），被保护而不参与DTNB成色反应的半胱氨酸分别为2.38和2.14。说明对半胱氨酸影响的大小顺序为苯菌灵 > GTP > 秋水仙碱[26]。

目前半胱氨酸巯基化法多应用在研究微管蛋白的构象即微管蛋白是否发生正确折叠上，在微管蛋白同药剂结合方面应用不多[8]。

6 微管蛋白靶标类药物的离体筛选

微管一直是人们研究的热点，这不仅是因为微管在细胞内具有重要的生理功能，更为关键的是以微管蛋白为靶的药物筛选在新药开发中具有非常重要的意义。目前，已开发出的微管蛋白靶标类药物包括抗肿瘤药物、除草剂、杀菌剂、驱虫剂等。人们一直试图将微管蛋白作为药物筛选的分子靶标，在体外实现对药物的筛选，并朝高通量筛选（high-throughput screening）的方向努力。

以微管为靶点的药物研发工作包括对原有药物的结构改造以提高抗有丝分裂活性、降低毒性；联合应用作用于不同位点的药物增强疗效；广泛筛选天热产物及合成半合成化合物以获得有活性的先导化合物；运用现代分子与细胞生物学手段进行作用机制与作用位点的研究[3]。

6.1 利用荧光淬灭进行药物离体筛选

Yakovich 等利用荧光淬灭并结合体外聚合法研究了从 *Leishmania tarentolae* 提取的微管蛋白同氨磺乐灵类似物的体外结合活性。结果表明氨磺乐灵和其类似物 GB-II-5 对 *L. tarentolaede* 微管蛋白的解离速率同对寄生虫 *L. amazonensis* 的解离速率相似，从而表明非致病性的寄生虫 *L. tarentolaede* 的微管蛋白在体外也可应用在药剂筛选中[9]。

Giles 等在研究原核表达的布氏锥虫（*T. brucei*）α-微管蛋白同氟乐灵及其类似物体外结合活性时，采用了荧光淬灭法评价药物的活性大小。4 个氟乐灵类似物（1007，1008，1016 和 1017）对布氏锥虫 α-微管蛋白有很高的亲和力，但同哺乳动物微管蛋白的亲和力较小，表明这 4 个化合物可选择性地结合到锥虫 α-微管蛋白上[21]。

利用荧光淬灭技术研究的多是药剂对单个微管蛋白（α-或 β-微管蛋白）荧光强度的影响，无法评判药剂对 α-、β-微管蛋白聚合的影响（从生物体内直接提取的微管蛋白除外）。尽管药剂可能会导致微管蛋白荧光强度的降低，但并不一定对聚合产生影响，在评价药物活性方面不够全面。

6.2 利用微管蛋白的体外聚合进行药物的离体筛选

由于从生物体内提取的或通过原核表达制备的微管蛋白，在体外适当的条件下均可发生聚合，而药物分子对微管蛋白的体外聚合有影响，故可以用微管蛋白的体外聚合动力学分析微管蛋白类药物的活性及筛选微管蛋白类药物。96 孔板和酶标仪的运用大大加速了微管蛋白聚合活性的检测，使抗微管类药物的高通量筛选成为可能[45]。

6.2.1 无荧光分子的加入

Oxberry 等在研究原核表达的捻转血矛线虫 *H. contortus*α-、β-微管蛋白的体外聚合时，发现阿苯达唑对两者的体外聚合有影响，表明了可应用微管蛋白的体外聚合进行药物结合及离体筛选研究[22]。

Dong 等在制备微管蛋白时采用一次微管聚合/解聚循环，然后进行离子交换层析，在此过程中应用高浓度的谷氨酸钠使微管蛋白获得较高的纯度，而且还能排除微管相关蛋白 MAPs 的影响。将这种含高浓度谷氨酸钠的微管蛋白溶液直接用于微管蛋白的体外聚合分析，表明这样的一个系统对于研究微管蛋白的体外聚合是可行的，并可在药物的高通量筛选中应用[2]。

随后 Hu 等采用两步筛选法进行了促进微管聚合的药物筛选。第一步是在含 GTP 的微管溶液中进行 1 500个样本的大规模筛选，根据最后的光度值评价药物的活性；第二步是根据第一步的实验结果，在无 GTP 的情况下进行样本对微管蛋白聚合影响的动力学分析。结果表明：在不到 3h 的时间就完成了 1 500个样本的分析；从 1 500个样本中筛选到了 108 个样本在 10mg/L 浓度时对聚合有 20% 的促进活性；从 108 个样本中又选取了 5 个进行下一步的动力学验证，其中3个是埃波霉素及其类似物，其他两个化合物同稳定微管的细胞毒性天然产物具有相同的药效基团。两步筛选法是一种高通量、成本小、有效的筛选微管稳定因子的手段[45]。

6.2.2　荧光分子的加入

利用浊度法检测α-、β-微管蛋白的体外聚合反应有些局限性，在体外进行药物筛选时会有许多限制，尤其是在体外进行药物的高通量筛选，如对微管蛋白的浓度要求较高，待检药物对微管蛋白的聚合影响明显。总之，灵敏度不高。后来，开发了用荧光染料标记的方法检测α-、β-微管蛋白的体外聚合，大大提高了检测的灵敏度。

Koo等在研究原核表达的辣椒疫霉（*P. capsici*）α-、β-微管蛋白的体外聚合时利用荧光共振能量转移技术（Fluorescence Resonance Energy Transfer technique，FRET），即在聚合前向α-微管蛋白溶液里加入荧光探针Alexa 488，向β-微管蛋白溶液里加入荧光探针Alexa 514，25℃孵育2h后，在荧光分析仪上测定荧光大小。研究发现，两个荧光探针的荧光值明显高于秋水仙碱，从而表明此方法可以应用在药物的高通量筛选中[13]。

Barron等根据α-、β-微管蛋白的体外聚合提出了一个高通量分析的方法，即在聚合中利用荧光监测微管聚合的策略。此法可以评价抑制聚合药物及促进聚合药物的活性，结果也很精确。Barron应用的荧光染料是4，6-二脒基-2-苯基吲哚（4，6-diamidino-2-phenylindole，DAPI）和DCVJ［9-（2，2-Dicyanovinyl）julolidine］。在微管蛋白解聚液聚合前将荧光染料加入，检测荧光染料发出的荧光值变化情况判断化合物对微管蛋白聚合的影响。在进行药物筛选时，一般在96孔酶标板上进行，这样，一次可以检测出许多样，基本能达到药物高通量筛选的要求。在没有化合物时，微管蛋白聚合动力学表现出典型的“抛物线”形状；而当有化合物存在时，由于化合物会对微管蛋白聚合产生影响，曲线就出现变化，不同化合物对聚合影响的程度不同，根据曲线的形态可判断化合物对聚合影响的大小，进而可判断与微管蛋白亲和力的强弱以及其潜在的应用价值[4]。

孙婉等以常规的免疫组化方法为基础优化实验条件，在96孔板上建立了以微管蛋白为靶点的高通量药物筛选模型；抗肿瘤药物紫杉醇和秋水仙碱作用于人肝癌$HepG_2$细胞后，细胞的免疫荧光强度发生了明显的可检测到的变化，间接反映药物对细胞微管蛋白聚合/解聚作用的影响，与理论预测结果一致。应用建立的模型对28种待测化合物进行HTS筛选，发现3种化合物具有明显地促进微管聚合/解聚作用，表明此模型可应用在化合物活性的精确测定及活性的快速筛选方面[3]。由此可以看出，基于人肿瘤细胞的以微管蛋白为靶点的高通量筛选方法可应用于抗肿瘤化合物的筛选。

Bane等对基于荧光染料的药剂筛选方法进行了改进，在DAPI的参与下，Bane等的方法不仅可以应用在筛选抑制微管聚合的药物中，如秋水仙碱及其类似物，而且还能应用在促进微管聚合的药物筛选中，如紫杉醇[46]。

荧光分子标记法检测α-/β-微管蛋白的体外聚合有许多优点，在有任何能检测荧光的设备（荧光分析仪、荧光分光光度计、荧光酶标仪等）的实验室均可进行。试验非常高效，对微管蛋白的浓度要求低，并能获得高度精确的数据。

寻找与发现新的抗微管蛋白类药物是件非常有意义的工作。微管蛋白不失为一种有效的靶蛋白。因此，利用编码微管蛋白的基因、这些基因的产物以及微管蛋白动力学不稳定性可研发多种高通量药物筛选的方法，促进以微管蛋白为靶点的抗有丝分裂剂和抗肿瘤药物的开发和应用。

参考文献

[1] Downing K H. Structural basis for the interaction of tubulin with proteins and drugs [J]. Annual Review of Cell and Developmental Biology, 2000, 16: 89-111.

[2] Dong H, Li Y Z, Hu W. Analysis of purified tubulin in high concentration of glutamate for application in high throughput screening for microtubule-stabilizing agents [J]. Assay and Drug Development Technologies, 2004, 2 (6): 621-628.

[3] 孙婉，李敏，魏少萌，等. 以微管蛋白为靶的高通量药物筛选方法的建立与应用 [J]. 中国新药杂志，2006，15（21）：1 828-1 831.

[4] Barron D M, Chatterjee S K, Ravindra R, et al. A fluorescence-based high-throughput assay for antimicrotubule drugs [J]. Analytical Biochemistry, 2003, 315: 49-56.

[5] 王霞. 一个新的拟南芥微管结合蛋白——AtMAP18 的功能分析 [D]. 北京：中国农业大学，2005.

[6] 毛同林. 两种拟南芥 65kDa 微管结合蛋白的功能分析 [D]. 北京：中国农业大学，2005.

[7] Castoldia M, Popova A. Purification of brain tubulin through two cycles of polymerization-depolymerization in a high-molarity buffer [J]. Protein Expression and Purification, 2003, 32: 83-88.

[8] Werbovetz K A, Brendle J J, Sackett D L. Purification, characterization, and drug susceptibility of tubulin from Leishmania [J]. Molecular and Biochemical Parasitology, 1999, 98: 53-65.

[9] Yakovich A J, Ragone F L, Alfonzo J D, et al. *Leishmania tarentolae*: Purification and characterization of tubulin and its suitability for antileishmanial drug screening [J]. Experimental Parasitology, 2006, 114: 289-296.

[10] Bellocq C, Andrey-Tornare I, PaunierDoret A M, et al. Purification of assembly-competent tubulin from *Saccharomyces cerevisiae* [J]. European Journal of Biochemistry, 1992, 210: 343-349.

[11] Yoon Y, Oakley B R. Purification and characterization of assembly-competent tubulin from *Aspergillus nidulans* [J]. Biochemistry, 1995, 34: 6 373-6 381.

[12] Hollomon D W, Butters J A, Barker H, et al. Fungal β-tubulin, expressed as a fusion protein, binds benzimidazole and phenylcarbamate fungicides [J]. Antimicrobial Agents and Chemotherapy, 1998, 42 (9): 2 171-2 173.

[13] Koo B S, Park H, Kalme S, et al. α-and β-tubulin from *Phytophthora capsici* KACC 40483: molecular cloning, biochemical characterization, and antimicrotubule screening [J]. Applied Microbiology and Biotechnology, 2009, 82: 513-524.

[14] Huang S J, Ren H Y, Yuan M. *In vitro* assembly of plant tubulin in the absence of microtubule-stabilizing reagents [J]. Chinese Science Bulletin, 2000, 45 (24): 2 258-2 263.

[15] Xu C H, Huang S J, Yuan M. Dimethyl sulfoxide is feasible for plant tubulin assembly *in vitro*: a comprehensive analysis [J]. Journal of Integrative Plant Biology, 2005, 47 (4): 457-466.

[16] MacDonald L M, Armson A, Andrew Thompson R C, et al. Characterisation of benzimidazole binding with recombinant tubulin from *Giardia duodenalis*, *Encephalitozoon intestinalis*, and *Cryptosporidium parvum* [J]. Molecular and Biochemical Parasitology, 2004, 138, 89-96.

[17] MacDonald L M, Armson A, Andrew Thompson R C, et al. Expression of *Giardia duodenalis* β-Tubulin as a soluble protein in *Escherichia coli* [J]. Protein Expression and Purification, 2001, 22: 25-30.

[18] MacDonald L M, Armson A, Andrew Thompson R C, et al. Characterization of factors favoring the expression of soluble protozoan tubulin proteins in *Escherichia coli* [J]. Protein Expression and Purification, 2003, 29: 117-122.

[19] Linder S, Schliwa M, Kube-Granderath E. Expression of *Reticulomyxafilosa* α-and β-tubulins in *Escherichia coli* yields soluble and partially correctly folded material [J]. Gene, 1998, 212: 87-94.

[20] Blackhall W J, Drogemuller M, Schnieder T, et al. Expression of recombinant β-tubulin alleles from *Cylicocyclus nassatus* (Cyathostominae) [J]. Parasitology Research, 2006, 99: 687-693.

[21] Giles N L, Armson A, Reid S A. Characterization of trifluralin binding with recombinant tubulin from *Trypanosoma brucei* [J]. Parasitology Research, 2009, 104: 893-903.

[22] Oxberry M E, Geary T G, Winterrowd C A, et al. Individual expression of recombinant α-and β-Tubulin from *Haemonchus contortus*: polymerization and drug effects [J]. Protein Expression and Purification, 2001, 21: 30-39.

[23] Katzenback B A, Plouffe D A, Haddad G, et al. Administration of recombinant parasite β-tubulin to goldfish (*Carassiusauratus* L.) confers partial protection against challenge infection with *Trypanosoma danilewskyi* Laveran and Mesnil, 1904 [J]. Veterinary Parasitology, 2008, 151: 36-45.

[24] Lubega G W, Geary T G, Klein R D, et al. Expression of cloned β-tubulin genes of *Haemonchus contortus* in *Escherichia coli*: interaction of recombinant β-tubulin with native tubulin and mebendazole [J]. Molecular and Biochemical Parasitolology, 1993, 62 (2): 281-92.

[25] Jang M H, Kim J, Kalme S, et al. Cloning, purification, and polymerization of *Capsicum annuum* recombinant α and β Tubulin [J]. Bioscience, Biotechnology, and Biochemistry, 2008, 72 (4): 1 048-1 055.

[26] Koo B S, Kalme S, Yeo S H, et al. Molecular cloning and biochemical characterization of α-and β-tubulin from potato plants (*Solanum tuberosum* L.) [J]. Plant Physiology and Biochemistry, 2009, 47: 761-768.

[27] Yoshida M, Narusaka Y, Minami E, et al. Expression of *Neurosporacrassa* β-tubulin, target protein of benzimidazole fungicides, in *Escherichia coli* [J]. Pesticide Science, 1999, 55: 362-364.

[28] 张聪，徐建强，于俊杰，等. 禾谷镰孢菌β-微管蛋白基因的克隆及在原核细胞中的表达 [J]. 南京农业大学学报，2011，34 (1): 51-56.

[29] Bell A, Wernli B, Franklin R M. Expression and secretion of malarial parasite β-tubulin in *Bacillus brevis* [J]. Biochimie, 1995, 77: 256-261.

[30] Vats-Mehta S, Yarbrough L R. Expression of chick and yeast β-tubulin-encoding genes in insect cells [J]. Gene, 1993, 128 (2): 263-267.

[31] Linder S, Schliwa M, Kube-Granderath E. Expression of *Reticulomyxa filosa* tubulins in *Pichia pastoris*: regulation of tubulin pools [J]. FEBS Letters, 1997, 417: 33-37.

[32] Nogales E, Wolf S G, Downing K H. Structure of the alpha beta tubulin dimer by electron crystallography [J]. Nature, 1998, 391: 199-203.

[33] Oakley B R. Tubulins in *Aspergillus nidulans* [J]. Fungal Genetics and Biology, 2004, 41: 420-427.

[34] Li M, Yang Q. Isolation and characterization of a *β*-tubulin Gene from *Trichoderma harzianum* [J]. Biochemical Genetics, 2007, 45: 529-534.

[35] Qiu J B, Xu J Q, Yu Y J, et al. Localisation of the benzimidazole fungicide binding site of *Gibberella zeae β2*-tubulin studied by site-directed mutagenesis [J]. Pest Management Science, 2010, 67: 191-198.

[36] Zou G, Ying S H, Shen Z C, et al. Multi-sited mutations of beta-tubulin are involved in benzimidazole resistance and thermotolerance of fungal biocontrol agent *Beauveria bassiana* [J]. Environmental Microbiology, 2006, 8 (12): 2 096-2 105.

[37] BlumeY B, Nyporkoa AY, Yemetsa A I, et al. Structural modeling of the interaction of plant α-tubulin

with dinitroaniline and phosphoroamidate herbicides [J]. Cell Biology International, 2003, 27: 171 - 174.

[38] Delye C, Menchari Y, Michel S, et al. Molecular bases for sensitivity to tubulin-binding herbicides in green foxtail [J]. Plant Physiology, 2004, 136: 3 920 - 3 932.

[39] Morrissette N, Sept D. Dinitroaniline interactions with tubulin: genetics and computational approaches to define the mechanisms of action and resistance [C] //Blume Y B et al. eds. The Plant Cytoskeleton: a Key Tool for Agro-Biotechnology. Springer Science and Business Media B. V., 2008: 327 - 349.

[40] Morrissette N S, Mitra A, Sept D, et al. Dinitroanilines bind α-tubulin to disrupt microtubules [J]. Molecular Biology of the Cell, 2004, 15: 1 960 - 1 968.

[41] Prichard R K. Genetic variability following selection of *Haemonchus contortus* with anthelmintics [J]. Trends in Parasitology, 2001, 17 (9): 445 - 453.

[42] Robinson M W, Trudgett A, Fairweather I. Benzimidazole binding to *Haemonchus contortus* tubulin: a question of structure [J]. Trends in Parasitology, 2002, 18 (4): 153.

[43] Robinson M W, McFerran N, Trudgett A, et al. A possible model of benzimidazole binding to β-tubulin disclosed by invoking an inter-domain movement [J]. Journal of Molecular Graphics and Modelling, 2004, 23: 275 - 284.

[44] Koo B S, Jang M H, Park H, et al. Characterization of *Capsicum annuum* recombinant α-and β-tubulin [J]. Applied Biochemistry and Biotechnology, 2010, 160 (1): 122 - 128.

[45] Hu W, Dong H, Li Y Z, et al. A high-throughput model for screening anti-tumor agents capable of promoting polymerization of tubulin *in vitro* [J]. Acta Pharmaceutica Sinica, 2004, 25 (6): 775 - 782.

[46] Bane S L, Ravindra R, Zaydman A A. High-throughput screening of microtubule-interacting drugs [C] //Zhou J ed. Microtubule Protocols. Totowa N J, Humana Press Inc: 281 - 288.

中国主要蔬菜细菌病害及其防治研究进展*

关　巍[1**]，杨玉文[1]，王铁霖[2]，赵廷昌[1***]
（1. 中国农业科学院植物保护研究所，植物病虫害国家重点实验室，北京　100193；
2. 中国中医科学院中药资源中心，北京　100700）

摘要：随着我国蔬菜产业的不断扩大，种植面积的不断增加，蔬菜病害的发生也更为频繁，给我国蔬菜产业造成严重危害。细菌性病害由于其传播速度快、影响范围广、传播媒介多等特点，给此类病害的防治带来极大困难。本文针对此问题，从发生为害情况、发生规律、检测方法以及防止手段等方面，介绍了西甜瓜、番茄、辣椒、芹菜和菜豆等主要蔬菜上的细菌性果斑病、叶斑病、疮痂病、软腐病、晕疫病、疫病等主要病害。旨在为蔬菜生产上对这些病害的防治提供理论指导。

关键词：蔬菜；细菌性病害；发生规律；检测方法；防治措施

蔬菜产业是我国农业的重要组成部分，不但反映国民生活水平，更是贫困地区脱贫的手段之一。2012 年蔬菜播种面积约 2 000万 hm^2（3 亿亩），总产量约 7 亿 t。据联合国粮农组织（FAO）统计，中国蔬菜播种面积和产量分别占世界的 43% 和 49%，均居世界第一。蔬菜生产具有较高的经济效益。据农业部初步统计，2010 年蔬菜对全国农民人均纯收入贡献逾 830 元，占农民人均收入的 14%，种植蔬菜是一些地区农民致富的重要手段[1]。

随着蔬菜栽培面积的扩大，蔬菜病害向更多地区扩展，如冬季南方北运蔬菜、夏季高山蔬菜与高原夏菜的发展，大大扩展了我国蔬菜病害发生的范围；随着栽培品种的多样化，以及世界和全国范围内引种的增加，病害的种类也随之增加，寄主新记录病害与国内新记录病害增多。尤其是全世界范围内的种子流通，使细菌性种传病害传播更广泛、为害更严重，给蔬菜生产造成严重威胁。

目前，蔬菜细菌性病害防治上还存在着一些难点，如连年轮作、种子检疫落后、种子消毒不力、缺少抗病品种、病菌抗药性上升、生防制剂效果不佳以及化学药剂过量使用等问题。针对以上问题，本文综述了 6 种我国主要蔬菜作物生产中常见的细菌性病害，从为害情况、发生规律、检测手段和防治措施等方面进行阐述，旨在为防治蔬菜细菌性病害提供理论参考，为我国蔬菜产业健康发展提供指导。

1　西甜瓜细菌性果斑病

1.1　为害发生情况

细菌性果斑病（Bacterial fruit blotch，BFB）是葫芦科作物生产上常见的细菌性种传病

*　基金项目：国家西甜瓜产业技术体系（CARS－26）；中国农业科学院科技创新工程项目资助；北京市自然基金（6162023）；国家支撑计划（2012BAD19B06）；公益性行业（农业）科研专项（201003066）

**　作者简介：关巍，男，助理研究员，研究方向：植物病害生物学

***　通讯作者：赵廷昌，男，研究员，研究方向：分子植物病理学；E-mail：zhaotgcg@163. com

害，在世界范围内多有发生，严重威胁各国的西甜瓜种植业的发展。在我国，果斑病的发生也逐年上升。目前，已在我国海南省[2]、新疆维吾尔自治区和内蒙古自治区[3]、中国台湾[4]、吉林省[5]、福建省[6]、山东省和河北省[7]、湖北省[8]及广东省[9]等多个省份检测出该病害，给我国的西甜瓜产业造成了严重的影响。

瓜类细菌性果斑病的病原菌是西瓜食酸菌（*Pseudonomas pseudoalcaligenes* subsp. *citrulli* = *Acidovorax avenae* subsp. *citrulli* = *Acidovorax citrulli*），主要侵染西瓜、甜瓜、南瓜、黄瓜等葫芦科作物，也可以侵染萎叶等其他植物。

1.2 发生规律

西瓜嗜酸菌主要依靠种子传播[10-11]，带病原菌的种子是该病害的主要初侵染源之一。病原菌可以在种子表面附着，也可侵入种子的内部。西瓜嗜酸菌抗逆性强，在种子内部可长时间存活，主要附着在种子的胚乳表层[12]。研究证明，将保存34年的西瓜种子和40年的甜瓜种子种植发芽后，用ELISA检测叶片，结果为阳性的病组织中富集果斑病菌。病原菌在在病残体上越冬的西瓜嗜酸菌也可成为次年初侵染的来源之一[13]。大棚条件下，喷灌和幼苗移植时，西瓜嗜酸菌也可侵染邻近幼苗，最终导致果斑病的大面积暴发[14]。感病的叶片及果实上的菌脓可借助昆虫、风力、雨水或农事操作等方式传播，成为果斑病的再侵染源[15]。高温高湿是瓜类果斑病大规模发生的有利条件，特别是强光、炎热及暴风雨过后，西瓜嗜酸菌的繁殖速度和传播速度加快，该病害大规模流行发生。

1.3 防治方法

BFB是我国的检疫性病害之一，进口种子时，应杜绝带菌的种子进入[16]。另一方面，从无病区引种，测定用于生产的种子的带菌率。

另外，制种时应使用无菌种子进行原种及商业化种子的生产；将制种田与其他瓜类田隔离。不能在有疑似病害发生的田块采种；与发病地块相邻的田块，即使本身未发病，也不能作为制种田。种子播种前，可用3% HCl处理，15min后用水清洗，随后用47%的加瑞农600倍液浸种，过夜处理后播种[17]。

在农业防治措施方面，需进行倒茬轮作。曾发生过瓜类果斑病的田块，至少3年不能种植西瓜或者其他葫芦科作物。田间灌溉宜采用滴灌取代喷灌。田块一旦发生病害，应及时清除病株及病果，并彻底清除田间的杂草残体。不宜在叶片露水未干的田块中工作，避免将在发病地块使用过的农事工具拿到未发病的田块中使用[17-18]。做好苗床清洁处理工作，瓜苗移植后，需及时彻底清理温室中的瓜苗残体及杂草等[16]，不同育苗室的工具不能相互交换使用[18]。

防治果斑病主要化学药剂有53.8%氢氧化铜悬浮剂（可杀得）800倍液、50%氯溴异氰尿酸（消菌灵）800倍液、47%春·王铜（加瑞农）800倍液等。因部分西瓜嗜酸菌株有耐铜性，因此应谨慎使用铜制剂；用47%的加瑞农或者90%的新植霉素，其苗期防效均可超过80%[19]。

在生物防治措施上，防治果斑病的生防菌主要有荧光假单胞菌（*Pesudomonas fluorescens*）工程菌[20]、酵母菌（*Pichia anomala*）[8]、部分葫芦科内生芽孢杆菌（*Bacillus* spp.）[21]等。

使用抗病品种，是防治果斑病的有效措施之一。Hopkins等研究证明，三倍体西瓜较与二倍体西瓜相比，三倍体西瓜更为抗病；抗性强的西瓜品种，果皮通常坚硬且颜色较深，而感病西瓜品种的果皮较浅[14]。但迄今未开发出具有商业价值的免疫或高抗西瓜品种。

2 番茄、辣椒疮痂病

2.1 为害发生情况

番茄－辣椒细菌性疮痂病，俗称“疱病”，是番茄和辣椒上普遍发生的一种病害。近年来，随着番茄和辣椒新品种的引进与栽培面积的不断扩大，病害发生日趋严重。一般病田发病率为20%左右，严重的达80%，常引起早期落叶、落花、落果，对产量影响较大。特别是南方6月，北方7—8月高温多雨或暴雨后，发病尤为严重。

病菌主要为害番茄和辣椒的叶片、茎蔓、果实，尤以叶片上发生普遍。叶片发病，初期形成水渍状、黄绿色的小斑点，扩大后变成圆形或不规则形，暗褐色，边缘隆起，中央凹陷的病斑，粗糙呈疮痂状。病斑大小为0.5～1.5mm。多个病斑结合在一起，所以在叶片上有的仅有几个大病斑，直径达6mm。严重时叶片发黄、干枯、破裂，早期脱落。茎部和果梗发病，初期形成水渍状斑点，逐渐发展成褐色短条斑。病斑木栓化隆起，纵裂呈溃疡状疮痂斑。果实发病，形成圆形或长圆形的黑色疮痂斑。潮湿时病斑上有菌脓溢出。如防治不及时，常引起落叶、落花、落果，甚至植株死亡。

细菌性疮痂病菌（*X. campestris* pv. *vesicatoria*）呈短杆状，两端钝圆，大小为（1.0～1.5）μm×（0.6～0.7）μm；单极生鞭毛，能游动；菌体排列成链状，有荚膜、无芽孢；革兰氏染色阴性，好气，不能发酵利用葡萄糖；在YDC平板培养基上，菌落呈圆形、有凸起，黄色黏稠状；在CKTM选择性培养基上的菌落比在YDC平板培养基上的要小一些，但在黄色菌落周围有一圈清晰的菌环；病原菌发育温度范围为5～40℃，最适温度为27～30℃，致死温度为59℃/10min[22-23]；能够利用半胱氨酸产H_2S，无氧化酶、脲酶活性，有过氧化氢酶活性，不能还原硝酸盐，能水解七叶苷、Tween80，能够液化明胶，能够降解蛋白，不能产吲哚，在0.1% TTC条件下不能生长，能够利用D－阿拉伯糖、D－葡萄糖、D－海藻糖、D－半乳糖和D－甘露糖产少量酸。

2.2 发生规律

病原菌细菌性疮痂病菌主要在种子表面或随病残体在土壤中越冬，成为病害初侵染来源，病菌从气孔或水孔侵入，在叶片上潜育期为3～6d，果实上5～6d。病菌与寄主叶片接触后从气孔进入，在细胞间隙繁殖，致使表皮组织增厚呈疮痂状。病残组织中的病菌在灭菌土壤中可存活9个月。种子带菌是病害远距离传播的重要途径。条件适宜时，病斑上溢出的菌脓借助雨水、昆虫及农事操作传播，并引起多次再侵染。病害多发生于7—8月高温多雨季节，尤其在暴雨过后，伤口增多，有利于病菌的侵染和传播，病害易发生和流行。在这一时期叶片上病斑不形成疮痂而迅速扩展至叶缘，或在叶片上形成许多小斑点而脱落。品种抗病性也有差异。氮肥过量，磷、钾肥不足会加重发病[23]。

2.3 防治方法

到目前为止，还未发现对辣椒或番茄上的所有细菌性疮痂病菌都具有抗性的商业品种。因此，选用不带菌的种子、实行作物轮作制、加强田间卫生管理、实施精确灌溉、一些杀细菌剂（如：铜制杀菌剂、苯并噻二唑等）组合使用，以上措施是目前防治该病害所广泛采用的防治措施。

辣椒细菌性疮痂病是种传病害，种子带菌率很高，一旦病原菌侵入未曾发病的地区，病原菌就会源源不断地侵染。需要对种子进行带菌检测来确定种子的健康状况，在一些疑似发病田，检测土壤中携带的细菌量也是必要的。4 000粒种子中有1粒种子带菌，PCR技术就能检测出来。因此可以应用 PCR 技术来定性和定量地检测种子上和土壤中的细菌。同时还要做好种子消毒工作，用 3% 中生菌素可湿性粉剂 1 000倍液浸种 30min，取出冷水冲洗后催芽播种，可以有效地消除种子表面携带的病原菌。可采用 55℃温水浸种 10min 或在 1∶10 的农用链霉素中浸种 30min 或用 0.1% 的高锰酸钾溶液浸种 15min，清水清洗后催芽播种。也可冷水浸种 10～12 h，再用 1% 硫酸铜溶液浸种 5min，捞出后用少量草木灰或生石灰中和酸性，即可播种。

在云南地区，辣椒疮痂病发病期一般在 5 月上旬，危害盛期在 6 月中旬至 7 月中旬，7 月下旬老病叶大量脱落，新病叶增加缓慢，病情不再发展，另外，在发病初期用 77% 可杀得可湿性粉剂或 72% 农用链霉素可溶性粉剂连续防治两次，可取得较好防效。也可使用 20% 叶枯唑可湿性粉剂 800 倍液喷雾，隔 7 天喷 1 次，连续喷 2～3 次。

3　番茄、辣椒青枯病

3.1　为害发生情况

番茄、辣椒青枯病是由茄科雷尔氏菌（*Ralstonia solanacearum* = *Pseudomonas solanacearum*）引起的细菌性维管束土传病害。侵入木质部维管束并且快速通过维管系统进入到植物地上部分。典型的病害症状包括木质部的褐变，叶片的偏上性生长和致死性枯萎。枯萎症状可能来自于大量细菌在木质部的定殖和大量可以导致维管功能障碍的胞外多糖的产生。几天之内，大量的青枯菌就可以在感病的寄主中生长并最终导致植物的萎蔫死亡。

青枯病是热带、亚热带和温带地区番茄生产的重要病害，发病严重时可使植株整片毁灭。发病国家包括亚洲的印度、朝鲜、埃及、孟加拉国、尼泊尔、日本、印度尼西亚、菲律宾、泰国和中国等。欧洲的德国、波兰、法国，非洲的南非、尼日利亚等。美洲的巴西、美国、巴拿马，澳洲的澳大利亚等。我国台湾及长江流域各省均有发生，尤以四川、浙江、福建、江西、湖南、广东和广西等地发病严重。

3.2　发生规律

青枯病菌一般在寄主植株的伤口或裂缝侵入，吸附在根系伸长区、新生梢或侧根。病原菌入侵作物根系，进入根系组织的细胞间隙中，同时增殖填充整个间隙，病菌通过破坏作物细胞胞间层的果胶聚合物，从中获取养分来进行增殖。进入作物根系之后，青枯病菌进入内皮层，穿过表皮细胞，进入维管束，最后进入木质部，在此进行大量繁殖，同时在植物体内向四处移动，病菌大量的繁殖和分泌的胞外多糖物质阻塞木质部导管，从而阻碍了水分在作物体内的运输，导致植株萎蔫死亡，这样大量的青枯病菌进入到土壤中或寄居于作物残体和水体等环境中腐生生活，直到遇到新的寄主再次引发病害。

3.3　防治方法

药剂防治主要是在发病初期进行灌根和喷雾处理。在发病初期多次施用菌杀清、双效灵、抗枯宁等药剂可将青枯病控制在一定的程度。Methyl Bromide、Chlorlpicrin、Metam Sodium 和 Clazomet 这 4 种药剂防治青枯病也可取得较好的效果。同时，抗生素对防治青枯病也有良好的效果。青枯病菌对四环素、链霉素、氯霉素、利福霉素、螺旋霉素、万古霉素都比较敏感，其中最有效的是四环素。

4 细菌性叶斑病

4.1 为害发生情况

细菌性叶斑病亦称细菌性斑点病或斑疹病，由丁香假单胞番茄致病变种（*Pseudomonas syringae* pv. *tomato*）引起，是番茄等蔬菜生产上的一种重要细菌病害，能够为害番茄的叶、茎、花、叶柄和果实，可造成极大的损失。

番茄细菌性斑点病主要为害番茄叶、茎、花、叶柄和果实。叶片感染，产生深褐色至黑色不规则斑点，直径 2～4mm，斑点周围有或无黄色晕圈。叶柄和茎秆症状和叶部症状相似，产生黑色斑点，但病斑周围无黄色晕圈。病斑易连成斑块，严重时可使一段茎部变黑。为害花蕾时，在萼片上形成许多黑点，连片时，使萼片干枯，不能正常开花。幼嫩果实初期的小斑点稍隆起，果实近成熟时病斑周围往往仍保持较长时间的绿色。病斑附近果肉略凹陷，病斑周围黑色，中间色浅并有轻微凹陷。细菌性斑点病症状与番茄细菌性疮痂病的症状相似，应注意区分。番茄细菌性疮痂病发病时，叶片上病斑圆形或不规则形，但边缘暗褐色，稍隆起，中部色淡，稍凹陷，表面粗糙，叶背早期出现水渍状小斑，逐渐扩展，近圆形或连接成不规则形黄色病斑，隆起较明显。病斑周围有黄色晕圈，后期干枯质脆茎上先于茎沟处出现褪绿水渍状小斑点，然后沿茎沟上下扩展，形成长椭圆形短条斑，中间稍凹陷，褐色，以后木栓化隆起，可以裂开溃疡状果实表面先出现水渍状褪绿斑点，逐渐扩大，病斑褐色到黑褐色，初期带有油渍亮光，后呈现黄褐或黑褐色木栓化，直径 0.2～0.5cm 大小近圆形粗糙枯死斑，带有黄绿色晕圈，病斑稍隆起成疮痂斑。有的病斑相互连接成不规则形大斑块，若果柄与果实连接处受害，易引起落果。

自 1993 年首次报道以来，番茄细菌性斑点病在摩洛哥、南非、印度、澳大利亚、新西兰、奥地利、保加利亚、捷克、斯洛伐克、法国、匈牙利、意大利、罗马尼亚、瑞士、英国、前苏联、南斯拉夫、巴西、约旦、希腊、委内瑞拉、葡萄牙、美国、以色列、加拿大、土耳其等 26 个国家和中国的台湾省有报道发生，该病可造成 5%～75% 的产量损失[24]，1978 年番茄细菌性斑点病造成佐治亚州 160 hm^2 的番茄歉收；1979 年安大略省种植当地番茄苗的 26 个农场中发病的占 61.5%，株发病率达 19%，种植引进番茄苗的个农场中发病的占 36.9%，株发病率为 19%。1998—2004 年，在吉林、辽宁、黑龙江、河北、甘肃、山西、新疆和天津也发现有该病发生[25]。

4.2 发生规律

番茄细菌性斑点病的病菌可在番茄植株、种子、病残体、土壤和杂草上越冬不显症，也可在拟南芥等多种植物的叶和根上存活，病菌在干燥的种子上可存活 20 年等，可随种子远距离传播。播种带菌种子，幼苗即可发病，幼苗发病后传入大田，并通过雨水、昆虫、农事操作等进一步传播，以致造成流行。由于该菌在我国北方冬季保护地番茄上可以平安越冬，因此病菌往往来源于邻作的病田。25℃以下的温度和相对湿度 80% 以上的条件有利发病，高于 30℃时，这种病害发病轻。病菌从开花到直径 3 cm 的幼果时期番茄的果实最感病。

4.3 防治方法

对番茄细菌性斑点病的防治，主要应采取以下方法。

（1）加强检疫，防止带菌种子传入非疫区。

（2）选用抗病品种。

（3）建立无病种子田，采用无病种苗。

（4）温汤56℃浸种30min，或用0.8%醋酸溶液浸种18 h，或用5%酸浸种5～10 h，或用1.05%次氯酸钠浸种20～30min，浸种后用清水冲洗掉药液，稍晾干后再催芽，可有效减少种子上的病原菌基数。

（5）与非茄科蔬菜实行3年以上的轮作。

（6）灌溉、整枝、打杈、采收等农事操作中要注意避免病害的传播。尽量不使用喷灌进行灌溉。

（7）药剂防治。可选用77%可杀得可湿性粉剂400～500倍液，或53.8%可杀得2 000干悬浮剂1 000倍液，或噻菌灵（龙克菌）悬浮剂500倍液，或14%络氨铜水剂300倍液，或0.3%～0.5%氢氧化铜，或200μL/L链霉素或新植霉素，0.05% Cryptonal（chinosol）和0.5% Miceram（copper oxychloride + zineb）进行防治，10 d喷1次，共喷3～4次。

5　芹菜细菌性软腐病

5.1　发生和为害情况

细菌性软腐病是芹菜上常见病害之一，在我国广西、河北、河南、山东、浙江、内蒙古等多个省（区）普遍发生。近年来在北京地区也发现该病害的发生，在病害发病早期，个别植株外围茎秆腐烂变褐，严重时成片的植株整体腐烂死亡，造成绝产。

早期的研究初步确定引起芹菜细菌性软腐病的病原菌是一种寄主范围广泛、致病能力极强的异质性病原细菌：胡萝卜软腐果胶杆菌（*Pectobacterium carotovorum*），其原命名为软腐欧文氏菌（*Erwinia carotovora*）。目前该病原菌在种单位以下具有3个亚种，即 *P. carotovorum* subsp. *carotovorum*，*P. carotovorum* subsp. *odoriferum* 及 *P. carotovorum* subsp. *brasiliensis*。最近，北京地区引起芹菜细菌性软腐病的病原菌被鉴定为胡萝卜软腐果胶杆菌的3个亚种，其中 *P. carotovorum* subsp. *odoriferum* 为优势种[26]。

5.2　发病规律

芹菜细菌性软腐病在设施及陆地栽培条件下均有大面积的发生，以成片集中分布为主。在苗期，植株间距适中、通风好、湿度小病害发生较轻。在近成熟期，植株各茎秆间的相互遮挡使得根部附近局部密闭，形成相对高温高湿的微环境，促进了病原菌的侵染，导致病害在成熟期发生严重，影响芹菜产量及品质。病害发生初期，植株外围叶柄上出现水浸状、浅褐色、不规则形的凹陷斑。当温度在24～28℃、相对湿度为70%～85%时病原菌向植株茎秆内部迅速扩展，发病部位由最初的浅褐色逐渐变为深褐色至黑色，并伴有一定的臭味。病害发生严重时薄壁细胞组织全部解体，植株整体腐烂死亡[27]。

病原菌随病残体在土壤中或留种株以及保护地的植株上越冬，借雨水或灌溉水、昆虫、夹杂着病残体的肥料等传播，病原菌可以由虫口或农事操作导致的伤口侵入。由于芹菜种植密度较大，该病在生长后期田间小环境湿度大的条件下发病重，且可与冻害或其他病害混发[28]。

5.3　防治方法

（1）种子干热处理。带菌种子经85℃处理3 d后，病原菌可被完全抑制；经75℃处理3 d后，病原菌致病力完全丧失[27]。

（2）传统防治方法。在栽培措施上，做好苗床管理，进行合理轮作，降低土壤中的含菌量，同时拔除病株病苗，并撒入生石灰等进行消毒，配合喷施72%农用硫酸链霉素可溶性粉剂1 200倍液，或20%叶枯唑可湿性粉剂500倍液，或80%乙蒜素乳油1 000倍液等防治软腐病；防治腐霉根腐病用 58% 甲霜灵·锰锌可湿性粉剂，或 50% 甲霜·铜可湿性粉剂 8 ~ 10 g/m^2，与半干细土 4 ~ 5kg 混拌均匀，在苗床浇足底水的前提下，先取 1/3 毒土撒在床面上，播种后再覆盖剩余的 2/3 毒土。幼苗出土一周后开始用多菌灵 ＋代森锌或甲霜灵·锰锌等喷雾，轮换用药，每周 1 次，共 2 ~ 3 次；防治菌核病可喷施 40% 菌核净可湿性粉剂 1 000倍液，或 50% 多菌灵可湿性粉剂 500 倍液等，或用石灰氮进行土壤消毒预防[28]。

（3）新型防治技术。使用纸筒育苗的新型防治技术，在芹菜移苗期采用专用含药育苗基质进行移栽。具体操作方法：首先配制含药基质，将育苗纸筒（规格为 4 cm × 7 cm）伸展开，4 个角固定好，向纸筒内装满含药育苗基质，去除纸筒表面多余的基质，用喷壶或带有花洒的水管浇水，待纸筒内的基质充分吸水后，开始移栽 1 月龄芹菜幼苗，每个纸筒栽入 1 ~ 2 株芹菜苗，之后按照正常的水分管理即可。定植前浇一遍水，使育苗纸筒能独立分开，定植时先按传统定植方法开穴，然后将幼苗直接带纸筒放入定植穴，使纸筒上沿高出土面 1 cm。定植后按照正常的管理进行浇水、缓苗。该技术对芹菜苗期及定植前期的腐霉根腐病、软腐病及菌核病均有较好的预防效果，纸筒育苗的幼苗生长健壮，而常规育苗生长较差，出现叶色变黄现象。定植后纸筒育苗无缓苗过程，与常规育苗相比植株生长翠绿、健壮，达到培育无病壮苗的目的。

6　莴苣黑腐病

6.1　发生和为害情况

莴苣黑腐病又称腐败病、细菌性叶斑病。主要为害肉质茎，也为害叶片。肉质茎染病，受害处先变浅绿色，后转为蓝绿色至褐色，病部逐渐崩溃，从近地面处脱落，全株矮化或茎部中空；叶片染病生不规则形水渍状褐色角斑，后变淡褐色干枯呈薄纸状，条件适宜时可扩展到大半个叶子，周围组织变褐枯死，但不软腐。

莴苣黑腐病的病原菌为油菜黄单胞菌葡萄蔓致病变种（*X. campestris* pv. *vitians*）。菌体杆状，短链生，大小（0.42 ~ 0.83）μm ×（0.65 ~ 1.25）μm。有荚膜，无芽孢，单极鞭毛，革兰氏染色阴性，好气性。肉汁胨琼脂平面上菌落乳黄色圆形，平滑且薄，边缘整齐。在肉汁胨液中培养呈轻云雾状，具不完整的菌膜，有片段的薄膜飘浮，在马铃薯柱上生长旺盛，呈亮黄色。最适生长温度 26 ~ 28℃，最高 35℃，最低 0℃，51 ~ 52℃经 10min 致死。

6.2　发生规律

病菌在病残体上或种子内越冬，翌年从幼苗叶片的气孔或叶缘水孔、伤口处侵入，细菌侵入后形成系统侵染。远距离传播主要靠种子，在田间借雨水、昆虫、肥料传播蔓延，高温高湿条件下易发病，地势低洼、重茬及害虫为害重的地块发病重。

6.3　防治方法

农业防治方法。与葱蒜类、禾本科作物实行 2 ~ 3 年以上轮作。施用堆肥。选用无病种子，雨后及时排水，注意防治地下害虫。

化学防治方法。发病初期开始喷洒 30% 氧氯化铜悬浮剂 800 倍液或 30% 绿得保悬浮剂 300 ~ 400 倍液或 50% 琥胶肥酸铜（DT）可湿性粉剂 500 倍液、70% 琥珀乙膦铝（DTM）可湿性粉剂 500 倍液、25% 噻枯唑可湿性粉剂 500 ~ 1 000倍液、72% 农用硫酸链霉素 3 500 ~ 4 000倍液，47% 加瑞农可湿性粉剂 1 000倍液，每亩喷对好的药液 50L，隔 10 天左右 1 次，防治 1 次或 2 次。采收前 3 天停止用药。

7 莴苣叶焦病

7.1 发生和为害情况

莴苣叶焦病可引起莴苣外侧叶片或心叶边缘产生褐色区，有的坏死，有的波及叶脉。组织坏死后，易被腐生菌寄生。叶片失水过多表现叶色淡、脉焦或叶脉间坏死，叶片水分严重不足，出现叶焦或叶缘烧焦或干枯。

莴苣叶焦病的病原菌为菊苣假单胞菌（*Psedomonas cichorii*），属荧光假单胞菌。菌体杆状，具多根极生鞭毛，生长适温30℃，寄主范围广，除为害菊苣外，还可侵染莴苣、卷心菜、花椰菜、番茄、芹菜、茼蒿等。

7.2 发生规律

莴苣、莴笋在叶片失水情况下，被该菌侵染后，引起健康组织的病变，但它们毕竟是后来腐生上来的，在坏死组织里繁殖、蔓延。尤其是遇有低湿高温时，会使叶片中水分耗尽，致叶片边缘细胞死亡；根系吸收水分过少时，也会产生类似的症状，使水分运输受到抑制。此外，根系生长弱、土壤干燥和低温、盐分浓度高等诸因素，均可使植株水分吸收受阻，尤其是成株，由于叶片多且大，失水多易出现上述症状。

7.3 防治方法

（1）保持土壤湿润和含水量适宜，避免温度过高、过低，可防止该病发生和蔓延。

（2）保护根系功能正常，相对湿度不宜长时间过高，尽量保持湿度正常，增加空气流通，有助于阻止叶片受到伤害。通风适当，使叶中水分散失适宜。

（3）土壤盐分含量不宜过高。

（4）提倡施用酵素菌沤制的堆肥或腐熟有机肥。

（5）必要时喷施惠满丰、促丰宝和保丰收多元叶面肥。

8 菜豆常见种传细菌性病害

目前，严重威胁我国豆类安全生产和影响粮食安全的植物细菌病害包括菜豆细菌性萎蔫病（*C. flaccumfaciens*）、菜豆晕疫病（*P. syringae* pv. *phaseolicola*）、菜豆普通细菌性疫病（*X. campestris* pv. *phaseoli*）、豌豆细菌性疫病（*P. syringae* pv. *syringae*）等病害，其中菜豆细菌性萎蔫病、菜豆晕疫病和豌豆细菌性疫病是我国进境植物检疫性有害生物，这些病原菌都可通过种子携带进行远距离传播。

8.1 菜豆细菌性萎蔫病

菜豆细菌性萎蔫病的病原菌为短小杆菌属菜豆萎蔫致病变种（*C. flaccumfaciens* pv. *flaccumfaciens*）该病主要为害豆类作物维管束引发系统发病，其典型症状为种苗叶片有不规则黄色病斑，并由叶边缘往中脉发展逐渐萎蔫；茎上一般有锈色病斑，严重时造成植株矮化，幼枝枯死[29]。

该病害主要寄主为菜豆、大豆、利马豆、赤豆、绿豆、豇豆、扁豆等豆科作物，其病原菌主要依靠种子进行远距离传播，能在种子内存活20年以上；目前主要分布于美国、加拿大、澳大利亚、土耳其、比利时、意大利、非洲南部和南美洲部分地区和国家，现被列为我国进境植物检疫性有害生物。

8.2 菜豆晕疫病

病原菌为丁香假单胞菌菜豆致病型（*Pseudomonas savastanoi* pv. *phaseolicola*，异名：

Pseudomonas syringae pv. *phaseolicola*）。该病害发病初期叶片上产生暗绿色水渍状不规则病斑，周围常伴有黄色褪绿晕圈，发病组织呈半透明状；后期叶片脱落，在潮湿的环境中，病斑表面常有菌脓产生[30]。

该病害主要为害菜豆、扁豆、豌豆、绿豆、豇豆、大豆、月豆、多花菜豆等多种豆科作物，其病原菌也主要依靠种子进行远距离传播；目前主要分布在印度、日本、中国、以色列、澳大利亚、比利时、英国、美国、阿根廷、哥伦比亚等国家，现也被列为我国进境植物检疫性有害生物。

8.3　菜豆普通细菌性疫病

病原菌为地毯草黄单胞菌菜豆变种（*Xanthomonas axonopodis* pv. *phaseoli*，异名：油菜黄单胞菌菜豆变种 *Xanthomonas campestris* pv. *phaseoli*）或（和）褐色黄单胞褐色亚种（*Xanthomonas fuscans* subsp. *fuscans*）[31-32]。

该病害在菜豆整个生育期均可发生，常危害叶、茎、豆荚和种子。病斑呈红褐色溃疡状条带，叶尖或叶缘上产生暗绿色水渍状无规则斑点，周围有黄色晕圈，病斑后期呈深褐色并凹陷，病部常溢出淡黄色菌脓[33]。感染种子是菜豆普通细菌性疫病初侵染来源和主要传播载体。目前该病害主要分布在美国、墨西哥、古巴、牙买加、多米尼加共和国、巴西、荷兰、意大利、中国、日本、印度、韩国等国家主要菜豆种植区，被列为 EPPO A2 类有害生物，其为害可导致 20% ~60% 的产量损失，严重时可高达 80%[34]。

8.4　豌豆细菌性疫病

病原菌为丁香假单胞菌豌豆致病型（*Pseudomonas syringae* pv. *pisi*）。

豌豆细菌性疫病主要为害茎秆、叶片和豆荚，系统侵染可导致整个植株萎蔫枯死，对产量和种子质量影响较大。主要寄主为豌豆、扁豆、香豌豆、野豌豆、紫花豌豆、豇豆、广叶山藜豆等，其病原菌主要通过带菌种子远距离传播，并可在种子表面或内部存活 10 个月以上[35]；目前在欧洲、亚洲、美洲和大洋洲均有发生，但在我国未见报道，并被列为我国《中华人民共和国进境植物检疫性有害生物名录》中的检疫性植物病原细菌。

8.5　防治方法

目前，对该病害的防治措施首先要加强种子检疫。已发病的国家和地区通常采用的防治措施：在无病田块种植无病种子，在发病田块选用抗病品种。常见的检测方法有如下几种。

8.5.1　传统检测方法

目前，豆类重要植物病原细菌传统检测方法主要有分离培养检测法、幼苗试种法和生理生化试验，其基本特点为：操作简单、灵敏度低、检测时间周期较长。主要有分离培养检测法：通常使用半选择性培养基可以更好地提高分离的靶向性以及减少其他杂菌的干扰，而且制备半选择性培养基需要添加放线菌酮、头孢克洛或者硼酸等抗生素和甲基紫 B、酚红或者溴甲酚紫等染料[36]。幼苗试种法：在温室适合发病条件下（27 ~30℃，相对湿度在 85% 以上），播种 2 ~3 周后即可检测出种子的带菌率。该方法比较直接，并最能反映种子带菌情况，也得到了国际种子检测协会的认可，但是受环境和气候影响较大，例如夏季比较敏感，而冬季容易产生假阳性，不能保证种子批次真实带菌情况。生理生化检测法：随着生理生化技术的快速发展，不断衍生出其他新的检测方法，如 Biolog、脂肪酸（FAME）等[37]。Biolog 鉴定系统主要是根据细菌对 95 种碳氮源的利用情况，以四唑紫（Tetrazolim）为指示剂，鉴定数据经过电脑处理并匹配数据库信息得出最终结果，实现待测物的快速检测。

8.5.2 血清学检测方法

血清学检测方法主要包括酶联免疫法（Enzyme-Linked Immunosorbance Assays，ELA-SA）、免疫荧光法（Immuno-fluorescence，IF）、直接琼脂双扩散（Direct A-gar Double Diffusion，DADD）、免疫分离法（Imnuno-isolation，IIS）和免疫试纸条（Immuno-strip Tests）等。

8.5.3 分子生物学检测方法

分子生物学检测方法主要有传统 PCR 法：该方法相比分离培养、生理生化法等传统的检测方法，PCR 具有快速、特异性强、灵敏度高等点（李志峰 等，2015）。通常利用 16S rDNA、质粒 DNA、扩增转录间隔区（ITS）、毒素基因片段（toxin genes）、无毒性基因（Avirulence genes）等靶标片段设计引物探针；BIO-PCR 法：利用选择性或半选择性培养基生物富集结合 PCR 方法，进行病原细菌检测的技术称之为 BIO-PCR[38]。该技术运用培养基能抑制非靶标菌的生长，但对靶标菌起到富集作用，从而较大程度上提高检测灵敏度，可以检测低于 10 cfu/mL 的靶标菌[39]；同时，在一定程度上还能减少植物组织提取液中的 PCR 抑制因子对检测结果的影响；另外还有实时荧光 PCR 检测技术的基本原理是在 PCR 反应体系中加入荧光信号基团，利用荧光信号积累实时监测整个 PCR 进程，最后通过每个样品的荧光信号到达并高于设定阈值所经历的循环数即 Ct 值；此外，环介导等温扩增技术（loop-mediated isothermal amplification，LAMP）目前也用来检测相关病害，主要是利用两对特殊设计的引物和具有置换活性的 DNA 聚合酶，在等温条件下与模板快速结合、特异的置换扩增，其产物是一系列大小不一的 DNA 片段混合物，可以电泳分析和荧光定量检测，也可以荧光目测比色和焦磷酸镁浊度肉眼观察检测；多重 PCR 法：该方法在菜豆相关检疫性病害的检测中也应用广泛，是在同一反应体系采用多对特异性引物进行扩增大小不同的靶标片段，可在电泳检测下鉴别出来，也可通过实时荧光 PCR 进行区分。

8.5.4 其他检测方法

红外光谱检测技术原理是通过连续变化的红外光照射物体后，致使分子吸收某些光的频率辐射，而引起分子振动和能级由基态到激发态的跃迁，并通过产生的特异性峰位对检测物进行鉴定。徐晓鸥等[40]首次利用红外光谱技术对菜豆晕疫病菌与豌豆细菌性疫病菌进行检测鉴定，发现这两种病原菌在 3 000～4 000/cm 内具有特异性峰位，可作为这两种菌的生物标记。该方法的优点在于检测步骤简单、周期短，仅耗时 2 h 左右。

参考文献

［1］ 李宝聚. 蔬菜主要病害 2013 年发生概况及 2014 年发生趋势［J］. 中国蔬菜，2014，2：5－8.

［2］ 张荣意，谭志琼. 西瓜细菌性果斑病症状描述和病原菌鉴定［J］. 热带作物学报，1998，19（1）：70－76.

［3］ 赵廷昌，孙福在. 哈密瓜细菌性果斑病病原菌鉴定［J］. 植物病理学报，2001，31（4）：357－364.

［4］ Cheng A H，Hsu Y L，Huang T C，et al. Susceptibility of cucurbits to *Acidovorax avenae* subsp. *citrulli* and control of fruit blotch on melon［J］. Plant Pathology Bulletin，2000，9（4）：151－156.

［5］ 金岩，张俊杰，吴燕华，等. 西瓜细菌性果斑病的发生与病原菌鉴定［J］. 吉林农业大学学报，2004，26（3）：263－266.

［6］ 蔡学清，黄月英，杨建珍，等. 福建省西瓜细菌性果斑病的病原鉴定［J］. 福建农林大学学报：自然科学版，2006，34（4）：434－437.

［7］ 赵廷昌，赵洪海，王怀松. 山东省西瓜、甜瓜发生瓜类细菌性果斑病［J］. 植物保护，2009，35

(5): 170 - 171.

[8] 王晓东，孙玉宏，葛米红，等. 武汉地区保护地甜瓜细菌性斑点病病原鉴定 [J]. 湖北农业科学，2010，49（8）：1 883 - 1 886.

[9] 任小平，李小妮，王琳，等. 广东西瓜果斑病的病原鉴定 [J]. 华南农业大学学报，2010，31（4）：40 - 43.

[10] Wall G C. Control of watermelon fruit blotch by seed heat-treatments [J]. Phytopathology 1989，79：1 191（Abstract）.

[11] Sowell G Jr，Schaad N W. *Pseudomonas pseudoalcaligenes* subsp. *citrulli* on watermelon：Seed transmission and resistance of plant introductions [J]. Plant Disease Report，1979，63：437 - 441.

[12] Dutta B，Genzlinger L L，Walcott R R. Localization of *Acidovorax avenae* subsp. *citrulli*（Aac），the bacterial fruitblotch pathogen in naturally infested watermelon seed [J]. Phytopathology，2008，98（6）：suppl，p. S49.

[13] Kucharek T，Perez Y，Hodge C. Transmission of watermelon fruit blotch bacterium from infested seed to seedlings [J]. Phytopathology，1993，83：467.

[14] Hopkins D L，Thompson C M，Elmstrom G W. Resistance of watermelon seedlings and fruit to the fruit blotch bacterium [J]. HortScience，1993，28（2）：122 - 123.

[15] 张祥林，莫桂花. 西瓜上的一种新病害——细菌性果斑病 [J]. 新疆农业科学，1996，4：183 - 184.

[16] 侯建雄，方雯霞. 西瓜细菌性果腐病及其检疫对策 [J]. 中国进出境动植检，1997，1：36 - 37.

[17] 赵廷昌，孙福在，刘双平，等. 哈密瓜细菌性果斑病及其防治 [J]. 植物保护，2001a，27（1）：46 - 47.

[18] 赵廷昌，王建荣，孙福在，等. 哈密瓜细菌性果斑病综合治理指南 [J]. 植保技术与推广，2003b，23（4）：17 - 18.

[19] 赵廷昌，孙福在，王建荣，等. 药剂处理种子防治哈密瓜细菌性果斑病的研究 [J]. 植物保护，2003a，29（4）：50 - 53.

[20] 周洪友，杨静，宋娟. 荧光假单胞工程菌株对甜瓜细菌性果斑病的生物防治 [J]. 中国植保导刊，2009，29（1）：9 - 12.

[21] 蔡学清，鄢凤娇，林玉，等. 西瓜细菌性果斑病拮抗内生细菌的分离和筛选 [J]. 福建农林大学学报，2009，38（5）：465 - 470.

[22] Sahin F，Miller S A. Characterization of Ohio strains of *Xanthomonas campestris* pv. *vesicatoria*，causal agent of bacterial spot of pepper [J]. Plant Disease，1996，80：773 - 778.

[23] 董金皋. 农业植物病理学（北方本）[M]. 北京：中国农业出版社，2001：401 - 402.

[24] Yunis H，Bashan Y，Okon Y，et al. Chemical control of bacterial speck of tomato and its effect on tomato yield [J]. Hassadeh，1980，60（5）：1 004 - 1 007.

[25] 赵廷昌，孙福在，李明远，等. 番茄细菌性斑点病的发生与防治 [J]. 中国蔬菜，2004（4）：64.

[26] 田宇，马亚丽，何付新，等. 北京地区芹菜细菌性软腐病菌鉴定及其致病力分析 [J]. 植物病理学报，2016，4：433 - 442.

[27] 晋知文，宋加伟，谢学文，等. 芹菜细菌性软腐病病原的分离与鉴定 [J]. 植物病理学报，2016，3：304 - 312.

[28] 石延霞，孟姗姗，陈璐，等. 李宝聚博士诊病手记（七十一）芹菜根腐类病害的病原菌鉴定及新型防治技术 [J]. 中国蔬菜，2014，6：71 - 73.

[29] 赵友福. 菜豆细菌性萎蔫病菌 [J]. 中国进出境动植检，1994（3）：38 - 39.

[30] 陈云兰，周丽洪. 菜豆晕疫病病原菌鉴定 [J]. 湖南农业科学，2013（1）：91－93，100.

[31] Mkandawire A B C，Mabagala R B，Guzmán P，et al. Genetic diversity and pathogenic variation of common blight bacteria（*Xanthomonas campestris* pv. *phaseoli* and *X. campestris* pv. *phaseoli* var. *fuscans*）suggests pathogen co-evolution with the common bean [J]. Phytopathology，2004，94（6）：593－603.

[32] Schaad N W，Postnikova E，Lacy G H，et al. Reclassifi-cation of *Xanthomonas campestris* pv. *citri*（ex *Hasse* 1915）Dye 1978 forms A，B/C/D，and E as *X. smithii* subsp. *citri*（ex *Hasse*）sp. *nov.* nom. rev. comb. nov.，*X. fuscans* subsp. *aurantifolii*（ex Gabriel 1989）sp. nov. nom. rev. comb. nov.，and *X. alfalfae* subsp. *citrumelo*（ex Riker and Jones）Gabriel et al.，1989 sp. nov. nom. rev. comb. nov.；*X. campestris* pv. *malvacearum*（ex Smith 1901）Dye 1978 as *X. smithii* subsp. *smithii* nov. comb. nov. nom. nov.；*X. campestris* pv. *alfalfae*（ex Riker and Jones，1935）Dye 1978 as *X. alfalfae* subsp. *alfalfae*（ex Riker et al.，1935）sp. nov. nom. rev.；and "var. fus-cans" of *X. campestris* pv. *phaseoli*（ex Smith，1987）Dye 1978 as *X. fuscans* subsp. *fuscans* sp. nov. Systematic and Applied Microbiology，2005，28：494－518.

[33] 陈泓宇，徐新新，段灿星，等. 菜豆普通细菌性疫病病原菌鉴定 [J]. 中国农业科学，2012，45（13）：2 618－2 627.

[34] Zapata M，Beaver J S，Porch T G. Dominant gene for common bean resistance to common bacterial blight caused by *Xanthomonas axonopodis* pv. *phaseoli* [J]. Euphytica，2011，179（3）：373－382.

[35] 封立平，尼秀媚，厉艳，等. LAMP 方法检测豌豆细菌性疫病菌. 植物检疫，2013，27（3）：80－84.

[36] Jones J B，Gitaitis R D，Schaad N W. Gram-negative bac-teria：*Acidovorax* and *Xylophilus*//Schaad N W，Jones J B，Chen W. Laboratory guide for identification of plant pathogenic bacteria（3rd edition）[M]. St Paul，Minnesota：APS Press，2001：121－138.

[37] Calzolari A，Tomesani M，Mazzucchi U. Comparison of immunofluorescence staining and indirect isolation for the detection of Corynebacterium flaccumfaciens in bean seeds [J]. EP-PO Bulletin，1987，17（2）：157－163.

[38] Schaad N W，Cheong S S，Tamaki S C，et al. A combined biological and enzymatic amplification（BIO-PCR）technique to detect *Pseudomonas syringae* pv. *phaseolicola* in bean seed extracts [J]. Phytopathology，1995，85（2）：243－248.

[39] Zhao T，Feng J，Sechler A，et al. An improved assay for detection of *Acidovorax citrulli* in watermelon and melon seed [J]. Seed Science and Technology，2009，37：337－349.

[40] 徐晓鸥，吴志毅，陈曦，等. 2 种豆类植物病原细菌的红外光谱检测与鉴定 [J]. 浙江农业科学，2014（2）：233－235.

琥珀酸脱氢酶抑制剂类杀菌剂及其抗性研究进展

李　静，侯毅平，周明国
（南京农业大学植物保护学院，南京　210095）

摘要：琥珀酸脱氢酶抑制剂类（SDHIs）杀菌剂是一类作用于病原菌琥珀酸脱氢酶而抑制其呼吸作用的杀菌剂，能够有效防治多种植物病原菌，目前已成功开发了18个品种。然而随着其使用量的增加，该类杀菌剂导致的抗性问题也日益凸显。本文主要对这类杀菌剂的品种、作用机制、应用、抗性机制、抗性发生情况以及抗性治理措施进行综述。

关键词：琥珀酸脱氢酶抑制剂；作用机制；抗性治理

Progress on Research and Development of Succinate Dehydrogenase Inhibitor Fungicides and Its Resistance

Li Jing，Hou Yiping，Zhou Mingguo
（*College of Plant Protection*，*Nanjing Agricultural University*，*Nanjing* 210095）

Abstract：SDHIs were classified by inhibitioning the activity of succinate dehydrogenase to interfere the respiration of plant pathogens，which is effective in controling a variety of plant pathogens. Now at least 18 compounds of SDHIs have been successfully developed. However，SDHIs resistance is increasingly highlighted with the increasing usage of it. This paper is mainly subscribed the compounds，mode of action，resistance mechanism，application resistance situation and resistance management of this group of fungicides.

Key words：Development of Succinate Dehydrogenase Inhibitor Fungicides；the mode of action；resistance management

1　引言

琥珀酸脱氢酶（Succinatedehydrogenase，简称SDH）属于黄素酶类，是一种膜结合酶，能够与线粒体内膜进行结合。该酶作为连接氧化磷酸化与电子传递的枢纽，能够为真核细胞线粒体和多种原核细胞需氧和产能的呼吸链提供电子，为线粒体的一种标志酶[1]。琥珀酸脱氢酶与FAD的关系是以共价键相互连接，因此它是酶和辅基的关系。琥珀酸脱氢酶能够专一的催化琥珀酸的脱氢过程[2-3]。

琥珀酸脱氢酶抑制剂类（SDHIs）杀菌剂是杀菌剂抗性行动委员会（FRAC）新划分出来的一类作用机制和抗性机理相似的化合物[4]。该类杀菌剂的作用靶标是琥珀酸脱氢酶复合物，通过作用于蛋白复合体Ⅱ影响病原菌的呼吸链电子传递系统来阻碍其能量的代谢，从而达到抑制病原菌的生长的目的，并最终导致其死亡[5]。

目前，已成功开发的这类杀菌剂有18个品种，其中20世纪60年代开发的萎锈灵（carboxin）是该类杀菌剂最早的品种，随后又开发了氧化萎锈灵（oxycarboxin），但这些品种防治谱较窄，只能用于防治菊花属锈病（*Puccinia oriana*）和大麦散黑穗病（*Ustilago nu da*）等[6-7]有限的病害种类。随后又成功研制出一些广谱性的 SDHIs 类杀菌剂，如灭锈胺（mepronil）、氟酰胺（flutolanil）、麦锈灵（benodanil）、甲呋酰胺（fenfuram）、啶酰菌胺（boscalid）、噻呋酰胺（thifluzamide）、呋吡菌胺（furametpyr）、吡噻酰胺（penthiopyrad）以及近 2 年新研发的氟吡菌酰胺（fluopyram）、联苯吡啶胺（bixafen）、吡唑萘菌胺（isopyrazam）、氟唑环菌胺（sedaxane）和氟唑菌苯胺（penflufen），它们可被用于防治许多作物上的多种病害[8-9]，这使得该类药剂不断的发展成为一类重要的杀菌剂。但这类杀菌剂因作用位点单一，使得这种药剂在被不断的广泛使用的同时抗药性问题也日益凸显[10]。因此，该类药剂的抗药性相关问题也成为其目前研究的热点。SDHIs 作用于电子传递链的复合物 II，抗性的发生主要与 SdhB、SdhC 或 SdhD 的点突变有关，大部分病原真菌对 SDHIs 的抗性与 SdhB 点突变有关，SdhB 点突变发生位置比较单一，在多种病原菌中突变均发生在相同的组氨酸上即 H272，而 SdhC 和 SdhD 突变位点比较多[11]。

2 SDHIs 杀菌剂品种及其应用

自 1966 年报道萎锈灵以来，SDHI 类杀菌剂根据其开发时间的先后共划分为三代，药剂的生物活性随着技术的发展不断提高，防治谱也在不断扩大。根据开发时间可将该类杀菌剂分为三代，第一代为 20 世纪 60—80 年代开发的 6 个产品；第二代为 20 世纪 90 年代开发的 2 个产品；第三代为 2000 年后开发的 10 个产品[12]。这些产品根据其化学结构可分为八类，其中吡唑酰胺类产品有 8 种，苯基苯甲酰胺类产品有 3 种，氧硫杂环己二烯酰胺类产品 2 种，吡啶乙基苯甲酰胺类、呋喃酰胺类、吡唑酰胺类、噻唑酰胺类及吡啶酰胺类产品各一种。详见表 1。

表 1　琥珀酸脱氢酶抑制剂类杀菌剂种类

序号	英文通用名	中文通用名	结构类型
第一代	*carboxin*	萎锈灵	氧硫杂环己二烯酰胺类
	oxycarboxin	氧化萎锈灵	氧硫杂环己二烯酰胺类
第二代	*mepronil*	灭锈胺	苯基苯甲酰胺类
	flutolanil	氟酰胺	苯基苯甲酰胺类
	benodanil	麦锈灵	苯基苯甲酰胺类
	fenfuram	甲呋酰胺	呋喃酰胺类
	furametpyr	呋吡菌胺	吡唑酰胺类
	thifluzamide	噻呋酰胺	噻唑酰胺类
第三代	*boscalid*	啶酰菌胺	吡啶酰胺类
	penthiopyrad	吡噻菌胺	吡唑酰胺类
	isopyrazam	吡唑萘菌胺	吡唑酰胺类
	fluxapyroxad	氟唑菌酰胺	吡唑酰胺类
	bixafen	联苯吡菌胺	吡唑酰胺类
	sedaxane	氟唑环菌胺	吡唑酰胺类
	penflufen	氟唑菌苯胺	吡唑酰胺类
	fluopyram	氟吡菌酰胺	吡啶乙基苯甲酰胺类
	benzovindiflupyr	苯并烯氟菌唑	吡唑酰胺类
	isofetamid	*isofetamid*	吡唑酰胺类

2.1　第一代 SDHIs 杀菌剂

2.1.1　萎锈灵（carboxin）

1966 年，萎锈灵由美国有利来路（现科聚亚）创制，并于 1969 年上市，是一种选择性较强的内吸型杀菌剂。生产中可采用拌种、闷种和浸种等方法防治大麦、小麦、燕麦、玉米、高粱、谷子等禾谷类黑穗病，亦可用于叶面喷洒防治小麦、豆类、梨等锈病以及棉花苗期病害及黄萎病、立枯病，也可作木材防腐剂。本品对丝核菌有效，特别适用于棉花、花生、蔬菜和甜菜的种子处理。对作物具有生长刺激作用，能够起到一定的增产作用。

2.1.2　氧化萎锈灵（oxycarboxin）

1973 年，有利来路（现科聚亚）公司在萎锈灵的基础上又开发了氧化萎锈灵，其与萎锈灵同属氧硫杂环己二烯酰胺类杀菌剂，并于 1975 年上市。氧化萎锈灵同萎锈灵一样都属于内吸型杀菌剂，主要作用于谷物和蔬菜上的锈病，兼具预防和治疗作用。

2.2　第二代 SDHIs 杀菌剂

2.2.1　灭锈胺（mepronil）

1980 年，灭锈胺由日本组合化学工业株会公社研发，并于 1981 年成功上市。灭锈胺也是一种内吸型杀菌剂，具有预防和治疗作用，能够有效的阻止和抑制病原菌的入侵。在生产中可以用于防治由担子菌亚门真菌引起的水稻、谷物、蔬菜、果树、烟草以及观赏植物上的病害。田间药效试验表明，对水稻纹枯（*Rhizoctonia solani*）有较好防效。灭锈胺用作种子和土壤处理剂时，能够有效的防治烟草及蔬菜上的猝倒病。此外，灭锈胺还可以用作各领域的抗菌防霉剂。

2.2.2　氟酰胺（flutolanil）

氟酰胺由日本农药株式会社创制，1981 年在英国布赖顿植保会议上首次报道，1986 年上市。是一种内吸型杀菌剂，兼具治疗和保护作用。能够阻碍植物体表菌丝的生长和穿透，引起菌丝的消解，主要用于防治由担子菌纲真菌引起的各种作物病害，如谷物雪腐病（*Typhula* spp.）、小麦纹枯病（*Rhizoctonia cerealis*）和马铃薯黑痣病（*Rhizoctonia solani*），甜菜、花生等作物上的白绢病（*Corticium rolfsii*），蔬菜及观赏植物上由立枯丝核（*Rhizoctonia solani*）等，其中对水稻纹枯病（*Rhizoctonia solani*）有特效。

2.2.3　麦锈灵（benodanil）

麦锈灵由巴斯夫公司研制，并于 1986 年上市，是一种内吸型杀菌剂，具有保护和治疗作用，主要用来防治谷物、咖啡、蔬菜和观赏植物上的黑粉病、锈病及棉花、马铃薯、大豆和烟草等作物上由丝核菌引起的其他病害，并有一定的增产作用。

2.2.4　甲呋酰胺（fenfuram）

甲呋酰胺是由 Shell 公司发现、安万特（现拜耳作物科学）公司开发的一种内吸性的拌种剂，可用于防治谷物种子胚内带菌引起的腥黑穗病（*Tilletia*）和黑粉病（*Ustilago* spp.），是汞制剂的新的替代品。

2.2.5　呋吡菌胺（furametpyr）

呋吡菌胺是由住友化学工业株式会社于 1989 年发现并开发，并于 1997 年上市。呋吡菌胺为内吸型杀菌剂，具有优良的传导性能，渗透作用强，兼具保护和治疗作用。该类杀菌剂主要作用于担子菌纲病原菌，且对丝核菌属、伏革菌属引起的水稻纹枯病、多种水稻菌核病以及白绢病等都有较好的治疗效果。

2.2.6　噻呋酰胺（thifluzamide）

噻呋酰胺由美国孟山都公司研制，是一种较为广谱的杀菌剂，并于1997年成功上市。噻呋酰胺属于内吸型杀菌剂，具有良好的传导性，兼具保护和治疗作用。该杀菌剂主要用于防治由担子菌引起的病害，尤其是对特别丝核菌引起的水稻、马铃薯、玉米等作物病害具有较好的效果。此外，噻呋酰胺也可防治由立枯丝核菌引起的纹枯病，对治疗水稻纹枯病有特效，且对水稻安全性好，能促进水稻生长。

2.3　第三代SDHIs杀菌剂

2.3.1　啶酰菌胺（boscalid）

1922年，啶酰菌胺由德国巴斯夫公司发现，并于2003年正式上市，是一种广谱的内吸型杀菌剂，兼具良好的渗透性。啶酰菌胺具有优异的预防作用，并有一定的治疗效果。啶酰菌胺主要通过抑制孢子萌发、芽管伸长及附着器形成来达到抑制病原菌生长的作用。主要用于防治由子囊菌和半知菌引起的褐腐（*Monilinia* spp.）、叶斑（*Mycosphaerella* spp.）以及由链格孢菌（*Alternaria* spp.）、灰霉菌（*Botrytis* spp.）、菌核病菌（*Sclerotinia* spp.）等引起的病害，对马铃薯早疫病、葡萄灰霉病等具有优异防效。

2.3.2　吡噻菌胺（penthiopyrad）

1996年，日本三井化学株式会社发现并研制了吡噻菌胺，并于2009年上市。吡噻菌胺具有渗透性和内吸性，具有保护和治疗作用。吡噻菌胺可用于防治链格孢属（*Alternaria* spp.）、壳二孢属（*Ascochyta* spp.）、葡萄孢属（*Botrytis* spp.）、白粉菌属（*Erysiphe* spp.）、丝核菌属（*Rhizoctonia* spp.）、核盘菌属（*Sclerotinia* spp.）、壳针孢属（*Septoria* spp.）、单囊丝壳属（*Sphaerotheca* spp.）和黑星菌属（*Venturia* spp.）等病原菌引起的病害。

2.3.3　吡唑萘菌胺（isopyrazam）

吡唑萘菌胺，由先正达公司研制，并于2010年上市，是一种广谱的内吸型杀菌剂，以保护作用为主，具有一定的治疗作用。吡唑萘菌胺主要用于防治小麦叶斑病（*Septoria tritici*）、褐锈病（*Puccinia recondita*）、条锈病（*Puccinia striiformis*），大麦网斑病（*Pyrenophora teres*）、云纹病（*Rhynchosporium secalis*）和柱隔孢叶斑病（*Ramularia collo-cygni*）等；还可用于防治梨果上的黑星病（*Venturia inaequalis*）和白粉病（*Podosphaera leucotricha*），蔬菜上的白粉病、叶斑病和锈病，油菜上的菌核病（*Sclerotinia*）和黑茎病（*Phoma*），以及香蕉上的黑条叶斑病（*Mycosphaerella fijiensis*）等。吡唑萘菌胺对小麦锈病和大麦锈腐病具有优异的防效。

2.3.4　氟唑菌酰胺（fluxapyroxad）

氟唑菌酰胺由巴斯夫公司研制，并于2011年上市，是一种光谱的内吸型杀菌剂，具有保护和治疗作用。氟唑菌酰胺主要通过抑制孢子发芽、芽孢管伸长、菌丝体生长以及孢子的形成来达到抑制病原菌生长的目的，可有效防治谷类作物及一些经济作物上由壳针孢菌（*Septoria*）、灰葡萄孢菌（*Botrytis*）、白粉菌（*Erysiphe*）、尾孢菌（*Cercospora*）、柄锈菌（*Puccinia*）、丝核菌（*Rhizoctonia*）、核腔菌（*Pyrenophora* spp.）等引起的病害。

2.3.5　联苯吡菌胺（bixafen）

联苯吡菌胺由拜耳作物科学公司研制，并于2011年上市，是一种广谱的内吸型杀菌剂，可有效防治谷类作物上由子囊菌、担子菌和半知菌引起的小麦叶枯病（*Septoria tritici*）、叶锈病（*Puccinia triticina*）、条锈病（*Puccinia striiformis*）、眼斑病（*Oculimacula* spp.）和黄斑病（*Pyrenophora triticirepentis*）等，以及大麦网斑病（*Pyrenophora teres*）、柱隔孢叶斑（*Ramularia collo-cygni*）、云纹病（*Rhynchosporium secalis*）和叶锈病（*Puccinia hordei*）等。

2.3.6 氟唑环菌胺（*sedaxane*）

氟唑环菌胺由先正达公司开发并生产，并于2011年上市，是先正达开发的专用种子处理剂中的第1个有效成分，能够用于防治多种作物中的土传和种传病害，以保护作用为主。

2.3.7 氟唑菌苯胺（*penflufen*）

氟唑菌苯胺由拜耳作物科学公司开发并生产，并于2012年正式上市，是一种内吸型杀菌剂，兼具保护和治疗作用。它可以通过渗透进入发芽的种子后通过木质部传导至整个植株，从而达到保护幼苗生长的作用。氟唑菌苯胺在生产中主要被用作种子处理剂，能够有效的防治由担子菌和子囊菌引起的各种土传和种传病害。如药剂处理马铃薯块茎或作物的种子可以有效的防治由立枯丝核菌（*Rhizoctonia solani*）引起的种传和土传病害。还可以有效防治谷类作物上由黑粉菌（*Tilletia Ustilago*）、丝核菌（*Rhizoctonia*）和旋孢腔菌（*Cochliobolus*）等引起的病害。

2.3.8 氟吡菌酰胺（fluopyram）

氟吡菌酰胺由拜耳作物科学公司研制，并于2012年上市，是一种广谱的内吸型杀菌剂，兼具保护和治疗作用。可用于防治多种作物上的灰霉病（*Botrytis cinerea*）、白粉病、菌核病（*Sclerotinia* spp.）和褐腐病等（*Monilia* spp.）；对香蕉叶斑病也具有良好的防治效果。由于其独特的化学结构，氟吡菌酰胺能够与靶标稳定结合，与其他类型的杀菌剂以及其他相同类型的杀菌剂均不存在交互抗性。因此在生产中单独或与其他杀菌剂复配使用均有良好的防治效果，且不易产生抗性。

2.3.9 苯并烯氟菌唑（benzovindiflupyr）

苯丙烯氟菌唑由先正达公司发现，由先正达和杜邦公司共同开发。该药剂对小麦叶枯病、花生黑斑病、小麦全蚀病及小麦基腐病均有很好的防治效果，尤其对小麦白粉病（*Blumeria graminis*）、玉米小斑病及灰霉病有特效，对亚洲大豆锈病（*Phakopsora pachyrhizi*）具有杰出防效。

2.3.10 isofetamid

Isofetamide 由日本石原产业株式会社研发，能够有效防治以果蔬作物为主的多种作物上由灰葡萄孢引起的灰霉病菌及核盘菌引起的菌核病。对链核盘菌和黑星病也具有较高的防治效果，对不完全菌类的链格孢属和菌绒孢属（*Mycovellosiella*）的菌丝伸长也有很强的抑制作用。此外，该药剂在低浓度时对黄瓜白粉病和褐斑病有90%左右的防效，但对疫病和霜霉病等卵菌病害没有活性。与其他相同类型的杀菌剂无交互抗性。

3 SDHIs 杀菌剂的作用机制

呼吸电子传递链是由Ⅰ、Ⅱ、Ⅲ和Ⅳ四个酶复合物组成的质子或电子传递体，是需氧生物获取能量的主要途径。SDHI类杀菌剂主要是通过作用于病原菌线粒体呼吸电子传递链上的复合体Ⅱ（即琥珀酸脱氢酶[13-14]）来干扰其功能，从而阻止其产生能量，达到抑制病原菌生长的目的，最终导致病原菌死亡[9]。

琥珀酸脱氢酶也称为琥珀酸-泛醌还原酶（succinate ubiquinone reductase，SQR），由黄素蛋白（Fp，SdhA）、铁硫蛋白（Ip，SdhB）和另外2种嵌膜蛋白（SdhC和SdhD）等4个亚基组成。其中SdhA和SdhB组成琥珀酸脱氢酶的外周膜结构域，该结构域较为保守[11]，具有琥珀酸脱氢酶活性；SdhC 和 SdhD 将 SdhA、SdhB 固定在内膜上，同时还具有泛醌还原酶活性[15]。该酶复合体为三羧酸循环的功能部分，与线粒体电子传递链相连，催化从琥珀酸氧化到延胡索酸和从泛醌还原到泛醌的偶联反应[16]。

4 SDHIs 类杀菌剂抗性发生机制

抗药性机制的研究可以让我们了解靶标位点对药剂产生抗药性的根本原因，并以此为理论依据设计一系列的抗性治理策略并应用到实际生产中去。单作用位点杀菌剂作用靶标单一、易产生抗药性，因此该类杀菌剂一直是抗性机制研究的热点。

SDHIs 类杀菌剂因其作用位点单一导致的高抗性风险被 FRAC 归类为中等至高抗性风险药剂，抗性产生原因主要是组成琥珀酸脱氢酶的不同亚基的氨基酸发生点突变而导致蛋白结构变化[11]。此外，有些病原真菌如黄瓜褐斑病（*Corynespora cassiicola*），对啶酰菌胺的抗性与琥珀酸脱氢酶的亚基基因点突变无关，说明除基因点突变引起的琥珀酸脱氢酶亚基结构改变外，还有其他抗药机制存在[17]。Avenot 等[1,18]的研究表明，SdhB 点突变是导致多数病原真菌对 SDHIs 产生的抗性的主要原因，少数病原真菌与亚基 SdhC 和 SdhD 上的氨基酸改变有关（表 2）。其中，由于 SdhB 点突变导致的抗性发生的病原菌突变位点多发生在相同的组氨酸上，而 SdhC 和 SdhD 突变位点比较多。但无论是哪个亚基，其氨基酸点突变均发生在保守区域。

表 2　电子传递链上的蛋白复合物 Ⅱ 上引起 SDHI 抗性的点突变

病原菌种类	杀菌剂	表现型	突变位点	抗性发生情况
Ustilago maydis	carboxin	CbxR	B（H253L）	Lab
Mycosphaerella graminicola	carboxin	CbxR	B（H267Y）	Lab
Pleurotus ostreatus	carboxin	CbxR	B（H239L）	Lab
Paracoccus denitrificans	carboxin	CbxR	B（H228N）	Lab
Xanthomonas campestris	carboxin	CbxR	B（H229L）	Lab
Alternaria alternata	boscalid	BosR	B（H277Y，R）	Field
Corynespora cassiicola	boscalid	BosR	B（H278Y）	Field
Botrytis cinerea	boscalid	BosR	B（P225L，T，F；H272Y，R）	Lab andField
Didymella bryoniae	boscalid	BosR	B（H－＞Y）	Field
Podosphaera xanthii	boscalid	BosR	B（H－＞Y）	Field
Aspergillus oryzae	carboxin	CbxR	B（H249Y，L，N）	Lab
Coprinus cinereus	carboxin	CbxR	C（N80K）	Lab
Alternaria alternata	boscalid	BosR	C（H134R）	Field
Corynespora cassiicola	boscalid	BosR	C（S73P）	Field
Aspergillus oryzae	carboxin	CbxR	C（T90I）	Lab
Paracoccus denitrificans	carboxin	CbxR	D（D89G）	Lab
Alternaria alternata	boscalid	BosR	D（D123E，D133R）	Field

（续表）

病原菌种类	杀菌剂	表现型	突变位点	抗性发生情况
Corynespora cassiicola	boscalid	BosR	D（S89P）	Field
Sclerotinia sclerotiorum	boscalid	BosR	D（D132R）	Field
Aspergillus oryzae	carboxin	CbxR	D（D124E）	Lab

注：CbxR：Carboxin resistant；BosR：Boscalid resistant；B：SdhB；C：SdhC；D：SdhD

5 SDHIs类杀菌剂抗性发生情况

保护性杀菌剂一般具有多个靶标位点，不易产生抗药性，因此关于抗性的研究主要是集中在单位点杀菌剂上。琥珀酸脱氢酶杀菌剂作用位点单一，重复用药等选择压力使其抗性风险备受关注，因此针对该类杀菌剂的抗性研究报道较多，并在许多病菌中获得了田间或室内的抗性菌株，目前该类杀菌剂中出现抗性问题最多的是萎锈灵和啶酰菌胺两种药剂，在其他品种中也会存在少量的抗药性问题，如甲呋酰胺和氟酰胺。此外，该类杀菌剂的不同品种间也存在不同程度的交互抗性。

萎锈灵是开发最早的一类SDHIs类杀菌剂，作用位点单一，再加上长期大量的使用，其抗药性问题比较突出。1986年，Leroux[15]等在冬大麦上分离到抗萎锈灵的病原菌菌株。1988年，人们在田间发现了对SDHIs类杀菌剂萎锈灵产生抗药性的大麦散黑穗病菌抗性菌株（*Ustilago nuda*）[19]。1991年，Keon等[20]检测到对萎锈灵产生抗性的玉米瘤黑粉病菌（*Ustilago maydis*）抗性菌株；1992年Broomfield等[21]的研究表明第253位氨基酸由组氨酸突变成亮氨酸是导致该类杀菌剂抗药性产生的主要原因，而突变后的该基因可以用于抗性菌株的筛选[22]。Skinner等[23]通过对敏感菌株和人工诱变的抗性菌株的SdhB基因的克隆和测序比对发现：SdhB亚单位267位的组氨酸突变为酪氨酸或亮氨酸均能够导致萎锈灵抗药性问题的产生；1998年，Mattson等[24-25]研究发现脱氮副球菌（*paracoccus denitrificans*）中的抗性突变有两种类型，即B（H228N）和D（D89G）。白腐菌（*Pleurotus ostreatus*）对萎锈灵的抗性也是由于SdhB亚单位上的点突变引起的[26]。

啶酰菌胺（boscalid）的开发与使用较晚，是第三代SDHIs类杀菌剂的典型代表，但随着其使用量的增加，在生产中同样存在抗药性问题。2007年，Stammler等[27]发现灰霉病菌（*Botrytis cinerea*）对啶酰菌胺存在抗性；同年，巴斯夫公司也报道了由于百合会霉病菌（*Botrytis elliptica*）中SdhB亚单位中的第272位氨基酸由组氨酸突变为酪氨酸或精氨酸而对啶酰菌胺产生抗性；2008年，Avenot等[28]监测到对啶酰菌胺具极高水平抗性的互隔交链孢霉（*Alternaria alternata*）菌株；此外，在国外的一些研究中也发现了啶酰菌胺的抗性菌株黄瓜褐斑病菌菌株；有研究显示在田间检测到对啶酰菌胺产生抗性的西瓜蔓枯病菌的抗性菌株，且抗性菌株的抗性水平均较高；2008年又发现了对啶酰菌胺具有较高抗性的白粉病菌菌株。2009年，Shima等[29]检测了米曲霉（*Aspergillus oryzae*）对抗萎锈灵突变体的3个突变位点，这3个突变位点分别位于SdhB、SdhC和SdhD三个亚单位上。

1986年，Leroux等[30]在田间还发现了甲呋酰胺的抗性菌株；2004年，Ito等[31]对报道灰盖鬼伞（*Coprinus cinereus*）对氟酰胺的抗性，且进一步的研究发现其SdhC亚基上的第80位氨基酸突变为天冬酰胺是导致抗药性产生的根本原因。

多数情况下，同一类药剂之间存在交互抗性，如互隔交链孢霉（*Alternaria alternata*）[29]

的277位和灰霉病菌（*Botrytis cinerea*）[31]的272位等同于大麦散黑穗病菌（*Ustilago nuda*）[20]的257位和小麦叶枯病菌（*Mycosphaerella graminicola*）[23]的267位，因此两者存在交互抗药性是正常的现象。但由于该类杀菌剂的抗性菌株中抗性产生的点突变位点存在差异，因此并不是所有的该类药剂间都存在交互抗性。如马铃薯早疫链格孢菌（*Alternalia solani*）[32]对啶酰菌胺表现为抗性，但所有抗性菌株对氟吡菌酰胺表现为敏感，大部分抗性菌株对吡噻菌胺表现为敏感。

在SDHIs类杀菌剂交互抗性问题的研究中，Avenot等[33]对3个啶酰菌胺抗性突变体和野生型敏感菌株互隔交链孢霉（*Alternaria alternata*）的敏感性进行测定，结果显示野生型敏感菌株对吡噻菌胺和氟吡菌酰胺均表现敏感；而抗性突变体对吡噻菌胺同样表现出不同程度的抗性，但对氟吡菌酰胺无抗性。同时Avenot发现田间对啶酰菌胺有抗性的西瓜蔓枯病菌对吡噻菌胺也具有抗性，说明啶酰菌胺与吡噻菌胺间存在交互抗性。而氟吡菌酰胺在离体条件下对互隔交链孢霉（*Alternaria alternata*）的琥珀酸脱氢酶突变体和野生型具有同样的抑制作用。但进一步的研究发现对啶酰菌胺有抗性的突变型SDHB－H277Y对氟吡菌酰胺敏感，而突变型SDHC－H134R对氟吡菌酰胺表现为低水平的抗性。氟吡菌酰胺在交互抗药性中的差异性说明它与其他SDHIs类杀菌剂在蛋白体复合物Ⅱ上的靶标位点存在差异。此外，在灰霉病菌中，啶酰菌胺与甲氧基类杀菌剂存在负交互抗药性，与萎锈灵等存在正交互抗药性，这与本实验室获得的油菜菌核病菌的抗性菌株结果相符。但本实验获得的抗甲氧基类杀菌剂的油菜菌核病菌中发现。甲氧基类杀菌剂与啶酰菌胺不存在负交互抗药性。

综上所述，导致该类杀菌剂抗性产生的点突变主要是发生在复合物Ⅱ的三个亚基中，其中以SdhB为主，SdhC、SdhD中则较少。此外SdhB点突变发生位置比较单一，在多种病原菌中突变均发生在相同的组氨酸上，而SdhC和SdhD突变位点比较多，不同氨基酸的点突变能够导致不同程度抗药性问题的产生[8]。此外，病原菌中还存在其他能够导致对该类药剂产生抗性突变的作用机制，如病原真菌如黄瓜褐斑病菌（*Corynespora cassiicola*）对啶酰菌胺的抗性突变的产生就与琥珀酸脱氢酶的亚基基因点突变无关。

6　SDHIs类杀菌剂抗性治理措施

随着SDHIs类杀菌剂类产品的广泛开发和频繁使用，抗性问题不可避免。根据国内外已有的相关药剂的田间抗性发生情况的报道，该类药剂已被FRAC归为中度抗性风险杀菌剂，为避免或延缓该类杀菌剂的抗性发生，FRAC对这类药剂的使用提出以下建议：按照生产商推荐的有效剂量和安全间隔期使用；限制SDHI类杀菌剂的用药次数，每生长季最多用药3次；使用复配产品来延缓抗性时，其桶混或复配的配伍通常要满足两方面的条件，使用的单剂对靶标病害也应该提供满意的防效，同时必须具有不同的作用机理，应该预防性使用SDHI类杀菌剂，即在病害发生早期施药[34]；新药剂大面积应用前及时建立靶标病原菌的敏感基线，且对田间抗性情况进行实时监测，并及时治理。2个或多个SDHI类杀菌剂复配时，它们虽然可以提供好的生物防效，但并不能作为抗性治理策略，必须视作单一的SDHI类杀菌剂来处理。

参考文献

[1]　Avenot H，Sellam A，Michailides T. Characterization of mutations in the membrane anchored subunits AaSDHC and AaSDHD of succinate dehydrogenase from *Alternaria alternata* isolates conferring field re-

sistance to the fungicide coscalid [J]. Plant Pathol, 2009, 58: 1 134 -1 143.

[2] Ackerll B A C. Progressin understanding structure function relationships in respiratory chain complexII [J]. FEBS Lett, 2000, 466: 1 -5.

[3] Miyadera H, Shiomi K, Ui H, et al. Atpenins, potent and specific inhibitors of mitochondrial complex II (succinate-ubiquinone oxidoreductase) [J]. Proc. Natl. Acad. Sci, 2003, 100: 473 -477.

[4] Russell P E P E. Fundicide resistance action committee (FRAC): a resistance activity update [J]. Ourlooks Pest Manage, 2009, 20 (3): 122 -125.

[5] Dubos T, Pasquali M, Pogoda F, et al. Differences between the succinate dehydrogenase sequences of isopyrazam sensitive *Zymoserptroria tritici* and insensitive *Fusarium graminearums* trains [J]. Pesticide biochemistry and physiology, 2013, 105 (1): 28 -35.

[6] Klappach K. SDHI Working Group of FRAC [OL]. [2010 -03 -03]. Http: //www. frac. info/frac/index. htm.

[7] Ulrich J T, Mathre D E. Mode of Action of Oxathiin Systemic Fungicides. V. Effect on Electron Transport System of *Ustilago maydis* and *Saccharomyces cerevisiae* [J]. J Bacteriol, 1972, 110: 628 -632.

[8] Stammler G, Brix H D, Glattli A, et al. Biological Properties of the Carboxamide Boscalid Including Recent Studies on Its Mode of Action [C] //Proceedings XVI International Plant Protection Congress Glasgow, 2007: 40 -45.

[9] Yanase Y, Yoshikawa Y, Kishi J, et al. The History of Complex II Inhibitors and the Discovery of Penthiopyrad [C] //Ohkawa H, Miyagawa H, Lee P W (eds.). Pesticide Chemisty Crop Protection, Public Health, Environmental Safety. WILEY-VCH, Weinheim, Germany, 2007: 295 -303.

[10] Dekker J. Development of Resistance to Modern Fungicides and Strategies for Its Avoidance [C] //Lyr H. Modern Selective Fungicides-Properties, Applications, Mechanisms of Action. Gustav Fisher Verlag, New York, 1995: 23 -38.

[11] 詹家绥，吴娥娇，刘西莉，等. 植物病原真菌对几类重要单位点杀菌剂的抗药性分子机制 [J]. 中国农业科学，2014，47 (17): 3 392 -3 404.

[12] 詹儒林，李伟，郑服丛. 杧果炭疽病菌对多菌灵的抗药性 [J]. 植物保护报，2005，32 (1): 71 -76.

[13] Hagerhall C. Succinate: Quinine Oxidoreductases. Variations on a Conserved Theme [J]. Biochem Biophys Acta, 1997, 1320: 107 -141.

[14] Kuhn P J. Mode of Action of Carboximides [J]. Symp Ser Br Mycol Soc, 1984, 9: 155 -183.

[15] 李良孔，袁善奎，潘洪玉，等. 琥珀酸脱氢酶抑制剂类 (SDHIs) 杀菌剂及其抗性研究进展 [J]. 农药，2011，50 (3): 165 -169.

[16] Keon J P R, White G A, Hargreaaves J A. Characterization and Sequence of a Gene Conferring Resistance to the Systemic Fungicide Carboxin from the Maize Smut Pathogen, *Ustilago maydis* [J]. Current Genetics, 1991, 19: 475 -481.

[17] Miyamoto T, Ishii H, Stammler G, et al. Distribution and molecular characterization of *Corynespora cassiicola* isolates resistant to boscalid [J]. Plant Pathology, 2010, 59 (5): 873 -881.

[18] Avenot H F, Sellam A, Karaoglanidis G, et al. Characterization of mutations in the iron-sulphur subunit of succinate dehydrogenase correlating with boscalid resistance in *Alternaria alternata* from California pistachio [J]. Phytopathology, 2008, 98 (6): 736 -742.

[19] Leroux P, Berthier G. Resistance to Carboxin and Fenfuramin Ustilago nuda Jens Rostr the Causal Agent of Barley Loose Smut [J]. Crop Protection, 1988, 7: 16 -19.

[20] Keon J P R, White G A, Hargreaves J A. Isolation, Characterization and Sequence of a Gene Conferring Resistance to the Systemic Fungicide Carboxin from the Maize Smut Pathogen, *Ustilago maydis*

[J]. Current Genetics, 1991, 19: 475 - 481.

[21] Brommfield P L E, Hargreaves J A. A Single Amino-acid Change in the Ironsu-lphur Protein Subunit of Succinate Dehydrogenase Confers Resistance to Carboxin in *Ustilago maydis* [J]. Current Genetics, 1992, 22: 117 - 121.

[22] Kinal H, Park C M, Bruenm J A. A Family of *Ustilago maydis* Expression Vectors: New Selectable Markers and Promoters [J]. Gene, 1993, 127: 151 - 152.

[23] Skinner W, Bailey A, Renwick A, et al. A Single Amino-acid Substitution in the Iron-sulphur Protein Subunit of Succinate Dehydrogenase Determines Resistance to Carboxin in *Mycosphaerella graminicola* [J]. Current Genetics, 1998, 34: 393 - 398.

[24] Matsson M, Ackrell B A, Cochran B, et al. Carboxin Resistance in Paracoccus denitrificans Conferred by a Mutation in the Membrane-anchor Domain of Succinate: Quinine Reductase [J]. Arch Microbiol, 1998, 170: 27 - 37.

[25] Matsson M, Hederstedt L. The Carboxin binding Site on Paracoccus denitrificans Succinate: Quinine Reductase Identified by Mutations [J]. Bioenerg Biomembr, 2001, 33: 99 - 105.

[26] Honda Y, Matsuyama T, Irie T, et al. Carboxin Resistance Transformation of the Homobasidiomycete Fungus *Pleurotus ostreatus* [J]. Current Genet, 2000, 37: 209 - 212.

[27] Stsmmle G, Brix H D, Glattli A, et al. Biological Properties of the Carboxamide Boscalid Including Recent Studies on Its Mode of Action [C] //Proceedings XVI International Plant Protection Congress Glasgow, 2007: 40 - 45.

[28] Aernot H F, Sellam A, Karaoglanidis G, et al. Characterization of Mutations in the Ironsulphur Subunit of Succinate Dehydrogenase Correlating with Boscalid Resistance in *Alternaria alternata* from California Pistachio [J]. Phytopathology, 2008, 98 (6): 736 - 742.

[29] Shima Y, Ito Y, Kaneko S, et al. Identification of Three Mutant Loci Conferring Carboxin resistance and Development of a Novel Transformation System in *Aspergillus oryzae* [J]. Fungal Genetics and Biology, 2009, 46: 67 - 76.

[30] Ito Y, Muraguchi H, Seshime Y, et al. Flutolanil and Carboxin Resistance in Coprinus cinereus Conferred by a Mutation in the Cytochrome b560 Subunit of Succinate Dehydrogenase Complex (Complex II) [J]. Mol Gen Genomics, 2004, 272: 328 - 335.

[31] Fernandez-Ortuno D, Chen F, Schnabel G. Resistance to pyraclostrobin and boscalid in *Botrytis cinerea* isolates from strawberry fields in the Carolinas [J]. Plant Disease, 2012, 96 (8): 1 198 - 1 203.

[32] Gudmestad N C, Arabiat S, Miller J S, Pasche J S. Prevalence and impact of SDHI fungicide resistance in *Alternaria solani* [J]. Plant Disease, 2013, 97 (7): 952 - 960.

[33] Avenot H F, Michailidesi T J. Progress in Understanding Molecular Mechanisms and Evolution of Resistance to Succinate Dehydrogenase Inhibiting (SDHI) Fungicides in Phytopat-hogenic Fungi [J]. Crop Protection, 2010, 29: 643 - 651.

[34] FRAC. Protocol of the discussions and use recommendations of the SDHI Working Group of the Fungicide Resistance Action Committee (FRAC) [EB/OL]. [2014 - 09 - 24]. http: //www. Frac. info/work/ Minutes. pdf.

猕猴桃溃疡病的发生与化学防治

陈　亮[1*]，刘君丽[1]，冯　聪[1]，颜克成[1]，张信旺[2]，金锡萱[2]
（1. 沈阳中化农药化工研发有限公司，新农药创制与开发国家重点实验室，
沈阳　110021；2. 四川省猕猴桃工程技术研究中心，成都　611630）

摘要：猕猴桃是一种原产于中国的经济价值很高的水果，近年来在国内的种植面积不断增加，已经初步形成规模化经营。猕猴桃溃疡病（*Pseudomonas syringae* pv. *actinidiae*）是危害猕猴桃产业发展的毁灭性细菌病害，国内主要猕猴桃产区均受到溃疡病危害，每年造成的产量损失在 10 万 t 以上。化学防治是控制猕猴桃溃疡病的重要措施，抗生素类、铜制剂类，以及一些保护性杀菌剂可以用于防治溃疡病。由于可用药剂种类少，作用机制单一，长期大量施用少数几种药剂，溃疡病菌抗性风险增加，导致无法对其进行有效防治。
关键词：猕猴桃溃疡病；细菌性病害；化学防治

1　中国的猕猴桃产业

猕猴桃（kiwifruit）是一种原产于中国的经济价值很高的水果，因其富含糖类和维生素，尤其是维生素 C 含量比柑橘高 5～10 倍，比苹果、梨高 30 倍，加之味道甜美，使其享有“果中之王”的称号，深受消费者喜爱[1]。中国猕猴桃产业已经形成较大规模，种植面积世界第一，产量也位居世界前列，远销欧美、日本等地区。栽培猕猴桃主要包括美味猕猴桃和中华猕猴桃两大类，美味猕猴桃表皮毛多而硬，中华猕猴桃表皮毛少而稀疏，常脱落[2]。目前猕猴桃在国内已经形成五大产区，一是陕西秦岭北麓（宝鸡市眉县和西安市周至县）；二是大别山区，河南伏牛山、桐柏山；三是广东河源和平县；四是贵州高原和湖南省西部；五是四川省的西北地区和湖北省的西南地区[3]。其中，陕西省西安市周至县、宝鸡市眉县、四川省的苍溪县因盛产猕猴桃成为猕猴桃之乡，其中，仅陕西省眉县猕猴桃种植面积就接近 30 万亩。

2　猕猴桃溃疡病的发现及其造成的损失

猕猴桃溃疡病首先于20 世纪80 年代年在日本静冈县[4]和美国加州[5]被发现，随后在韩国、意大利等国家陆续发现猕猴桃溃疡病，近年来溃疡病呈现加速扩展的趋势，中国[6]、日本[4]、韩国[7-8]、新西兰[9]、澳大利亚、法国[10]、意大利[11]、西班牙[12]、葡萄牙[13]、瑞士、美国[14]、土耳其[15]、智利等国家的猕猴桃种植都受到溃疡病的影响。

国内于 1985 年首次报道在湖南省常德市石门县东山峰农场发现猕猴桃溃疡病，当时感病猕猴桃面积 13hm^2，在短时间内造成猕猴桃植株成片死亡。随后于 1987 年在四川、1990 年在安徽、1991 年在陕西都有溃疡病发生[16]。近年来猕猴桃溃疡病分布面积不断扩大，在

---

* 作者简介：陈亮，男，高级工程师，主要从事杀菌剂高通量筛选与应用技术研究

多个猕猴桃种植区，包括陕西[17]、四川[18]、湖南[19]、广东、福建[20]、安徽[21]、江苏[22]、浙江[23]、江西等地区均有溃疡病发生危害的报道。

溃疡病菌侵染猕猴桃后病害发展快、流行性强、危害重，同时缺乏有效的防控药剂，目前已经对猕猴桃产业构成严重威胁。猕猴桃树龄越长，尤其达到十年以上树龄发病越严重。随着猕猴桃溃疡病发病面积的不断扩大，造成产量损失也逐年增加，以 2012 年为例，全国由于猕猴桃溃疡病造成的产量损失达到 10 万 t 以上（表 1）[24]。从单个猕猴桃产区的情况来看，由于溃疡病造成的产量损失几乎能够摧毁当地的猕猴桃产业。根据文献报道，1989 年四川省苍溪县三溪口林场大面积暴发猕猴桃溃疡病，损失惨重，产量由 1988 年的 15 万 kg 猛跌到 5 万 kg，因无法有效控制病害，只能清除病树，到 1990 年产量仅有 2.5 万 kg[25]。另据文献报道，陕西省眉县 2015 年猕猴桃栽植面积 1.99 万 hm^2，溃疡病平均病园率达到 62.5%，其中，当地主要栽培品种之一的“红阳”病园率 100%，病菌在树体内部形成系统侵染，迅速蔓延至整个果园，导致毁园现象的出现[26]。

由于猕猴桃溃疡病危害严重，造成的经济损失巨大，1996 年其被列入中国森林植物检疫对象名单，2009 年国家质量监督检验检疫总局关于印发《意大利猕猴桃进境植物检疫要求》的通知，明确把猕猴桃溃疡病菌列入关注的检疫性有害生物名单中。2013 年，国家林业局将其列入全国林业危险性有害生物名单。

表 1　中国主要省份猕猴桃产量以及溃疡病危害损失（2012）

地区	面积（hm^2）	产量（t）	溃疡病造成损失（t）
陕西	30 000	340 000	70 000
四川	13 800	60 000	13 000
湖南	7 200	25 168	6 200
河南	6 670	36 000	7 000
贵州	5 500	12 057	2 000
浙江	3 000	7 403	—
江西	2 400	8 523	1 600
重庆	1 600	2 453	800
安徽	900	2 235	1 000
山东	600	1 935	700
广西	600	1 823	—
河北	600	1 347	400
福建	500	1 980	—
云南	300	1 642	—
江苏	300	1 349	300
合计	73 970	503 915	103 000

3 猕猴桃溃疡病病原菌的确定及其病害症状

猕猴桃溃疡病是近年在农业生产中最具毁灭性的细菌性病害之一，也是一种检疫性病害。当前对于猕猴桃溃疡病的病原菌存在争议，日本、韩国、意大利、法国、新西兰等国的研究人员根据溃疡病的危害症状、寄主范围、细菌的生理生化特征，认为猕猴桃溃疡病的病原细菌为丁香假单胞杆菌猕猴桃致病变种（*Pseudomonas syringae* pv. *actinidiae*），该观点也成为国际主流意见，被大多数国家的研究人员所认可。也有一些研究结果表明其他病原细菌能引起猕猴桃溃疡病，例如美国的 Opgennorth D C 认为丁香假单胞杆菌李致病变种（*P. syringae* pv. *morsprunorum*）引起猕猴桃溃疡病[5]，日本、伊朗、意大利等国家的研究人员报道丁香假单胞杆菌丁香致病变种（*P. syringae* pv. *syringae*）引起猕猴桃枝干溃疡[27]。国内的植物病理学家对中国主要猕猴桃产区的溃疡病菌进行了相关的鉴定研究，结果与国际主流意见相同，认为中国的猕猴桃溃疡病菌为丁香假单胞杆菌猕猴桃致病变种（*P. syringae* pv. *actinidiae*）。

猕猴桃溃疡病（*P. syringae* pv. *actinidiae*）是一种低温高湿病害，春季气温偏低、连续阴雨易导致溃疡病大发生。病菌在田间借风雨、昆虫或农事活动传播，主要从植株体表伤口或者自然孔口侵入，在染病枝蔓和土壤中病残体上越冬。溃疡病菌主要危害猕猴桃的主干、枝蔓、新梢、叶片、花蕾和花等部位。溃疡病典型症状为叶片上布满多角形的褐色斑点，周围形成黄色晕圈，常导致叶片翻卷；嫩芽发生焦枯，花蕾呈现褐色，不能开放而枯萎；枝干的伤口、皮孔、芽眼、叶痕等处溢出乳白色黏液，后期氧化变为红褐色，病部组织下陷造成溃疡腐烂，枝蔓枯死[28]；果实不易感染，但是溃疡病发生后，不仅产量下降，而且果实品质也受到严重影响，主要表现为果皮增厚，口感和果色变差，果实畸形[29]。猕猴桃溃疡病一年有两个发病期，第一个在春季果树萌芽期至谢花期，初期主要危害枝干，后期随气温升高，叶片、花蕾等部位开始发病。第二个在秋季果实成熟期，主要危害叶片[30]。

4 猕猴桃溃疡病的化学防治

化学防治是控制猕猴桃溃疡病的有效措施，但是猕猴桃溃疡病属于细菌性病害，与真菌病害相比，防治细菌病害药剂较少，主要以抗生素类、铜制剂类以及一些保护性杀菌剂为主，具体的防治药剂包括春雷霉素、噻菌铜、噻霉酮、硫酸铜、可杀得（氢氧化铜）、叶枯唑、噻唑锌等[31-32]，施用方法主要有病斑刮除与药剂涂抹、整株喷雾、枝干注射[33]等。各种药剂施用方法中，在防治适期进行整株喷雾效果较为明显，而刮除病斑结合药剂涂抹以及枝蔓注射的方式效果不如喷雾明显，尤其是注射的处理方式，受到药剂在植株体内传导特性的影响非常明显[31,33]。猕猴桃溃疡病防治适期为春季植株萌芽前施用化学药剂对溃疡病进行预防，以及在发病期对溃疡病蔓延的控制；秋季采果后入冬前在施用化学药剂的同时对果园内发病枝蔓、杂草等进行清理，以铲除越冬菌源，减轻来年春季病害压力。

参考文献

[1] 文国琴，石大兴，吴雪梅，等. 猕猴桃组织培养研究的现状与进展［J］. 北方果树，2004（3）：1-4.

[2] 李莎莎. 猕猴桃溃疡病相关细菌的鉴定及致病性研究［D］. 合肥：安徽农业大学，2013.

[3] 邵宝林. 猕猴桃溃疡病风险分析及其病原鉴定检测和生物防治研究［D］. 成都：四川农业大

学，2013.
[4] Serizawa S，Ichikawa T，Takikawa Y，et al. Occurrence of bacterial canker of kiwifruit in Japan：description of symptoms，isolation of the pathogen and screening of bactericides [J]. Japanese Journal of Phytopathology，1989，55 (4)：427－436.
[5] Dan CO. Pseudomonas Canker of Kiwifruit [J]. Plant Disease，1983，67 (11)：1 283－1 284.
[6] 高小宁，赵志博，黄其玲，等. 猕猴桃细菌性溃疡病研究进展 [J]. 果树学报，2012 (2).
[7] Koh Y，Kim G，Jung J，et al . Outbreak of bacterial canker on Hort16A (Planchon) caused by pvin Korea [J]. New Zealand Journal of Crop & Horticultural Science，2010，38 (4)：275－282.
[8] Koh Y J，Lee D H. Canker of kiwifruit by *Pseudomonas syringae* pv. *morsprunorum* [J]. Korean Journal of Plant Pathology，1992.
[9] 黄其玲，高小宁，赵志博，等. GFPuv 标记猕猴桃溃疡病菌的生物学特性及其在土壤、根系中的定殖 [J]. 中国农业科学，2013，46 (2)：282－291.
[10] Liu X，Park M，Kim MG，Gupta S，Wu G，Cho J. First report of Pseudomonas syringae pv. actinidiae，the causal agent of bacterial canker of kiwifruit in Slovenia [J]. Plant Disease，2014，98 (11)：S131－137.
[11] Gaohui，Wu，Meihui，Song，Ziyang，Xiu，et al. Identification of Pseudomonas syringae pv. actinidiae as Causal Agent of Bacterial Canker of Yellow Kiwifruit (Actinidia chinensis Planchon) in Central Italy [J]. Journal of Phytopathology，2009，157 (11－12)：768－770.
[12] Abelleira A，López M M，Peñalver J，et al. First report of bacterial canker of kiwifruit caused by *Pseudomonas syringae* pv. *actinidiae* in Spain [J]. Plant Disease，2011，95 (12)：1 583－1 584.
[13] Balestra G M，Renzi M，Mazzaglia A. First report of bacterial canker of Actinidia deliciosa caused by *Pseudomonas syringae* pv. *actinidiae* in Portugal [J]. New Disease Reports，2010，22 (11)：2 510－3.
[14] 李黎，钟彩虹，李大卫，等. 猕猴桃细菌性溃疡病的研究进展 [J]. 华中农业大学学报，2013，32 (5)：124－133.
[15] Bastas K K，Karakaya A. First report of bacterial canker of kiwifruit caused by *Pseudomonas syringae* pv. *actinidiae* in Turkey [J]. Plant Disease，2012，96 (3)：452－454.
[16] 方炎祖，朱晓湘，王宇道. 湖南猕猴桃病害调查研究初报 [J]. 四川果树科技，1990 (1)：28－29.
[17] 梁英梅，张星耀，田呈明，等. 陕西省猕猴桃枝干溃疡病病原菌鉴定 [J]. 西北林学院学报，2000，15 (1)：37－39.
[18] 刘瑶，朱天辉，樊芳冰，等. 四川猕猴桃溃疡病的发生与病原研究 [J]. 湖北农业科学，2013，52 (20)：4 937－4 942.
[19] 尹春峰，林文力，肖伏莲，等. 湖南猕猴桃溃疡病菌 16S rDNA 和 16S－23S rDNA 间隔区序列的克隆与分析 [J]. 植物检疫，2015，29 (3)：57－61.
[20] 宋晓斌，张学武，马松涛. 猕猴桃溃疡病研究现状与前景展望 [J]. 陕西林业科技，1997 (4)：62－64.
[21] 承河元，李瑶，万嗣，等. 安徽省猕猴桃溃疡病菌鉴定 [J]. 安徽农业大学学报，1995 (3)：219－223.
[22] 杜永章，王永兰，施正伟. 猕猴桃溃疡病的发生与防治 [J]. 现代农业科技，2009 (24)：175－177.
[23] 张慧琴，李和孟，冯健君，等. 浙江省猕猴桃溃疡病发病现状调查及影响因子分析 [J]. 浙江农业学报，2013，25 (4)：832－835.
[24] 胡家勇. 猕猴桃溃疡病菌株收集及其寄主和传播媒介的鉴定 [D]. 合肥：安徽农业大学，2014.

[25] 刘绍基，唐显富，王忠肃，等. 四川省苍溪猕猴桃溃疡病的发生规律 [J]. 中国果树，1996 (1).

[26] 王鑫，李小晶. 猕猴桃溃疡病防治技术 [J]. 北方果树，2015 (4)：40－41.

[27] Balestra G M，Varvaro L. *Pseudomonas syringae* pv. *syringae* causal agent of disease on floral buds of Actinidia deliciosa (A. Chev) Liang et Ferguson in Italy [J]. Journal of Phytopathology，2008，145 (8－9)：375－378.

[28] 胡锦，杨义成. 猕猴桃溃疡病的发生与防治对策 [J]. 四川农业科技，2007 (9)：50－51.

[29] Balestra GM，Mazzaglia A，Quattrucci A，et al. Current status of bacterial canker spread on kiwifruit in Italy [J]. Australasian Plant Disease Notes，2009，4 (1)：34－36.

[30] Li Y，Song X，Zhang X. Studies on Laws of Occurrence of Bacterial Canker in Kiwifruit [J]. Journal of Northwest Forestry College，2000.

[31] 龙友华，夏锦书. 猕猴桃溃疡病防治药剂室内筛选及田间药效试验 [J]. 贵州农业科学，2010，38 (10)：84－86.

[32] 王西锐，李艳红. 噻霉酮防治猕猴桃溃疡病药效试验 [J]. 烟台果树，2013 (1)：18－19.

[33] 李泉厂，陈金焕，王西红. 不同药剂防治猕猴桃溃疡病效果研究 [J]. 现代农业科技，2013 (11)：130－131.

三唑类杀菌剂的研究进展

罗舜文，陈东明，周明国
（南京农业大学植物保护学院，南京 210095）

摘要：三唑类（triazole）化合物是20世纪70年代初开发的一类非常重要的药剂，被广泛应用于农用杀菌剂、除草剂、植物生长调节剂、杀虫杀螨剂、抗病毒药剂以及临床抗癌药物。三唑类药剂中最大的一类为农用杀菌剂，对大多数子囊菌、担子菌和半知菌引起的病害均有防治效果，不仅具有高效、低毒和持续时间长的优点，还有良好的内吸性，对病害具有保护和治疗作用，使其在农药分子设计领域越来越受到人们的重视。三唑类化合物是重要的内吸性杀菌剂，并已开发出几十种产品；我国从20世纪80年代仿制三唑类农药以来，该类药剂一直被广泛应用于一些重要农作物主要病害的防控。该类化合物是甾醇生物合成C－14脱甲基化酶抑制剂（DMI），对植物体内麦角甾醇合成具有抑制作用，属于麦角甾醇生物合成抑制剂（Ergosterol Biosynthesis Inhibitors，EBIs）。麦角甾醇存在于大多数病原菌体中，是细胞膜的组成成分之一，具有重要的生物功能。早期研究人员发现病原菌对三唑类杀菌剂产生抗药性受多基因控制，抗性菌株适应性差，在自然界中难以形成抗药性群体，因此普遍认为植物病原真菌在田间对该类杀菌剂产生抗药性的风险较小[1-2]。但近几十年来随着三唑类杀菌剂被大量推广应用，却逐渐发现植物病原真菌对该类杀菌剂的抗药性风险并不乐观。由于其应用广泛且作用位点单一，田间许多重要的靶标病原菌已逐渐对其产生了不同程度的抗药性[3-5]。因此，本文拟就目前有关三唑类杀菌剂的品种、作用机制、抗药性的发生现状和抗药机制进行简要综述。

关键词：三唑类；麦角甾醇生物合成抑制剂；抗药性；抗药机制

Abstract: The triazole compounds are very important kinds of agents from the early 1970s, which are widely used in agricultural fungicide, herbicide, plant growth regulator, insecticidal acaricide, antiviral agents and clinical anti-cancer drugs. The largest category of triazole is agricultural fungicide, they can prevent and cure of diseases which are most caused by ascomycetes, basidiomycetes and deuteromycetes, not only high efficiency, low toxicity and long duration, but also good systemic action, having protection and treatment effect, to make it more and more get people's attention in the field of pesticide molecular design. Triazole compounds are important systemic fungicide, and have developed dozens of products; in our country from the generic three azole fungicides since the 1980s, the fungicides had been widely used in the prevention and control some diseases of main crops. This kind of compound is sterol Biosynthesis of C－14 to demethylation Inhibitors (DMI), which has an inhibitory effect on Ergosterol synthesis in plants, belonging to the Ergosterol Biosynthesis inhibitor (EBIs). Ergosterol exists in most of the pathogenic bacteria in the body, which is one of the components of cell membranes and has important biological function. Early researchers found pathogens resistant to triazole fungicide controlled by polygene, poor resistance strain adaptation, difficult to form drug resistance groups in nature, so is generally believed that plant pathogenic fungi in the field of this kind of fungicide resistant to less risk[1-2]. But in recent decades as the triazole fungicide widely applied, but gradually found that plant pathogenic fungi of the fungicide resistance risk are not optimistic. Because of its wide application and site of a single, many important target pathogens had been gradually on the field and have a different degree of drug resistance[3-5]. Therefore, this thesis is about the triazole fungicide resistance of variety, mechanism of action and current status of and resistant mechanism is reviewed.

1 三唑类杀菌剂

三唑类化合物的母体结构中都含有一个三唑氮杂环，其杂环上的氮原子可与甾醇14α－脱甲基酶P450（sterol 14α-demethylase P450，简称P450－14DM，在真菌中称CYP51）的血红素－铁活性中心以配位键结合，从而抑制酶的活性。CYP51是麦角甾醇生物合成途径中的关键性酶，一旦其活性受到抑制，会阻碍麦角甾醇的合成，导致真菌细胞膜结构破坏和细胞死亡。目前，三唑类杀菌剂已有三十几个品种上市。1974年拜耳公司研制成功的三唑酮（Triadimefon）是三唑类杀菌剂第一个商品化的产品[6]，以及该公司于20世纪70年代开发的三唑醇（Triadimenol）、20世纪80年代开发的烯唑醇（Diniconazole）[7]和丙环唑（Propiconazole）[8]、20世纪90年代初期研发的戊唑醇（Tebuconazole）[9]都是目前国内常用的防治小麦病害的三唑类杀菌剂。最近研发出来的氟醚唑（Tetraconazole，意大利Isagro）[10]、羟菌唑（Metconazole，美国氰胺）[11]、丙硫菌唑（Prothioconazole，德国拜尔）[12]、氟硅唑（Flusilazole，美国杜邦）[13]等新型的三唑类化合物，与常用的三唑酮等三唑类杀菌剂相比，分子结构变化很大，且大多含氟，除了对禾谷类作物的锈病、白粉病有活性外，对纹枯病等病害亦有很好的活性且持效期长。特别是丙硫菌唑代表了三唑啉硫酮这类新化合物，具有保护、治疗和铲除作用。几乎对所有小麦病害都可以防治，并能防治非谷物的多种病害，如油菜粉和花生的土壤传播病害与叶面病害。

2 三唑类杀菌剂的作用机制

三唑类化合物的含氮杂环部分的氮原子与细胞色素P－450的铁离子结合，显示抗菌活性和植物生长调节活性。麦角甾醇在许多菌类中是组成细胞膜的主要甾醇，其生物合成以乙酰CoA为先导化合物，经火落酸、鲨烯、羊毛甾醇变为麦角甾醇。羊毛甾醇在24位甲基化、14位和4位的脱甲基化，其后甾醇核和侧链上的双键发生移动后，变为麦角甾醇。由于依赖于细胞色素P－450的氧化酶系统，14位的a－甲基为羟甲基，然后变为甲酰基，以甲酸形式脱离，接着14位的双键饱和[15]。

R. Gadners[16]以酵母和鼠肝均浆为酶，麦角甾醇生物合成抑制剂尤其抑制甾醇生物合成的脱甲基化反应，此时明显地与细胞色素P－450结合。接着T. E. Wiggins[16]用鼠肝微粒件和酵母细胞色素P－450研究了三唑类化合物作为生长调节剂阻碍植物中赤霉素的生物合成，在抑制生长的同时，有抗菌活性。研究表明：有关化合物与细胞色素P－450的结合性，化合物与细胞色素P－450的结合程度与杀菌活性和抑制脱甲基活性一致，这表明化合物中的三唑环的4位氮原子与叶啉的铁离子结合。

三唑类化合物降低了植物中的赤霉素的活性，添加赤霉素可恢复其生长抑制作用。另外，发现经三唑类化合物处理的植物中有甲基甾醇的富集。由此来看，三唑类化合物阻碍植物的赤霉素和甾醇的生物合成，抑制其生长。

3 三唑类杀菌剂的抗药性发生现状

20世纪80年代初期，德国和英国分别在田间发现了大麦白粉病菌（*Erysiphe graminis* f. sp. *hordei*）对环丙唑醇（cyproconazole）的抗药性菌株；80年代后期，欧洲和美国加利福尼亚州同时在田间发现了葡萄白粉病菌（*Uncinula necator*）对三唑酮的抗药性菌株。随后，有关柑橘绿霉病菌（*Penicillium digitatum*）、大麦网斑病菌（*Pyrenophora teres*）、大麦云纹病菌（*Rhynchosporium secalis*）、甜菜叶斑病菌（*Cercospora beticola*）、苹果黑星病菌（*Venturis inaepualis*）、小麦纹枯病菌（*Rhizoctonia cerealis*）、小麦白粉病菌（*Erysiphe graminis* f. sp. *tritici*）、黄瓜白粉病菌（*Spherotheca fulilginea*）和番茄叶霉病菌（*Fluvia fulva*）等重要植物病原真菌在田间对三唑类杀菌剂产生不同程度抗药性的报道陆续出现。同时还发现三唑类杀菌剂对梨疮痂病菌（*Venturia nashicola*）、小麦壳针孢叶枯病菌（*Mycosphaerella graminicola*）、香蕉黑条叶斑病菌（*Mycosphaerella fijiensis*）、小麦叶锈病菌（*Puccinia triticina*）和苹果轮纹病菌（*Botryosphaeria dothidea*）等植物病原真菌的田间防效明显下降，并且经室内诱导发现禾谷镰孢菌（*Fusarium graminearum*）对戊唑醇也产生了抗药性[17]。

4 三唑类杀菌剂的抗药机制

4.1 抗药性的遗传机制

一般而言，1 个或几个基因可以控制病原真菌对某一类杀菌剂产生抗药性[18]。据现有的研究报道，植物病原真菌对三唑类杀菌剂的抗药性遗传机制较为复杂，不同病原菌对同一种三唑类杀菌剂或同种病原菌对不同三唑类杀菌剂都有着不同的抗性遗传机制，因此很难用某一种简单的模式来解释病原菌对该类杀菌剂的抗性遗传机制[19]。植物病原真菌对三唑类杀菌剂的抗药性遗传机制通常被认为受多基因控制，而这些基因属于微效基因，在田间表现为连续性数量遗传性状[20]。室内抗药性突变体诱导研究发现，构巢曲霉（*Aspergillus nidulans*）对抑霉唑[21]、红粒丛赤壳菌（*Nectria haematococca*）对氯苯嘧啶醇[22]和戊唑醇[23]、玉蜀黑粉菌（*Ustilago maydis*）对三唑酮[24]、番茄叶霉病菌和苹果黑星病菌对氟硅唑[25]等抗药性的遗传都受多基因控制，其抗性倍数相对较低，且具有累加效应。有些抗性突变体表现出多基因效应，其适合度（如菌落生长速率、孢子萌发率、产孢量等）和致病力都有明显下降的现象。而红粒丛赤壳菌南瓜变种（*Nectria haematococca* var. *cucurbitae*）对三唑醇[26]的抗性则表现为单基因遗传，其抗性倍数较高。抗性遗传的复杂性不仅表现在室内诱变抗性突变体上，还存在于田间分离的抗性病原真菌中。如大麦白粉病菌对三唑醇[27]的抗性表现为主效基因遗传；苹果黑星病菌对氯苯嘧啶醇的抗性表现为单基因遗传[28]，而对氟硅唑和腈菌唑的抗性则表现为多基因遗传[29]；大麦网斑病菌对三唑醇的抗性表现为主效基因遗传，而对丙环唑的抗性则表现为多效基因遗传。

4.2 抗药性的生理生化机制

4.2.1 降低药剂与作用位点酶的亲和力

由于病原菌自身可以改变杀菌剂作用位点的结构，使杀菌剂对该作用位点的亲和力下降，从而无法发挥其杀菌作用。近年来，人们在柑橘青霉病菌上成功研究出一套分离细胞色素 P450 同工酶的方法，该方法有可能成为研究药剂与作用位点酶亲和力下降这一抗药性机制的有效手段。例如，白假丝酵母（*Candida albicans*）对三唑类杀菌剂的抗药性机制即属于这一类。研究还表明，抗药性菌株中细胞色素 P450 同工酶之一的细胞色素 P450 - 14DM 可能有 1 个脱辅基蛋白分子发生了变异，从而使得该酶与三唑类杀菌剂的亲和力下降，导致抗药性的产生。

4.2.2 增加对药剂的排泄

有些病原菌虽然能吸收大量的药剂，但由于其能够很快将这些药剂再排出体外，因而不会中毒。构巢曲霉的敏感菌株和抗性菌株均能吸收氯嘧啶醇和抑霉唑，然而 2 种菌株体内药物的含量却不同，进一步研究发现，这是由于2种菌株对上述药剂的排泄速率不同所致：抗性菌株可能由于体内运输体蛋白基因的过量表达而促进了其排泄作用，对药物的排泄速率快于敏感菌株，使得药物分子在与作用位点结合之前就被排出菌体外，从而表现出较高的抗药性。意大利青霉菌（*Penicillium italicum*）对三唑类杀菌剂的抗性突变体也表现为对药剂的积累量减少，致使到达作用靶点的药量不足而使得杀菌活性下降。通过紫外诱变的方法，周明国等[30]获得了脉孢霉（*Neurospora crassa*）对三唑醇的抗性突变体，对其抗性机制的研究发现，抗性突变体对药剂的吸收能力与敏感菌株相当，但排泄能力有所增强。总之，菌株可通过降低药剂在细胞内的积累，降低菌体内药剂的实际作用浓度，从而减轻对抗性突变菌株体内甾醇生物合成过程的抑制作用，最终表现出抗药性。

4.2.3 脱毒作用

其一为解毒代谢作用。有些病原真菌可通过一系列代谢途径将体内有毒药剂转化成无毒物质，使药剂失去作用而表现出抗药性。立枯丝核菌（*Rhizoctonia solani*）能够将抑霉唑代谢为无毒化合物，从而对其产生抗药性。瓜枝孢菌（*Cladosporium cucumerinum*）和意大利青霉菌对三唑类杀菌剂的抗性也与该作用有关[31]。将散黑穗病菌 *Ustilago avena* 对三唑醇的抗性菌株通过体内标记4，4-二甲基甾醇后在非标记培养基中培养，结果发现存在未标记的甲基甾醇，说明对生命活动有毒害作用的甲基甾醇被快速更新（脱毒）可能是其抗药性机制之一。此外，还有些抗药性病原真菌则是由于丧失了将杀菌剂转化为较高活性化合物的能力，如三唑酮在敏感菌株中需代谢成三唑醇才能发挥杀菌作用，而在抗药性菌株中这种代谢作用被阻止，从而也表现出抗药性。

4.2.4 对药剂的通透性下降

由于病原菌细胞膜通透性发生改变，导致药剂无法进入细胞内而不能到达其作用位点，从而使杀菌剂无法发挥作用。这种改变可以被看做是病原菌自我保护机制的增强，从而表现为其抗药能力的提高。抗戊唑醇的禾谷丝核菌菌株对戊唑醇的适应能力明显高于敏感菌株，其细胞膜受损伤的程度相对较小，通透性低于敏感菌株。Dahmen 等[32]指出，某些三唑类杀菌剂能够增大小麦秆锈病菌（*Puccinia graminis* f. sp. *tritici*）细胞膜的通透性，导致细胞内电解质外渗，使细胞内药剂积累浓度下降而表现出抗药性。

4.3 抗药性的分子机制

4.3.1 ABC 运输蛋白基因控制的抗三唑类杀菌剂分子机制

ABC 运输蛋白（ATP-binding cassette transporter）是目前已知的、最大的蛋白质家族之一，从细菌到人类已鉴定出共计150多个不同的 ABC 蛋白。ABC 运输蛋白是呈镶嵌状态存在于真菌细胞膜上的膜转运蛋白，是细胞膜上的外排机能泵，含有一个 ATP 结合区，通过结合并水解 ATP 而为膜转运提供能量，以运输质膜上的糖、氨基酸、磷脂和肽类物质，具有广泛的运输范围[33]。现有的研究证明，ABC 运输蛋白表达量的高低可能是植物病原真菌对三唑类杀菌剂具有抗药性或敏感的关键因素之一[34]。Cools 等[35]认为，桃疮痂病菌（*Cladosporium carpophilum* Thun）对三唑类杀菌剂的抗药性机制可能与其依赖 ABC 运输蛋白增强了对有毒物质的排泄有关。Delsorbo 等[36]从构巢曲霉中克隆了编码丝状真菌 ABC 运输蛋白的 artA 和 artB 基因，将 artB 基因的 cDNA 转入酵母菌 PDR5 基因缺失的突变体中，所获得的转化子对三唑类杀菌剂可产生抗药性。同时也发现采用少数三唑类药剂、抗生素和植物防御素（豌豆素）处理构巢曲霉可明显增强其 atrA 和 atrB 基因的表达，并且在药物处理仅几分钟后基因表达就开始上升，这与依赖能量的药物排出活性基因表达的启动情形相一致，也进一步说明 atrA 和 atrB 基因在三唑类杀菌剂的外排中具有作用。构巢曲霉 atrB 基因缺失的突变体对苯胺基嘧啶类、苯并咪唑类、苯基吡咯类、苯基吡啶胺类、Strobirulin 类和一些三唑类，天然化合物如喜树碱、植保素白藜芦醇，以及突变剂 4 - 硝基喹啉氧化物的敏感性都有所增加；与此相反，atrB 基因超量表达的突变体对这些有毒药剂的敏感性均下降，其敏感性与超量表达的水平呈负显著相关[37]。三唑类杀菌剂氯苯嘧啶醇在 atrB 基因超量表达的突变体中的积累量比其野生型中的少；用抑制能量生成的代谢型抑制剂处理这些突变体可使其对氯苯嘧啶醇的积累量恢复至与野生型一致的水平[37]。由此可见，atrB 基因超量表达可导致编码蛋白的量上升，使菌体对药剂的排出量增加，最终减少了药剂在病菌体内的累积，表现为具有抗药性。

4.3.2　与 CYP51 蛋白基因相关的抗三唑类杀菌剂分子机制

甾醇 14α - 脱甲基化酶（CYP51，P450 - 14DM）属于细胞色素 P450 基因家族中的成员，广泛存在于各种生物体内，是生物甾醇合成途径中的关键酶，迄今已有 30 多年的研究历史。在真菌中，CYP51 催化 14α - 甾醇经 14α - 脱甲基化反应生成麦角甾醇。由于三唑类杀菌剂的作用机制即是其含氮杂环氮原子与 14α - 甾醇经 CYP51 蛋白基因的血红素绑定区（血红素 - 铁活性中心）以配位键结合，从而抑制 14α - 脱甲基酶的活性[38]。因此，植物病原真菌对三唑类杀菌剂的抗药性机制之一可能是由于 CYP51 蛋白基因发生点突变导致氨基酸的取代，从而引起酶与药剂亲和力的下降而产生的。这种假设现已得到证实，如意大利青霉菌、指状青霉菌、大麦白粉病菌、苹果黑星病菌、葡萄白粉病菌等植物病原真菌对三唑类杀菌剂的抗药性都与 CYP51 基因的突变有关[3]。其中有报道表明，大麦白粉病菌、葡萄白粉病菌和指状青霉菌对三唑醇的抗药性主要是由于菌体内 CYP51 基因序列中的第 458 位碱基由 A 突变成了 T，该突变使得 136 位的酪氨酸（Tyr）被苯丙氨酸（Phe）取代，从而导致病原菌对三唑醇产生了抗药性[39]。李红叶等[18]在研究指状青霉菌对抑霉唑的抗药性机制时，分别克隆了 1 个敏感菌株和 2 个抗药性菌株的 CYP51 基因，并确定了其序列；通过序列比较发现，抗药性菌株 CYP51 基因编码区有 4 个核苷酸被取代，分别为 967、1028、1263 和 1650 位核苷酸碱基由敏感菌株的 G、T、T、C 突变成了 C、A、C、T，碱基的取代导致 209 位的甘氨酸（Gly）变成了丙氨酸（Ala），308 位的酪氨酸（Tyr）变为组氨酸（His），437 位的组氨酸（His）变为酪氨酸（Tyr）；同时还发现，指状青霉菌对抑霉唑的抗药性并非完全由基因突变引起，可能还与 CYP51 蛋白基因的超表达有关。Delye 等[4]研究了与大麦白粉病菌对三唑类杀菌剂抗药性产生相关的 CYP51 蛋白基因点突变机制，结果发现，与敏感菌株相比，抗性菌株 CYP51 蛋白基因的 F136Y 氨基酸序列发生了突变。F136Y 氨基酸突变现已被作为多种植物病原真菌对三唑类杀菌剂抗药性突变的研究热点。Leroux 等[40]就对三唑类杀菌剂抗药性水平不同的野生型小麦壳针孢叶枯病菌进行了 CYP51 蛋白基因型序列分析，发现了多个与三唑类杀菌剂抗药性产生相关的点突变。与敏感菌株相比，Y459S/D/N、G460D 或者 Y461S/H 点突变决定了对 SBIs 杀菌剂的低抗水平，而 I381V 点突变决定了对三唑类杀菌剂的高抗性水平。

5　总结

植物病原菌对三唑类杀菌剂产生抗药性是其适应环境的结果，在病原菌的长期进化过程中是一种必然事件。我们应积极探索植物病原菌对杀菌剂产生抗药性的原因进而来规避其产生抗药性的风险，同时应积极采取合理措施来降低抗药性的风险，比如和其他靶标的杀菌剂混配使用。

参考文献

[1] 何秀萍，张博润. 微生物麦角固醇的研究进展 [J]. 微生物学通报，1998，25 (3)：166-169.

[2] 刁春玲，刘芳，宋宝安. 农用杀菌剂作用机理的研究进展 [J]. 农药，2006，45 (6)：374-377.

[3] 周明国，刘经芬，叶钟音. 关于杀菌剂抗药性研究方法的综述 [J]. 南京农业大学学报，1987 (4，增刊)：128-134.

[4] Delye C，Laigret F，Corio-Costet M F. A multation in the 14α-demethylase gene (CYP51) of Unicinula necator that correlates with resistance to a sterol biosynthesis inhibitor [J]. Appl Environ Microb，1997b，63 (8)：2 966-2 970.

[5] Golembievski R C，Vargas J R，Jones A L，et al. Detection of demethylation inhibitor (DMI) resistance in Sclerotinia homeocarpa populations [J]. Plant Dis，1995，79 (5)：491-493.

[6] 李海屏. 杀菌剂新品种开发进展及发展趋势 [J]. 江苏化工，2004，32 (6)：7-12.

[7] Haiyan Liu，Gengliang Yang，Shubin Liu，et al. Molecular Recognition Properties and Adsorption Isotherms of Diniconazole-Imprinted Polymers [J]. Journal of Liquid Chromatography & Related Technologies，2007，28 (15)：2 315-2 323.

[8] Katsuhiko Sekimata，Sunyoung Han，Koichi Yoneyama，et al. A Specific and Potent Inhibitor of Brassinosteroid Biosynthesis Possessing a Dioxolane Ring [J]. Journal of Agricultural & Food Chemistry，2002，50 (12)：3 486-90.

[9] Horsley R D，Pederson J D，Schwarz P B，et al. Integrated Use of Tebuconazole and Fusarium Head Blight-Resistant Barley Genotypes [J]. Agronomy Journal，2006，98 (1)：194-197.

[10] Khan M F R，Smith L J. Evaluating fungicides for controlling Cercospora leaf spot on sugar beet [J]. Crop Protection，2005，24 (1)：79-86.

[11] Ito A，Saishoji T，Kumazawa S，et al. Structure-Activity Relationships of the Azole Fungicide Metconazole and Its Related Azolylmethylcycloalkanols [J]. Journal of Pesticide Science，1999，24 (3)：262-269.

[12] Blackbeard J，Kilburn D. No resistance to new mildewicide [J]. Arable Farming，1996.

[13] Garandeau J M，Chollet J F，Miginiac L. Synthesis of (3-aminopropyl) arylsilanes，including one or two heterocyclic patterns，potential fungicides [J]. Helvetica Chimica Acta，1997，80：706-718.

[14] 周子燕，李昌春，高同春，等. 三唑类杀菌剂的研究进展 [J]. 安徽农业科学，2008，36 (27)：11 842-11 844.

[15] 白林. 张应年，李生英. 三唑类化合物的杀菌活性和植物生长调节作用 [J]. 甘肃高师学报，2000，5 (2)：51-55.

[16] 王进贤译. 三唑类化合物的杀菌和植物生长调节作用 [J]. 农药，1988，27 (5)：50-55.

[17] 叶滔，马志强，王文桥，等. 禾谷镰孢菌对戊唑醇抗药性的诱导及抗性菌株特性研究 [J]. 农药学学报，2011，13 (3)：261-266.

[18] 迟玉杰，杨谦. 植物病原菌对杀菌剂产生抗性的机制与抗性的利用 [J]. 农业系统科学与综合研究，2001，17 (4)：313-316.

[19] 李红叶，谢清云，宋爱环. 抑霉唑敏感与抗性指状青霉菌株 CYP51 基因序列比较 [J]. 菌物系统，2003，22 (1)：153-156.

[20] 杨谦. 植物病原菌抗药性分子生物学 [M]. 北京：科学出版社，2003：69-80.

[21] VANTUYL J M. Genetics of fungal resistance to systemic fungicides [J]. Meded Landbouwhogesch Wageningen，1977，77 (2)：136-137.

[22] KALAMARAISLAE，DEMOPOULOSVP，ZIOGASBN，et al. A high mutable major gene for tradimenol resistance in *Nectria haematococcca* var. cucurbitae [J]. New J Plant Pathol，1989 (95，Suppl)：

109 - 115.

[23] Akakkal R, Debieu D, Lanen C, et al. Inheritance and mechanisms of resistance to tebuconazole, a sterol C14 - demethylation inhibitor in *Nectria haematococca* [J]. Pestic Biochem Physiol, 1998, 60 (3): 147 - 166.

[24] Wellmanh, Schauzk. DMI-resistance in *Ustilago maydis*: 1. haracterization and genetics analysis of triadimefon-resistant laboratory mutants [J]. Pestic Biochem Physiol, 1992, 43 (3): 171 - 181.

[25] Koller W, Smtthf D, REYNOLDSKL. Phenotypic instability of flusilazole sensitivity in *Venturia inaequalis* [J]. *Plant Pathol*, 1991, 40 (4): 608 - 611.

[26] Brown J K M, Jessop A C, Thoma S S, et al. Genetics control of the response of *Erysiphe graminis* f. sp. *hordei* to ethirimol and triadimenol [J]. Plant Pathol, 1992, 41 (2): 126 - 135.

[27] Stanis V F, Jonos A L. Reduced sensitivity to sterol inhibiting fungicides in field isolates of *Venturia inaequalis* [J]. Phytopathol, 1985, 75 (10): 1 098 - 1 101.

[28] Sholberg P L, Haag P D. Sensitivity of *Venturia inaequalis* isolates from British Columbia to flusilazole and myclobutanil [J]. Can J Plant Pathol, 1993, 15: 102 - 107.

[29] Peever T L, Milgroom M G. Inheritance of triadimenol resistance in Pyrenophorateres [J]. *Phytopathology*, 1992, 82 (8): 821 - 828.

[30] 周明国, Hollomon D W. *Neurospora crssa* 对三唑醇的抗药性分子机制 [J]. 植物病理学报, 1997, 27 (3): 275 - 280.

[31] Fuchs A, De Vries F W. Diastereomer-selective resistance in *Cladosporium cucumerinum* to triazole-type fungicides [J]. Pestic Sci, 1984, 15 (1): 90 - 96.

[32] Dahmen H, Hoch H C, Staub T. Differential effects of sterol inhibitors on growth, cell membrane permeability and ultra structure in two target fungi [J]. Phytopathology, 1988, 78 (8): 1 033 - 1 042.

[33] Del Sorbo G, Schoonbeek H, Dewaard M A. Fungal transporters involved in efflux of natural toxic compounds and fungicides [J]. Fungal Genet Biol, 2000, 30 (1): 1 - 15.

[34] Dewaard M A. Significance of ABC transporters in fungicide sensitivity and resistance [J]. Pectic Sci, 1997, 51 (3): 271 - 275.

[35] Cools H J, Ishii J, Butters J A, et al. Cloning and sequence analysis of the eburicol 14α-demethylase encoding gene (CYP51) from the Japanse pear scab fungus *Venturia inashicola* [J]. J Phytopathology, 2002, 150 (8 - 9): 444 - 450.

[36] Delsorbo G, Andrade A C, Vannistelrooy J G, et al. Multidrug resistance in Aspergillus nidulans involves novel ATP-binding cassette transporters [J]. Mol Gen Genet, 1997, 254 (4): 417 - 426.

[37] Andrade A C, Delsorbo G, Vannistelrooy J G, et al. The ABC transporter AtrB from Aspergillus nidulans mediates resistance to all major classes of fungicides and some natural toxic compounds [J]. Microbiology, 2000, 146 (8): 1 987 - 1 997.

[38] Ji H, Zhang W, Zhang M, et al. Structure-based de novo design, synthesis, and biological evaluation of non-azole inhibitors specific for lanosterol 14*alpha*-demethylase of fungi [J]. J Med Chem, 2003, 46 (4): 474 - 485.

[39] Delye C, Bousset L, Corio-Costet M F. PCR cloning and detection of point mutations in the eburicol 14*alpha*-demethylase (CYP51) gene from *Erysiphe graminis* f. sp. *hordei*, a "recalcitrant" fungus [J]. Curr Genet, 1998, 34 (5): 399 - 403.

[40] Leroux P, Albertini C, Gautier A, et al. Mutations in the CYP51 gene correlated with changes in sensitivity to sterol 14*alpha*-demethylation inhibitors in field isolates of *Mycosphaerelia graminicola* [J]. Pest Manag Sci, 2007, 63 (7): 688 - 698.

活性氧与呼吸链氧化磷酸化损伤研究概述

潘夏艳，武　健，段亚冰，侯毅平，王建新，周明国
（南京农业大学植物保护学院，南京　210095）

摘要：呼吸链在细胞生长、增殖分化和死亡等生命活动中扮演着十分重要的调控者的角色，也是许多药物的靶标。活性氧（ROS）是细胞代谢不可避免的产物，在呼吸链中由复合物Ⅰ和复合物Ⅲ产生超氧根阴离子（O_2^-），在不同的超氧化物歧化酶（SOD）作用和芬顿反应下生成 H_2O_2 和 OH^-。活性氧是引起呼吸链损伤的主要途径之一。本文介绍了呼吸链的结构和功能，重点从呼吸链氧化磷酸化功能的破坏阐述了 ROS 引起呼吸链损伤的机制。

关键词：呼吸链；氧化磷酸化；活性氧

Review of Reactive Oxygen Species and Oxidative Phosphorylation Damage of Respiratory Chain

Pan Xiayan, Wu Jian, Duan Yabing, Hou Yiping, Wang Jianxin, Zhou Mingguo
(*Plant Protection College*, *Nanjing Agricultural University*, *Nanjing* 210095, *China*)

Abstract: Respiratory chain plays vital role in cell growth, cell proliferation differentiation and cell death and other life activities. It is also the target of many drugs. Reactive oxygen species (ROS) is the inevitable product of cell metabolism. O_2^- is produced by complex Ⅰ and Ⅲ, and then generates H_2O_2 and OH^- under the action of superoxide dismutase (SOD). ROS is the main way to cause damage to the respiratory chain. This study introduces the structure and function of the respiratory chain and mainly elaborates damage mechanism of respiratory chain from function damage of oxidative phosphorylation.

Key words: respiratory chain; oxidative phosphorylation; reactive oxygen species

呼吸链氧化磷酸化是指细胞内伴随着有机物氧化，利用生物氧化过程中释放的自由能，促使 ADP 与无机磷酸结合生成 ATP 的过程。一切生命活动都是通过能量来驱动的，而氧化磷酸化被认为是细胞内的“动力室”[1]。真核细胞中，氧化磷酸化是在线粒体内膜上进行的；而原核生物中由于没有线粒体结构的分化，氧化磷酸化是在细胞膜上进行的。不管是在真核生物还是原核生物中，氧化磷酸化都参与许多细胞生命活动，如生物合成通路、细胞凋亡、氧化还原调控信号转导途径和维持体内钙平衡等[2]。越来越多的报道证实了呼吸链结构的复杂性和氧化磷酸化功能的重要性，一旦真核细胞线粒体的损伤或者原核细胞细胞膜被破坏，都会对这些生物造成极大的伤害，甚至直接被杀死。因此，开发一系列以真核原核呼吸链复合物为作用靶标的药剂，能够有效的控制人体和植物病害。本文主要介绍了整个呼吸链氧化磷酸化，并初步描述了氧化磷酸化与 ROS 之间的关系，为后续进一步研究开发呼吸链为靶标的药物提供了一定的理论依据。

1 呼吸链氧化磷酸化系统

1.1 呼吸链的组成

氧化磷酸化作用包括电子传递和 ATP 的偶联机制，整个过程被称为呼吸链。呼吸链主要由复合物Ⅰ、复合物Ⅱ、复合物Ⅲ和复合物Ⅳ组成[3]。复合物Ⅰ，又称 NADH 脱氢酶，是呼吸链上最大、最复杂的酶复合物，由 30 条以上的多肽链组成，10 个以上亚基组成 L 形的复合物横跨线粒体内膜或细胞膜。复合物Ⅱ，又称琥珀酸脱氢酶，由 4 条多肽链组成，含有一个以 FAD 为辅基的黄素蛋白、2 个铁硫蛋白和一个细胞色素 b。复合物Ⅲ，又称细胞色素 c 还原酶，由 10 多条多肽链组成，包括 2 个细胞色素 b，一个细胞色素 c1 和一个铁硫蛋白。复合物Ⅳ，又称细胞色素氧化酶，是呼吸链末端氧化酶，包括细胞色素 aa3 和含铜蛋白。每个复合物都有对应的抑制剂，鱼藤酮和杀粉蝶菌素 A 抑制复合物Ⅰ，阻断电子由 NADH 像 CoQ 的传递，但不影响 $FADH_2$ 到 CoQ 的氢传递。TTFA 抑制复合物Ⅱ，抗霉素和粘噻唑抑制复合物Ⅲ，氰化物和一氧化碳等则抑制复合物Ⅳ的电子传递（图 1）。

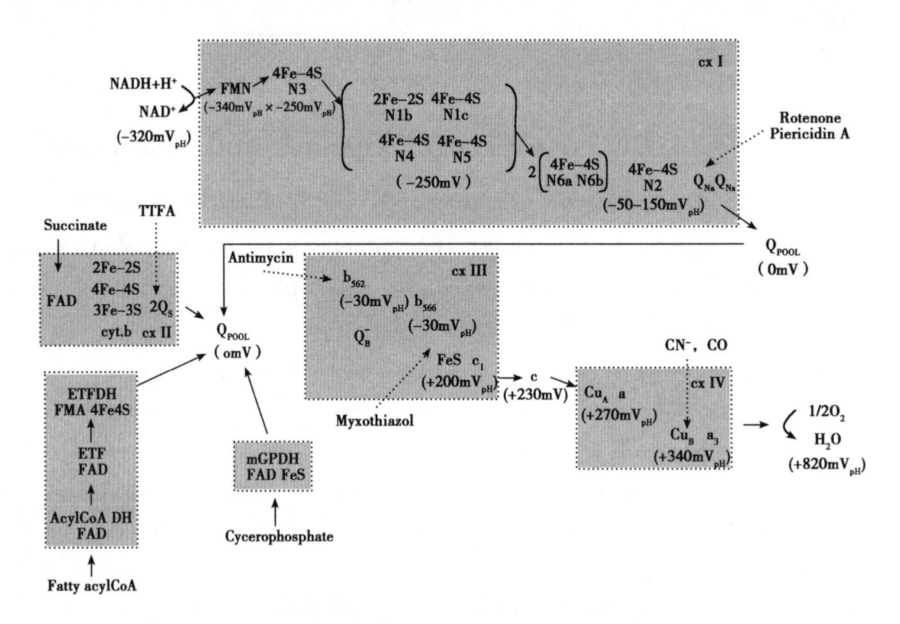

图 1 呼吸链电子传递中心

Fig. 1 Electron transfer centers of the respiratory chain in mammalian mitochondria

Qpool：泛醌池；Q_B：蛋白结合的泛醌

Qpool：ubiquinone of the pool，QB：protein bound ubiquinone

1.2 氧化磷酸化复合物生物合成和组装

以线粒体呼吸链为例，线粒体的分化涉及整个细胞生命周期组织的形成。在有丝分裂中，线粒体前体生成和分化，形成有正常组件的线粒体子细胞。线粒体的分化由两大信号网络来调控表达两大细胞基因组：核 DNA 和线粒体 DNA 实现的。线粒体基因组编码呼吸链和 ATP 合成酶复合物的蛋白组件。但是，上百种核编码蛋白参与呼吸链和内质网合成以及线粒体蛋白的输入。线粒体蛋白的输入途径非常复杂，由膜上的转运蛋白和内膜或基质上的分子伴侣介导运输。核编码的线粒体蛋白分为两大类，一类是 N 端导肽的前体蛋白，这些导肽作用于信号分子从而与线粒体输入受体互作；另一类是非导肽蛋白，携带各种各样的信号因子。在某些情况下，导肽会嵌入线粒体外膜移位酶（TOM）装置中，然后输入蛋白从4个途径进行运输线粒体前体蛋白。第一个途径是外膜上的不含靶标导肽的前体，像孔蛋白等，是通过小线粒体内膜间隙中的分子伴侣（Tim蛋白）从TOM复合物运输到排列组装蛋白的装置（SAM复合物）。第二条途径是通过MIA40/ERV1通路将蛋白直接运输到内膜间隙。第三条途径是内膜疏水性的前体，从TOM复合物开始，后结合到分子伴侣Tim9～Tim10或者Tim8～Tinm13复合物，然后分子伴侣将该蛋白运输到内膜上的TIM22复合物。第四条途径是含导肽的前体蛋白，直接被运输到基质中的TIM23复合物上[4]（图2）。

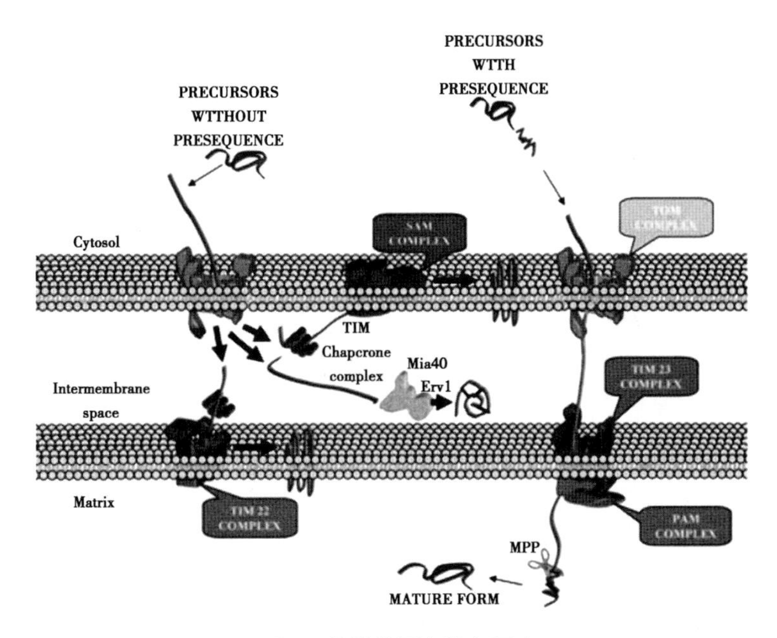

图2　线粒体蛋白输入通路

Fig. 2　Protein import pathways in mitochondria

呼吸链上的氧化磷酸化复合物都是由不同的亚基组成的，因此前体通过上述不同的途径到达指定位置之后，会组装成完整的复合物。比如复合物Ⅰ，外围臂组件和膜臂组件先组合形成中间络合物，之后再组装成完整的复合物Ⅰ[5]（图3）。

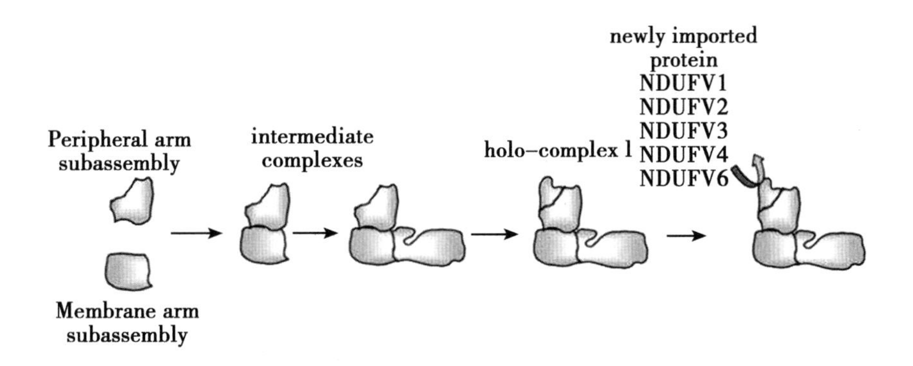

图3　复合物Ⅰ组装

Fig. 3　Assembly scheme of complex Ⅰ

1.3 转录因子调控氧化磷酸化复合物的组装

氧化磷酸化复合物的组装主要是通过转录因子调控完成的。环磷酸腺苷（cAMP）和钙离子（Ca^{2+}）信号能激活环磷腺苷效应元件结合蛋白（CREB）和传感器来调控 CREB 结合蛋白家族（TORCs）[6]。CREB 和 TORC 一起激活下游的转录共激活因子 PGC-1α 和细胞色素 c。PGC-1α 的激活和表达和很多正负信号通路相关，PGC-1α 能被去乙酰化酶（SIRT1）正调控，被腺苷酸活化蛋白激酶（AMPK）和分裂原激活的蛋白激酶（P38MAPK）磷酸化，或者被 AKT 抑制磷酸化[7]。PGC-1α 被激活后，与 NRF1 和 NRF2 转录因子共同作用，激活线粒体转录和复制相关因子的表达，如线粒体转录因子（TFAM）；也能激活结构蛋白和组装的转录因子氧化磷酸化（OXPHOS）复合物的表达。最后，在线粒体蛋白输入到指定位置之后，mTFA 和 CREB 蛋白作用激活 mtDNA 的表达[8]（图 4）。

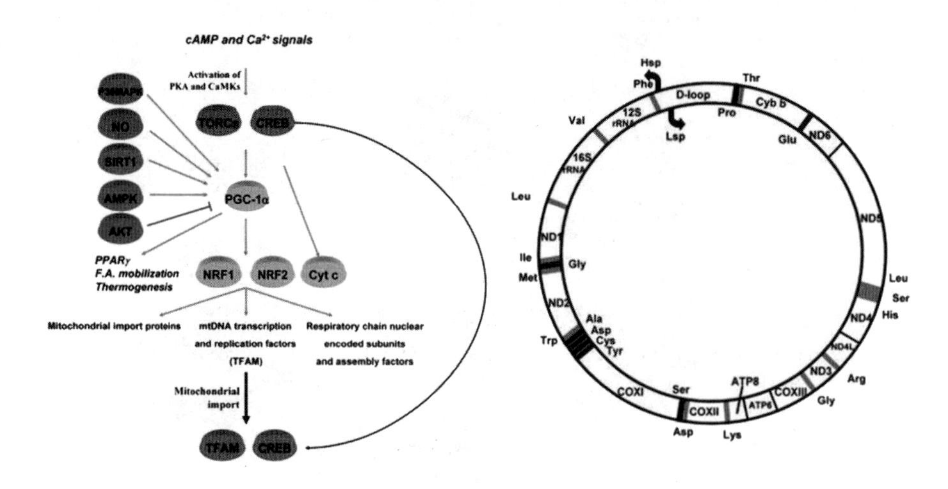

图 4　转录因子调控线粒体生物合成示意图

Fig. 4　Schematic representation of transcription factors which control mitochondrial biogenesis

2　活性氧（ROS）和氧化磷酸化

2.1 ROS 化学过程

一旦呼吸链氧化磷酸化过程被破坏，最直接的结果是会在胞内积累 ROS。氧化磷酸化功能受损，导致电子转运链异常，ROS 大量积累，导致蛋白质、膜脂、DNA 及其他细胞组分的严重损伤[9]。ROS 种类非常多，有超氧根阴离子（O_2^-），过氧化氢（H_2O_2），羟基自由基（OH^-）等。O_2^- 不能移动，只能在产生部位被代谢或生成 H_2O_2[10]；H_2O_2 可以自由地在细胞内穿梭，而且非常稳定，是许多信号通路的调控因子；OH^- 是伤害力最强的一种 ROS，但是其半衰期非常短，极其不稳定。这几种 ROS 之间可以相互转换，O_2 和 O_2^- 在 H^+ 作用下能生成 H_2O_2；H_2O_2 能被直接代谢生成 H_2O，或者在芬顿反应下生成 OH^-[11]（图 5）。

2.2 ROS 产生和代谢

在完整的细胞内，到处分布着 Ca^{2+} 和 H^+。一旦 Ca^{2+} 积累，就会激活 TCA 循环，TCA 循环中也会产生一定的 ROS。之后激活呼吸链复合物。整个呼吸链中，只有复合物 I 和复

$$O_2 \cdot 1e^- \rightarrow O_2 \cdot^-$$

$$O_2 \cdot^- + 1e^- + 2H^- \rightarrow H_2O_2$$

$$O_2 + 2e^- + 2H^- \rightarrow H_2O_2$$

$$H_2O_2 + 2e^- + 2H^- \rightarrow 2H_2O$$

$$Fe^{2+} + H_2O_2 \rightarrow Fe^{3+} + OH \cdot + OH^- \text{ (Fenton reaction)}$$

图 5　活性氧的化学过程

Fig. 5　chemical process of ROS

合物Ⅲ能产生 O_2^-，之后再转换成 H_2O_2 和 OH^-[12-13]。细胞内有效清除活性氧的保护机制分为酶促和非酶促两类。酶促脱毒系统包括超氧化物歧化酶（SOD）、过氧化氢酶（CAT）和谷胱甘肽过氧化物酶（GPX）等。非酶类抗氧化剂包括抗坏血酸、谷胱甘肽、甘露醇和类黄酮。SOD 将 O_2^- 代谢生成 H_2O_2，CAT 和 GPX 等能代谢 H_2O_2 生成 H_2O[14]（图6）。

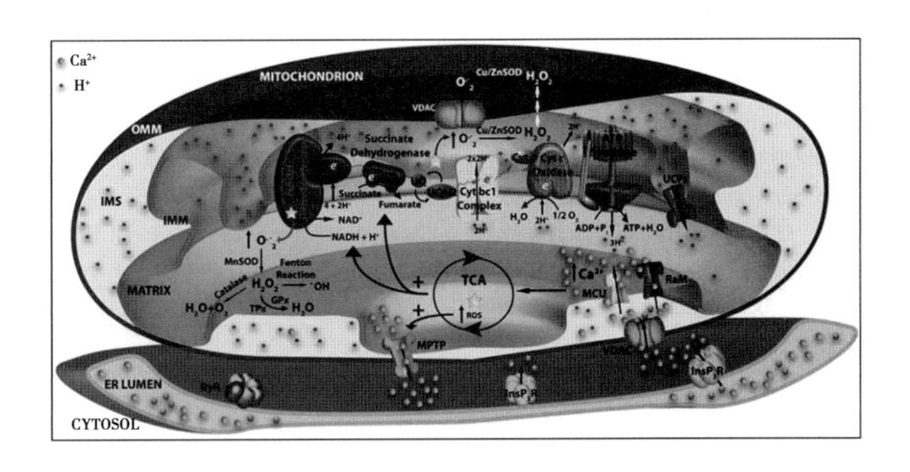

图 6　钙离子激活氧化磷酸化和活性氧产生代谢

Fig. 6　Ca^{2+} stimulation of oxidative-phosphorylation and ROS production/ metabolism

3　ROS 增加对细胞造成的危害

过量的 ROS 会降低细胞膜电位，导致细胞凋亡相关蛋白的释放从而启动细胞凋亡过程，导致 DNA 裂解，诱导细胞凋亡[15]。ROS 积累会影响细胞的代谢速率，影响细胞的结构和生长。ROS 能够修饰氨基酸，使蛋白肽链断裂，形成蛋白质的交联聚合物，改变蛋白的构象等对蛋白质造成一系列的损伤，从而影响正常细胞的生命活动[16]。

4　总结

呼吸链氧化磷酸化功能受损，电子传递通路发生障碍，导致 ROS 积累。呼吸链中产生 ROS 的部位主要为氧化磷酸化通路中的复合物复合物Ⅰ（NADH 脱氢酶）和复合物Ⅲ（泛醌-细胞色素 c 还原酶）。如果以呼吸链为靶标，能使细胞内 ROS 积累，导致细胞死亡，从而起到治疗人体疾病或植物病害的作用。未来特异性地针对呼吸链氧化磷酸化功能障碍，对病害防控有着重要的研究前景。

参考文献

[1] Paz M L, González Maglio D H, Weill F S, et al. Mitochondrial dysfunction and cellular stress progression after ultraviolet B irradiation in human keratinocytes [J]. Photodermatol Photoimmunol Photomed, 2008, 24 (3): 115 - 122.

[2] Wagner B K, Kitami T, Gilbert T J, et al. Large-scale chemical dissection of mitochondrial function [J]. Nat Biotechnol, 2008, 26 (3): 343 - 351.

[3] Leonard J V, Schapira A H. Mitochondrial respiratory chain disorders I: mitochondrial DNA defects [J]. Lancet, 2000, 355 (9200): 2 299 - 2 304.

[4] Neupert W, Herrmann J M. Translocation of proteins into mitochondria [J]. Annu Rev Biochem, 2007, 76: 723 - 749.

[5] Lazarou M, McKenzie M, Ohtake A, et al. Analysis of the assembly profiles for mitochondrial and nuclear-DNA-encoded subunits into Complex I. Mol Cell Biol, 2007, 27: 4 228 - 4 237.

[6] Herzig S, Long F, Jhala U S, et al. CREB regulates hepatic gluconeogenesis through the coactivator PGC-1 [J]. Nature, 2001, 413: 179 - 183.

[7] Screaton R A, Conkright M D, Katoh Y, et al. The CREB coactivator TORC2 functions as a calcium- and cAMP-sensitive coincidence detector [J]. Cell, 2004, 119: 61 - 74.

[8] De Rasmo D, Signorile A, Roca E, et al. cAMP response element-binding protein (CREB) is imported into mitochondria and promotes protein synthesis [J]. FEBS J, 2009, 276: 4 325 - 4 333.

[9] Soeur J, Eilstein J, Léreaux G, et al. Skin resistance to oxidative stress induced by resveratrol: from Nrf2 activation to GSH biosynthesis [J]. Free Radic Biol Med, 2015, 78: 213 - 223.

[10] Riley P A. 1994. Free radicals in biology: oxidative stress and the effects of ionizing radiation. Int. J. radiat. Biol. 65: 27 - 33.

[11] Pridmore R D, Pittet A C, Praplan F, et al. Hydrogen peroxide production by *Lactobacillus johnsonii* NCC 533 and its role in anti-Salmonella activity [J]. Fems Microbiol. Lett, 2008, 283: 210 - 215.

[12] Miwa S, St-Pierre J, Partridge L, et al. Superoxide and hydrogen peroxide production by Drosophila mitochondria [J]. Free Radic Biol, 2003, 35 (18): 938 - 948.

[13] Murphy M P. How mitochondria produce reactive oxygen species [J]. Biochem J, 2009, 417 (1): 1 - 13.

[14] Wang, L, Duan, Q, Wang, T, et al. Mitochondrial Respiratory Chain Inhibitors Involved in ROS Production Induced by Acute High Concentrations of Iodide and the Effects of SOD as a Protective Factor [J]. Oxid. Med. Cell. Longev., 2015: 217 670.

[15] Hildeman D A, Mitchell T, Teague T K, et al. Reactive oxygen species regulate activation-induced T cell apoptosis [J]. Immunity, 1999, 10 (6): 735 - 744.

[16] Fruehauf J P, Meyskens F L Jr. Reactive oxygen species: a breath of life or death? [J]. Clin Cancer Res, 2007, 13: 789 - 794.

The Fungicide Resistance Action Committee-structure, Objectives and its Contribution to Fungicide Resistance Management in Agriculture

Dietrich Hermann*

(*Syngenta Crop Protection AG*, *Schwarzwaldallee* 215, 4058 *Basel*, *Switzerland.*)

Modern agricultural production is dependent on the use of fungicides to avoid yield and quality losses due to fungal diseases; thus fungicides are important to complement other means of control like genetic resistance and cultural techniques. Intensive use of chemical control measures has however led to its own challenges, including selection for fungicide resistance in many pathogens.

As the ability of the industry to discover and develop fungicides of new modes of action is limited and the regulatory hurdles are constantly rising, sustainable use of existing technologies is key. The implementation of resistance management strategies is an important element in maintaining a sufficient choice of tools to the grower and prolonging the effectiveness and usefulness of modern fungicides. The Fungicide Resistance Action Committee (FRAC) and its global network play a vital role in understanding resistance issues and designing and supporting these strategies.

FRAC is a specialist industry group of CropLife International (CLI), consisting of the main fungicide manufacturers developing and supplying products to the market. Current member companies of global FRAC are ADAMA, BASF, Bayer CropScience, Dow Agrosciences, Du Pont, FMC, Isagro, KI Chemical, Sumitomo and Syngenta. In addition, several other companies are represented in technical and regional FRAC working groups. Its purpose is to identify potential and existing resistance problems and their relevance to the grower, evaluate scientific approaches and knowledge, and provide global fungicide resistance management guidelines as well as education to prolong the effectiveness of fungicides and limit crop losses should resistance occur.

Fungicide resistance management strategies are based around good agronomic practice, optimum fungicide timing, the use of appropriate dose rates, and optimizing the use of alternative modes of action in programs. Effective dissemination of information and implementation of resistance management strategies is made much easier by the presence of local organisations. FRAC local groups, consisting of expert industry representatives, are very active in North America, Japan, Brazil, South Africa and are being encouraged in other countries of the world.

FRAC does not work in isolation-there is excellent communication with country Fungicide Resistance Action Groups (FRAGs, which usually consist of university and official experts as well as

* Corresponding author: Dietrich Hermann; E-mail: dietrich. hermann@ syngenta. com

industry members), universities and research institutions, advisory bodies, EPPO and many country regulatory authorities. This forms a network of resistance experts and scientists to perform appropriate research and to help make the best practical recommendations.

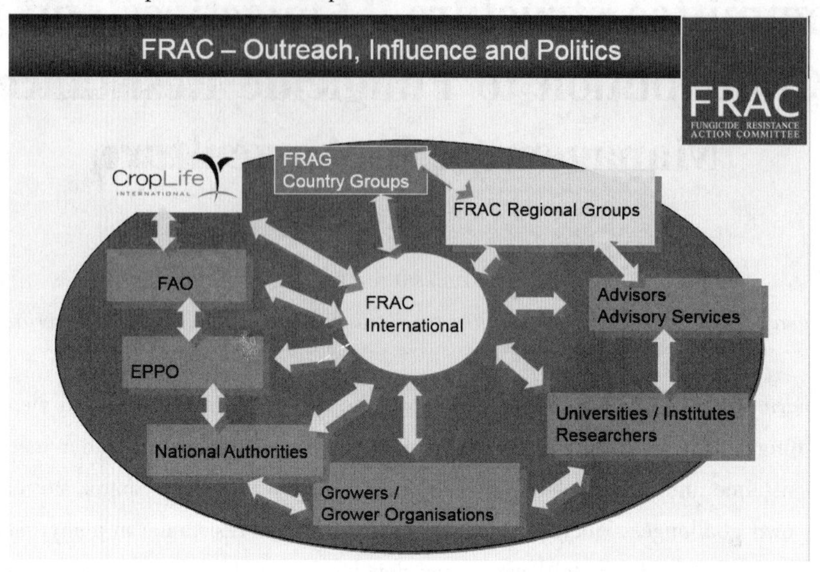

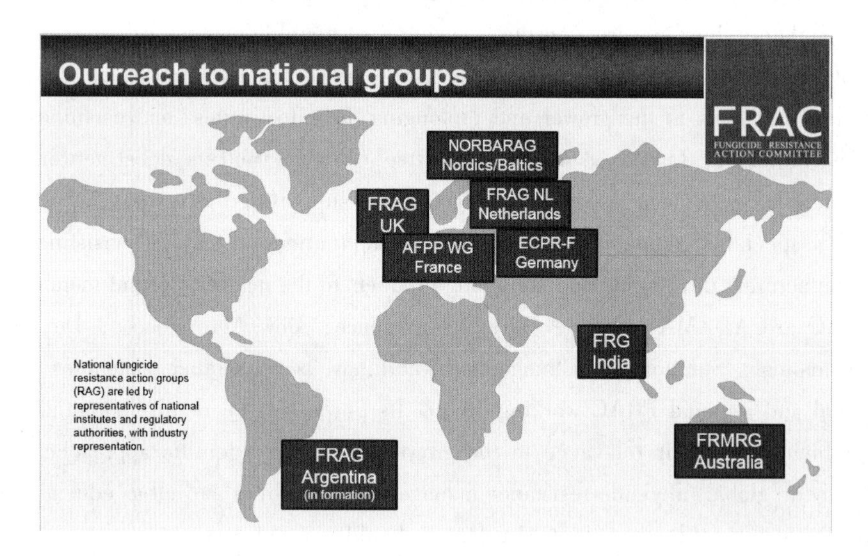

The detailed technical work of FRAC is done in Working Groups. If fungicides from different manufacturers have the same mode of action and if this mode of action bears at the same time a significant resistance risk, a FRAC Working Group is usually established to analyze the resistance risk and to develop and publish common resistance management recommendations. There are currently FRAC Working Groups for sterol biosynthesis inhibitors (SBIs), QoI fungicides, anilinopyrimidines (APs), SDHI fungicides, carboxylic acid amides (CAAs) and azanaphthalenes (AZNs), and a Task Force has recently been established to work on resistance management principles for the QiI fungicides. Formation of new working groups e. g. Arylphenyl-ketones or OSBPI (Oxysterol-binding protein inhibitors) is anticipated. These groups meet regularly and publish annually updated reports on the resistance status and suitable resistance management recommendations. In addition, the FRAC Banana Working Group, composed of fungicide manufacturers and fruit companies, coordinates resistance management recommendations for all specific fungicides used in banana production. For older modes of action for which regular monitoring programs are no longer performed (benzimidazoles, phenylamides and dicarboximides), so-called Expert Fora are available at the FRAC website (www. frac. info) to give advice and collect important published literature on resistance monitoring methods and resistance management.

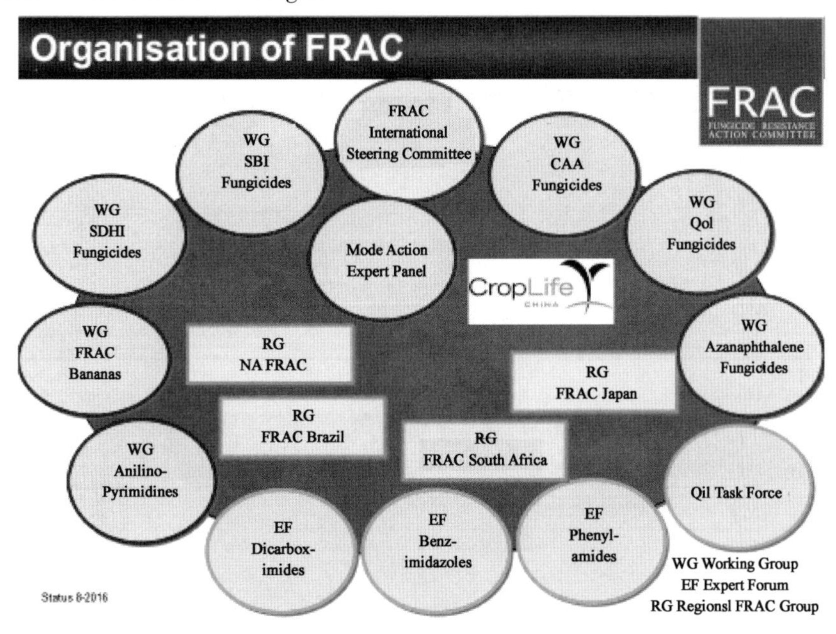

Since the classification of fungicides according to their mode of action and cross resistance pattern is a fundamental basis for resistance management under practical agronomic conditions, a major contribution of FRAC is its work in assessing the mode of action and resistance risk of fungicides and in compiling the annually updated Fungicide Mode of Action list. This list has become the standard reference for the classification of fungicides according to mode of action and cross resistance risk. In order to support the FRAC Steering Committee in the scientific evaluation of matters related to inclusion or changes of antifungal agents on the FRAC Mode of Action code list and the Mode of Action Poster, a technical subgroup of FRAC was established in 2012, the FRAC Mode of Action Expert Panel. Proposals for inclusions typically come from the industry or are triggered by the FRAC steering committee based on publication or registration of new fungicides. The panel then works closely with scientists of the suppliers or their nominated collaborators in evaluating the proposals and defining the information needed for inclusion to the list. The members of this Panel are recognised industrial experts in the field of mode of action. Of course, experts outside industry are consulted where necessary, to ensure agreement on a scientific basis.

The Mode of Action code list however gives no information on specific products or their disease control efficacy. Inclusion of a compound is not a regulatory process and thus gives no "approval" for a product and its value in disease control or resistance management.

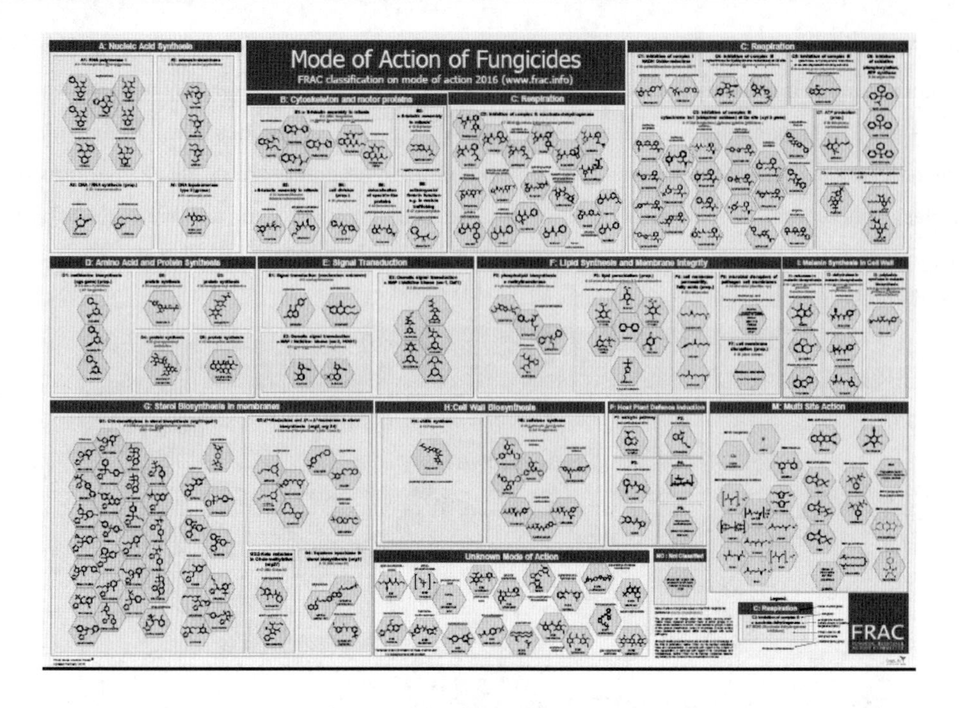

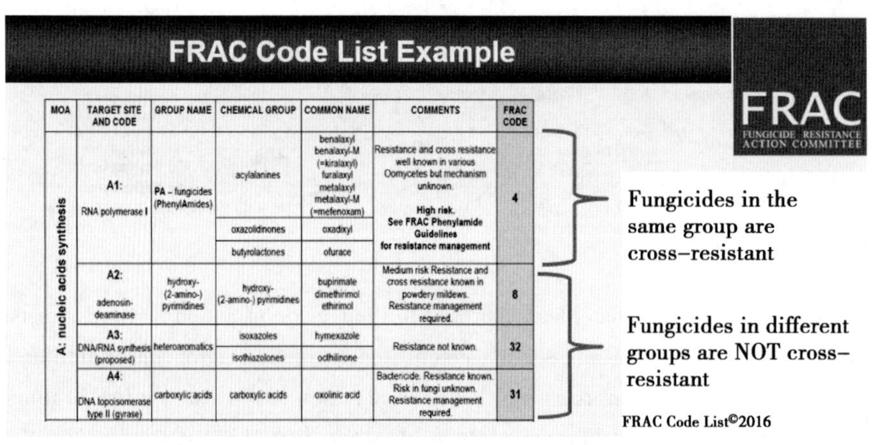

FRAC Code List Example

MOA	TARGET SITE AND CODE	GROUP NAME	CHEMICAL GROUP	COMMON NAME	COMMENTS	FRAC CODE
A: nucleic acids synthesis	A1: RNA polymerase I	PA – fungicides (PhenylAmides)	acylalanines	benalaxyl benalaxyl-M (=kiralaxyl) furalaxyl metalaxyl metalaxyl-M (=mefenoxam)	Resistance and cross resistance well known in various Oomycetes but mechanism unknown. **High risk. See FRAC Phenylamide Guidelines for resistance management**	4
			oxazolidinones	oxadixyl		
			butyrolactones	ofurace		
	A2: adenosin-deaminase	hydroxy-(2-amino-) pyrimidines	hydroxy-(2-amino-) pyrimidines	bupirimate dimethirimol ethirimol	Medium risk Resistance and cross resistance known in powdery mildews. Resistance management required.	8
	A3: DNA/RNA synthesis (proposed)	heteroaromatics	isoxazoles	hymexazole	Resistance not known.	32
			isothiazolones	octhilinone		
	A4: DNA topoisomerase type II (gyrase)	carboxylic acids	carboxylic acids	oxolinic acid	Bactericide. Resistance known. Risk in fungi unknown. Resistance management required.	31

The overall activities of FRAC are managed by the Steering Committee. The FRAC Steering Committee is currently chaired by Dietrich Hermann (Syngenta), and comprises senior, technical people from the Research and Development functions of agrochemical companies. All committee members are elected to their positions. The chairpersons of the Working Groups and Expert Fora are members of the Steering Committee to ensure coordination and peer review of the groups. The members of the Steering Committee make a commitment to undertake an active and constructive role in resistance management matters and all members of FRAC are required to actively contribute with scientific data and discussions.

A key responsibility of FRAC is communication, education and training in the area of fungicide resistance management. The core platform to share information from FRAC is the website (www.frac.info). Here interested stakeholders can find:

- The publication of FRAC Working Group findings and recommendations, updated at least annually. Recommendations are also published for some of the fungicide groups without active FRAC Working Groups.
- The publication of reference and educational material such as the FRAC monographs, the FRAC Mode of Action list and wall chart, the FRAC list of pathogens resistant to fungicides and FRAC recommended sensitivity monitoring methods.
- Other papers and documents such as the FRAC recommendations for mixtures and their use.

In order to emphasize the impact of grower decisions and practises on resistance management, FRAC also contributes to CLI training materials focusing on sustainable and responsible use of pesticides. FRAC encourages scientists in academia and extension to develop tools helping the advisor and grower in making fungicide decisions based on resistance risk and cross resistance information.

FRAC is also proactive, along with the other Resistance Management Committees, in discussions with key bodies such as the European Commission, EPPO, FAO and EPA to try to bring resistance management science and practice into consideration with policy makers and to raise understanding on the need to maintain a broad toolset for modern agriculture.

In conclusion, FRAC, within the global network to local RAGs, officials and advisors, supports proactive resistance management using all the tools available in an integrated approach. We actively support and invest in research on resistance risk and the effectiveness of management strategies. FRAC encourages the use of FRAC codes in product labelling.

Although it cannot be claimed that the resistance management strategies defined and supported by FRAC and other experts in the field have prevented the occurrence of fungicide resistance, it is clear that they have reduced the impact of resistance in practical situations and have certainly limited or slowed down the rate of spread.

Effect of Azoxystrobin and Kresoxim-Methyl on Rice Blast and Rice Grain Yield in China*

Chen Yu[1,3,4]**, Yang Xue [1,3,4], Li Yunfei[2],
Zhang Aifang[1,3,4], Yao Jian[2], Gao Tongchun[1,3,4,]***
(1. *Institute of Plant Protection and Aro-Products Safety*, *Anhui Academy of Agricultural Sciences*, *Hefei* 230031, *China*; 2. *Anhui Entry-Exit Inspention and Quarantine Bureau*, *Hefei* 230022, *China*; 3. *Scientific Observing and Experimental Station of Crop Pests in Hefei*, *Ministry of Agriculture*, *China*; 4. *Laboratory of Quality & Safety Risk Assessment for Agro-Products* (*Hefei*), *Ministry of Agriculture*, *China*)

Abstract: Sensitivity to azoxystrobin and kresoxim-methyl of 80 single-spore isolates of *Magnaporthe oryzae* was determined. The EC_{50} values for azoxystrobin and kresoxim-methyl in inhibiting mycelial growth of the 80 *M. oryzae* isolates were 0.006 ~ 0.056μg/mL and 0.024 ~ 0.287μg/mL, respectively. The EC_{50} values for azoxystrobin and kresoxim-methyl in inhibiting conidia germination of the *M. oryzae* populations were 0.004 ~ 0.051μg/mL and 0.012 ~0.105μg/mL, respectively. There was significant difference of sensitivity to azoxystrobin or kresoxim-methyl between isolates respresenting differential sensitivity to carbendazim (MBC) and kitazin P (IBP), however, there was no correlation between this difference in sensitivity to azoxystrobin or kresoxim-methyl and sensitivity to MBC or IBP, indicating that there was no cross-resistance between azoxystrobin or kresoxim-methyl and MBC or IBP. In the protective and curative tests, kresoxim-methyl exhibited higher protective and curative control efficacy than azoxystrobin when applied at 150 and 250μg/mL accordingly, while azoxystrobin exhibited stronger inhibitory activity against *M. oryzae* isolates than that of kresoxim-methyl in the *in vitro* test. The results of field trials also showed that both azoxystrobin and kresoxim-methyl at 187.5g. a. i/hm^2 provided over 73% control efficacy in 2010 and 2011 in both two sites, reepectively, exhibiting excellent activity against rice blast. Taken together, azoxystrobin and kresoxim-methyl should be good alternatives to MBC or IBP for the control of rice blast in China, but should be carefully used as they were both at risk fungicides.

Key words: *Magnaporthe oryzae*; Azoxystrobin; Kresoxim-Methyl; Baseline sensitivity; Cross-resistance; Rice blast

Rice blast, caused by *Magnaporthe oryzae* Couch (anamorph *Pyricularia oryzae* Cavara, formerly known as *P. grisea* (Cook) Sacc.), previously referred to in the literature as *M. grisea* (Hebert) Barr, is one of the most destructive diseases of rice (*Oryza sativa* L.) and is a serious constraint to production in most of the world's rice growing areas [1-2]. Under favorable conditions,

---

* Foundation item: Special Fund for Agro-Scientific Research in the Public Interest from the Ministry of Agriculture, China (201303023)

** First auther: Chen Yu, associate researvher, major in plant pathology; E-mail: chenyu66891@ sina. com

*** Corresponding author: Gao Tongchun, major in plant pathology; E-mail: gtczbs@ sina. com it is a very important rice disease in China, Japan and the United States, and causes severe damage to rice yield[1,3-4]. Severe blast has been attributted to susceptible cultivar, loss of flood water, high nitrogen fertilization, sandy light soils and fields surrounded by trees[1,5-6]. The *M. oryzae* pathogen can infect rice plants at any growth stage from seedling through grain formation, and causes leaf blast, collar rot, nodal blast, neck blast, or panicle blast[1,7], among which the neck blast is the most destructive in terms of yield loss[2,8]. Rice blast is especially problematic in temperate mountain areas with high humidity and precipitation. This disease could be managed by changing planting date, avoiding excessive inputs of nitrogen, maintaining high levels of soil moisture, planting blast-resistant cultivars and using fungicides. However, farmers preferred to applying fungicides for disease control and to prevent significant reductions in grain yield when the symptoms of blast appeared[1-2].

In the early 20th century, copper-based fungicides were used in China, but were discontinued due to low efficacy and phytotoxicity[9-10]. During 1980s organophosphorus fungicides such as kitazin (EBP), kitazin P (IBP) and isoprothiolane (FJ-one) with different chemical structures but similar mode of action were widely used[10-11]. Tricyclazole, a melanin biosynthesis inhibitor[12-13], was introduced as an alternative for IBP in the early 1990's because resistance to organophosphate fungicides became widespread[14]. Carbendazim (MBC), a benzimidazole fungicide, was used as an auxiliary chemical against blast disease in practice by mixing with other chemicals such as sulphur. However, resistance of *M. oryzae* to carbendazim has also been reported in China[15]. Therefore, assessment of new chemical classes of fungicides, which have different mode of action and cross-resistance with these commonly used blasticides, is one of the preferred strategies[16].

The strobilurin-based (QoI) fungicides have proven very effective in controlling rice blast in the United States[1]. The specific target for the QoI fungicides is the quinol-oxidizing (Qo) site of the mitochondrial enzyme cytochrome b[17], as these chemicals at the Qo site blocks electron transport, thereby inhibiting respiration[18]. Azoxystrobin and kresoxim-methyl, belonging to QoI fungicides, are relatively new for the control of rice blast in China. However, in our knowledge, no reports on the baseline sensitivity to azoxystrobin and kresoxim-methyl of field isolates of *M. oryzae* from China for are available till now. This provides an opportunity to study the baseline sensitivity of *M. oryzae* isolates to azoxystrobin and kresoxim-methyl although they have recently been used for the control of rice diseases in China. Construction of the baseline sensitivity is a tool for monitoring sensitivity shifts and effectiveness of resistance management strategies and is one of the major tasks undertaken by scientists working in the chemical control of pests, forming a significant part of the registration process for pesticides[19]. Moreover, the novel mode of action of azoxystrobin and kresoxim-methyl makes them excellent candidates for management of fungicide resistance in *M. oryzae* populations in China.

The objectives of this current study were to (i) establish the baseline sensitivity of *M. oryzae* isolates to azoxystrobin and kresoxim-methyl using field isolates from China, (ii) determine the cross-resistance patterns of azoxystrobin or kresoxim-methyl with fungicides belonging to other chemical classes, (iii) evaluate the protective and curative activity of azoxystrobin and kresoxim-methyl against rice blast and (iv) test the efficacy of azoxystrobin or kresoxim-methyl in controlling rice blast and their effect on rice grain yield.

1 Materials and methods

1.1 Origin and collection of *M. oryzae* isolates

Field single-spore isolates of *M. oryzae* were isolated from diseased rice samples with typical symptoms collected in Jiangsu and Anhui Province of China between 2010 and 2011. The rice fields were separated from each other by more than 10 km. In each field several plants with symptoms of rice blast were randomly collected from a designated 5.0m^2 ×5.0m^2 area, air-dried, placed in paper envelopes, and stored at -4℃. To enhance sporulation, diseased samples were disinfested with 95% ethyl alcohol, and placed in 90 mm Petri dishes on moistened filter paper under black light (BL, 20W), then incubated at 25℃ for 2~5 days. Individual spore was transferred onto 2% water agar (WA) plates, and incubated at 25℃ for 6 h. Single germ tube of a germinated spore was individually removed from the WA plates, and placed onto potato dextrose agar (200 g potato, 20 g dextrose, 35 g agar and 1 l distilled water) (PDA) plates amended with 0.5g/L of streptomycin to avoid any contaminated bacteria in dark. Growing mycelia were then transferred to PDA slants at 4℃ in dark with 5~12 isolates collected from each rice field[9]. A total of 80 single-spore isolates were collected from 13 rice fields in Anhui Province (Table 1). QoI fungicides had never been used to control rich blast before sampling.

Table 1 Cross-resistance between carbendazim/kitazin P and azoxystrobin/ kresoxim-methyl in *Magnaporthe oryzae*

Isolates	Sensitivity to		EC_{50}[1] for azoxystrobin (μg/mL)		EC_{50} for kresoxim-methyl (μg/mL)	
	Carbendazim	Kitazin P	Inhibiting growth	Inhibiting germination	Inhibiting growth	Inhibiting germination
Cun01	S[2]	S	0.019 c[3]	0.007 d	0.223 a	0.025 f
Cun02	S	R	0.017 d	0.007 d	0.081 d	0.081 b
Cun05	R	S	0.016 d	0.006 d	0.223 a	0.022 f
QS07	R	S	0.019 bc	0.008 c	0.263 a	0.063 c
QS09	R	S	0.026 a	0.011 b	0.135 c	0.035 e
QS13	R	S	0.018 c	0.009 c	0.152 bc	0.052 c
TP01	S	S	0.027 a	0.014 a	0.227 a	0.027 ef
TP03	S	R	0.020 b	0.015 a	0.093 d	0.093 a
TP06	S	R	0.018 d	0.006 d	0.115 c	0.015 c
XC02	R	S	0.017 d	0.007 d	0.177 b	0.077 b
XC06	R	S	0.011 e	0.006 d	0.212 a	0.012 g
XC08	S	R	0.017 d	0.007 d	0.246 a	0.046 c

[1] EC_{50}, 50% effective concentration;

[2] R, resistant to this fungicide; S, sensitive to this fungicide;

[3] Mean values with the same letters within the same column were not significantly different with least significant difference test at $P=0.05$.

1.2 Fungicides

Azoxystrobin (93%), kresoxim-methyl (95.3%), IBP (68%) in technical grade were provided by Syngenta, BASF and Shanghai Jiangnan Chemical Company, respectively, and dissolved in methanol to 10mg/mL for stock solution. MBC (98%, supplied by Shenyang Chemical Research Institute) was dissolved in 0.1 mol/mL HCl to 10mg/mL for stock solution. Salicylhydroxamic acid (SHAM), which strongly inhibits the alternative respiration of fungi, in technical grade (99%) was purchased from Sigma-Aldrich and dissolved in methanol to 50mg/mL for stock solution. These stock solutions were stored at 4℃ in the dark and were added to molten media, when they were cooled to approximately 50℃.

1.3 Determination of sensitivity of growth to azoxystrobin and kresoxim-methyl

To evaluate the sensitivity of mycelial growth to azoxystrobin and kresoxim-methyl, the 80 *M. oryzae* isolates were used to determine the EC_{50} values (the concentration of the fungicide causing a 50% reduction in the growth rate compared to an unamended control). Autoclaved yeast glycerol agar medium (5 g yeast extract, 20mL glycerol, 0.25 g $MgSO_4$, 6 g $NaNO_3$, 0.5 g KCl, 1.5 g KH_2PO_4, 20 g agar and 1 L distilled H_2O) was amended with azoxystrobin or kresoxim-methyl to obtain final concentrations of 0, 0.0125, 0.025, 0.05, 0.1, 0.2, 0.4, and 0.8μg/mL. In addition, sensitivity tests for azoxystrobin and kresoxim-methyl were usually conducted in the presence of SHAM, a specific inhibitor of alternative oxidase and the final concentration of SHAM was 100μg/mL for each fungicide treatment as described by studies[20]. An inverted mycelial plug (5 mm in diameter), cut from the edge of a 3 d-old colony was transferred to 9 cm Petri dishes containing the amended media. Three replicates per concentration were used and all the tests were repeated twice. The mycelial growth was measured after 7 ~ 10 days of incubation at 25℃ in a growth chamber (12 h photoperiod). Two perpendicular diameters of each fungal colony were measured and averaged (the diameter of the plug was subtracted). The EC_{50} value, which was the fungicide concentration that resulted in 50% mycelia grwoth inhibition, was determined by probit analysis[21-22].

1.4 Determination of sensitivity of conidia germination to azoxystrobin and kresoxim-methyl

Azoxystrobin or kresoxim-methyl was added to WA after sterilization with final concentrations of 0, 0.01, 0.02, 0.04, 0.1, 0.2 and 0.8μg/mL. Sensitivity tests were also conducted in the presence of SHAM at a concentration of 100μg/mL. Conidia were obtained from colonies grown on tomato potato sucrose agar (TPSA, 150 g tomato, 200 g potato, 20 g sucrose, 35 g agar and 1 liter of water) as described by a previous study. Conidia suspensions containing 10^5 conidia per millilitre were poured onto the surface of WA plates. Conidia were allowed to germinate at 25℃ for 8 h. Germination was quantified at 3 sites by counting 100 conidia per site. A conidium was scored germinated if the germ tube had reached at least half the length of the conidium. Three plates for each concentration were used, and the experiment was performed twice. For each isolate, a linear regression of the percent inhibition related to the control of spore germination versus the $\log_{10}$ transformation for each of the seven concentrations was obtained. EC_{50}, which was the fungicide concentration that resulted in 50% spore germination inhibition, was calculated with the regression equation for each isolate as described previously[23].

1.5 Cross-resistance of azoxystrobin or kresoxim-methyl with IBP and MBC

Sensitivity of the isolates used in this study to IBP or MBC was tested through a discriminatory dose test according to a previous study[15]. Twelve *M. oryzae* isolates with differential sensitivity to IBP or MBC were arbitrarily chosen for the cross-resistance analysis. The 12 isolates which were sensitive or resistant to IBP or MBC, were compared regarding their sensitivity to azoxystrobin and kresoxim-methyl (EC_{50}) to determine cross-resistance of azoxystrobin and kresoxim-methyl with IBP or MBC as described previously[20]. This experiment was performed twice.

1.6 Evaluation of protective and curative activity of azoxystrobin and kresoxim-methyl against rice blast

The *M. oryzae* isolate Cun05, which was resistant to MBC from the field of Anqing, Anhui Province, with stable pathogenicity on rice (unpublished data) was used for inoculation on rice plants in the protective and curative activity tests. Conidia suspensions were prepared as described above as inoculums and adjusted to 10^5 conidia per ml by centrifugation. The experiment was performed in a rice experimental field in Anhui Academy of Agricultural Sciences. The rice field was divided into 24 plots and each plot was $0.5 \times 0.5\ m^2$ including at least 40 rice plants. Each of the plots was separated by a 30cm interval with untreated rice plants.

To evaluate the protective and curative activity of azoxystrobin and kresoxim-methyl, rice plants (yuanfengzao, a local susceptible rice cultivar) were inoculated by spraying conidia suspension to rice plants at four-leaf stage and at least 40 plants per treatment were approximately equally inoculated. The six treatments were: (1 ~ 2) azoxystrobin SC (25% Amistar) commercial formulation applied at 150 and 250μg/mL; (3 ~ 4) kresoxim-methyl WG (50% Stroby) commercial formulation applied at 150 and 250μg/mL; (5) 50% MBC WG commercial formulation applied at 750μg/mL; (6) no treatment control (CK). Each treatment was conducted with four replicates and a total of 24 plots were treated and the experiment was performed twice. No other fungicides were applied to the experimental plots. All other treatments, such as fertilizers, were used in accordance with standard farm products. The rice plants were at same growth stage and sprayed with fungicides 24, 48 and 72h before (protective treatments) and after (curative treatments) inoculation. Visual disease assessment was made 10 days after the inoculation. The severity of blast symptoms was evaluated and recorded using a 0 ~ 9 scale standard based on the type and size of lesions described by the International Rice Research Institute[24]. Disease severity was calculated as follows:

Disease severity (%) = [Σ (The number of diseased plants in this index × Disease index) / (Total number of plants investigated × The highest disease index)] ×100 [25]

The protective and curative efficacies (%) of azoxystrobin, kresoxim-methyl and MBC were calculated using the following formula: (SC-ST) /SC × 100, where SC was the average severity of diseased rice leaves from over 40 untreated control plants and ST was that from over 40 fungicide-treated rice plants. The experiment was performed twice.

1.7 Field trial for azoxystrobin and kresoxim-methyl in controlling rice blast and their effect on rice yield

Rice cultivar yuanfengzao, a susceptible cultivar to rice blast, was sown in fields located in Anqing city and Lu'an city, China, during rice growth seasons of 2010 and 2011. The two fields were known to be naturally infested with *M. oryzae* isolates. Each of the fields was divided into 24 plots and each plot had an area of 40 m^2 (5 m × 8 m). Each of the plots was separated by a 50cm interval with untreated rice plants. The seven treatments were: (1 ~ 2) azoxystrobin SC (25% Amistar) commercial formulation applied at 112.5 and 187.5 g a. i./hm^2; (3 ~ 4) kresoxim-methyl WG (50% Stroby) commercial formulation applied at 112.5 and 187.5 g a. i./hm^2; (5) 50% MBC WG commercial formulation applied at 562.5 g a. i./hm^2; (6) no treatment control (CK). The volume of water delivered was 750 L/hm^2. The cultivar was planted following normal agronomic practice. Fungicides were applied at booting and heading stage. Each treatment was conducted with four replicates and a total of 24 plots were treated. No other fungicides were applied to the experimental plots. All the other treatments, such as herbicides and fertilizers, were used in accordance with standard farm products. Visual disease assessment was made at least 14 days after fungicide application. Activity of fungicides controlling rice blast in the field was evaluated based on at least 100 rice plants collected at random from each plot. The disease severity was evaluated and recorded using a 0 ~ 9 scale standard based on the type and size of lesions described by the International Rice Research Institute for each plot[24]. Disease severity was calculated as follows:

Disease severity (%) = [Σ (The number of diseased plants in this index × Disease index) / (Total number of plants investigated × The highest disease index)] ×100. The control efficacies (%) of azoxystrobin, kresoxim-methyl and MBC were calculated using the following formula: (SC-ST) /SC × 100, where SC was the average disease index of four replicates of untreated plots and ST was that from the four replicates of the fungicide-treated plots. The rice yield was evaluated according to the previous study for each treatment[1].

1.8 Data analysis

Data from repeated experiments were combined for analysis because variances between experiments were homogeneous. All data were processed with the SIGMASTAT Statistical Software Package (SPSS Science, version 11). EC_{50} values were calculated from the sensitivity tests described above from the fitted regression line of the log-transformed percent inhibition plotted against the log-transformed fungicide concentration[26].

2 Results

2.1 Sensitivity of growth to azoxystrobin and kresoxim-methyl

The EC_{50} values for azoxystrobin and kresoxim-methyl in inhibiting mycelial growth of the 80 *M. oryzae* isolates were 0.006 ~ 0.056μg/mL and 0.024 ~ 0.287μg/mL, with the average EC_{50} values of (0.026 ± 0.006) μg/mL and (0.082 ± 0.019) μg/mL, respectively (Fig. 1 A & B). Both of the two baseline sensitivity curves were unimodal (Fig 1 A & B), representing range-of-variation factors of 7.2 and 12.0, respectively.

In the presence of SHAM, azoxystrobin exhibited stronger inhibitory activity against mycelial growth of *M. oryzae* isolates than that of kresoxim-methyl. The sensitivity data can be used as baselines for monitoring any future changes in sensitivity to azoxystrobin and kresoxim-methyl by *M. oryzae* populations in China. Such baseline data are essential for assessing resistance risk, managing field-resistance, and monitoring resistance levels in the pathogen populations.

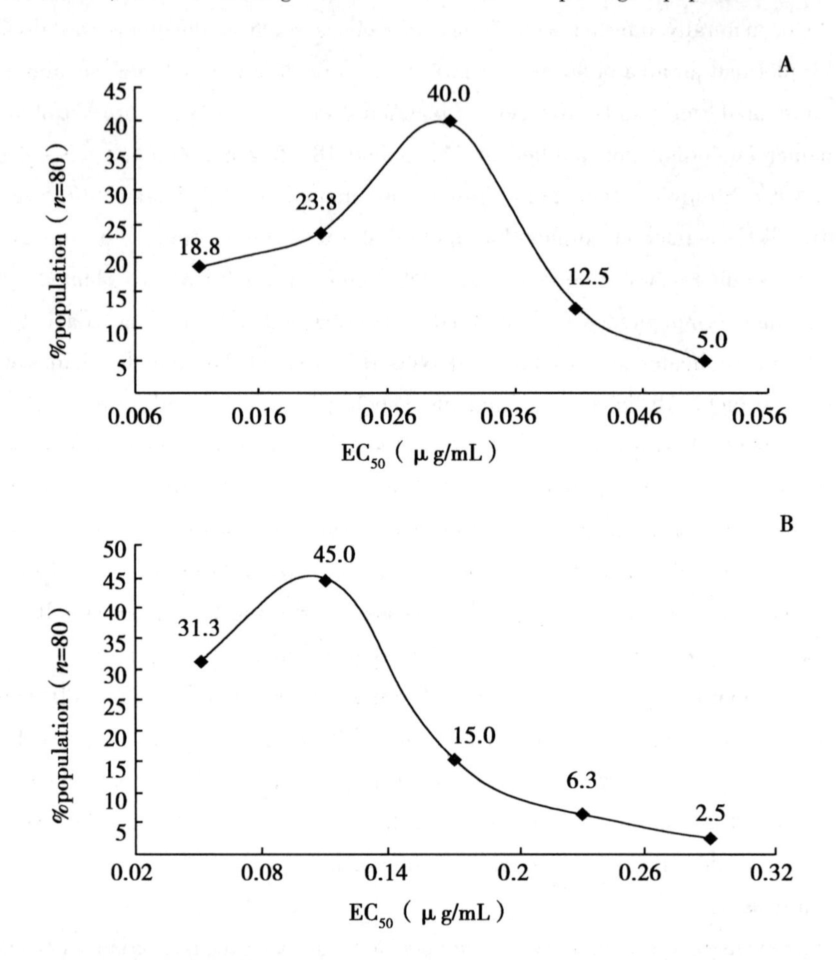

Fig. 1 Frequency distribution of sensitivity of mycelial growth (base-line sensitivity) of *Magnaporthe oryzae* isolates from China to azoxystrobin (A) and kresoxim-methyl (B) (in presence of salicylhydroxamic acid)

2.2 Sensitivity of conidia germination to azoxystrobin and kresoxim-methyl

The EC_{50} values for azoxystrobin and kresoxim-methyl in inhibiting conidia germination of the *M. oryzae* populations were 0.004 ~ 0.051μg/mL and 0.012 ~ 0.105μg/mL, with the average EC_{50} values of (0.012 ± 0.004) μg/mL and (0.042 ± 0.015) μg/mL, respectively (Fig. 2 A & B). Both of the two baseline sensitivity curves were unimodal (Fig. 2 A & B), representing a range-of-variation factors of 17.6 and 8.8, respectively. These data indicated that conidia germination of *M. oryzae* isolates was very sensitive to azoxystrobin and kresoxim-methyl. This baseline data would also be essential for monitoring any further insensitivity or resistance to azoxystrobin and kresoxim-methyl in *M. oryzae* populations in China.

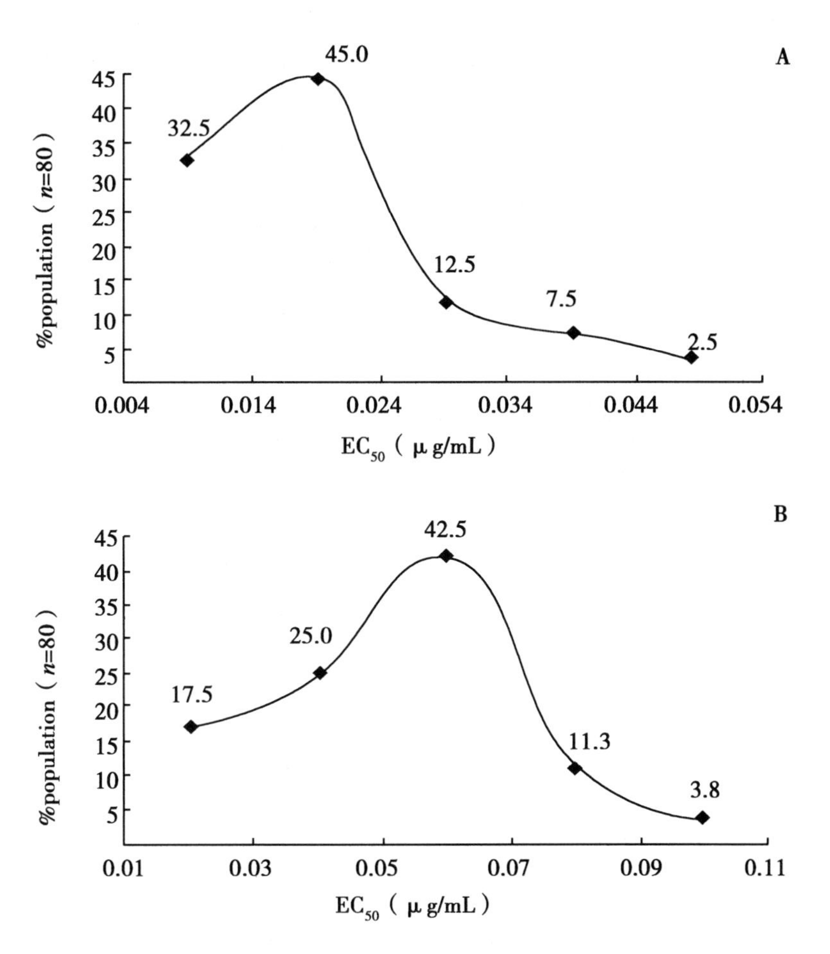

Fig. 2 Frequency distribution of sensitivity of conidia germination (base-line sensitivity) of *Magnaporthe oryzae* isolates from China to azoxystrobin (A) and kresoxim-methyl (B) (in presence of salicylhydroxamic acid)

2. 3 Cross-resistance patterns

There was significant difference of sensitivity to azoxystrobin or kresoxim-methyl between isolates (Table 1). However, there was no correlation between this difference in sensitivity to azoxystrobin or kresoxim-methyl and sensitivity to the benzimidazole fungicide MBC or the organophosphorus fungicide IBP. This suggested that there was no cross-resistance between azoxystrobin or kresoxim-methyl and MBC or IBP. Moreover, azoxystrobin and kresoxim-methyl showed significantly stronger activity of inhibiting germination than inhibiting growth. The EC_{50} values of the 12 tested isolates were 0. 008 ~ 0. 027μg/mL and 0. 093 ~ 0. 263μg/mL for inhibiting growth, and 0. 006 ~ 0. 015μg/mL and 0. 012 ~ 0. 093μg/mL for inhibiting germination, for azoxystrobin or kresoxim-methyl, respectively.

2. 4 Protective and curative activity of azoxystrobin and kresoxim-methyl against rice blast

When the fungicides were applied 1, 3 and 5 d before inoculations, the average control efficacy of azoxystrobin or kresoxim-methyl at 250μg/mL were more than 80% in all cases, indicating excellent protective activity against rice blast. Moreover, kresoxim-methyl applied at 150 or 250μg/

mL provided higher control efficacy than that of azoxystrobin in corresponding treatments (Table 2). MBC applied at 750μg/mL provided significant lower protective control efficacy than those of azoxystrobin or kresoxim-methyl in corresponding treatments (Table 2).

Also, both azoxystrobin and kresoxim-methyl applied 1 and 3 d after inoculations at 250μg/mL exhibited over 80% curative control efficacy against rice blast, but less than 80% curative control efficacy when applied 5 d after inoculations (Table 2). However, in all cases, kresoxim-methyl applied at 150 or 250μg/mL provided higher curative control efficacies in corresponding treatments (Table 2). Worse still, MBC applied at 750μg/mL exhibited less than 70% curative control efficacy in each treatment (Table 2).

Table 2 Protective and curative activity of azoxystrobin and kresoxim-methyl against rice blast

Treatment	Dosage (μg mL⁻¹)	Protective activity[2]						Curative activity[2]					
		1 d		3 d		5 d		1 d		3 d		5 d	
		Average disease severity (%)[3]	Control efficacy (%)[3]	Average disease severity (%)	Control efficacy (%)	Average disease severity (%)	Control efficacy (%)	Average disease severity (%)	Control efficacy (%)	Average disease severity (%)	Control efficacy (%)	Average disease severity (%)	Control efficacy (%)
25% azoxystrobin SC[1]	150 μg/mL	1.52 c	77.1 c	1.53 c	77.0 b	1.93 d	72.1 c	1.61 c	73.7 c	1.92 c	72.1 c	1.90 c	70.1 b
	250 μg/mL	0.91 d	86.5 b	0.96 d	85.5 a	1.36 c	80.3 b	1.18 d	80.8 b	1.51 d	78.0 b	1.88 c	70.4 b
50% kresoxim-methyl WG	150 μg/mL	1.45 c	78.1 c	1.05 d	84.2 a	1.31 c	81.1 b	1.55 c	74.7 c	1.57 d	77.1 b	1.88 c	70.4 b
	250 μg/mL	0.57 e	91.4 a	0.87 d	86.9 a	1.02 e	85.3 a	0.78 e	87.3 a	1.05 e	84.7 a	1.31 d	79.4 a
50% carbendazim WG	750 μg/mL	2.36 b	64.4 d	2.72 b	59.0 c	3.08 b	55.5 d	2.23 b	63.6 d	2.63 b	63.3 d	2.97 b	53.2 c
CK	—	6.63 a	—	6.64 a	—	6.92 a	—	6.13 a	—	6.86 a	—	6.35 a	—

[1] SC and WG represented Suspension Concentrates and Water Dispersible Granules, respectively.

[2] Visual disease assessment was made 10 days after the inoculation. The severity of blast symptoms was evaluated and recorded using a 0 ~ 9 scale standard based on the type and size of lesions described by the International Rice Research Institute (International Rice Research Institute, 1996). Disease severity was calculated as follows:

Disease severity (%) = [Σ (The number of diseased plants in this index × Disease index) / (Total number of plants investigated × The highest disease index)] ×100. And the protective (1, 3 and 5 d before inoculation) and curative (1, 3 and 5 d after inoculation) control efficacies were calculated as follows:

Control efficacy (%) = [(Average disease severity of control-Average disease severity of treated group) /Average disease severity of control] ×100.

[3] Mean vales with the same letters within the same column were not significantly different ($P > 0.05$, Fisher's LSD).

Taken together, azoxystrobin and kresoxim-methyl exhibited excellent protective and curative control efficacies against rice blast, and moreover, kresoxim-methyl was more effective than azoxystrobin in this *in vivo* test.

2.5 Effect of azoxystrobin and kresoxim-methyl on rice blast and rice yield

The incidence of rice blast at both sites was correlated with moisture period during the booting and heading stage. At both two sites, kresoxim-methyl at 187.5 g a. i. /hm^2 provided over 80% control efficacy in 2010 and 2011, exhibiting excellent activity against rice blast (Table 3). Also, control efficacy of azoxystrobin at 187.5 g a. i. /hm^2 at both sites were all over 70%, which were significantly higher than that of MBC applied at 562.5 g a. i. /hm^2 (Table 3). Interestingly, kresoxim-methyl, which exhibited less inhibitory activity than azoxystrobin *in vitro*, provided higher control efficacy in the field trials in corresponding treatments (Table 3).

Rice blast caused significant yield reductions to rice plants in unsprayed plots in both two sites. When applied with azoxystrobin or kresoxim-methyl, the rice yield was significantly improved, and moreover, the rice grain yield following the best fungicide treatment (kresoxim-methyl at 187.5 g a. i. /hm^2) was 935 kg/hm^2 more than the untreated control in Lu'an city in 2010 (Table 3).

Table 3 Effect of azoxystrobin and kresoxim-methyl on rice blast and rice blast and rice yield

Site	Treatment	Dosage (g. a. i/ha)	Disease index[2]		Control efficacy (%)[2]		Yield (kg/ha)	
			2010	2011	2010	2011	2010	2011
Anqing City	25% azoxystrobin SC[1]	112.5	2.97	2.40	66.9 d[3]	71.3 c	8 322 c	8 275 b
		187.5	1.92	1.86	78.6 b	77.8 b	8 468 b	8 493 a
	50% kresoxim-methyl WG	112.5	2.38	1.64	73.4 c	80.4 b	8 399 c	8 275 b
		187.5	1.49	1.22	83.4 a	85.4 a	8 576 a	8 446 a
	50% carbendazim WG	562.5	3.47	3.32	61.2 e	60.3 d	8 101 d	8 026 c
	CK		8.96	8.36			7 660 e	7 812 d
Lu'an City	25% azoxystrobin SC[1]	112.5	3.18	2.81	64.3 c	70.6 c	8 266 c	8 301 b
		187.5	2.34	2.06	73.5 b	78.5 ab	8 441 b	8 468 a
	50% kresoxim-methyl WG	112.5	2.88	2.15	67.7 c	77.5 b	8 387 bc	8 373 b
		187.5	1.59	1.80	82.5 a	81.2 a	8 668 a	8 525 a
	50% carbendazim WG	562.5	3.65	3.72	59.1 d	61.1 d	8 144 d	8 075 c
	CK		8.93	9.56			7 733 e	7 698 d

[1] SC and WG represented Suspension Concentrates and Water Dispersible Granules, respectively.

[2] Visual disease assessment was made 10 days after the inoculation. The severity of blast symptoms was evaluated and recorded using a 0 ~ 9 scale standard based on the type and size of lesions described by the International Rice Research Institute (International Rice Research Institute, 1996). Disease severity was calculated as follows:

Disease severity (%) = [Σ (The number of diseased plants in this index × Disease index) / (Total number of plants investigated × The highest disease index)] ×100. And the protective (1, 3 and 5 d before inoculation) and curative (1, 3 and 5 d after inoculation) control efficacies were calculated as follows:

Control efficacy (%) = [(Average disease severity of control-Average disease severity of treated group) /Average disease severity of control] ×100%.

[3] Mean vales with the same letters within the same column were not significantly different (P>0.05, Fisher's LSD).

3 Discussion

Application of fungicides for managing rice blast has been relied upon over the last few decades, because few cultivars with effective genetic resistance are available. The benzimidazole fungicides, particularly MBC, have been routinely used during each period of booting and heading in areas with warm and moist weather to control rice blast in China for more than 30 years[27]. A distinct advantage of MBC was its systemic activity which not only protected plants from infection but also provided disease control when applied after the early stages of infection[26]. However, MBC differed from other protective fungicides in its site-specific mode of action which made resistance problems appear within a few years and resulted in control failures. Among the blasticides described above, IBP almost had not been used in China but FJ-one which is cross-resistant with organophosphorus fungicides has been used in some regions[15]. Because of upcoming resistance problem, China now is facing a challenge to find alternative fungicides for managing rice blast. Therefore, introduction of an alternative fungicide with different mode of action for consistent control of rice blast is urgent. In theory, when fungicides are introduced with specific targets to control, each of the targets requires a baseline sensitivity to be established so that product use strategies can be monitored and possible resistance can be detected[19]. The mode of actions of QoI fungicides, such as azoxystrobin and kresoxim-methyl, which were not widely used for the control of rice blast in China, have been proved to be different from that of MBC and IBP[15]. This study also suggested that there was no cross-resistance between QoIs and MBC or IBP. This could be explained by the difference in mode of action in these fungicides (Table 1). This study shows that azoxystrobin and kresoxim-methyl have strong activity against *M. oryzae* by inhibiting both mycelial growth and conidia germination (Table 1).

Base-line sensitivities of *M. oryzae* isolates to azoxystrobin and kresoxim-methyl were established with two different methods. The results showed that the selected isolates were very sensitive to these two fungicides and the low EC-values and the relatively narrow range of sensitivity variation and unimodal frequency distribution of EC-values indicated the absence of resistant subpopulations among the isolates used in this study. Therefore, the sensitivity data can be used as baseline sensitivity for monitoring any future changes in sensitivity to azoxystrobin and kresoxim-methyl in *M. oryzae* populations in China. Such baseline data are essential for assessing resistance risk, managing field-resistance, and monitoring resistance levels in the pathogen populations.

Rice blast could cause significant grain yield reduction when no fungicides were sprayed, and result in at least a 40% grain yield loss where this disease severely occurred in susceptible cultivars in previous studies[1]. Fungicide applications have been commonly used in the United States for the control of rice blast for rice production[5]. Previous study also showed that good control of rice blast was possible with a single application of azoxystrobin made at heading stage, but two applications at both booting and heading stage were more effective. In present study, kresoxim-methyl applied at 150 and 250μg/mL exhibited higher protective and curative control efficacy than that of azoxystrobin at 150 and 250μg/mL accordingly (Table 2), while azoxystrobin exhibited stronger inhibitory activity against *M. oryzae* isolates than that of kresoxim-methyl in the *in vitro* tests. Nevertheless, both azoxystrobin and kresoxim-methyl provided excellent protective and curative activity against rice blast. The results of field trials also showed that both, azoxystrobin and kresoxim-methyl, at 187.5 g a. i. /hm^2 provided over 73% control efficacy in 2010 and 2011 in both two sites, respectively, exhibiting excellent activity against rice blast. Moreover, kresoxim-methyl provided higher control efficacy than azoxystrobin in the field trials in corresponding treatments as well.

In our knowledge, the present study is the first report on the baseline sensitivity of Chinese *M. oryzae* populations to azoxystrobin and kresoxim-methyl and the excellent protective and curative efficacy of these two fungicides in controlling rice blast. Taken together, azoxystrobin and kresoxim-methyl should be good alternatives to the commonly used blasticides for the control of rice blast. But this does not mean that azoxystrobin and kresoxim-methyl could be extensively used for the disease control as these two QoIs are also at risk fungicides, and resistance to QoIs has already reported in many previous studies[17,28-34]. Therefore, these QoI fungicides should be carefully used in spray programmes in alternation or tank mixing with other effective fungicides with different modes of action.

参考文献

[1] Groth D E. Azoxystrobin rate and timing effects on rice head blast incidence and rice grain and milling yields [J]. Plant Disease, 2006, 90: 1 055 - 1 058.

[2] Seebold K W, Datnoff L E, Correa-Victoria F J, et al. Effect of silicon rate and host resistance on blast, scald, and yield of upland rice [J]. Plant Disease, 2000, 84: 871 - 876.

[3] Noguchi M T, Yasuda N, Fujita Y. Evidence of genetic exchange by parasexual recombination and genetic analysis of pathogenicity and mating type of parasexual recombinants in rice blast fungus, *Magnaporthe oryzae* [J]. Phytopathology, 2006, 96: 746 - 750.

[4] Zeng J, Feng S, Cai J, et al. Distribution of mating type and sexual status in Chinese rice blast populations [J]. Plant Disease, 2009, 93: 238 - 242.

[5] Greer C A, Webster R K. Occurrence, distribution, epidemiology, cultivar reaction, and management of rice blast disease in California [J]. Plant Disease, 2001, 85: 1 096 - 1 102.

[6] Long D H, Lee F N, TeBeest D O. Effect of nitrogen fertilization on disease progress of rice blast on susceptible and resistant cultivars [J]. Plant Disease, 2000, 84: 403 - 409.

[7] Jia Y, Valent B, Lee F N. Determination of host responses to *Magnaporthe grisea* on detached rice leaves using a spot inoculation method [J]. Plant Disease, 2003, 87: 129 - 133.

[8] Bonman J M, Estrada B A, Bandong J M. Leaf and neck blast resistance in tropical lowland rice cultivars [J]. Plant Disease, 1989, 73: 388 - 390.

[9] Zhang C Q, Zhu G N, Ma Z H, et al. Isolation, characterization and preliminary genetic analysis of laboratory tricyclazole-resistant mutants of the rice blast fungus, *Magnaporthe grisea* [J]. Phytopatholgy, 2006, 154: 392 - 397.

[10] Zhang C, Huang X, Wang J, et al. Resistance development in rice blast disease caused by *Magnaporthe grisea* to tricyclazole [J]. Pestic Biochem Physiol, 2009, 94: 43 - 47.

[11] Katagiri M, Uesugi Y. Similarities between the fungicidal action of isoprothiolane and organophosphosphorus thiolate fungicides [J]. Phytopathology, 1977, 67: 1 415 - 1 417.

[12] Chrysayi M T, Sisler H D. Effect of tricyclazole on growth and secondary metabolism in *Pyricularia oryzae* [J]. Pestic Biochem Physiol, 1978, 8: 26 - 32.

[13] Woloshuk C P, Sisler H D, Tokousbalides M C, et al. Melanin biosynthesis in *Pyricularia oryzae*: site

of tricyclazole inhibition and pathogenicity of melanin-deficient mutants [J]. Pestic Biochem Physiol, 1980, 14: 256 - 264.

[14] Peng Y L, Liu J F, Ye Z Y. Studies on the resistance to kitazin p in *Pyricularia oryzae* [J]. J Nanjing Agric Univ, 1990, 13: 45 - 48.

[15] Zhang C Q, Zhou M G, Shao Z R, et al. Detection of sensitivity and Resistance variation of *Magnaporthe grisea* to kitazin P, carbendazim and tricyclazole Rice Sci, 2004, 18: 317 - 323.

[16] FRAC (Fungicide Resistance Action Committee), FRAC Code List: Fungicides Sorted by Modes of Action. 2007. Available from: www. frac. info.

[17] Kim Y S, Dixon E W, Vincelli P, et al. Field resistance to strobilurin (QoI) fungicides in *Pyricularia grisea* caused by mutations in the mitochondrial *cytochrome* b gene [J]. Phytopathology, 2003, 93: 891 - 900.

[18] Bartlett D W, Clough J M, Godwin J R, et al. The strobilurin fungicides [J]. Pest Manage Sci, 2000, 58: 649 - 662.

[19] Russell P E. Sensitivity baselines in fungicide resistance research and management. In: FRAC Monograph, 3. CropLife International, Brussels. 2004. Available from: www. frac. Info.

[20] Zhang C Q, Zhang Y, Zhu G N. The mixture of kresoxim-methyl and boscalid, an excellent alternative controlling grey mould caused by *Botrytis cinerea* [J]. Ann Appl Biol, 2008, 153: 205 - 213.

[21] Chen Y, Chen C J, Zhou M G, et al. Monogenic resistance to a new fungicide JS399-19 in *Gibberella zeae* [J]. Plant Pathol, 2009, 58: 565 - 570.

[22] Li H, Diao Y, Wang J X, et al. JS399-19, a new fungicide against wheat scab [J]. Crop Prot, 2008, 27: 90 - 95.

[23] Zhang C, Liu Y, Ding L, et al. Shift of sensitivity of *Botrytis cinerea* to azoxystrobin in greenhouse vegetables before and after exposure to the fungicide [J]. Phytoparasitica, 2011, 39: 293 - 302.

[24] International Rice Research Institute (1996) Standard Evaluation System for Rice. 4th edn. Manila: IRRI.

[25] Wang J X, Ma H X, Chen Y, et al. Characterization of sensitivity of Rhizoctonia solani, causing rice sheath blight, to mepronil and boscalid [J]. Crop Prot, 2009, 28: 882 - 886.

[26] Chen Y, Li H, Chen C, et al. Sensitivity of *Fusarium graminearum* to fungicide JS399-19: *in vitro* determination of baseline sensitivity and the risk of developing fungicide resistance [J]. Phytoparasitica, 2008, 36: 326 - 337.

[27] Zhou M G, Ye Z Y, Liu J F. Progress of Fungicide Resistance [J]. J Nanjing Agric Univ, 1994, 17: 33 - 41.

[28] Gisi U, Chin K M, Knapova G, et al. Recent developments in elucidating modes of resistance to phenylamide, DMI and strobilurin fungicides [J]. Crop Prot, 2000, 19: 863 - 872.

[29] Ishii H, Fraaije B A, Noguchi K, et al. Occurrence and molecular characterization of strobilurin resistance in cucumber powdery mildew and downy mildew [J]. Phytopathology, 2001, 91: 1 166 - 1 171.

[30] Jiang J, Ding L, Michailides T J, et al. Molecular characterization of field azoxystrobin-resistant isolates of *Botrytis cinerea* [J]. Pestic Biochem Phys, 2009, 93: 72 - 76.

[31] Ma Z, Michailides T. Advances in understanding molecular mechanisms of fungicide resistance and molecular detection of resistant genotypes in phytopathogenic fungi [J]. Crop Prot, 2005, 24: 853 - 863.

[32] Sierotzki H, Wullschleger J, Gisi U. Point mutation in cytochrome b gene conferring resistance to strobilurin fungicides in *Erysiphe graminis* f. sp. *tritici* field isolates [J]. Pestic Biochem Physiol, 2000, 68:

107 – 112.

[33] Vincelli P, Dixon E. Resistance to QoI (strobilurin-like) fungicides in isolates of *Pyricularia grisea* from perennial ryegrass [J]. Plant Disease, 2001, 86: 235 – 240.

[34] Ziogas B N, Baldwin B C, Young J E. Alternative respiration: A biochemical mechanism of resistance to azoxystrobin (ICIA5504) in *Septoria tritici* [J]. Pestic Sci, 1997, 50: 28 – 34.

寡雄腐霉对水稻发芽率的影响与其对水稻恶苗病菌的抑制及杀灭作用

何　玲*，袁会珠，杨代斌，闫晓静**
（中国农业科学院植物保护研究所，农业部作物有害生物综合治理综合性重点实验室，北京　100193）

摘要：采用浸种法，以咪鲜胺为对照药剂，考察寡雄腐霉对不同水稻品种种子发芽率与其对恶苗病菌的抑制及杀灭效果进行了研究，并结合室内盆栽试验，分析结果并筛选出最佳的浸种药剂。结果表明，寡雄腐霉制剂2 000倍稀释液不仅能明显促进室内水稻种子发芽和幼苗生长，且能很好的抑制及杀灭恶苗病菌，是最佳的防治水稻恶苗病的浸种处理药剂。

关键词：寡雄腐霉；水稻恶苗病；浸种

The Effect of *Pythium oligandrum* on Rice Germination Rate and the Inhibition and Killing Effect Against *Fusarium moniliforme*

He Ling，Yuan Huizhu，Yang Daibin，Yan Xiaojing
（*Key Laboratory of Integrated Pest Management in Crops*，*Ministry of Agriculture*，*Institute of Plant Protection*，*Chinese Academy of Agricultural Sciences*，*Beijing* 100193，*China*）

Abstract：Adopting the soaking method to explore the effect of germination rate on four varieties of rice seeds with *Fusarium moniliforme* and the inhibition and killing effect on *Fusarium moniliforme* of Polyversum of *Pythium oligandrum* preparation，and in combination with pot experiment indoor，analyzing the results and selecting the best soaking agents，comparing with that of prochloraz. The results showed that *Pythium oligandrum* preparation 2 000 – fold dilution not only promote the germination and the growth of rice seeds indoor，and can suppress and kill *Fusarium moniliforme* splendidly. To sum up，Polyversum of *Pythium oligandrum* preparation 2 000 – fold dilution is the best soaking treatment agents to control *Fusarium moniliforme sheld.*

Key words：*Pythium oligandrum*；*Fusarium moniliforme sheld*；Seed soaking

水稻恶苗病俗称徒长病、白秆病、抢先稻等，是一种系统性侵染的真菌病害。病原菌为*Fusarium moniliforme*，为半知菌亚门真菌，为害水稻全株。受恶苗病菌侵染的水稻谷粒播种后常常不能正常发芽或者无法出土。水稻苗期发病常表现为“病苗细高且细长，夜色淡黄，根系发育不良，有些在移栽前后死去”[1]。水稻恶苗病对亚洲的水稻产量造成了巨大的影响，在亚洲地区最高损失据报道已达70%[2]。水稻恶苗病在我国各稻区发生较多且严重，水稻浸种是防治水稻恶苗病的一种有效方法，但近年来随着化学药剂的长时间使用，对水稻恶苗病的防治效果下降，其发病率呈上升趋势，且由于化学药剂的残效期长、残留量高，影

* 第一作者：何玲，女，硕士研究生；E-mail：1193676003@qq. com
** 通讯作者：闫晓静，女，副研究员，主要从事杀菌剂作用机理与使用技术研究；E-mail：yanxiaojing@caas. cn

响水稻安全性，对大米农艺性状造成了巨大的影响。所以，寻找一种安全高效的防治水稻恶苗病的生防制剂是解决此难题的重要途径。

寡雄腐霉 *Pythium oligandrum* 属于茸边生物界 Chromista 卵菌门 Oomycota[3]，是自然界中存在的一种寄生真菌，可以在水稻、番茄、烟草等作物根围定殖，通过寄生作用达到杀灭病菌的效果[4]。已有研究证明，寡雄腐霉产生的寡雄蛋白不仅对病原真菌的繁殖与生长起到抑制作用，还能诱导植物产生抗病性，促进植物的生长[5]。目前，多利维生（Polyversum）是目前唯一以寡雄腐霉登记注册的微生物农药产品，鉴于其对水稻恶苗病菌抑制作用显著[6]，且能有助于水稻生长[7]，本研究以咪鲜胺为对照药剂，详细研究了用寡雄腐霉制剂浸种后，带菌水稻种子的发芽率、带菌率以及幼苗长势的变化，以期选择出防治恶苗病的最佳浸种药剂。

1　材料与方法

1.1　供试材料

供试带恶苗病菌水稻品种为中龙香粳 1 号、松粳香 1 号、北粳 1 号和阳光 600。25% 咪鲜胺乳油购自东莞市瑞德丰生物科技有限公司，寡雄腐霉可湿性粉剂由北京比奥瑞生物科技有限公司提供。

超净工作台（北京东联哈尔仪器制造有限公司）；MLS－3781L 高压灭菌锅（北京百立隆生物技术有限公司）；MJX 智能型霉菌培养箱（宁波江南仪器厂）。

1.2　试验方法

1.2.1　选种

用比重 1.01 的盐水挑选子粒成熟且饱满、大小一致、谷壳完整无破损的水稻种子用于本试验。

1.2.2　配制药液

用蒸馏水分别将寡雄腐霉可湿性粉剂和 25% 咪鲜胺乳油稀释成 500 倍液、1 000倍液、1 500倍液、2 000倍液和 2 500倍液待用。

1.2.3　药液浸种

将 1 500粒中龙香粳一号水稻种子放入 500mL 烧杯中，加入适量配制好的不同浓度的寡雄腐霉可湿性粉剂，设置 3 次重复。按此方法用咪鲜胺 500 倍液、1 000倍液、1 500倍液、2 000倍液和 2 500倍液分别处理中龙香粳 1 号。试验设置清水对照，并放在 15℃的霉菌培养箱中进行培养。按照上述步骤处理其余 3 个品种的种子。

1.2.4　发芽试验

用药剂浸种第 4、6、8、10、12、14 天后，分别从上述烧杯中各取出 180 粒种子放入培养皿中进行培养，培养皿中倒入适量提前融化的琼脂，每个处理 3 个重复，每个培养皿中放入 60 粒水稻种子。培养 4d 后，统计发芽率。培养条件为：设定光照度为 8 500lx；温度设定为 30℃；光循环周期为 16 h 光照处理和 8 h 黑暗处理[8]。

1.2.5　带菌率测定试验

用药剂浸种第 4、6、8、10、12、14 天分别从上述烧杯中各取出 60 粒种子，用体积分数为 0.01 的次氯酸钠浸泡 2min，灭菌水冲洗 3 次后，均匀摆在含硫酸链霉素（200Lg/mL）的 PSA 培养皿中，每个处理 3 次重复，每个重复放入 20 粒种子，并置于 25℃霉菌培养箱中培养 5d，记录种子带菌率[9]。

1.2.6　盆栽试验

待水稻种子中龙香粳1号、松粳香1号、北粳1号和阳光600浸种10d后，从各烧杯中取出100粒种子种在10个管中，在温室中培养。观察发病情况并测量水稻植株的根长、根重、地上部长度和地上部重量。

2　结果与分析

2.1　寡雄腐霉可湿性粉剂浸种后对水稻发芽率的影响

Table 1　The effect of Polyversum and prochloraz against rice germination rate of Zhonglong Xiangjing No. 1

Treatment	发芽率（%）					
	4d	6d	8d	10d	12d	14d
Polyversum 500	97.8 ±0.01a	98.3 ±0.00a	98.9 ±0.01a	99.4 ±0.01ab	99.4 ±0.01a	99.4 ±0.01a
Polyversum 1 000	97.2 ±0.02a	98.3 ±0.02a	98.9 ±0.02a	99.4 ±0.01ab	99.4 ±0.01a	98.9 ±0.01ab
Polyversum 1 500	96.7 ±0.06a	97.8 ±0.02ab	98.3 ±0.02ab	98.9 ±0.01ab	99.4 ±0.01a	99.4 ±0.01a
Polyversum 2 000	98.3 ±0.02a	97.8 ±0.01ab	99.4 ±0.01a	100.0 ±0.00a	100.0 ±0.00a	99.4 ±0.01a
Polyversum 2 500	96.1 ±0.03a	98.9 ±0.02a	98.9 ±0.02a	98.9 ±0.02ab	98.9 ±0.02a	98.9 ±0.01ab
Prochloraz 500	88.9 ±0.03b	90.6 ±0.05cd	92.2 ±0.03b	93.3 ±0.04b	94.4 ±0.06a	95.6 ±0.03b
Prochloraz 1 000	91.7 ±0.04ab	91.7 ±0.06bcd	93.3 ±0.04ab	93.9 ±0.06ab	96.7 ±0.02a	95.6 ±0.03b
Prochloraz 1 500	94.4 ±0.04ab	93.9 ±0.03abc	95.6 ±0.05ab	96.1 ±0.01ab	97.8 ±0.03a	97.8 ±0.01ab
Prochloraz 2 000	95.0 ±0.03ab	95.6 ±0.03abc	95.0 ±0.04ab	96.7 ±0.04ab	98.3 ±0.02a	98.3 ±0.02ab
Prochloraz 2 500	95.0 ±0.03ab	95.6 ±0.04abc	95.0 ±0.03ab	96.7 ±0.03ab	98.3 ±0.02a	98.3 ±0.00ab
CK	88.9 ±0.03b	87.8 ±0.04d	86.1 ±0.05c	84.4 ±0.05c	83.9 ±0.08b	83.3 ±0.04c

Note：Data in a column followed by the same letters are not significantly different at $P_{0.05}$ by Duncan's multiple range test

试验结果见表1。使用寡雄腐霉可湿性粉剂与咪鲜胺乳油稀释液对中龙香粳1号水稻浸种后：①浸种4d，寡雄腐霉处理发芽率显著高于清水处理，咪鲜胺处理与清水处理不存在显著性差异。除咪鲜胺500倍稀释液处理外，两种药剂处理之间差异不显著；②浸种6d，寡雄腐霉处理发芽率显著高于清水处理，咪鲜胺500倍和1 000倍稀释液处理与清水处理不存在显著性差异，其余3个处理发芽率显著高于清水处理。除咪鲜胺500倍稀释液处理外，两药剂处理间差异不显著；③浸种8、10、12和14d后，两药剂处理之间差异不显著，但两药剂处理发芽率均显著高于清水处理。

Table 2　The effect of Polyversum and prochloraz against rice germination rate of Songjing Xiang No. 1

Treatment	4d（%）	6d（%）	8d（%）	10d（%）	12d（%）	14d（%）
Polyversum 500	96.7 ±0.02a	97.2 ±0.01a	97.8 ±0.03a	98.3 ±0.02a	98.9 ±0.04a	98.9 ±0.01a
Polyversum 1 000	96.1 ±0.03a	96.7 ±0.02a	97.2 ±0.02a	98.9 ±0.02a	98.9 ±0.01a	98.3 ±0.02a
Polyversum 1 500	95.0 ±0.03a	96.7 ±0.02a	97.2 ±0.03a	98.9 ±0.01a	98.3 ±0.00a	98.9 ±0.01a

(续表)

Treatment	4d (%)	6d (%)	8d (%)	10d (%)	12d (%)	14d (%)
Polyversum 2 000	97.2 ±0.03a	97.8 ±0.02a	98.3 ±0.02a	99.4 ±0.01a	99.4 ±0.01a	99.4 ±0.01a
Polyversum 2 500	95.6 ±0.04a	96.1 ±0.03ab	97.8 ±0.03a	98.3 ±0.03a	98.9 ±0.01a	99.4 ±0.01a
Prochloraz 500	87.8 ±0.06a	88.9 ±0.03cd	90.6 ±0.01bc	92.2 ±0.08a	94.4 ±0.04a	95.6 ±0.04a
Prochloraz 1 000	90.0 ±.07a	89.4 ±0.05bcd	91.1 ±0.02bc	92.8 ±0.03a	95.0 ±0.04a	96.1 ±0.01a
Prochloraz 1 500	93.3 ±0.06a	92.8 ±0.07abcd	92.8 ±0.02ab	95.6 ±0.04a	97.2 ±0.03a	97.8 ±0.03a
Prochloraz 2 000	94.4 ±0.01a	94.4 ±0.03abc	93.3 ±0.02ab	95.6 ±0.03a	97.2 ±0.05a	98.3 ±0.02a
Prochloraz 2 500	94.4 ±0.05a	95.0 ±0.03abc	93.9 ±0.01ab	96.1 ±0.02a	96.7 ±0.02a	98.9 ±0.01a
CK	87.8 ±0.09a	86.1 ±0.05d	85.6 ±0.08c	83.9 ±0.06b	82.8 ±0.06b	81.7 ±0.03b

Note: Data in a column followed by the same letters are not significantly different at $P_{0.05}$ by Duncan's multiple range test

结果见表2。使用寡雄腐霉可湿性粉剂与咪鲜胺乳油稀释液处理松粳香1号后：①浸种4d后，两药剂处理间差异不显著，但两药剂处理发芽率显著高于清水处理；②浸种6d后，除咪鲜胺500倍液、1 000倍液和1 500倍液处理外，其余处理发芽率显著高于清水处理；③浸种8d后，除咪鲜胺500倍液和1 000倍液处理外，两药剂处理发芽率显著高于清水处理；④浸种10、12和14d后，两药剂处理之间差异不显著，但两种药剂处理发芽率均显著高于清水处理。

Table 3　The effect of Polyversum and prochloraz against rice germination rate of Yangguang 600

Treatment	4d (%)	6d (%)	8d (%)	10d (%)	12d (%)	14d (%)
Polyversum 500	97.2 ±0.01a	98.3 ±0.02a	98.9 ±0.01a	98.9 ±0.01a	98.3 ±0.02ab	98.9 ±0.01ab
Polyversum 1 000	96.7 ±0.02a	97.8 ±0.01a	98.3 ±0.03a	98.9 ±0.01a	98.9 ±0.02a	99.4 ±.01a
Polyversum 1 500	96.7 ±0.02a	97.8 ±0.01a	98.3 ±0.02a	99.4 ±0.01a	98.9 ±0.01a	98.9 ±0.02ab
Polyversum 2 000	97.8 ±0.01a	98.9 ±0.01a	98.9 ±0.01a	99.4 ±0.01a	99.4 ±0.01a	99.4 ±0.01a
Polyversum 2 500	96.1 ±0.01a	97.2 ±0.02a	98.3 ±0.02a	98.3 ±0.02a	99.4 ±.01a	98.3 ±0.03ab
Prochloraz 500	88.3 ±0.03b	89.4 ±0.01b	90.6 ±0.01b	92.2 ±0.01c	92.8 ±0.01d	93.3 ±0.04b
Prochloraz 1 000	89.4 ±0.03b	90.6 ±0.01b	91.1 ±0.05b	92.8 ±0.02bc	92.8 ±0.03d	93.3 ±0.04b
Prochloraz 1 500	91.7 ±0.02b	92.8 ±0.03b	92.8 ±0.05ab	95.0 ±0.02bc	95.6 ±0.01c	96.1 ±0.01ab
Prochloraz 2 000	92.2 ±0.03b	93.3 ±.03b	93.9 ±0.06ab	95.0 ±0.03bc	96.1 ±0.01bc	96.7 ±0.03ab
Prochloraz 2 500	92.2 ±0.02b	93.3 ±0.03b	93.9 ±0.03ab	95.6 ±0.01b	96.1 ±0.01bc	97.2 ±0.01ab
CK	91.1 ±0.03b	90.0 ±0.03b	88.3 ±0.03b	86.7 ±0.02d	83.9 ±0.02e	83.9 ±0.05c

Note: Data in a column followed by the same letters are not significantly different at $P_{0.05}$ by Duncan's multiple range test

从表3数据可以得出，使用寡雄腐霉可湿性粉剂与咪鲜胺乳油稀释液处理阳光600后：①浸种4、6和8d后，寡雄腐霉处理发芽率显著高于清水处理，但咪鲜胺处理与清水处理不存在显著差异；②处理10d和12d后，寡雄腐霉处理发芽率显著高于咪鲜胺处理，且药剂处理发芽率均显著高于清水处理；③浸种14d后，两种药剂处理间差异不显著，但药剂处理发芽率均显著高于清水处理。

Table 4　The effect of Polyversum and prochloraz against rice germination rate of Beijing No. 1

Treatment	4d（%）	6d（%）	8d（%）	10d（%）	12d（%）	14d（%）
Polyversum 500	98.3 ±0.02a	98.9 ±0.02a	98.9 ±0.01a	98.3 ±0.00abc	97.8 ±0.02a	98.9 ±0.01abc
Polyversum 1 000	97.8 ±0.01a	98.3 ±0.02a	98.3 ±0.00a	98.9 ±0.01ab	98.3 ±0.02a	98.9 ±0.02abc
Polyversum 1 500	97.2 ±0.03a	98.3 ±0.02a	98.3 ±0.02a	98.9 ±0.02ab	98.9 ±0.02a	99.4 ±0.01ab
Polyversum 2 000	98.9 ±0.01a	99.4 ±0.01a	98.9 ±0.02a	99.4 ±0.01a	99.4 ±0.16a	100.0 ±0.00a
Polyversum 2 500	97.8 ±0.02a	98.3 ±0.02a	98.9 ±0.01a	99.4 ±0.01a	98.9 ±0.01a	99.4 ±0.01ab
Prochloraz 500	89.4 ±0.01d	90.0 ±0.02d	91.7 ±0.02bc	92.2 ±0.01f	92.8 ±0.02ab	93.3 ±0.00d
Prochloraz 1 000	90.6 ±0.01cd	91.1 ±0.02cd	91.1 ±0.03bc	92.8 ±0.01ef	92.2 ±0.01ab	93.9 ±0.02d
Prochloraz 1 500	92.2 ±0.01bcd	92.8 ±0.01bcd	92.2 ±0.03bc	94.4 ±0.03def	96.1 ±0.01a	96.7 ±0.02c
Prochloraz 2 000	92.8 ±0.01bc	93.3 ±0.02bc	92.8 ±0.03bc	96.1 ±0.02bcd	96.1 ±0.03a	97.2 ±0.01bc
Prochloraz 2 500	93.9 ±0.03b	94.4 ±0.03b	93.3 ±0.02b	95.6 ±0.03cde	96.7 ±0.02a	97.8 ±0.01abc
CK	92.2 ±0.03bcd	90.6 ±0.02cd	89.4 ±0.02c	88.3 ±0.03g	86.1 ±0.03b	85.0 ±0.02e

Note：Data in a column followed by the same letters are not significantly different at $P_{0.05}$ by Duncan's multiple range test

分析表 4 可以发现，使用寡雄腐霉可湿性粉剂与咪鲜胺乳油稀释液处理北粳 1 号后，寡雄腐霉处理发芽率显著高于清水处理。在药后 4、6 和 8d，寡雄腐霉处理发芽率显著高于咪鲜胺处理。

2.2　寡雄腐霉可湿性粉剂浸种对水稻恶苗病菌带菌率的影响

室内带菌率测定结果表明，整个浸种过程中，随着浸种时间的推移，用寡雄腐霉制剂稀释液和咪鲜胺乳油稀释液处理后的 4 个品种的种子带菌率均随着时间的延长而呈现出下降趋势。对于中龙香粳一号，在浸种 14d，除寡雄腐霉 2 000倍液和 2 500倍液，其余处理的种子带菌率均为零；对松粳香 1 号，浸种 14d 后，除寡雄腐霉 1 500倍液、2 000倍液和 2 500倍液，其余处理的种子带菌率均为零；对于阳光 600 和北粳 1 号，浸种 12d 后，除寡雄腐霉 2 000倍液和 2 500倍液，其余药剂处理的种子带菌率均为零。此外，咪鲜胺乳油稀释液处理带菌率低于寡雄腐霉制剂稀释液处理的带菌率，表明在一定程度上，咪鲜胺对水稻恶苗病菌的抑制的速率要优于寡雄腐霉处理。

Table 5　The effect of Polyversum and prochloraz against percentage of infected seed of Zhonglong Xiangjing No. 1

Treatment	4d（%）	6d（%）	8d（%）	10d（%）	12d（%）	14d（%）
Polyversum 500	55.0 ±0.00abc	46.7 ±0.06bc	35.0 ±0.05b	26.7 ±0.06bcde	16.7 ±0.08bcde	0.0 ±0.00c
Polyversum 1 000	56.7 ±0.03abc	51.7 ±0.06bc	41.7 ±0.03b	30.0 ±0.00bcd	18.3 ±0.03bcd	0.0 ±0.00c
Polyversum 1 500	60.0 ±0.10abc	53.3 ±0.08b	43.3 ±0.06b	31.7 ±0.06bcd	21.7 ±0.06bc	0.0 ±0.00c
Polyversum 2 000	61.7 ±0.06ab	55.0 ±0.00b	45.0 ±0.00b	35.0 ±0.00bc	21.7 ±0.08bc	5.0 ±0.00b
Polyversum 2 500	65.0 ±0.05ab	55.0 ±0.09b	45.0 ±0.05b	36.7 ±0.06b	23.3 ±0.03b	6.7 ±0.03b

（续表）

Treatment	4d (%)	6d (%)	8d (%)	10d (%)	12d (%)	14d (%)
Prochloraz 500	46.7 ±0.13c	40.0 ±0.10c	28.3 ±0.03b	18.3 ±0.06e	8.3 ±0.03e	0.0 ±0.00c
Prochloraz 1 000	51.7 ±0.06bc	43.3 ±0.03bc	33.3 ±0.06b	21.7 ±0.03de	11.7 ±0.06de	0.0 ±0.00c
Prochloraz 1 500	53.3 ±0.03abc	46.7 ±0.06bc	35.0 ±0.10b	23.3 ±0.08de	11.7 ±0.03de	0.0 ±0.00c
Prochloraz 2 000	58.3 ±0.13abc	48.3 ±0.03bc	38.3 ±0.15b	25.0 ±0.00cde	13.3 ±0.08cde	0.0 ±0.00c
Prochloraz 2 500	60.0 ±0.00abc	50.0 ±0.00bc	40.0 ±0.00b	26.7 ±0.06bcde	15.0 ±0.00bcde	0.0 ±0.00c
CK	66.7 ±0.08a	70.0 ±0.09a	73.3 ±0.21a	76.7 ±0.13a	80.0 ±0.00a	81.7 ±0.08a

Note: Data in a column followed by the same letters are not significantly different at $P_{0.05}$ by Duncan's multiple range test

Table 6 The effect of Polyversum and prochloraz against percentage of infected seed of Songjingxiang No. 1

Treatment	4d (%)	6d (%)	8d (%)	10d (%)	12d (%)	14d (%)
Polyversum 500	61.7 ±0.03abc	48.3 ±0.15b	35.0 ±0.00bc	26.7 ±0.03bc	15.0 ±0.05b	0.0 ±0.00b
Polyversum 1 000	65.0 ±0.10ab	51.7 ±0.20b	36.7 ±0.10bc	26.7 ±0.10bc	21.7 ±0.06b	0.0 ±0.00b
Polyversum 1 500	65.0 ±0.15ab	53.3 ±0.19ab	40.0 ±0.00bc	28.3 ±0.03bc	23.3 ±0.08b	3.3 ±0.03b
Polyversum 2 000	65.0 ±0.05ab	55.0 ±0.05ab	41.7 ±0.03bc	31.7 ±0.03bc	25.0 ±0.09b	8.3 ±0.08b
Polyversum 2 500	66.7 ±0.03ab	56.7 ±0.08ab	45.0 ±0.00b	33.3 ±0.12b	25.0 ±0.00b	8.3 ±0.03b
Prochloraz 500	51.7 ±0.03c	43.3 ±0.06b	31.7 ±0.08c	21.7 ±0.03c	11.7 ±0.06b	0.0 ±0.00b
Prochloraz 1 000	55.0 ±0.00bc	45.0 ±0.00b	33.3 ±0.06bc	25.0 ±0.00bc	13.3 ±0.06b	0.0 ±0.00b
Prochloraz 1 500	58.3 ±0.08abc	48.3 ±0.03b	38.3 ±0.12bc	26.7 ±0.06bc	15.0 ±0.00b	0.0 ±0.00b
Prochloraz 2 000	60.0 ±0.00abc	48.3 ±0.08b	40.0 ±0.00bc	28.3 ±0.08bc	18.3 ±0.03b	0.0 ±0.00b
Prochloraz 2 500	63.3 ±0.03abc	51.7 ±0.03b	43.3 ±0.08bc	31.7 ±0.03bc	21.7 ±0.08b	0.0 ±0.00b
CK	68.3 ±0.08a	73.3 ±0.16a	75.0 ±0.10a	76.7 ±0.03a	78.3 ±0.16a	83.3 ±0.15a

Note: Data in a column followed by the same letters are not significantly different at $P_{0.05}$ by Duncan's multiple range test

Table 7 The effect of Polyversum and prochloraz against percentage of infected seed of Yangguang 600

Treatment	4d (%)	6d (%)	8d (%)	10d (%)	12d (%)	14d (%)
Polyversum 500	45.0 ±0.05bc	40.0 ±0.00cde	26.7 ±0.03b	13.3 ±0.03def	0.0 ±0.00d	0.0 ±0.00b
Polyversum 1 000	48.3 ±0.03bc	40.0 ±0.05cde	28.3 ±0.03b	16.7 ±0.03cde	0.0 ±0.00d	0.0 ±0.00b
Polyversum 1 500	48.3 ±0.06bc	41.7 ±0.03bcd	31.7 ±0.03b	18.3 ±0.03cd	0.0 ±0.00d	0.0 ±0.00b
Polyversum 2 000	50.0 ±0.00bc	43.3 ±0.03bc	31.7 ±0.03b	20.0 ±0.00bc	6.7 ±0.03c	0.0 ±0.00b
Polyversum 2 500	53.3 ±0.03abc	48.3 ±0.03b	33.3 ±0.06b	23.3 ±0.03b	11.7 ±0.03b	0.0 ±0.00b
Prochloraz 500	40.0 ±0.05c	28.3 ±0.03e	16.7 ±0.03c	0.0 ±0.00g	0.0 ±0.00d	0.0 ±0.00b
Prochloraz 1 000	45.0 ±0.00bc	33.3 ±0.03ef	26.7 ±0.03b	10.0 ±0.00f	0.0 ±0.00d	0.0 ±0.00b
Prochloraz 1 500	50.0 ±0.00bc	35.0 ±0.05def	28.3 ±0.03b	11.7 ±0.03ef	0.0 ±0.00d	0.0 ±0.00b

（续表）

Treatment	4d（%）	6d（%）	8d（%）	10d（%）	12d（%）	14d（%）
Prochloraz 2 000	66. 7 ±0. 25a	38. 3 ±0. 03cde	31. 7 ±0. 03b	15. 0 ±0. 05cdef	0. 0 ±0. 00d	0. 0 ±0. 00b
Prochloraz 2 500	55. 0 ±0. 05abc	38. 3 ±0. 03cde	31. 7 ±0. 03b	16. 7 ±0. 03cde	0. 0 ±0. 00d	0. 0 ±0. 00b
CK	60. 0 ±0. 05ab	65. 0 ±0. 09a	70. 0 ±0. 05a	71. 7 ±0. 03a	75. 0 ±0. 05a	78. 3 ±0. 06a

Note：Data in a column followed by the same letters are not significantly different at $P_{0.05}$ by Duncan's multiple range test

Table 8　The effect of Polyversum and prochloraz against percentage of infected seed of Beijing No. 1

Treatment	4d（%）	6d（%）	8d（%）	10d（%）	12d（%）	14d（%）
Polyversum 500	48. 3 ±0. 03cde	38. 3 ±0. 03cd	33. 3 ±0. 03bc	15. 0 ±0. 00c	0. 0 ±0. 00c	0. 0 ±0. 00b
Polyversum 1 000	53. 3 ±0. 03bcd	40. 0 ±0. 00bcd	33. 3 ±0. 08bc	15. 0 ±0. 00c	0. 0 ±0. 00c	0. 0 ±0. 00b
Polyversum 1 500	55. 0 ±0. 00bcd	41. 7 ±0. 06bc	35. 0 ±0. 00bc	16. 7 ±0. 06bc	0. 0 ±0. 00c	0. 0 ±0. 00b
Polyversum 2 000	55. 0 ±0. 05bcd	45. 0 ±0. 00b	38. 3 ±0. 03b	18. 3 ±0. 03bc	11. 7 ±0. 08b	0. 0 ±0. 00b
Polyversum 2 500	58. 3 ±0. 03b	45. 0 ±0. 00b	38. 3 ±0. 03b	21. 7 ±0. 03b	13. 3 ±0. 03b	0. 0 ±0. 00b
Prochloraz 500	43. 3 ±0. 03e	31. 7 ±0. 03e	20. 0 ±0. 05e	0. 0 ±0. 00d	0. 0 ±0. 00c	0. 0 ±0. 00b
Prochloraz 1 000	46. 7 ±0. 03de	35. 0 ±0. 00de	23. 3 ±0. 03de	15. 0 ±0. 05c	0. 0 ±0. 00c	0. 0 ±0. 00b
Prochloraz 1 500	55. 0 ±0. 05bcd	38. 3 ±0. 03cd	25. 0 ±0. 05de	16. 7 ±0. 03bc	0. 0 ±0. 00c	0. 0 ±0. 00b
Prochloraz 2 000	55. 0 ±0. 05bcd	40. 0 ±0. 00bcd	30. 0 ±0. 00cd	18. 3 ±0. 03bc	0. 0 ±0. 00c	0. 0 ±0. 00b
Prochloraz 2 500	56. 7 ±0. 06bc	40. 0 ±0. 05bcd	30. 0 ±0. 00cd	18. 3 ±0. 03bc	0. 0 ±0. 00c	0. 0 ±0. 00b
CK	66. 7 ±0. 08a	71. 7 ±0. 03a	73. 3 ±0. 06a	75. 0 ±0. 00a	76. 7 ±0. 03a	81. 7 ±0. 13a

Note：Data in a column followed by the same letters are not significantly different at $P_{0.05}$ by Duncan's multiple range test

2. 3　室内温室盆栽结果

Table 9　The effect of Polyversum and prochloraz against the growth of rice of Zhonglong Xiangjing No. 1

Treatment	Root length	Root weight	Height	Fresh weight
Polyversum 500	13. 5 ±2. 86b	0. 03 ±0. 01a	17. 5 ±1. 89a	0. 03 ±0. 02b
Polyversum 1 000	13. 1 ±2. 37b	0. 03 ±0. 02a	16. 6 ±3. 06ab	0. 02 ±0. 01c
Polyversum 1 500	12. 9 ±1. 22b	0. 02 ±0. 01bc	16. 2 ±1. 99abc	0. 02 ±0. 01c
Polyversum 2 000	14. 5 ±1. 24a	0. 04 ±0. 01a	17. 6 ±3. 08a	0. 04 ±0. 02a
Polyversum 2 500	13. 7 ±2. 17ab	0. 03 ±0. 01a	16. 5 ±2. 60ab	0. 04 ±0. 01b
Prochloraz 500	7. 2 ±1. 31ef	0. 01 ±0. 00c	15. 6 ±1. 50bc	0. 02 ±0. 02c
Prochloraz 1 000	7. 9 ±0. 52de	0. 01 ±0. 01c	16. 5 ±2. 56ab	0. 02 ±0. 01c
Prochloraz 1 500	8. 8 ±0. 52cd	0. 02 ±0. 01b	16. 4 ±2. 56abc	0. 03 ±0. 01b
Prochloraz 2 000	9. 0 ±1. 98cd	0. 02 ±0. 01b	17. 0 ±1. 18ab	0. 04 ±0. 01a
Prochloraz 2 500	9. 5 ±1. 64c	0. 02 ±0. 01b	16. 7 ±0. 84ab	0. 03 ±0. 01b
CK	6. 3 ±1. 50f	0. 01 ±0. 01c	14. 5 ±1. 32c	0. 02 ±0. 01c

Note：Data in a column followed by the same letters are not significantly different at $P_{0.05}$ by Duncan's multiple range test

分析表9数据发现，对于中龙香粳一号，寡雄腐霉和咪鲜胺处理后，水稻幼苗根长显著高于对照，且寡雄腐霉处理显著高于咪鲜胺处理；除寡雄腐霉1 500倍液处理外，两种药剂处理后水稻幼苗根重显著高于对照。除寡雄腐霉1 500倍液、咪鲜胺500倍液和咪鲜胺1 000倍液，寡雄腐霉处理后水稻幼苗根重显著高于咪鲜胺处理；除寡雄腐霉1 500倍液、咪鲜胺500倍液和咪鲜胺1 500倍液，两药剂处理后幼苗鲜高显著高于对照。除咪鲜胺500倍液处理，寡雄腐霉处理后鲜高显著高于咪鲜胺处理；除寡雄腐霉1 000倍液、1 500倍液和咪鲜胺500倍液、1 000倍液，其余处理后幼苗鲜重显著高于对照。

Table 10 The effect of Polyversum and prochloraz against the growth of rice of Songjingxiang No. 1

Treatment	Root length	Root weight	Height	Fresh weight
Polyversum 500	15.4 ±2.63a	0.06 ±0.02a	17.5 ±2.29bcd	0.05 ±0.01bc
Polyversum 1 000	15.1 ±1.42a	0.05 ±0.01b	16.6 ±2.94cd	0.05 ±0.01bc
Polyversum 1 500	15.0 ±1.01a	0.05 ±0.01b	17.3 ±1.59bcd	0.06 ±0.01bc
Polyversum 2 000	15.9 ±2.81a	0.07 ±0.01a	19.1 ±2.56a	0.08 ±0.01a
Polyversum 2 500	14.8 ±1.56a	0.04 ±0.01c	16.9 ±2.26bcd	0.05 ±0.01de
Prochloraz 500	8.2 ±0.77cd	0.02 ±0.01d	16.9 ±1.02bcd	0.02 ±0.01g
Prochloraz 1 000	8.8 ±0.76bc	0.03 ±0.01d	17.3 ±0.81bcd	0.04 ±0.01ef
Prochloraz 1 500	9.44 ±4.49b	0.03 ±0.00d	17.5 ±3.81bcd	0.05 ±0.01cd
Prochloraz 2 000	9.7 ±0.88b	0.04 ±0.01c	18.4 ±1.25ab	0.05 ±0.01cde
Prochloraz 2 500	9.5 ±0.51b	0.04 ±0.01c	18.0 ±0.79abc	0.06 ±0.01b
CK	7.6 ±2.65d	0.02 ±0.01d	16.1 ±1.03d	0.03 ±0.01f

Note: Data in a column followed by the same letters are not significantly different at $P_{0.05}$ by Duncan's multiple range test

试验结果见表10。对于松粳香1号，除咪鲜胺500倍液，其余处理后幼苗根长显著高于对照，且寡雄腐霉处理显著高于咪鲜胺处理；除咪鲜胺500倍液、1 000倍液和1 500倍液，其余处理幼苗根重显著高于对照；除寡雄腐霉2 000倍液、咪鲜胺2 000倍液和咪鲜胺2 500倍液，其余处理幼苗鲜高显著高于对照；除咪鲜胺1 000倍液，两种药剂处理幼苗鲜重显著高于对照。

Table 11 The effect of Polyversum and prochloraz against the growth of rice of Yangguang 600

Treatment	Root length	Root weight	Height	Fresh weight
Polyversum 500	15.3 ±3.21bc	0.06 ±0.02a	19.8 ±4.82ab	0.05 ±0.01bc
Polyversum 1 000	15.2 ±3.78bc	0.04 ±0.01b	21.4 ±5.19a	0.05 ±0.02bc
Polyversum 1 500	16.4 ±4.16ab	0.06 ±0.02a	19.2 ±2.95ab	0.05 ±0.01bc
Polyversum 2 000	18.1 ±2.49a	0.06 ±0.02a	21.8 ±2.31a	0.06 ±0.01b
Polyversum 2 500	13.9 ±2.44cd	0.03 ±0.01d	20.0 ±4.10ab	0.04 ±0.01bcd
Prochloraz 500	10.6 ±4.66ef	0.02 ±0.00ef	19.3 ±4.83ab	0.03 ±0.01bcd

（续表）

Treatment	Root length	Root weight	Height	Fresh weight
Prochloraz 1 000	10.3 ±3.68f	0.02 ±0.00ef	18.0 ±4.35b	0.02 ±0.00d
Prochloraz 1 500	12.7 ±2.89de	0.02 ±0.01f	20.8 ±3.35ab	0.02 ±0.01cd
Prochloraz 2 000	13.1 ±2.76cd	0.03 ±0.01de	21.6 ±5.01a	0.01 ±0.01a
Prochloraz 2 500	16.8 ±1.27ab	0.04 ±0.01bc	20.7 ±3.32ab	0.03 ±0.01bcd
CK	9.3 ±5.27f	0.03 ±0.01cd	18.3 ±2.99b	0.03 ±0.01bcd

Note：Data in a column followed by the same letters are not significantly different at $P_{0.05}$ by Duncan's multiple range test

结果见表11。对于阳光600，除咪鲜胺500倍液和1 000倍液，其余处理幼苗根长显著高于对照；除寡雄腐霉2 500倍液和咪鲜胺2 000倍液，两种药剂处理幼苗根重显著高于对照；除寡雄腐霉1 000倍液、2 000倍液和咪鲜胺2 000倍液，其余处理幼苗鲜高与对照无差异；两种药剂处理后，寡雄腐霉2 500倍液、咪鲜胺500倍液、1 000倍液、1 500倍液和2 500倍液幼苗鲜重显著高于对照。

Table 12　The effect of Polyversum and prochloraz against the growth of rice of BeiJing No. 1

Treatment	Root length	Root weight	Height	Fresh weight
Polyversum 500	18.3 ±3.45ab	0.08 ±0.01abc	20.6 ±3.63c	0.06 ±0.02b
Polyversum 1 000	19.4 ±4.13a	0.08 ±0.02ab	24.6 ±2.48a	0.07 ±0.01a
Polyversum 1 500	19.5 ±3.29a	0.08 ±0.01bcd	21.8 ±3.01bc	0.06 ±0.01bc
Polyversum 2 000	20.1 ±2.81a	0.07 ±0.01cdef	24.1 ±2.52ab	0.05 ±0.01bc
Polyversum 2 500	16.4 ±4.15bc	0.03 ±0.01g	22.4 ±4.28abc	0.03 ±0.01cd
Prochloraz 500	14.5 ±2.79c	0.07 ±0.03def	17.2 ±3.24d	0.05 ±0.01cd
Prochloraz 1 000	15.1 ±3.50c	0.04 ±0.01g	21.0 ±4.21c	0.05 ±0.02bc
Prochloraz 1 500	18.7 ±3.40ab	0.07 ±0.01bcde	20.6 ±4.79c	0.05 ±0.01bc
Prochloraz 2 000	20.1 ±1.92a	0.06 ±0.01f	21.4 ±4.25c	0.06 ±0.01b
Prochloraz 2 500	19.8 ±3.80a	0.06 ±0.01ef	21.1 ±3.15c	0.06 ±0.01bc
CK	18.6 ±4.16ab	0.09 ±0.02a	22.9 ±3.47abc	0.07 ±0.01a

Note：Data in a column followed by the same letters are not significantly different at $P_{0.05}$ by Duncan's multiple range test

见表12。对于北粳1号，除咪鲜胺500倍液和1 000倍液，其余处理幼苗根长显著高于对照；除寡雄腐霉500倍液和1 000倍液，其余药剂处理幼苗根重显著高于对照；除咪鲜胺500倍液，其余处理鲜高与对照差异不显著；除寡雄腐霉1 000倍液，两药剂处理后幼苗鲜重均显著高于对照。

3　结果和讨论

对水稻种子进行浸种处理是目前防治水稻恶苗病的主要措施，主要浸种药剂有咪鲜胺乳油、恶线灵可湿性粉剂和使百克乳油等，但经这些药剂处理后，在一定程度上都会减弱种子活力，从而影响水稻苗的成苗率及长势[10]。本研究表明，在浸种过程中，咪鲜胺500倍液和1 000倍稀释液处理发芽率整体偏低，这可能是由于高浓度咪鲜胺处理对水稻种子活力有一定影响。随着浸种时间的延长，清水处理发芽率偏低，这可能是由于恶苗病菌在水中产生大量分生孢子[11]，使得整体带菌率提高，影响水稻发芽；浸种14d后，寡雄腐霉制剂2 000倍稀释液不仅可以很好地抑制水稻恶苗病菌，且保证95%以上的发芽率，且在一定程度上促进种子的生长，其根长、根重、鲜高和鲜重均显著高于清水处理。

参考文献

[1] Wang G C. Studies on the pathogens of the rice bakanae disease in ZheJiang [J]. Acta Phytopathologica Sinica, 1990, 20 (2): 93 - 97.

[2] Chen J F. Experiments on the control of bakanae disease of rice [J]. Acta Phytophylacica Sinica, 1985, 12 (2): 140 - 141.

[3] Ji Z J. Research progress of rice bakanae disease resistance [J]. Chinese Rice, 2008 (2): 24 - 25.

[4] Jiang B R. The soaking test of rice bakanae disease pesticides [J]. Shanghai Agricultural Science and Technology, 2012 (1): 107.

[5] Wang A Y. Inhibitory effect of the secretion of *Pythium oligandrum* on plant pathogenic fungi and the control effect against tomato gray mould [J]. Acta Phytophylacica Sinica, 2007, 34 (1): 57 - 60.

[6] OuYang Y N. The effect of "Polyversum" of *Pythium oligandrum* preparation on rice's pro - growth effect and disease prevention and yield increase [J]. Chinese Rice, 2007, 6: 48 - 50.

[7] D Cizkova. The effect of biological preparation polyversum and *befungin* against fungi of genus *Ophistoma* sp [C]. Slovak and Czech Plant Protection Conference, 1997, 343 - 344.

[8] Hua Z Y. The safety test of 16% prochloraz · cartap WP on rice soaking [J]. Guangxi plant protection, 2012, 25 (1): 10 - 11.

[9] Guan Hongdan, Liu Min, Sheng Haian, et al. The effect test of different seed treatment against rice bakanae disease [J]. Shanghai Agricultural Science and Technology, 2012 (2): 123.

[10] Cao D D. Effect of Seed Soaking in different chemicals to control rice bakanae disease in seedling stage [J]. Seed, 2014, 33 (4): 86 - 87.

[11] Pan Y L. Dispersal of *Fusarium moniliforme*, causing bakanae disease of rice, in seed soaking [J]. Journal of Anhui Agricultural Sciences, 2000, 28 (5): 616 - 618.

黑龙江省水稻恶苗病菌对咪鲜胺敏感基线的建立*

徐　瑶**，李　鹏，穆娟微***
（黑龙江省农垦科学院植物保护研究所，哈尔滨　150038）

摘要：利用菌丝生长速率法检测了来自黑龙江省 8 个县（市）的 32 个恶苗病菌株对咪鲜胺的敏感性，确定敏感基线值为 0.003 6μg/mL，为黑龙江省监测恶苗病菌对咪鲜胺的抗药性奠定基础。
关键词：恶苗病；咪鲜胺；抗药性；敏感基线

Establishment of a Sensitive Baseline for *Fusarium moniliforme* from Heilongjiang Rice to Prochloraz

Xu Yao，Li Peng，Mu Juanwei
（*Institute of Plant Protection*，*Heilongjiang Academy of Land Reclamation Sciences*，*Harbin* 150038，*China*）

Abstract：Mycelium growth rate method was used to detect the sensitivity of 32 *Fusarium moniliforme* strains to prochloraz from 8 counties（cities） in Heilongjiang province. The results showed that the value of baseline sensitivity is 0.003 6μg/mL，which lay the foundation for monitoring the resistance of *F. moniliforme* to prochloraz in Heilongjiang province.
Key words：Rice Bakanae；Prochloraz；Prevention；sensitivity baseline

水稻恶苗病是由串珠镰孢菌（*Fusarium moniliforme* Sheld）引起的真菌病害[1]，是严重影响黑龙江省水稻生产的主要病害之一，发病地块一般减产 10% ~20%，严重的可减产 50% 以上。近年来，旱育秧田的大面积推广使苗床通气性改善，给恶苗病病菌的生长繁殖提供了有利条件；种子生产环节上存在不足，导致病菌再侵染；加之单一化学药剂的长期使用，病菌逐渐产生抗药性，防治效果日益降低，使得水稻恶苗病的发生日趋严重。有关该病的研究得到人们日益重视。

自 20 世纪 90 年代咪鲜胺代替多菌灵防治恶苗病以来，咪鲜胺以防效高、对作物安全等优势倍受农民青睐。咪鲜胺浸种防治水稻恶苗病在黑龙江已应用长达 20 年，连续使用单一杀菌剂会导致抗药性的出现[2]，监测其抗药性对保护水稻安全生产意义重大。本文通过测定黑龙江省不同地区恶苗病菌对咪鲜胺的敏感性，确立该地区恶苗病菌对咪鲜胺的敏感性基线，为黑龙江省监测恶苗病菌对咪鲜胺的抗药性奠定基础。

---

* 基金项目：黑龙江垦区一戎水稻科技奖励基金会支持
** 第一作者：徐瑶，女，硕士，助理研究员，从事植物病害与综合防治研究；E-mail：xuyao20111@163.com
*** 通讯作者：穆娟微，女，硕士，研究员，从事水稻植保技术研究；E-mail：mujuanwei@126.com

1 材料与方法

1.1 供试菌株

于2013年8月在黑龙江省虎林、阿城、讷河、佳木斯、铁力、富裕、绥化和庆安等8个县（市）采集67个水稻恶苗病病株，采用组织分离方法进行病原菌的分离，通过单孢分离纯化培养获得42个恶苗病菌菌株，每个县（市）随机抽取4个菌株，共计32个供试菌株。

1.2 恶苗病菌对咪鲜胺敏感性测定

通过预备试验确定出药剂的5个有效浓度，配制含药平板。以室内生长速率法测定咪鲜胺对水稻恶苗病菌菌株的毒力大小[3-4]。在活化后的恶苗病菌菌落边缘上打取直径为5mm的菌碟，挑取菌碟置于含药平板培养皿的中央，将其置于25℃恒温培养箱中培养，7d后以十字交叉法测菌落直径，计算不同浓度药剂对各菌株菌丝生长抑制率。求出药剂对菌株的毒力公式，即回归方程 $y = ax + b$ 及 x 与 y 之间的相关系数 r，计算出咪鲜胺对各供试菌株的抑制中浓度（EC_{50}）。根据病菌对咪鲜胺的敏感性频率分布建立恶苗病菌对咪鲜胺的敏感基线。

2 结果与分析

2.1 供试菌株对咪鲜胺的敏感性

利用室内生长速率法测定了咪鲜胺对32个供试菌株的毒力大小，通过分析得出各个菌株的毒力回归方程和 EC_{50} 值，相关系数均在0.917 7以上。各菌株的 EC_{50} 值差异较大，最小 EC_{50} 值为0.001 5μg/mL，最大 EC_{50} 值为1.306 7μg/mL，相差871倍（表）。

表 恶苗病菌对咪鲜胺的敏感性

菌株编号	回归方程	相关系数（r）	抑制中浓度 EC_{50}（μg/mL）
NH-1	$y=6.4972+0.5276x$	0.971 2	0.001 5
NH-2	$y=4.8489+0.8887x$	0.939 3	0.110 9
NH-3	$y=5.1614+0.9517x$	0.996 0	0.676 7
NH-4	$y=6.3128+1.2559x$	0.990 5	0.090 1
HL-1	$y=6.0650+0.9657x$	0.972 2	0.078 9
HL-2	$y=6.5597+1.5288x$	0.988 6	0.095 5
HL-3	$y=6.8946+1.6113x$	0.995 1	0.066 7
HL-4	$y=6.2463+1.4879x$	0.999 2	0.145 3
AC-1	$y=6.5054+1.2276x$	0.953 3	0.059 4
AC-2	$y=5.2193+0.9631x$	0.987 6	0.382 5
AC-3	$y=5.8216+1.3977x$	0.976 3	0.258 3
AC-4	$y=6.7348+1.5648x$	0.995 2	0.077 9
FY-1	$y=6.9658+1.6748x$	0.933 6	0.067 0
FY-2	$y=6.8176+1.3059x$	0.992 2	0.040 6

（续表）

菌株编号	回归方程	相关系数（r）	抑制中浓度 EC_{50}（μg/mL）
FY－3	$y=5.6343+1.3447x$	0.994 7	0.337 5
FY－4	$y=5.8861+0.7267x$	0.927 8	0.060 4
JMS－1	$y=6.6156+0.7164x$	0.918 6	0.005 6
JMS－2	$y=6.8043+1.6514x$	0.991 7	0.080 8
JMS－3	$y=4.8695+1.1231x$	0.983 0	1.306 7
JMS－4	$y=6.3355+1.2011x$	0.989 0	0.077 3
TL－1	$y=7.1019+1.5442x$	0.998 0	0.043 5
TL－2	$y=6.8057+1.5492x$	0.970 9	0.068 3
TL－3	$y=7.0060+2.0426x$	0.995 7	0.104 2
TL－4	$y=6.7572+1.3585x$	0.992 8	0.048 3
SH－1	$y=6.6492+1.5130x$	0.986 7	0.081 3
SH－2	$y=6.2687+1.1461x$	0.989 0	0.078 2
SH－3	$y=6.8109+1.0713x$	0.991 0	0.020 4
SH－4	$y=6.3879+1.4884x$	0.996 1	0.116 8
QA－1	$y=7.0408+1.3585x$	0.976 8	0.031 5
QA－2	$y=6.6950+1.4483x$	0.984 2	0.067 6
QA－3	$y=6.4316+1.2585x$	0.994 2	0.072 8
QA－4	$y=6.4264+1.3735x$	0.917 7	0.206 9

2.2 恶苗病菌对咪鲜胺敏感基线的建立

采用类平均法（UPGMA）根据32个菌株对咪鲜胺的敏感性进行聚类分析（图1），将恶苗病菌对咪鲜胺的敏感性划分为 $EC_{50}<0.01$μg/mL（NH－1、JMS－1）、0.01μg/mL $<EC_{50}<0.2$μg/mL（SH－3、QA－1、FY－2、TL－1、TL－4、AC－1、FY－4、HL－3、FY－1、QA－2、TL－2、QA－3、JMS－4、AC－4、SH－2、HL－1、JMS－2、SH－1、NH－4、HL－2、TL－3、NH－2、SH－4、HL－4）、$EC_{50}>0.2$μg/mL（AC－2、FY－3、AC－3、QA－4、NH－3、JMS－3）3个类群。

在32个供试菌株中，$EC_{50}<0.01$μg/mL的2个菌株（NH－1、JMS－1）对咪鲜胺最敏感，EC_{50}值平均0.0036μg/mL，将其定为黑龙江省水稻恶苗病菌对咪鲜胺的敏感基线。

3 结论与讨论

目前为止，敏感基线的确定方法大致有3种[3-7]：第一种方法，FRAC建议的以没有接触过被测药剂且没有接触过被测药剂同类药剂地区的菌株敏感性（EC_{50}）作为敏感基线；第二种方法，将最敏感的一个菌株或几个菌株 EC_{50} 的平均值作为敏感基线；第三种方法，绘出菌株敏感性频率分布图，正态分布曲线 EC_{50} 的平均值就是敏感性基线。

由于咪鲜胺作为防治水稻恶苗病的药剂在黑龙江省已连续使用长达20年，要获得从未

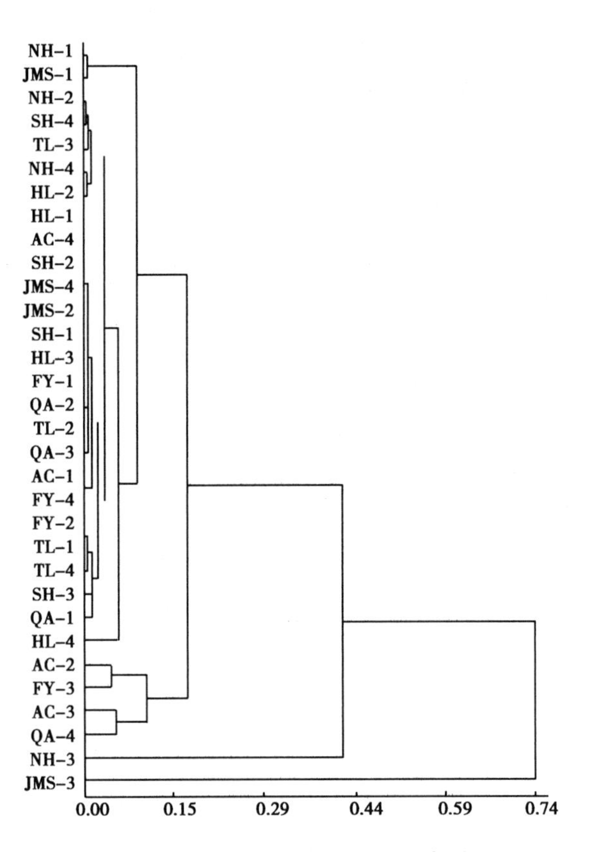

图 1　咪鲜胺对水稻恶苗病菌 EC_{50} 值聚类分析结果

接触被测药剂的野生菌株是不可能做到的。而 FRAC 建议方法中采集的水稻恶苗病菌必须是没有接触被测药剂或同类药剂的菌株，对野生敏感菌株的获得要求很严。所以第一种方法不适合作为确定黑龙江省水稻恶苗病菌对咪鲜胺敏感基线的方法。

第二种方法确定敏感基线，有可能会高估病原菌的抗药水平。本研究使用第二种方法将黑龙江省水稻恶苗病菌对咪鲜胺的敏感基线值确定为 0.003 6μg/mL，与刘永锋[5]确定的敏感基线值 0.005μg/mL 相近，比卢国新[9]确定的敏感基线值 0.000 746 6μg/mL 高，不会高估计病原菌的抗药水平。恶苗病菌对咪鲜胺敏感性基线的建立，为黑龙江省监测恶苗病菌对咪鲜胺的抗药性奠定基础。

参照 Takuo Wada 等[10]采用第三种方法，将本研究 0.01μg/mL < EC_{50} < 0.2μg/mL 类群（菌株频率最高的类群）的 24 个菌株绘出菌株敏感性频率分布图（图 2），菌株对咪鲜胺的敏感性呈连续的单峰曲线，符合正态分布。卢国新[6]将恶苗病菌对咪鲜胺的敏感基线确定为 0.000 746 6μg/mL，而本研究 0.01μg/mL < EC_{50} < 0.2μg/mL 组 EC_{50} 平均值为 0.074 3 μg/mL，对咪鲜胺敏感性差。因此，此区组的 EC_{50} 平均值不能作为黑龙江省水稻恶苗病菌对咪鲜胺的敏感基线。

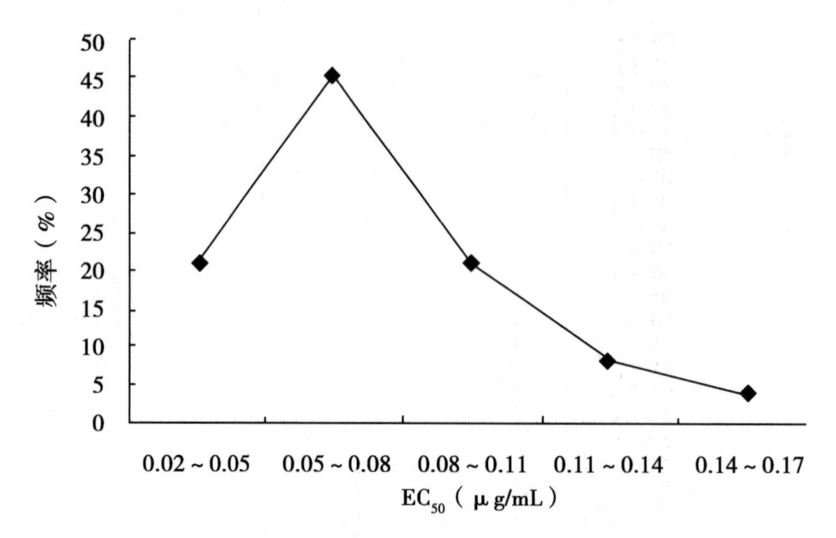

图 2　24 个恶苗病菌株对咪鲜胺敏感性频率分布

参考文献

［1］ 产祝龙，丁克坚，檀根甲. 水稻恶苗病的研究进展［J］. 安徽农业科学，2002，30（6）：880 - 883.

［2］ 陈夕军，卢国新，童蕴慧，等. 水稻恶苗病菌对三种浸种剂的抗性及抗药菌株的竞争力［J］. 植物保护学报，2007，34（4）：425 - 430.

［3］ 范子耀，孟润杰，韩秀英，等. 马铃薯早疫病菌对咯菌腈的敏感基线及其对不同药剂的交互抗性［J］. 植物保护学报，2012，39（2）：153 - 158.

［4］ 李恒奎，陈长军，王建新，等. 禾谷镰孢菌对氰烯菌酯的敏感性基线及室内抗药性风险初步评估［J］. 植物病理学报，2006，36（3）：273 - 278.

［5］ 刘永锋，陈志谊，周保华，等. 江苏省部分稻区恶苗病菌对水稻浸种剂的抗药性检测［J］. 江苏农业学报，2002，18（3）：190 - 192.

［6］ 卢国新. 江苏水稻恶苗病种类和抗药性研究［D］. 江苏，扬州大学，2005.

［7］ 潘洪玉，杜红军，郭金鹏，等. 东北春麦区小麦赤霉病菌对多菌灵敏感性的测定［J］. 吉林农业科学，2002，27（增刊）：44 - 45.

［8］ 郑丽娜，靳学慧，张亚玲，等. 黑龙江省稻瘟病菌对施保克敏感性分析［J］. 黑龙江八一农垦大学学报，2009，21（2）：13 - 16.

［9］ 郑睿，聂亚锋，于俊杰，等. 江苏省水稻恶苗病菌对咪鲜胺和氰烯菌酯的敏感性［J］. 农药学学报，2014，16（3）：693 - 698.

［10］ Takuo Wada，Seiichi Kuzuma，Mitsuki Takenaka. Sensitivity of *fursarium monilifore* Isolates to pefurazote［J］. Ann Phytopath Soc，1990，56（4）：449 - 456.

施药时期及方式对杀线剂防治小麦孢囊线虫病效果的影响

任玉鹏[1,2]，张 杰[3]，王晓坤[1]，慕 卫[1]，刘 峰[1]
（1. 山东农业大学植物保护学院，泰安 271018；2. 山东省农药科学研究院，济南 250100；3. 山东省莱州市农业局，烟台 261400）

摘要：为优化阿维菌素、甲氨基阿维菌素苯甲酸盐和噻唑膦防治小麦孢囊线虫病的使用技术，在室内条件下不同温度阶段浸渍法处理孢囊，测定了3种杀线剂对线虫孵化的影响，并在田间小麦播种期、齐苗期、冬前期和返青期采用不同施药方式分别比较了6种药剂对孢囊线虫病的防治效果。结果表明：阿维菌素和甲维盐在4℃和15℃两个阶段处理孢囊均可显著降低线虫的孵化率，15℃条件下处理孵化率显著低于4℃处理，噻唑膦仅在15℃处理孢囊可显著降低线虫孵化率。在田间，2%阿维菌素乳油、2%甲维盐乳油和40%噻唑膦乳油在小麦返青期分别施用15kg/hm^2、15kg/hm^2、5kg/hm^{2}灌根对小麦孢囊线虫病的防治效果可达50%以上，3%阿维悬浮种衣剂2.5kg/100kg种子处理下防效为41.65%，0.5%阿维菌素颗粒剂及10%噻唑膦颗粒剂在28.35kg/hm^2、30kg/hm^{2}防效分别为43.01%、57.17%。因此，施药时期及方式对药剂防治小麦孢囊线虫病效果存在较大的影响，为轻简化施药及保证效果，建议在发病较轻地块播种期拌种处理或颗粒剂种—肥—药同播，重病田结合播种期种衣剂处理或颗粒剂种—肥—药同播与返青期随灌溉水补充施药进行防治。

关键词：禾谷孢囊线虫；杀线剂；施药时期；防治效果

Effects of Three Kinds of Nematicides on Controlling Cereal Cyst Nematode (*Heterodera avenae*) in Different Application Periods

Ren Yupeng[1,2], Zhang Jie[3], Wang Xiaokun[1], Mu Wei[1], Liu Feng[1]
(1. *Collegeof Plant Protection*, *Shandong Agricultural University*, *Tai'an* 271018, *China*; 2. *Shandong Academy of Pesticide Sciences*, *Jinan* 250100, *China*; 3. *Agricultural Bureau of Laizhou*, *Yantai* 261400, *China*)

Abstract: The bioassay and field trials were conducted to evaluate the effects of abermectin, emamectin benzoate and fosthiazate on the hatching of cereal cyst nematode (*Heterodera avenae* Wollenweber) and on the controlling effect in different growth period of wheat anddifferent applying mode. The results indicated that, in the bioassay of the hatching of cereal cyst nematode, it was significantly decreased in the treatment of abermectin and emamectin benzoate both under the temperature of 4℃ and 15℃; and the hatching percentage under the temperature of 15℃ was lower than it under the temperature of 4℃, fosthiazate can also inhibit the hatching of cereal cyst nematode at 15℃. In the field trials, by using the doses of 15kg/hm^2, 15kg/hm^2, 5kg/hm^2 respectively atgreen stage, 2% abamectin EC, 2% emamectin benzoate EC and 40% fosthiazate EC had decent effects with the value more than 50%, 3% abamectin seed coating had control effect of 41.65% by using the doses of 2.5 kg/100 kg seedand by using the doses of 28.35kg/hm^2、30kg/hm^2 respectively atsowing date, 0.5% abamectin GR and 10% fosthiazate GR had control effect of 43.01%、57.17%. Overall, considering the great influence of different application periods on the control effect of nematicides, in order to make the nematicides application simplify and ensure that the control effect, nematicides seed coating，GR or EC should be applied according to the incidence in different field.

Key words：*Heterodera avenae* Wollenweber；Nematicides；Application time；Control effect

禾谷孢囊线虫病（Cereal Cyst Nematodes，CCNs）是一类主要为害燕麦、小麦、大麦和黑麦等麦类作物及多种禾本科牧草的世界性植物寄生线虫病害，其主要是由禾谷孢囊线虫（*Heterodera avenae* Wollenweber）[1]引起。1908 年英国首次在小麦上发现，在世界上的 32 个国家已有报道。我国最早于 1987 年在湖北省天门县报道其为害小麦，1989 年鉴定为小麦禾谷孢囊线虫（*H. avenae*）[2-3]。目前，该病害在我国 16 个省市小麦主产区均有发生，并呈逐年上升趋势，严重威胁着我国粮食安全[4-7]。孢囊线虫一年发生一代，主要侵染小麦根部，影响植株对土壤中水肥的吸收，进而影响小麦的产量。在小麦成熟期，雌成虫形成孢囊，从小麦根部脱落于土壤进入滞育状态[8]。禾谷孢囊线虫的孵化是其生活史中的重要阶段，其孵化期及孵化量直接影响着线虫田间群体密度及病害发生的早晚及程度。我国禾谷孢囊线虫群体一般须经过 2 个月或以上时间低温阶段，二龄幼虫才能自孢囊中孵出，且低温后一定幅度的温度升高可使二龄幼虫在短期内大量孵化[9]，若遇到小麦等寄主的根系，建立侵染点并侵入[10]，如果无法找到寄主，则只能在土壤中存活几天或数周[3]。

由于我国当前生产上广泛种植的小麦多为感病品种，轮作等农业防治措施难以推广，生物防治仅处于起步阶段，因此对小麦孢囊线虫病的防治仍以化学防治为主。目前，化学防治的工作主要集中在杀线剂室内毒力和敏感性测定及田间药效方面。不过由于孢囊线虫雌成虫死后在土壤中会形成较厚孢囊壳保护体内卵，提高抗逆性[11]，增加了药剂防治的难度，所以选择合适的施药时期对发挥药效有重要影响，而对于杀线剂对禾谷孢囊线虫孢囊孵化的影响以及防治适期方面目前少见报道。本文通过阿维菌素、甲氨基阿维菌素苯甲酸盐（甲维盐）、噻唑膦以及阿维种衣剂、阿维颗粒剂、噻唑膦颗粒剂 6 种对环境较为安全且对线虫毒力较高的药剂进行试验，研究了 3 种药剂在室内条件下对禾谷孢囊线虫孢囊孵化的影响以及在及田间条件下 6 种药剂不同施药时期及方式对小麦禾谷孢囊线虫的防治效果，初步明确 6 种杀线剂的最佳施药时期，为指导小麦生产和禾谷孢囊线虫的防治提供理论依据。

1　材料与方法

1.1　材料

供试线虫：线虫土样于 2014 年 8 月采自山东省济南市济阳县孙耿镇郑家村禾谷孢囊线虫重病田，用 Fenwick 漂浮过筛法分离孢囊样品，并通过蔗糖液离心法分离样品中孢囊[12]，在体视显微镜下选择当年新形成的大小相对一致的饱满孢囊，放置灭菌培养皿内，用灭菌水冲洗 3 次，用于不同处理下的孵化试验。

供试药剂：选取对小麦孢囊线虫防治效果较好的 3 种杀线剂，2% 甲维盐乳油、2% 阿维菌素乳油、40% 噻唑膦乳油。同时，配制相同配方但不含原药的 3 种空白助剂乳油，剩余部分用溶剂补足。3% 阿维阿维菌浮种衣剂，0.5% 阿维菌素颗粒剂，10% 噻唑膦颗粒剂。试验中各制剂均为本实验室加工，各剂型加工质量经检测均复合国家同类剂型质量标准。

供试小麦品种：矮抗 58（感病品种），河南省益农种业有限公司。

试验地：山东省济南市济阳县孙耿镇郑家村。多年种植小麦，与玉米轮作，经调查小麦孢囊线虫发生严重。土壤类型为壤质土，水肥条件良好。小麦拌种时间为 2014 年 10 月 9 日，小麦播种时间为 2014 年 10 月 10 日，机播，播种量 225kg/hm^2，播种后田间进行正常施肥和除草。

1.2 方法

1.2.1 室内不同杀线剂对禾谷孢囊线虫孢囊孵化的影响

将供试3种乳油及溶剂用去离子水配制成一定浓度梯度的溶液。噻唑膦的浓度分别为10、100、1 000mg/L，阿维菌素的浓度分别为0.05、0.5、5mg/L，甲维盐的浓度分别为0.05、0.5、5mg/L，噻唑膦溶剂的浓度分别为15、150、1 500mg/L，阿维菌素溶剂的浓度分别为2.5、25、250mg/L，甲维盐溶剂的浓度分别为2.5、25、250mg/L。

1.2.1.1 未经杀线剂处理的禾谷孢囊线虫孵化特点

采用Ioannis O等的浸渍法[13]，将禾谷孢囊放于24孔细胞培养板内，每孔放10个孢囊并加入2mL无菌水。在4℃条件下放置8周之后，再转至15℃条件下培养，重复4次。每隔2 d检查1次孵出的线虫数量，查后用移液器将孵化出的幼虫吸出，并加入等量的无菌水，直至线虫孵化结束，最后将孢囊压破，统计孢囊内的卵数和已形成卷曲但未孵出的幼虫数量以及孵化的幼虫数量，计算孢囊孵化率。

1.2.1.2 杀线剂在4℃预处理阶段处理孢囊对线虫孵化的影响

将禾谷孢囊放于24孔细胞培养板内，每孔放10个孢囊并加入2mL一定浓度的杀线剂溶液，并设清水空白对照，每3d换一次新药液。在4℃条件下放置8周之后，用无菌水冲洗孢囊，并将杀线剂溶液换成无菌水，再转至15℃条件下培养，每浓度重复4次。按照1.2.1.1的方法，每2d统计线虫孵出的数量，直至线虫孵化结束，最后将孢囊压破，统计线虫孵化率。

1.2.1.3 杀线剂在15℃孵化阶段处理孢囊对线虫孵化的影响

将禾谷孢囊放于24孔细胞培养板内，每孔放10个孢囊并加入2mL无菌水。在4℃条件下放置8周之后，将无菌水换成一定浓度的杀线剂溶液，浓度同1.2.1.2方法，并设清水空白对照，再转至15℃条件下培养，每浓度重复4次。按照1.2.1.1的方法，每2 d统计线虫孵化的数量，直至线虫孵化结束，最后将孢囊压破，统计线虫孵化率。

1.2.2 不同施药时期对田间条件下小麦禾谷孢囊线虫病的防治效果

田间试验分别对甲维盐、阿维菌素、噻唑膦3种药剂设置2个施药剂量，3个施药时期。其中，2%甲维盐乳油用量15、30kg/hm^2，2%阿维菌素乳油用量15、30kg/hm^2，40%噻唑膦乳油用量5、7.5kg/hm^2，3种药剂分别于齐苗期（2014年10月23日）、冬前期（2014年12月18日）、返青期（2015年3月10日）进行灌根施药，设清水处理对照，每处理重复3次，每小区6m^2，各小区采用随机区组排列。3%阿维阿维菌素悬浮种衣剂用量2、2.5kg/100kg种子于播种前一天，0.5%阿维菌素颗粒剂用量9.45、18.9、28.35kg/hm^2，10%噻唑膦颗粒剂9、15、30kg/hm^2，利用种肥同播技术将颗粒剂与肥料混匀施用，每处理重复3次，每小区60m^2，各小区随机排列。田间病情调查：分别于小麦播种前与小麦收获后每小区五点取样，用取土器挖取0～20cm的土壤，风干混匀，称取500g土壤采用过筛漂浮法（黄文坤等，2011）分离土壤中孢囊，并于体视显微镜下统计每个取样点土壤中饱满孢囊数量，根据公式（1）、（2）、（3）计算孢囊减退率、防治效果[14]、繁殖系数[15]。

$$\text{孢囊减退率}(\%) = (\text{播种前孢囊数量} - \text{收获后孢囊数量}) / \text{播种前孢囊数量} \times 100 \quad (1)$$

防治效果（%）＝（防治区孢囊减退率－空白区孢囊减退率）/（100－空白区孢囊减退率）（2）

繁殖系数＝收获后孢囊数量/播种前孢囊数量（3）

1.2.3 统计分析方法

数据分析采用DPS数据处理系统软件[16]，计算数据平均值与标准差、标准误，并用Duncan's新复极差法检验差异显著性（$P\leq0.05$）。

2 结果与分析

2.1 不同杀线剂不同阶段处理孢囊对禾谷孢囊线虫孵化的影响

2.1.1 未经杀线剂处理的禾谷孢囊线虫孵化的特点

由图1可知，未经杀线剂处理的禾谷孢囊，在经过4℃条件下放置8周之后，转至15℃条件下培养，第4天开始有二龄幼虫孵化，并呈上升趋势，第10天达到高峰，此后呈下降趋势，在第32天接近0，结束孵化，平均单孢囊孵出88.37条线虫。

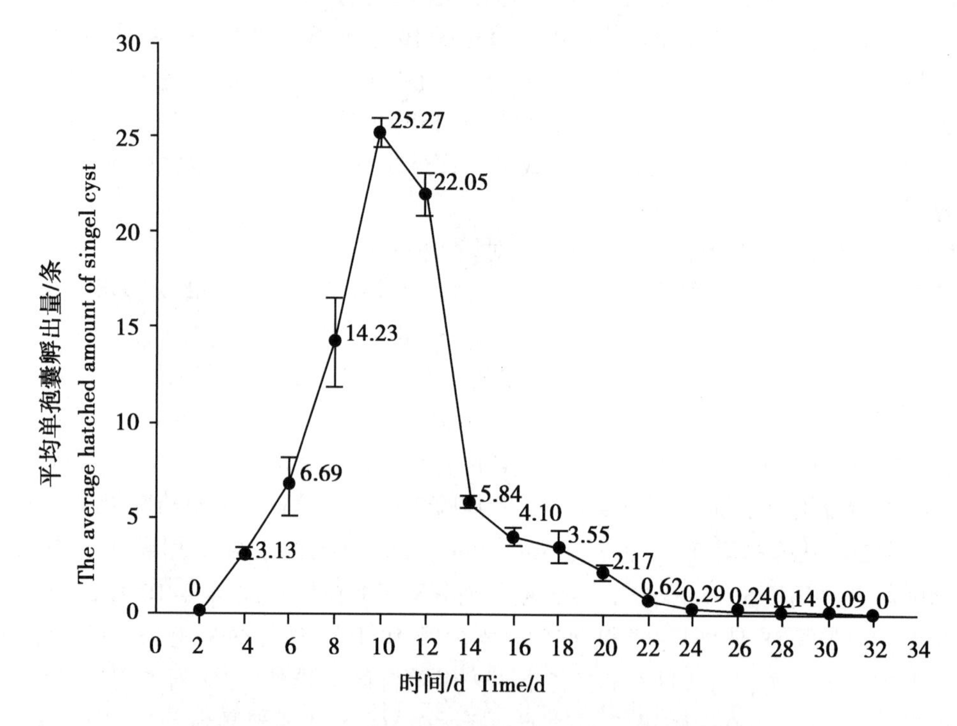

图1 未经杀线剂处理的孢囊4℃处理8周后的孵化特点

Fig. 1 Hatching characteristic of cysts after treatment eight weeks in 4℃ without nematicide treating

2.1.2 4℃预处理阶段杀线剂处理孢囊对线虫孵化的影响

由图2可知，在4℃低温处理时用40%噻唑膦乳油处理孢囊，各浓度处理线虫孵化率为26.17%～31.50%，空白对照处理线虫孵化率为33.14%，1 000、100mg/L处理线虫孵化率显著低于对照处理与10mg/L处理；噻唑膦溶剂对照处理各浓度处理线虫孵化率为30.21%～31.19%，1 500mg/L处理线虫孵化率显著低于100、10mg/L及空白对照处理，各溶剂浓度处理与对照处理线虫孵化率33.14%差异不大。说明在4℃低温预处理时40%噻唑膦乳油对孢囊线虫的孵化有一定的抑制作用。

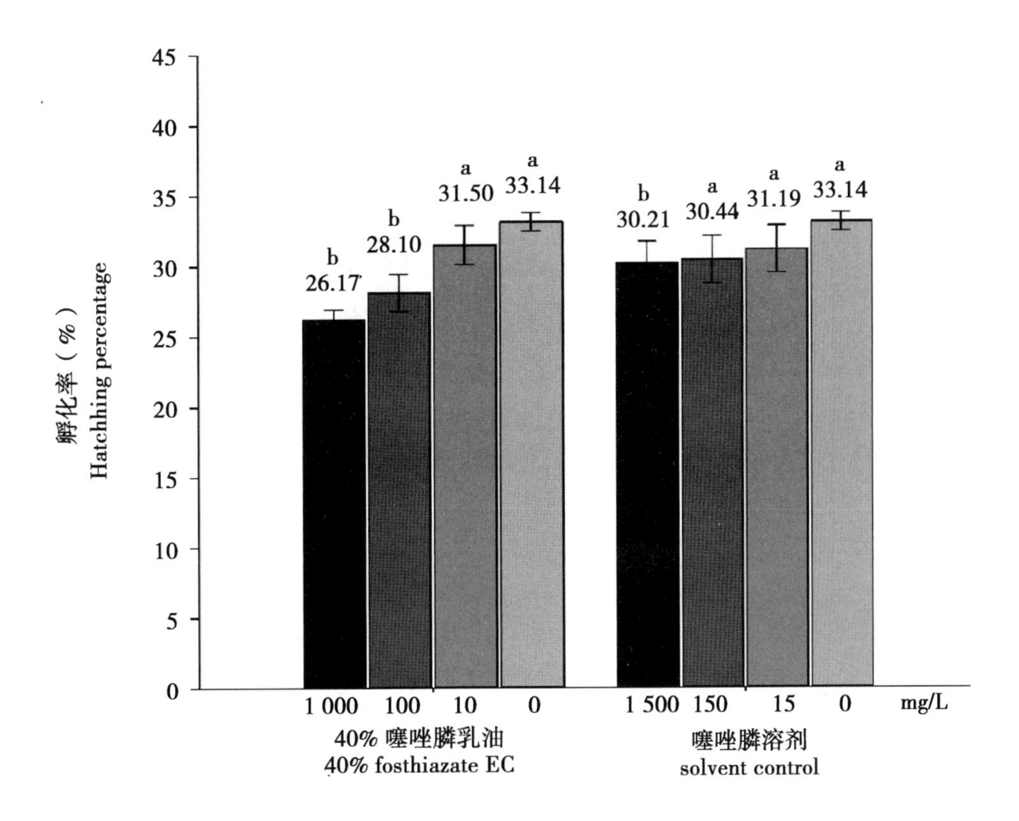

图2 噻唑膦在4℃预处理对禾谷孢囊线虫孵化的影响

Fig. 2 Hatching characteristic of *Heterodera avenae* in 4℃ with pretreatment by fosthiazate

由图3可知，用2%阿维菌素乳油在4℃低温处理时处理孢囊，各浓度处理线虫孵化率为3.20%～19.33%，空白对照处理线虫孵化率为33.14%，阿维菌素各浓度处理线虫孵化率均显著低于对照处理，并随阿维菌素浓度的降低线虫孵化率显著增加；阿维菌素溶剂对照处理各浓度线虫孵化率为25.48%～30.86%，与空白对照处理线虫孵化率33.14%差异不显著。说明2%阿维菌素乳油在4℃低温预处理时对孢囊线虫的孵化有显著地抑制作用。

由图4可知，用2%甲维盐乳油在4℃低温处理时处理孢囊，各浓度处理线虫孵化率为4.15%～18.56%，空白对照处理线虫孵化率为33.14%，甲维盐各浓度处理线虫孵化率均显著低于对照处理，并随甲维盐浓度的降低线虫孵化率显著增加；甲维盐溶剂对照处理各浓度线虫孵化率为21.66%～33.63%，且随溶剂浓度升高线虫孵化率降低，空白对照线虫孵化孵化率为33.14%，溶剂处理0.05mg/L处理与空白对照差异不显著，0.5mg/L、5mg/L处理显著低于空白对照处理。说明2%甲维盐乳油在4℃低温预处理时对孢囊线虫的孵化有显著地抑制作用，且其溶剂对线虫的孵化也有一定的抑制作用。

2.1.3 15℃孵化阶段杀线剂处理孢囊对禾谷孢囊线虫孵化的影响

由图5可知，用40%噻唑膦乳油在15℃孵化阶段处理孢囊，各浓度处理线虫孵化率为0.07%～0.27%，且各浓度处理间差异不显著，空白对照处理线虫孵化率为32.03%，显著

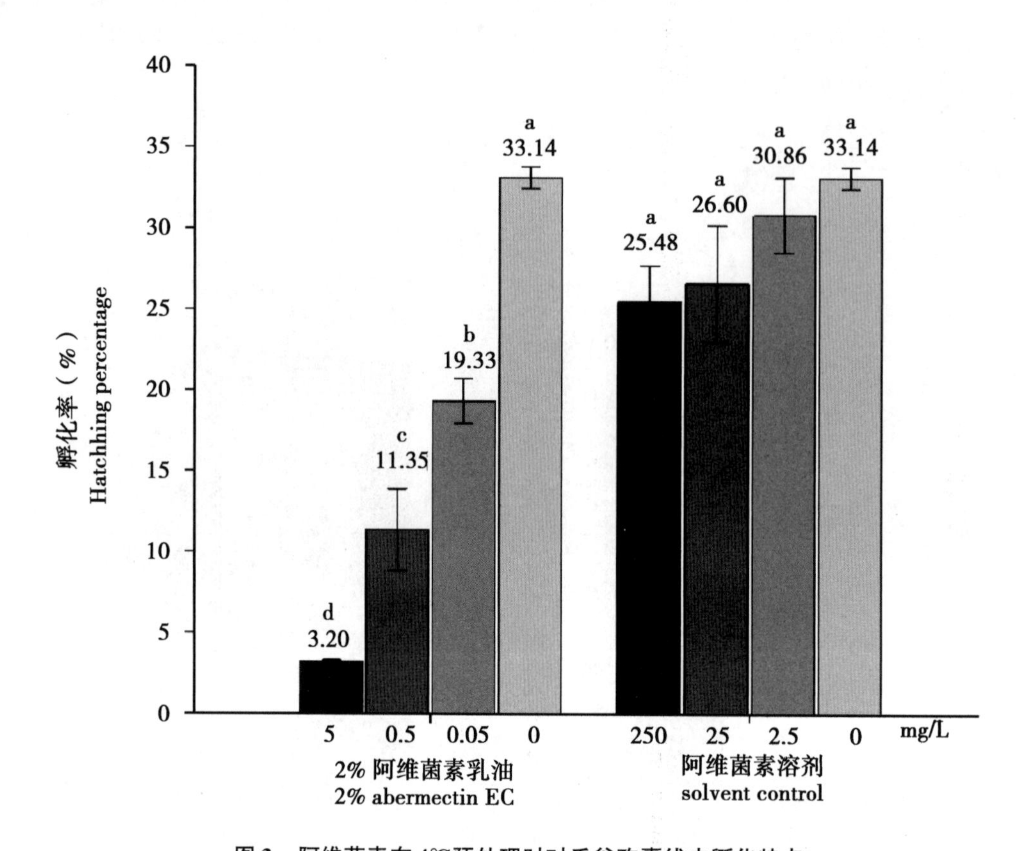

图 3　阿维菌素在 4℃预处理时对禾谷孢囊线虫孵化特点

Fig. 3　Hatching characteristic of *Heterodera avenae* in 4℃ with pretreatment by abermectin

高于噻唑膦各浓度处理线虫孵化率；噻唑膦溶剂处理各浓度线虫孵化率为 23.07% ~ 27.30%，且随浓度升高线虫孵化率降低，1 000mg/L 显著低于 100mg/L、10mg/L 处理，空白对照处理线虫孵化率为 32.03%，显著高于溶剂各浓度处理线虫孵化率。由此说明，在 15℃孵化时 40% 噻唑膦乳油对孢囊线虫孵化有显著地抑制作用。

由图 6 可知，用 2% 阿维菌素乳油在 15℃孵化阶段处理孢囊，各药剂浓度处理线虫孵化率为 0.22% ~1.01%，且随药剂浓度升高线虫孵化率降低，空白对照处理线虫孵化率为 32.03%，药剂各浓度处理显著低于空白对照处理；阿维菌素乳油溶剂各浓度处理线虫孵化率为 24.57% ~ 28.45%，空白对照处理线虫孵化率为 32.03%，溶剂各浓度处理间线虫孵化率差异不显著，且与空白对照处理差异不显著。由此说明：在 15℃孵化时 2% 阿维菌素乳油对孢囊线虫的孵化有显著地抑制作用。

由图 7 可知，用 2% 甲维盐乳油在 15℃孵化阶段处理孢囊，各药剂浓度处理线虫孵化率为 0.12% ~ 1.14%，且随药剂浓度升高线虫孵化率下降，空白对照处理线虫孵化率 32.03%，药剂处理各浓度线虫孵化率显著低于空白对照处理；甲维盐乳油溶剂各浓度处理线虫孵化率为 25.22% ~32.59%，随溶剂浓度升高呈下降趋势，空白对照处理线虫孵化率为 32.02%，溶剂 250mg/L 处理组线虫孵化率显著低于空白对照处理，25mg/L、2.5mg/L 处理与空白对照差异不显著。由此说明，在 15℃孵化时 2% 甲维盐乳油及其溶剂对孢囊线虫孵化有显著抑制作用。

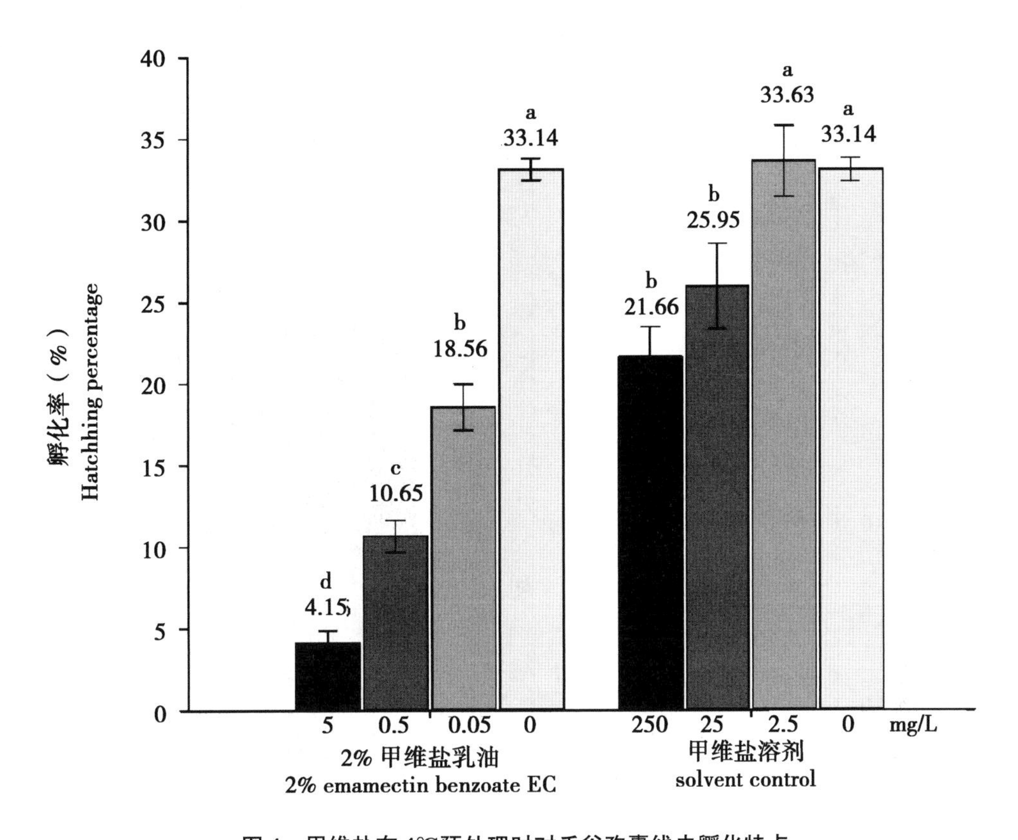

图4　甲维盐在4℃预处理时对禾谷孢囊线虫孵化特点

Fig. 4　Hatching characteristic of *Heterodera avenae* in 4℃ with pretreatment by emamectin benzoate

2.1.4　田间不同时期施药对小麦孢囊线虫病的防治效果

由下表可以看出，灌根施药处理在施药前试验地各小区孢囊基数存在差异，但其群体密度整体上分布比较均匀，空白对照处理孢囊基数显著低于2%阿维菌素乳油15kg/hm^{2}返青期处理、2%甲维盐乳油30kg/hm^{2}冬前期、返青期处理以及40%噻唑膦乳油5、7.5kg/hm^{2}返青期处理，且2%甲维盐乳油30kg/hm^{2}返青期处理以及40%噻唑膦乳油5、7.5kg/hm^{2}返青期处理孢囊基数显著高于2%甲维盐15kg/hm^{2}齐苗期处理，其他各处理间孢囊基数差异不显著。施药后，各处理孢囊数量发生较大变化，2%甲维盐30kg/hm^{2}冬前期处理孢囊数最多，与其齐苗期及空白对照处理差异不显著且显著高于其他处理，40%噻唑膦5kg/hm^{2}冬前期处理孢囊数最少，与2%阿维菌素30kg/hm^{2}齐苗期、返青期处理、2%甲维盐15kg/hm^{2}齐苗期、返青期处理、2%甲维盐30kg/hm^{2}返青期处理、40%噻唑膦5kg/hm^{2}返青期处理、7.5kg/hm^{2}冬前期、返青期处理差异不显著，且显著低于其他各处理，其他各处理间差异不显著。在孢囊减退率上，2%阿维菌素乳油30kg/hm^{2}返青期处理、2%甲维盐乳油30kg/hm^{2}返青期处理、40%噻唑膦乳油5kg/hm^{2}冬前期、返青期处理、7.5kg/hm^{2}冬前期、返青期处理孢囊减退率分别为57.64%、60.30%、59.34%、59.22%、58.40%、62.55%，孢囊减退率均高于50%，对孢囊增长有较显著的抑制效果；2%阿维菌素15kg/hm^2、2%甲维盐15kg/hm^{2}对孢囊也有较好的抑制效果，空白处理在小麦收获后土壤中孢囊数量增加了20.72%。比较各处理间的防效（校正孢囊减退率）后发现，2%阿维菌素乳油、2%甲维盐

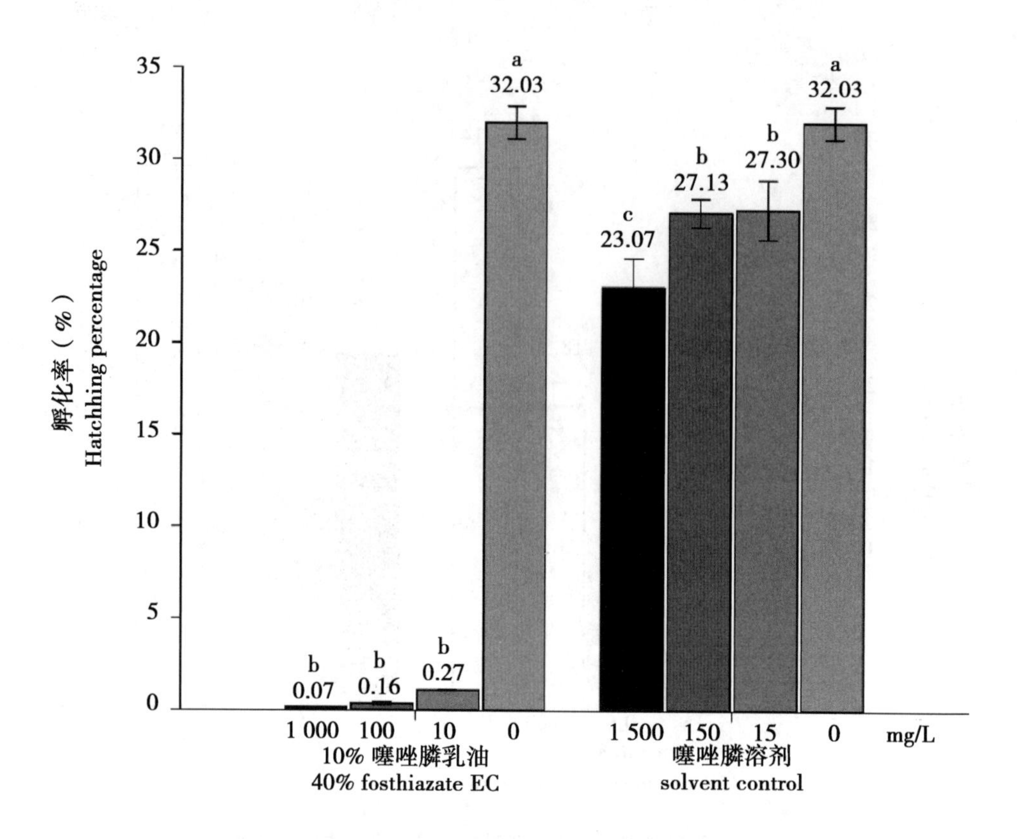

图 5　噻唑膦在 15℃孵化时处理对禾谷孢囊线虫孵化的影响

Fig. 5　Hatching characteristic of *Heterodera avenae* in 15℃ with treatment by fosthiazate

乳油、40% 噻唑膦乳油 3 种药剂在小麦返青期施药，2 个剂量均能达到较好的防治效果，其防效分别为 51.98%、63.57%、54.35%、65.49%、64.18%、67.02%，均大于 50%，而 40% 噻唑膦乳油 2 个剂量在小麦冬前期施药也能达到较好的防治效果，其防效为 64.77%、63.47%，且 2% 阿维菌素乳油 30kg/hm^2，2% 甲维盐乳油 30kg/hm^2 返青期施药及 40% 噻唑膦乳油 2 个剂量冬前期、返青期防效显著高于其他各处理。

由表 2 可以看出，种衣剂处理下在施药前田间孢囊分布比较均匀，阿维菌素颗粒剂 18.9kg/hm^2、28.35kg/hm^2 两个处理孢囊基数显著低于对照处理，其他各处理间差异不显著；施药后最终孢囊数量产生较大变化，空白对照处理孢囊数量显著高于药剂处理孢囊数，其他各处理间差异不显著；在孢囊减退率上，噻唑膦颗粒剂 30kg/hm^2 处理显著高于其他各处理，可达 48.21%，除此之外，噻唑膦颗粒剂 15kg/hm^2，阿维菌素颗粒剂 28.35kg/hm^2，阿维菌素种衣剂 2.5kg/100kg 种子处理之间差异不显著，且显著高于其他各处理，除空白对照外，阿维菌素颗粒剂 9.45kg/hm^2 孢囊减退率最低；在防效上，与孢囊减退率结果一致，防效最高的为噻唑膦颗粒剂 30kg/hm^2，可达 56.89%，其次为噻唑膦颗粒剂 15kg/hm^2，阿维菌素颗粒剂 28.35kg/hm^2，阿维菌素种衣剂 2.5kg/100kg 种子，其防效分别为 40.99%、43.01%、41.65%，防效最低的为阿维菌素颗粒剂 9.45kg/hm^2，仅为 22.75%。

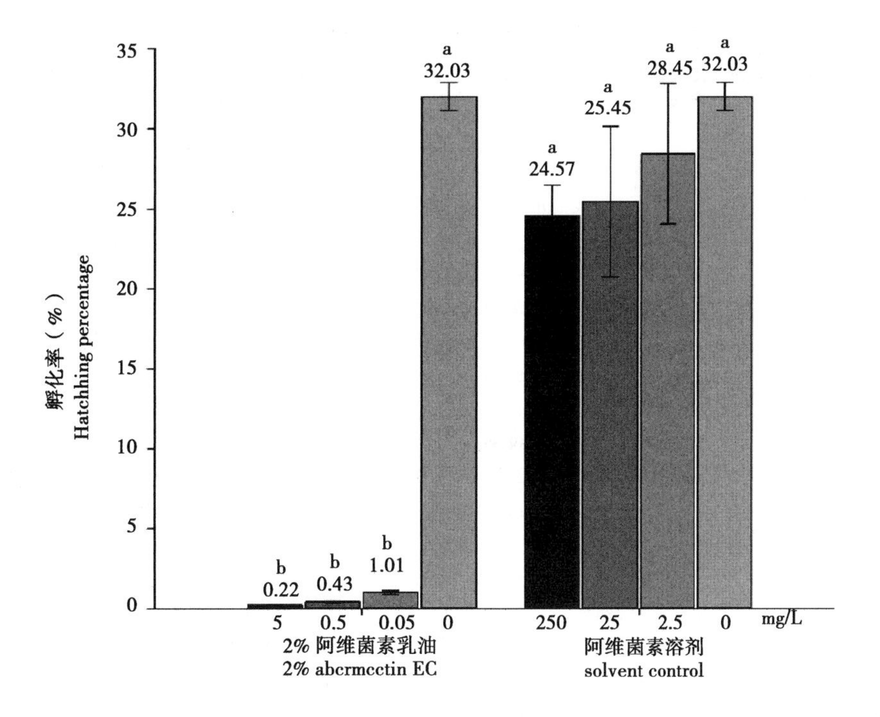

图6　阿维菌素在15℃孵化时处理对禾谷孢囊线虫孵化的影响

Fig. 6　Hatching characteristic of *Heterodera avenae* in 15℃ with treatment by abermectin

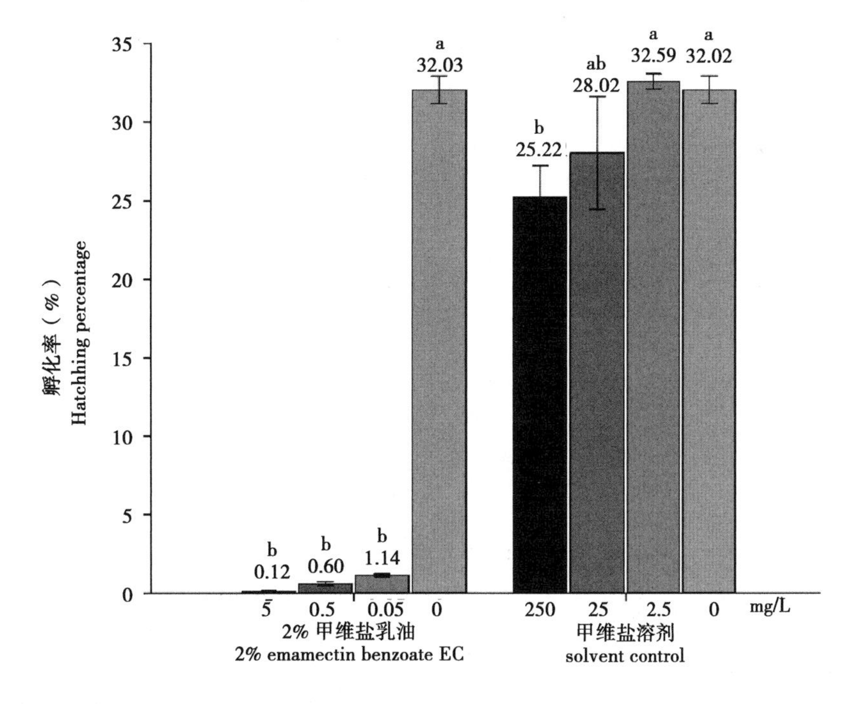

图7　甲维盐在15℃孵化时对禾谷孢囊线虫孵化特点

Fig. 7　Hatching characteristic of *Heterodera avenae* in 15℃ with treatment by emamectin benzoate

表 1　不同施药时期灌根处理对小麦孢囊线虫的防治效果

Table 1　Control of different time in the mode of root-irrigation to *Heterodera avenae* in the field

药剂 Nematicide	施药剂量 Dosage (kg/hm^2)	施药时期 Time	土壤中平均孢囊数（个/500g土）Average number of cysts in soil (cysts/500g · soil) 播种前 Preplanting	收获后 Post-harvest	繁殖系数 Reproduction index	孢囊减退率（%）Average cyst reducing rate	防治效果（%）Efficacy
2% 阿维菌素乳油 2% abermectin EC	15	齐苗期 Seeding stage	46.20 ± 5.22 abc	32.73 ± 5.34 bcdef	0.69 ± 0.02 e	31.42 ± 0.02 e	42.02 ± 0.94 d
		冬前期 Earlier winter stage	46.80 ± 12.80 abc	32.40 ± 5.81 bcdef	0.81 ± 0.02 bc	18.63 ± 0.02 gh	34.35 ± 0.55 e
		返青期 Green stage	58.80 ± 2.23 ab	31.80 ± 1.45 bcdef	0.56 ± 0.02 fg	44.14 ± 0.02 cd	51.98 ± 1.35 bc
	30	齐苗期 Seeding stage	49.20 ± 2.75 abc	29.00 ± 1.93 bcdefg	0.59 ± 0.01 f	40.65 ± 0.01 d	48.44 ± 1.28 c
		冬前期 Earlier winter stage	48.13 ± 7.08 abc	35.93 ± 4.52 bcd	0.78 ± 0.01 cd	21.86 ± 0.01 fg	33.88 ± 1.39 e
		返青期 Green stage	58.80 ± 2.23 ab	24.73 ± 2.08 cdefg	0.42 ± 0.01 h	57.64 ± 0.01 b	63.57 ± 0.62 a
2% 甲维盐乳油 2% emamectin benzoate EC	15	齐苗期 Seeding stage	36.07 ± 3.47 bc	27.93 ± 2.75 cdefg	0.76 ± 0.01 d	24.00 ± 0.01 f	34.94 ± 2.28 e
		冬前期 Earlier winter stage	42.07 ± 3.88 abc	33.53 ± 2.03 bcde	0.82 ± 0.01 bc	18.46 ± 0.01 gh	28.84 ± 3.53 f
		返青期 Green stage	47.67 ± 9.70 abc	23.87 ± 5.31 defg	0.52 ± 0.01 g	47.82 ± 0.01 c	54.35 ± 2.42 b
	30	齐苗期 Seeding stage	53.27 ± 10.57 abc	38.27 ± 7.07 abc	0.70 ± 0.01 e	30.32 ± 0.01 e	40.54 ± 0.44 d
		冬前期 Earlier winter stage	58.93 ± 7.58 ab	49.60 ± 6.30 a	0.85 ± 0.01 b	15.34 ± 0.01 h	26.71 ± 1.90 f
		返青期 Green stage	61.53 ± 5.81 a	26.00 ± 2.91 cdefg	0.40 ± 0.01 hi	60.30 ± 0.01 ab	65.49 ± 1.67 a
40% 噻唑膦乳油 40% fosthiazate EC	5	齐苗期 Seeding stage	48.80 ± 7.44 abc	34.33 ± 5.42 bcde	0.69 ± 0.03 e	30.58 ± 0.03 e	38.12 ± 0.82 de
		冬前期 Earlier winter stage	41.33 ± 6.68 abc	15.93 ± 2.60 g	0.41 ± 0.01 hi	59.34 ± 0.01 ab	64.77 ± 2.64 a
		返青期 Green stage	61.47 ± 8.89 a	21.87 ± 1.46 efg	0.41 ± 0.01 hi	59.22 ± 0.01 ab	64.18 ± 0.63 a
	7.5	齐苗期 Seeding stage	54.60 ± 1.51 abc	32.27 ± 1.20 bcdef	0.58 ± 0.01 f	41.62 ± 0.01 d	50.81 ± 0.68 bc
		冬前期 Earlier winter stage	46.47 ± 4.01 abc	19.27 ± 1.47 fg	0.42 ± 0.01 hi	58.40 ± 0.01 ab	63.47 ± 1.30 a
		返青期 Green stage	60.33 ± 5.66 a	20.87 ± 0.41 efg	0.37 ± 0.01 i	62.55 ± 0.01 a	67.02 ± 1.14 a
CK	—	—	34.87 ± 6.05 c	42.33 ± 6.67 ab	1.21 ± 0.01 a	-20.72 ± 0.01 i	—

注：表中数据为平均值 ± 标准误。同列数据后不同字母表示经 Duncan's 新复极差法检验差异显著（$P \leq 0.05$）。Note: The data are $\overline{X}$ ± SE. Data in the same column followed by the different levers indicate significant difference at $P \leq 0.05$ by DMRT.

表 2　不同施药方式处理对小麦孢囊线虫的防治效果

Table 2　Control of different applying mode to *Heterodera avenae* in the field

药剂 Nematicide	施药剂量 Dosage	土壤中平均孢囊数（个/500g 土）Average number of cyst in soil（cyst/500g · soil）		繁殖系数 Reproduction index	孢囊减退率（%）Average cyst reducing rate	防效（%）Efficacy
		播种前 Before sowing	收获后 Post-harvest			
3% 阿维菌素种衣剂	2kg/100kg 种子	30.93 ± 0.71 ab	21.20 ± 1.25 b	0.70 ± 0.02 d	30.00 ± 0.02 b	41.65 ± 1.44 b
	2.5kg/100kg 种子	26.47 ± 0.37 ab	20.40 ± 0.58 b	0.79 ± 0.01 c	20.33 ± 0.01 c	33.62 ± 1.59 c
0.5% 阿维菌素颗粒剂	9.45kg/hm^2	26.40 ± 3.42 ab	24.73 ± 3.23 b	0.93 ± 0.02 b	7.00 ± 0.02 d	22.75 ± 1.90 d
	18.9kg/hm^2	22.27 ± 2.68 b	17.53 ± 1.23 b	0.83 ± 0.01 c	17.00 ± 0.02 c	30.54 ± 1.20 c
	28.35kg/hm^2	22.80 ± 3.52 b	15.60 ± 2.05 b	0.68 ± 0.02 d	32.00 ± 0.02 b	43.01 ± 1.48 b
10% 噻唑膦颗粒剂	9kg/hm^2	30.00 ± 1.83 ab	23.93 ± 1.31 b	0.80 ± 0.01 c	20.43 ± 0.01 c	33.81 ± 1.08 c
	15kg/hm^2	30.53 ± 2.69 ab	21.40 ± 2.20 b	0.71 ± 0.02 d	28.99 ± 0.02 b	40.99 ± 1.11 b
	30kg/hm^2	31.47 ± 3.12 ab	16.13 ± 1.49 b	0.52 ± 0.01 e	48.21 ± 0.01 a	56.89 ± 1.27 a
CK	—	34.87 ± 6.05 a	42.33 ± 6.67 a	1.21 ± 0.01 a	-20.72 ± 0.01 e	—

注：表中数据为平均值 ± 标准误。同列数据后不同字母表示经 Duncan's 新复极差法检验差异显著($P \leq 0.05$)。Note: The data are $\bar{X} \pm SE$. Data in the same column followed by the different levers indicate significant difference at $P \leq 0.05$ by DMRT.

3　讨论

化学药剂的施用时机对药效的有效发挥具有较大的影响，选择关键时期用药可以在保证效果的同时达到减量施药的目的。Banyer & Fisher 将禾谷孢囊线虫的孵化过程分为 2 个阶段：一是幼虫在卵内的发育，二是破壳而出[17]。刘维志报道二龄幼虫可在秋季或初冬在小麦播种后少量侵染，越冬，但不能继续发育，次年春季继续发育[18]。彭德良等研究表明在华北麦区，仅有少量二龄幼虫侵入，翌年春天小麦返青后的低温可造成短时间内大量侵入[19]。因此，在华北麦区，理论上从小麦播种期到返青期均可施用药剂防治小麦孢囊线虫病。利用化学药剂破坏孢囊线虫孵化侵染过程中任何一个环节均能有效地降低其对小麦的为害[20]，但由于目前未见有报道对已经侵入根内的线虫有良好杀伤作用的药剂，因此，在华北防治小麦孢囊线虫施药时间不应晚于小麦返青线虫孵化侵染高峰期。

在施药的防治适期方面，已有研究只考察了单一时期施药对小麦孢囊线虫病的防治效果，而对于不同时期用药的效果是否存在差异未见报道。如吴绪金等在田间播种期施药，结果是 15% 涕灭威颗粒剂防病效果较好，其次是 5% 线敌颗粒剂和 10% 福气多颗粒剂[21]。裴世安等研究表明返青期施用 0.5% 阿维菌素颗粒剂对小麦孢囊线虫病有良好的防治效果[22]。本研究中，阿维菌素、甲维盐、噻唑膦在返青期灌根施药对孢囊线虫病均能够达到较好的防治效果，而噻唑膦在冬前期灌根施药也能够获得有效防治孢囊线虫病的效果，阿维种衣剂拌种处理以及阿维颗粒剂，噻唑膦颗粒剂处理下也有一定的防治效果。

同一药剂不同时期施药效果的差异可能与药剂在土壤中的分布和稳定性有关，同时也可能与孢囊的抵抗渗透能力变化有关。本研究设定了4℃低温时进行药剂处理和15℃孵化适温时进行药剂处理以模拟华北地区田间冬前施药与返青期施药，40%噻唑膦乳油及其助剂在4℃预处理时对孢囊线虫的孵化未见有较明显的影响，1.8%阿维菌素乳油、1%甲维盐乳油及其助剂明显抑制孢囊线虫的孵化，而在后期15℃处理时噻唑膦、阿维菌素、甲维盐对孢囊线虫的孵化均具有显著地抑制作用，这与田间试验中各药剂在返青期的防效普遍高于齐苗期和冬前期以及种衣剂拌种处理的结果是一致的，同时也表明孢囊这种结构前期对内部卵或幼虫的保护作用强，但到临近孵化时这种保护作用会减弱。由此，综合考虑轻简化施药以及减药增效的要求，建议在山东麦区发病较轻地块采用种衣剂处理种子的方式进行防治；在重病田，可结合播种前种衣剂处理或颗粒剂种—药—肥同播与返青期根据田间侵染情况采用阿维菌素、甲维盐或噻唑膦随返青水灌溉施药的施药方式进行防治。

田间防效往往是多种作用方式协同作用的结果。本研究虽然证实3种杀线剂可通过抑制孢囊的孵化来控制土壤中线虫的侵染繁殖，但它们对二龄幼虫的毒杀作用以及降低二龄幼虫侵染能力的作用也不容忽视。如有研究表明噻唑膦可以通过影响根结线虫搜寻侵染寄主的能力发挥作用[20]。返青期施药防效高于其他时期一方面与该时期二龄幼虫大量孵化，药剂可直接作用于二龄幼虫有关；另一方面可能与该时期二龄幼虫破囊而出时降低孢囊的抗逆性利于药剂进入发挥作用有关。五种药剂中，噻唑膦在冬前期施药也能够获得较好的防治效果，是否与其具有内吸活性，且具有干扰孢囊线虫化学感应、搜寻寄主的能力有关有必要进一步明确。此外，不同的施药时期，田间土壤的疏松程度不同，也会影响到灌根后药剂在土壤中的扩散与分布，对药剂能否顺利到达靶标所在土层发挥作用影响很大，如种衣剂可直接将药剂带入地下，更有利于药剂与靶标的接触，而灌根处理药剂存在被土壤吸附等原因，阻碍药剂药效发挥，需要更高的用量保证防效，这需要进一步研究与分析。另外，在生产中不应期待杀线剂完全控制孢囊线虫的发生，但在线虫侵染为害的薄弱时期，提前施药，可以降低化学农药的用量，抑制孢囊线虫的发生与为害，这对于提高农药利用率和节省用药成本，降低农药对环境造成的影响是十分有利的。

参考文献

[1] Meagher J W. World Dissemination of the Cereal-Cyst Nematode (*Heterodera avenae*) and Its Potential as a Pathogen of Wheat [J]. Journal of Nematology, 1977, 9-15.

[2] Wang M Z, Peng D L, Wu X Q. Study on a cyst nematode wheat disease, I. Identification of pathogen [J]. Journal of Huazhong Agricultural University (华中农业大学学报), 1991, 10 (4): 352-356.

[3] Chen P S, Peng D L. The Cereal-Cyst Nematode on wheat. [J] Plant Protection (植物保护), 1992, 18 (6): 37-38.

[4] Gao J, Wang Z H, Zhang S M, Research progress on cereal cyst nematode [J]. China Plant Protection (中国植保导刊), 2007, 27 (5): 10-13.

[5] Peng D L, Nicol J M, Li H M, et al. Current knowledge of cereal cyst nematode (*Heterodera avenae*) on wheat in China [C] //Cereal cyst nematodes: status, research and outlook. Proceedings of the First Workshop of the International Cereal Cyst Nematode Initiative. 2009: 29-34.

[6] Li H X, Liu Y E, Wei Z, et al. The detection of *Heteradera avenae* from the cereal field in Autonomous Regions of Tibet and Xinjiang [C]. Nematalagy Research in China, Vol. 4 (中国线虫学研究（第四卷)). 2012, 4: 164-165.

[7] Peng D L, Huang W K, Sun J H, et al. First report of cereal cyst nematode (*Heterodera avenae*) in Tianjin, China [C] //Nematalagy Research in China (Vol. 4) (中国线虫学研究(第四卷)), 2012, 4: 162 - 163.

[8] Li X H, Ma J, Chen S L. Effect of temperature on the hatch of *Heterodera avenge* field population [J]. Journal of Plant Protection (植物保护学报), 2012, 39 (3): 260 - 264.

[9] Wang M Z, Yan J K. Study on a cyst nematode wheat disease, Ⅱ. Hatching of *Heterodera avenae* Wollenweber. [J]. Journal of Huazhong Agricultural University (华中农业大学学报), 1993, 12 (6): 561 - 565.

[10] Prot J C. Migration of plant-parasitic nematodes towards plant roots [J]. Revuede Nématologie, 1980, 3 (2): 305 - 318.

[11] Feng Z Z. The key control technology and application of wheat cereal cyst nematode [D]. Zhengzhou: Henan Agricultural University, 2014.

[12] Zhen J W, Cheng H R, Fang Z D. Three kinds of separation methods on nematode cysts in soil and comprehensive evaluation [J]. Plant Protection (植物保护), 1995, 21 (1): 14 - 15.

[13] Giannakou Ioannis O, Karpouzas Dimitrios G, Anastasiades Ioannis et al. Factors affecting the efficacy of non-fumigant nematicides for controlling root-knot nematodes [J]. Pest Management Science, 2005, volos (10): 961 - 972 (12).

[14] Hao R, Huang W K, Liu C J, et al. Effect of seed-coatings on controlling cereal cyst nematode (*Heterodera avenae*) of wheat [J]. Plant Protection (植物保护), 2014 (1): 182 - 186.

[15] Zhang J, Li S X, Ren Y P. Effect of the cereal cyst nematode, *Heterodera avenge*, on wheat yield in Shandong Province [J], Journal of Plant Protection (植物保护学报), 2014, 41 (2): 242 - 247.

[16] Tang Q Y, Feng G M. Practical statistical analysis and DPS Data Processing System [M]. Beijing: Science Press (北京: 科学出版社), 2002.

[17] Banyer R J, Fisher J M. Effect of temperature on hatching of eggs of *Heterodera avenae* [J]. Nematologica, 1971, 17 (4): 519 - 534.

[18] Liu W Z. Plant Nematology [M]. Beijing: China Agriculture Press (北京: 中国农业出版社), 2000: 299 - 301.

[19] Zhang D S, Peng D L, Qi S H, et al. Characteristic of reproduce of cereal cyst nematode in wheat and its effect on development [J]. Plant Protection (植物保护), 1994, 20 (3): 4 - 6.

[20] Zou Y X, Li X H, Zhang X F, et al. Action mode of fosthiazate on southern root knot nematode, *Meloidogyne ihcognita* [J]. Journal of Hebei Agricultural University (河北农业大学学报), 2011, 34 (5): 78 - 81.

[21] Wu X J, Yang W X, Sun B J, et al. Effect of different nematicides on the controlling cereal cyst nematode and wheat growth [J]. Journal of Henan Agricultural Sciences (河南农业科学), 2007 (5): 57 - 60.

[22] Pei S A, Wang X, Geng L X, et al. Effects of different nematicides on cereal cyst nematode of wheat [J]. Plant Protection (植物保护), 2012, 38: 166 - 170.

三唑类与甲氧基丙烯酸酯类杀菌剂种子处理对小麦白粉病防效与产量的影响*

周洋洋**，杨　帅，杨代斌，袁会珠，闫晓静***
（中国农业科学院植物保护研究所，农业部有害生物综合治理重点实验室，北京　100193）

摘要：2014—2015 年，通过田间药效试验研究了三唑类杀菌剂与甲氧基丙烯酸酯类杀菌剂共计 13 种杀菌剂不同浓度处理下，对小麦白粉病防治效果和小麦产量的影响。结果表明，三唑类杀菌剂较低浓度处理而甲氧基丙烯酸酯类杀菌剂在中、高浓度有较好的防治效果和产量。其中 8% 吡唑醚菌酯微囊悬浮剂、6% 四氟醚唑水乳剂防效最佳分别为 64.06%、69.32%；8% 肟菌酯悬浮剂、3% 戊唑醇微囊悬浮种衣剂具有最佳增产率，分别为 30.48%、24.92%。8% 吡唑醚菌酯微囊悬浮剂与 3% 戊唑醇微囊悬浮剂与 8% 吡唑醚菌酯水乳剂、6% 戊唑醇悬浮剂比较有更高的防效与增产率。

关键词：三唑类杀菌剂；甲氧基丙烯酸酯类杀菌剂；防治效果；小麦白粉病；微囊悬浮剂

Influence of Seed - treatment with Fungicides of Triazole and Strobilurin on Control Effect against *Blumeria graminis* and Yield of Wheat

Zhou Yangyang, Yang Shuai, Yang Daibin, Yuan Huizhu, Yan Xiaojing
(*Institute of Plant Protection Chinese Academy of Agriculture Science. /Key Laboratory of Integrated Pest Management in Crops, MOA, Beijing* 100193, *China*)

Abstract: Influence of seed - treatment with triazole fungicides and strobilurin fungicides on control effect of wheat powdery mildew and yield of wheat were studied in this paper, 13 kinds of fungicides with different concentrations were used in this field trial, 2014—2015. The results generally shows that low dosage triazole fungicides, high and intermediate dosage strobilurin fungicides provided excellent control efficacy of wheat powdery mildew and higher rate of growth. 8% pyriproxyfen CS and 8% tetracoazole EW obtained the best control efficiency at 64.06% and 69.32 against wheat powdery mildew, respectively; 8% trifloxystrobin SC and 3% tebuconazole CS had the highest yield of wheat at the increase rate of 30.48% and 24.92%. Considering the influence of different formulation, 8% pyriproxyfen CS and 3% tebuconazole CS were better on efficiency and yield of wheat than that of 8% pyriproxyfen EW and 6% tebuconazole SC.

Key words: Triazole fungicides; Strobilurin fungicides; Control effect; Wheat powdery mildew; Capsule suspension

---

* 基金项目：农作物重大病虫害防控关键共性技术研发（2012BAD19B01）

** 第一作者：周洋洋，男，河南商丘人，硕士，主要从事热烟雾机对农药稳定性影响研究；E-mail：zhouyangyang0521@126.com

*** 通讯作者：闫晓静，副研究员，主要从事新型杀菌剂作用靶标研究以及杀菌剂使用技术研究；E-mail：xjyan@ippcaas.cn

20世纪70年代始末期以来，小麦白粉病逐渐发展为我国麦区的最重要病害之一[1]。白粉病在小麦幼苗期至成株期的均有发生，危害小麦叶片、叶鞘、茎秆与穗部[2]。病害发生可造成减产5%～30%[3]，目前三唑类、甲氧基丙烯酸酯类杀菌剂作为小麦白粉病防治药剂应用广泛[4]。三唑类杀菌剂可以通过叶面喷雾、拌种或撒施药土用于防治白粉病、锈病低量高效[5]，在实际生产中，以三唑类杀菌剂为活性成分的种衣剂已在防治植物病虫害中被广泛采用[6]，对禾谷类作物锈病、白粉病、纹枯病具有活性高和持效期长的特点，此外三唑类杀菌剂有植物生长调节作用，因此，合理严格的使用技术对于避免药害发生、充分发挥其功效，提高产量具有重要意义[7]。甲氧基丙烯酸酯类杀菌剂兼具保护和治疗作用，对禾谷类作物白粉病、锈病有良好的防治效果[8]。该类药剂具有内吸性强、杀菌谱广、良好的环境相容性特点，还能够延缓植物衰老，提高作物的产量和品质[9-10]。微胶囊农药作为一种新型剂型，由于壁材将有效成分与周围环境隔离，降低了降解速率，减轻药害延长了持效期[11]，对种子包衣处理长期有效防止病虫草害提供了有意义的手段[12]。本文以小麦为靶标作物，以三唑类与甲氧基丙烯酸的酯类杀菌剂为主要试验药剂，研究两类杀菌剂不同浓度、不同剂型种子包衣在大田条件下对小麦白粉病防治效果以及小麦产量的影响，为两类杀菌剂的科学合理使用及管理提供理论依据。

1 材料与方法

1.1 试验材料

1.1.1 小麦品种

矮抗58（河南省新乡市河南科技学院选育，冬小麦品种）。

1.1.2 试验药剂

12.5%硅噻菌胺悬浮剂（美国孟山都公司）；8%咪酰胺水乳剂（中国农业科学院植物保护研究所配制）；6%戊唑醇悬浮剂（拜耳作物科学（中国）有限公司）；3%戊唑醇微囊悬浮剂（中国农业科学院植物保护研究所配制）；6%四氟醚唑水乳剂（意大利意赛格公司）；3%苯醚甲环唑（南京红太阳集团有限公司）；12.5%粉唑醇悬浮剂（中国农业科学院植物保护研究所配制）；8%醚菌酯（青岛奥迪斯生物科技有限公司）；8%嘧菌酯（中国农业科学院植物保护研究所配制）；8%吡唑醚菌酯（中国农业科学院植物保护研究所配制）；8%吡唑醚菌酯水乳剂（中国农业科学院植物保护研究所配制）；8%丁香菌酯（吉林八达农药有限公司）；8%肟菌酯（拜耳作物科学（中国）有限公司）。

1.2 试验方法

1.2.1 种子包衣及药剂用量

于播种前进行小麦种子包衣，具体用量与药种比如表1。

Table 1 The ratio of medicament and seed and actual dosage of different dispose

序号	药剂	药种比	所需用量（g）
1	6%戊唑醇悬浮剂	1：500	30
		1：1 000	15
2	3%戊唑醇 微囊悬浮种衣剂	1：600	25
		1：1 200	12.5

（续表）

序号	药剂	药种比	所需用量（g）
3	6%四氟醚唑水乳剂	1∶500	30
		1∶1 000	15
4	3%苯醚甲环唑水乳剂	1∶200	75
		1∶400	37.5
5	12.5%粉唑醇悬浮剂	1∶500	30
		1∶1 000	15
6	8%醚菌酯悬浮剂	1∶100	150
		1∶200	75
		1∶400	37.5
7	8%嘧菌酯悬浮剂	1∶100	150
		1∶200	75
		1∶400	37.5
8	8%吡唑醚菌酯微囊悬浮种衣剂	1∶100	150
		1∶200	75
		1∶400	37.5
9	8%吡唑醚菌酯水乳剂	1∶100	150
		1∶200	75
		1∶400	37.5
10	8%丁香菌酯悬浮剂	1∶100	150
		1∶200	75
		1∶400	37.5
11	8%肟菌酯悬浮剂	1∶100	150
		1∶200	75
		1∶400	37.5
12	12.5%硅噻菌胺悬浮剂	1∶400	37.5
13	8%咪鲜胺水乳剂	1∶100	150
		1∶200	75

1.2.2　三唑类、甲氧基丙烯酸酯类包衣处理

试验在中国农业科学院植物保护研究所河南新乡综合试验基地进行，于2014年10月18日机播播种，亩播种量为15kg，每处理试验面积约500m^2（167m×3m）。

1.2.3　三唑类、甲氧基丙烯酸酯类种子包衣对小麦白粉病防治效果

在空白对照明显发病时，调查计算两类种衣剂不同包衣剂量下对白粉病的防治效果。

1.2.4　三唑类、甲氧基丙烯酸酯类包衣处理对小麦产量的影响

在小麦成熟时，测定不同处理的小麦产量，计算增产率。

2　结果与分析

2.1　三唑类、甲氧基丙烯酸酯类种子包衣对小麦白粉病的防治效果

Table 2　The control effect of coating under different medicament and dosage

序号	药剂	药种比	防治效果（%）
1	6% 戊唑醇悬浮剂	1∶500	41.28
		1∶1 000	64.20
2	3% 戊唑醇微囊悬浮种衣剂	1∶600	64.17
		1∶1 200	67.07
3	6% 四氟醚唑水乳剂	1∶500	67.62
		1∶1 000	69.32
4	3% 苯醚甲环唑水乳剂	1∶100	45.74
		1∶200	60.14
5	12.5% 粉唑醇悬浮剂	1∶500	45.70
		1∶1 000	56.67
6	8% 醚菌酯悬浮剂	1∶100	52.28
		1∶200	43.19
		1∶400	29.20
7	8% 嘧菌酯悬浮剂	1∶100	53.39
		1∶200	50.45
		1∶400	43.22
8	8% 吡唑醚菌酯微囊悬浮种衣剂	1∶100	64.06
		1∶200	45.21
		1∶400	37.46
9	8% 吡唑醚菌酯水乳剂	1∶100	53.56
		1∶200	41.93
		1∶400	26.21
10	8% 丁香菌酯悬浮剂	1∶100	41.43
		1∶200	38.35
		1∶400	27.12

（续表）

序号	药剂	药种比	防治效果（%）
11	8%肟菌酯悬浮剂	1：100	41.61
		1：200	39.28
		1：400	23.21
12	12.5%硅噻菌胺悬浮剂	1：400	40.06
13	8%咪鲜胺水乳剂	1：100	53.71
		1：200	46.14

由田间药效调查数据得知，虽然时间跨度较长，三唑类、甲氧基丙烯酸酯类种子包衣处理对小麦后期特别是抽穗到成株期白粉病仍具有一定的防治效果，总体而言，三唑类杀菌剂具有更高的防效。

甲氧基丙烯酸酯类杀菌剂防效均具有随着药剂浓度的增大而增大的特点。从表中得知，8%吡唑醚菌酯微囊悬浮剂具有最高防效，可达64.06%；8%醚菌酯悬浮剂、8%嘧菌酯悬浮剂、8%吡唑醚菌酯水乳剂高浓度处理下，防效均可达50%以上；8%吡唑醚菌酯水乳剂与8%吡唑醚菌酯微囊悬浮剂比较，后者防效低中高浓度处理分别提高11.25%、3.28%、10.5%，其原因可能是后者具有良好缓释作用从而具有更久持效期，在作物生长后期，药剂仍具有一定浓度防治病害。

三唑类杀菌剂防效均具有随着药剂浓度的降低而增大的特点。从表中得知，6%四氟醚唑水乳剂具有最高防效，可达69.32%；6%戊唑醇悬浮剂、3%戊唑醇微囊悬浮剂、3%苯醚甲环唑水乳剂、12.5%粉唑醇低浓度处理均可达到60%以上防效；6%戊唑醇悬浮剂与3%戊唑醇微囊悬浮剂处理比较，尽管戊唑醇含量降低一半，防效却有所提高，且3%戊唑醇微囊悬浮剂高浓度处理防效也可达到64.17%，与6%戊唑醇悬浮剂比较防效提高22.89%，与低浓度比较无显著差异，可能与其具有缓释作用的特点有关，因此，后者具有更高的安全性和更好的防治效果。

其他类别的杀菌剂，如12.5%硅噻菌胺悬浮剂防效为40.06%、8%咪酰胺水乳剂防效最高可达53.71%，对后期白粉病有一定的效果，需进一步研究。

2.2 三唑类、甲氧基丙烯酸酯类种子包衣对小麦产量的影响

Table 3 The wheat yield and rate of growth under the different medicament and dosage coating dispose

序号	药剂	药种比	亩产量（kg）	增产率（%）
1	6%戊唑醇悬浮种衣剂	1：500	447.9	3.26
		1：1 000	522.4	20.46
2	3%戊唑醇微囊悬浮种衣剂	1：600	533.3	22.96
		1：1 200	541.8	24.92
3	6%四氟醚唑水乳剂	1：500	425.3	-1.93
		1：1 000	465.5	7.33

（续表）

序号	药剂	药种比	亩产量（kg）	增产率（%）
4	3%苯醚甲环唑水乳剂	1：100	478.7	10.37
		1：200	532.9	22.87
5	12.5%粉唑醇悬浮剂	1：500	407.7	-5.99
		1：1 000	535.1	23.39
6	8%醚菌酯悬浮剂	1：100	509.0	17.36
		1：200	516.5	19.07
		1：400	438.5	1.09
7	8%嘧菌酯悬浮剂	1：100	469.7	8.30
		1：200	561.1	29.37
		1：400	442.4	2.01
8	8%吡唑醚菌酯微囊悬浮剂	1：100	489.4	12.84
		1：200	528.3	21.82
		1：400	462.7	6.69
9	8%吡唑醚菌酯水乳剂	1：100	416.8	-3.91
		1：200	487.1	12.31
		1：400	392.5	-9.50
10	8%丁香菌酯悬浮剂	1：100	431.8	-0.43
		1：200	479.4	10.53
		1：400	466.0	0.34
11	8%肟菌酯悬浮剂	1：100	420.8	-2.98
		1：200	487.3	12.35
		1：400	565.9	30.48
12	12.5%硅噻菌胺悬浮剂	1：400	495.5	14.26
13	8%咪鲜胺水乳剂	1：100	501.0	15.52
		1：200	457.2	5.41
14	CK		433.7	

小麦测产所得亩产量结果如表，总体看来，甲氧基丙烯酸酯类具有更好的增产效果。从表中得知，三唑类杀菌剂处理小麦亩产量范围则在407.7～541.8kg。各三唑类药剂处理均具有随药剂浓度的降低而产量增加的特点，较高浓度处理的对小麦产量没有较强的增产效应，而低浓度处理则均具有较为明显的增产作用。就药剂种类分析，戊唑醇、粉唑醇、苯醚甲环唑具有较好的增产效应，增产率可达20%以上；6%戊唑醇悬浮剂与3%戊唑醇微囊悬浮剂比较，后者在较高浓度处理下，也能达到20%以上增产率，具有较高的安全性与增产

效果，表现出了良好的剂型优势，其原因可能是微囊悬浮剂缓慢释放药剂的剂型特点，能够更加持久的发挥对白粉病抑菌杀菌作用的同时，可以更充分发挥其对植物生长的生理调节机制，抵抗不良环境伤害，协调植物地上部与地下部营养生长于生殖生长关系，从而提高经济产量。

甲氧基丙烯酸酯类杀菌剂处理小麦亩产量范围为392.5～561.1kg。其中8%醚菌酯悬浮剂、8%嘧菌酯悬浮剂、8%吡唑醚菌酯微囊悬浮剂、8%吡唑醚菌酯水乳剂、8%丁香菌酯悬浮剂均有随着药剂浓度的增加，小麦的亩产量有先增加后降低的规律，而中间浓度处理有最高的增产率；此外，肟菌酯则随浓度的降低而亩产量增大。各甲氧基丙烯酸酯类杀菌剂处理对比得知，8%肟菌酯悬浮剂具有最强的增产效果，增产率可达30%以上；8%醚菌酯悬浮剂、8%嘧菌酯悬浮剂、8%吡唑醚菌酯微囊悬浮剂则具有较好的增产效果，增产率可达20%左右；8%吡唑醚菌酯微囊悬浮剂与8%吡唑醚菌酯水乳剂两种不同剂型增产率比较得知，微囊悬浮种衣剂具有更好的增产作用，其原因可能是微囊悬浮种衣剂发挥了缓释特性，对外界病害（如：白粉病等）具有更持久的抵抗性，另一方面可以发挥甲氧基丙烯酸酯类杀菌剂诱导禾谷类作物生理变化的特性，抑制乙烯生物合成、提高内生细胞分裂素、增强植物对二氧化碳吸收、提高对氮的吸收利用率，从而延缓作物衰老，获得更好的增产效果。

其他类别的杀菌剂，如12.5%硅噻菌胺悬浮剂、8%咪酰胺水乳剂最高增产率在15%左右，有一定的增产效果，需进一步研究。

3 讨论

本试验以无药剂处理麦种作为空白对照，使用了三唑类杀菌剂、甲氧基丙烯酸酯类杀菌剂等不同剂型药剂共计13种，设定了不同药剂浓度处理麦种，对小麦生长后期白粉病防治效果、小麦产量进行了研究。从防效数据得知，三唑类杀菌剂低浓度处理与甲氧基丙烯酸酯类高浓度处理具有较好的防效；在较好的防治好白粉病影响的同时，保证了小麦生长健康、籽粒充实，小麦每平方米的生长密度、小麦麦穗数、千粒重等产量指标未受到更大的影响，进而为亩产量的增加，提供一定的依据。就小麦产量角度而言，三唑类杀菌剂在较低浓度处理下，增产效果更佳而甲氧基丙烯酸酯类杀菌剂则在高、中浓度有较高的增产率，其原因可能在于不同类别杀菌剂刺激小麦生长、增产的最佳浓度不同，三唑类较低而甲氧基丙烯酸酯类杀菌剂较高；就剂型特点而言，8%吡唑醚菌酯微囊悬浮剂、3%戊唑醇微囊悬浮剂各浓度梯度处理与同含量的吡唑醚菌酯水乳剂、更高含量的戊唑醇悬浮剂由于微囊悬浮剂控释性、稳定性的特点，能够在较长的小麦生长期内保持防治作用与刺激增产的作用，从而得到了比常规剂型更佳的防治效果与增产率。

参考文献

[1] Shao Z R, Liu W C. The present situation and countermeasures of wheat powdery mildew in our country [J]. Chinese Agricultural Science Bulletin, 1996, 12 (6): 21 – 23.

[2] Xia X Q, Yao G. The identification, trend and control method of wheat powdery mildew [J]. Technology and Market, 2007 (5): 34.

[3] Guo X S, Xiong Z Z, Fu Y S, et al. Research the harm of wheat powdery mildew and comprehensive prevention and control [J]. Shanghai Agricultural Science and Technology, 2006 (3): 88 – 89.

[4] Bi Q Y, Ma Z Q. Sensitivity of *Blumeria graminis* f. sp. *tritici* to strobilur in fungicides and their cross – resistance [J]. Acta Phytopathologica Sinica, 2012, 42 (3): 315 – 318.

[5] Zhou Z Y, Li C C, Gao T C, et al. Research progress on triazole fungicides [J]. Journal of AnHui Agricultural Sciences, 2008, 36 (27): 11 842 - 11 844.

[6] Yan X J, Xu Y, Gao Y Y, et al. Safety assessment of Seed - treatment with Triadimef on and Difenoconazole to Wheat Seeding Growth [J]. Pesticide Science and Amdinistration, 2013, 34 (1): 52 - 57.

[7] Liu C D, Wang P S, Wang J Q, etal. Research advances in triazole fungicides and used to control wheat diseases [J]. Journal of Shandong Agricultural University (natural science), 2005, 36 (1): 157 - 160.

[8] Zhang G S. Current status of application, development and prospect of strobin fungicides [J]. Pesticide Science and Administration, 2003, 24 (12): 30 - 34.

[9] Luo Y P, Li Y X, Zhao P L, et al. The progresses of research on strobilurin fungicides [J]. Sciencepaper Online, 2006, 8 (1): 20 - 26.

[10] Si B B, Yang Z. Studies on mechanism and resistance to strobilur in fungicides [J]. World Pesticides, 2007, 29 (6): 5 - 9.

[11] Zhang Y B. Review China's pesticide microcapsules [J]. China Agrochemicals, 2008, 4 (2): 36 - 39.

[12] Cao Y F, Zhong S L, Cao X M, et al. Research progress on pesticide capsule suspension [A]. 2009 pesticide preparation processing and fertilizer application technology exchange meeting.

Chemical control of *Botryosphaeria dothidea* causing canker disease on Chinese hickory (*Carya cathayensis*) according the Spore dispersal and canker development patterns*

D. J. Dai[a,b] · L P. Sun[c] · B. C. Xu[c] · Y. P. Wang[a] · H. D. Wang[b] · C. Q. Zhang[a]
(*a. Department of Crop Protection, Zhejiang Agriculture and Forest University, Lin; an* 311300, *China*; *b. Institute for the Control of Agrochemicals of Zhejiang Province, Hangzhou* 310020; *c. The Forest Pest Control Station of Hangzhou Municipal, Hangzhou* 310009, *China*)

Abstract: Understanding the biology and epidemiology is essential for developing appropriate management strategies to control canker caused by *Botryosphaeria dothidea* which is the most important disease that threaten the production of Chinese hickory. Ascospores were not captured throughout the study period in 2009—2012 and conidia are found to be the main source of inoculums on Chinese hickory. None conidium was observed until March 31, in 2009, 2011, 2012 and until April 7 in 2010. The late spring to early summer (May to June) is the main spore dispersal period. The number of spores trapped during May 12 to 26 May was 43. 9%, 52. 8%, 51. 4%, 49. 2% for 2009, 2010, 2011, and 2012, respectively. Through comparisons of the patterns of canker lesion multiplication and spore dispersal, we speculated that *B. dothidea* overwintered on hickory trees, including the retained lesions and infections remained latent until suitable conditions for symptom development. A total of 107 single-conidium isolates of *B. dothidea* were further tested for their sensitivities to trifloxystrobin. Trifloxystrobin had weak activity against the mycelial growth of *B. dothidea*, but during stage of spore germination, the frequency distribution of the EC_{50} values for 107 isolates was a unimodal curve, ranging from 0. 76mg/L to 36. 7mg/L with an average of 9. 73 ± 1. 13mg/L. Thus, these sensitivity data could be used as a baseline for monitoring the shift of sensitivity in *B. dothidea* populations. Control of canker disease on Chinese hickory by different fungicide programs indicated that the time of the first spray inhibiting the mycelial growth of *B. dothidea* is very important which should be done before the initiation of canker lesion expansion and far before the initiation of spore dispersal. And, the following two sprays should adopt fungicides to inhibit mainly the spore germination and somewhat the mycelial growth.

Key words: *Botryosphaeria dothidea*; *Carya cathayensis*; spore dispersal; canker development; control efficacy; difenoconazole; trifloxystrobin; baseline sensitivity

1 Introduction

Chinese hickory (*Carya cathayensis* Sarg.), endemic from China, is known for its distinctive fragrance and the high nutritional value of its nuts (Wang and Cao, 2005; Zhu et al., 2008; Ma et al., 2009). Currently, more than 15 000 hm^2 of Chinese hickory are cultivated, correspond-

* Corresponding author: Zhang Chunqing; E-mail: cqzhang@ zafu. edu. cn

ing to $ 70 000 000 per year in Zhejiang province alone. Canker disease caused by *Botryosphaeria dothidea* (Moug. ex Fr.) Ces. & De Not (Zhang and Xu, 2011) is the most important disease that threatens the production of Chinese hickory. Nearly 90% of orchards in Zhejiang and Anhui Provinces were seriously affected by this disease. Symptoms were similar to those of canker of *C. illinoinensis* (Sinclair and Lyon, 2005); small, elliptical lesions developed on bark at points of infection and then enlarged to form large, sunken, elongated cankers. The cankers coalesced forming large diffuse areas of blighted tissue, which turned black (Fig. 1).

Fungal species of the family *Botryosphaeriaceae* are long known as pathogens to cause cankers on a variety of woody hosts such as avocado (*Persea americana*), grapevine (*Vitis vinifera*), almond (*Prunus dulcis*), citrus (*Citrus* spp.), and plum (*Prunus* spp.) (McDonald and Eskalen, 2011; Michailides and Morgan, 1993; Urbez-Torres et al., 2006). They infect the host plants through wounds, breach the outer bark, and then colonize the phloem, vascular cambium and xylem, resulting in disruption in the flow of nutrients and water in the diseased plants. This blockage causes weakening and decay of the wood at the infection sites (Slippers et al., 2007). Members of the *Botryosphaeriaceae* often produce spores in two types of fruiting bodies called perithecia (ascospores) and pycnidia (conidia), representing the teleomorphic and anamorphic stages of the fungus, respectively. The anamorphic asexual stage of the *Botryosphaeriaceae* is commonly observed in nature and pycnidia can be observed protruding from the bark in or around the canker tissue as well as on surrounding dead bark and twigs (Jacobs and Rehner, 1998).

B. dothidea is an ascomycete fungus that is very difficult to control once it has been established in a pistachi orchard because it produces pycnidia within the bark where fungicides cannot reach. On pistachi, it overwinters on retained rachises, fruit mummies, petioles, and cankers of blighted shoots and panicles (Michailides, 1991). No ascocarps of the pathogen have been discovered in pistachio orchards and pycnidiospores are believed to be the main source of inoculum (Ntahimpera et al., 2002). On Chinese hickory, canker lesions caused by *B. dothidea* occur mainly on tree trunk and big branches but not observed on leaves, nuts, and panicles (Zhang and Xu, 2011). However, little is known about the epidemiology of this disease and the mode of inoculum dispersal. Also, there are no field-scale studies on dispersal patterns of spores of *B. dothidea* in China. These factors limit our understanding of the epidemiology of the Chinese hickory canker pathogen and make the control efficacy for *B. dothide* on Chinese hickory is very unsatisfied. Cultural practices, such as pruning and erasions of canker lesions are laborious and bad efficacy and therefore are unacceptable for farmers until present. The application of fungicides is the main strategy using in practices. However, the best acquired control efficacy was only about 60% through spraying of difenoconazole, a demethylation-inhibitor (DMI) fungicide (Claire et al., 2015), for three times from middle of April to early May (Zhang et al., 2011). In order to develop more effective management strategy, the current study was conducted: (a) to determine the dynamics of the canker disease incidence and the seasonal abundance of spores of *B. dothide* in Chinese hickory groves; (b) to build the baseline sensitivity of *B. dothide* from Chinese hickory to trifloxystrobin, a Qo inhibiting (QoIs) fungicide; and (c) assess the control efficacy in Chinese hickory through different fungicide application programs.

2 Materials and Methods

2.1 *Spore trapping*

Spore traps (JIADUO GROUP, Henan, China) were operated from 10 March through19 November during 2009 to 2012 in Tuankou county, Zhejiang province where is the main production region of Chinese hickory. The experimental sites were naturally infected. No fungicides were applied during the experiments and all the other practices, such as application of fertilizers, were done as normal. Spore traps were located in the middle of selected commercial Chinese hickory forests. Two JDBZ1 spore traps (1.05m × 1.05m × 2.0m) were operated at an air sampling speed of 0.5 m/sec with the intake orifice 1.6 m above the ground. Spore trap slides preparation and mounting was done according to the instructions for the JDBZ1 sampler. In each trap, four glass microscope slides (26 by 76 mm) coated with silicone oil (Merck) were played in and slides was changed weekly and returned to the laboratory. Slides were examined under a light microscope at 40′ magnification and spores were counted. *B. dothide* was identified based on morphological characters (shape, size, color, and absence or presence of septa) as previously described (Urbez-Torres et al., 2006).

2.2 *Multiplication dynamics of canker lesions*

Expansion dynamics of canker lesions were surveyed while spore trapping was carried out. In the experimental sites, the number and severity of canker lesions on 37 Chinese hickory plants randomly chosen was surveyed weekly. The canker lesion of the trees was scored 1, 3, 5, and 7 (Zhang et al., 2011), where class 1 represented the diameter of the canker lesion < 0.5cm, 3 represented the diameter of the canker lesion is 0.5 ~ 1.99 cm, 5 represented the diameter of the canker lesion is 2.0 ~ 4.99 cm, 7 represented the diameter of the canker lesion is > 5.0 cm. The canker lesion index (CLI) of each tree was calculated as following: CLI (%) = (Σ (number of lesions of each class × each evaluation class) /total number of lesions × 7) × 100. The CLI value of 1 March was adopted as the background value of this year (CLI_0). Increased CLI value = CLI value or a survey-CLI_0.

2.3 *Collection of isolates for determination of sensitivity to trifloxystrobin*

During 2010 to 2013, isolates from diseased trees were collected from Lin'an, Tonglu and Chun'an in Zhejiang province. A total of 107 single-conidium isolates of *B. dothide* were collected from 23 Chinese hickory orchards with no history of use of Q_0 center inhibitors (Q_0Is). Orchards were separated, at least, 20 km from each other. About 5 isolates from each orchard, obtained on potato dextrose agar (PDA), were identified on the basis of their conidium morphology and DNA sequencing of the internal transcribed spacer (ITS) region with primers ITS1/ITS4 as described previously (White et al., 1990; Zhang and Xu, 2011). Isolates were kept on PDA slants at 4℃.

2.4 *Determination of sensitivity to trifloxystrobin*

Technical-grade of trifloxystrobin provided by the Institute of Pesticide, Zhejiang University (Hangzhou, China) was dissolved in the analytical grade acetone (Sangon Company, Shanghai, China) to prepare the stock solutions which were stored at 4℃ in the dark. The sensitivity was assessed for the inhibition of both mycelial growth and spore germination respectively on PDA and

2.0% water agar (WA) amended with a series of concentrations of trifloxystrobin (Zhang et al., 2008). Salicylhydroxamic acid (SHAM), a specific inhibitor of alternative oxidase (Avila-Adame and Köller, 2003), at a concentration of 100mg/L was added to each trifloxystrobin concentration treatment. The PDA or WA plates without fungicides but 100mg/L SHAM were used as the controls. Three replication plates per isolate-fungicide-concentration combination were used. For inhibiting growth, the PDA plates were incubated for 5 days at 25℃, and the diameter of each colony was measured in two perpendicular directions, with the original mycelial plug diameter (5 mm) subtracted; For inhibiting germination, conidia were allowed to germinate at 25℃ for 14 h. Germination was quantified at three sites by counting 100 conidia per site. A conidium was scored germinated if the germ tube had reached at least half the length of the conidium. The effective fungicide concentration (mg/L) to inhibit 50% of mycelial growth or spore germination (EC_{50}) was calculated with the regression equation for each isolate (Zhang et al., 2008).

2.5 *Determination of the baseline sensitivity*

The baseline sensitivity of all the isolates to trifloxystrobin was built. A histogram of EC_{50} was plotted for all these isolates, and the shape of the frequency distribution was analyzed by examining the EC_{50} in regards to curve shape, range and mean values, as well as the ratio of the highest to lowest EC_{50} values (Russell, 2004).

2.6 *Evaluations of disease control efficacy through different fungicide programs*

In 2014 and 2015, the canker disease control experiments were carried out at two sites (Tuankou and Changhua county) with the following eight treatments: (A) 50% trifloxystrobin WDG (Bayer) applied at 100 g a. i. per ha for three times of sprays respectively on 10 March, 1 April, and 20 April; (B) 50% trifloxystrobin WDG applied at 100 g a. i. per ha for three times of sprays respectively on 20 March, 1 April, and 20 April; (C) 10% difenoconazole WDG (Syngenta) applied at 50 g a. i. per ha for three times of sprays respectively on 10 March, 1 April, and 20 April; (D) 10% difenoconazole WDG applied at 50 g a. i. per ha for three times of sprays respectively on 20 March, 1 April, and 20 April; (E) 10% difenoconazole WDG applied at 50 g a. i. per ha for once on 10 March + 50% trifloxystrobin WDG applied at 100 g a. i. per ha for twice respectively on 1 April and 20 April; (F) 50% trifloxystrobin WDG applied at 100 g a. i. per ha or once on 10 March + 10% difenoconazole WDG applied at 50 g a. i. per ha for twice respectively on 1 April and 20 April; (G) 10% difenoconazole WDG applied at 50 g a. i. per ha for twice on 10 March and 1 April + 50% trifloxystrobin WDG applied at 40 g a. i. per ha for once on 20 April; (CK) no fungicide treatment control. Treatment plots were laid out as a randomized complete block design with four replicates. Each plot had an area of about $50m^2$. All experimental sites were naturally infected. No other fungicides were applied to the experimental plots. All the other treatments, such as fertilizers, were used in accordance with standard farm products. Visual disease assessment was made twice, respectively on1 May and 20 May. Control efficacy of all the treatments was evaluated based on10 trees collected at random from each plot. The CLI value of 1 March was adopted as the background value of this year (CLI_0). Increased CLI value = CLI value of a survey-CLI_0. Percentage of efficacy was determined by applying the following formula: 100 × (Increased CLI value for control plot-increased CLI value for treatment plot) /Increased CLI value or control plot ((Zhang et al.,

2011).

3 Results

3.1 *Temporal dynamics of spore dispersal*

Throughout the study period in 2009—2012, ascospores (sexual spores) were not observed. Only conidia (asexual spores) were trapped. These spores were colorless, aseptate, thin-walled conidia, (17.2 ±0.8) μm in length and 4.3 ± 0.6μm in width. Neither conidium was observed until March 31, in 2009, 2011, and 2012 and until April 7 in 2010. Then, the number of trapped spores increased rapidly with the peak value on May 26, in 2009, 2010, and 2012 and on May 19 in 2011. Late spring to early summer was the main spore dispersal period. The number of spores trapped during May 12 to 26 May was 43.9%, 52.8%, 51.4%, 49.2% of total spores trapped for 2009, 2010, 2011, and 2012, respectively. No conidia were observed during July 22 to August 26 for all the four years. Although a weak dispersal was observed in early to middle autumn, no spore was trapped after October 15 (Fig. 1).

3.2 *Multiplication dynamics of canker lesions*

Throughout the study period in 2009—2012, canker lesions caused by *B. dothidea* occurred mainly on tree trunk and main branches but not on leaves and nuts. Expansions of canker lesions were observed since March 17 which was about 15 days in advance of the spore dispersal. Then, the lesions multiply rapidly especially during April 7 to May 19 with the peak value on May 19, in 2009, 2011, 2012 and on Jun 3 in 2010. Then, the decrease of CLI was observed except that a weak re-increase occurred during August 26 to September 1 for all the four years (Fig. 2).

background value of this year (CLI_0). Increased CLI value = CLI value or a survey-CLI_0.

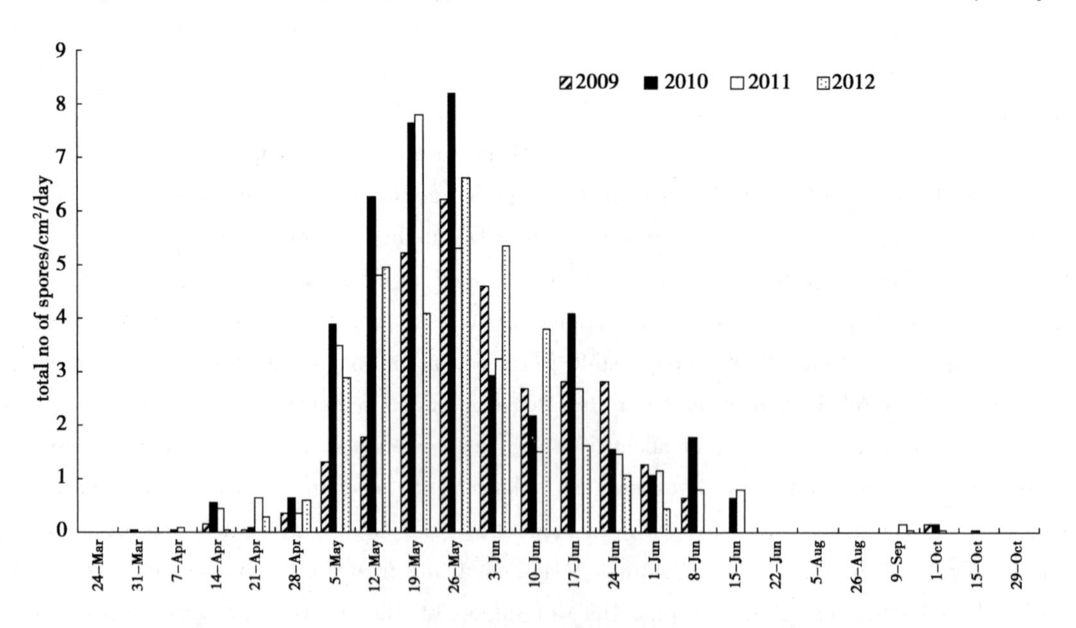

Fig. 1 Total number of *Botryosphaeria dothidea* spores trapped per day using JDBZ1 spore traps in Chinese hickory forests

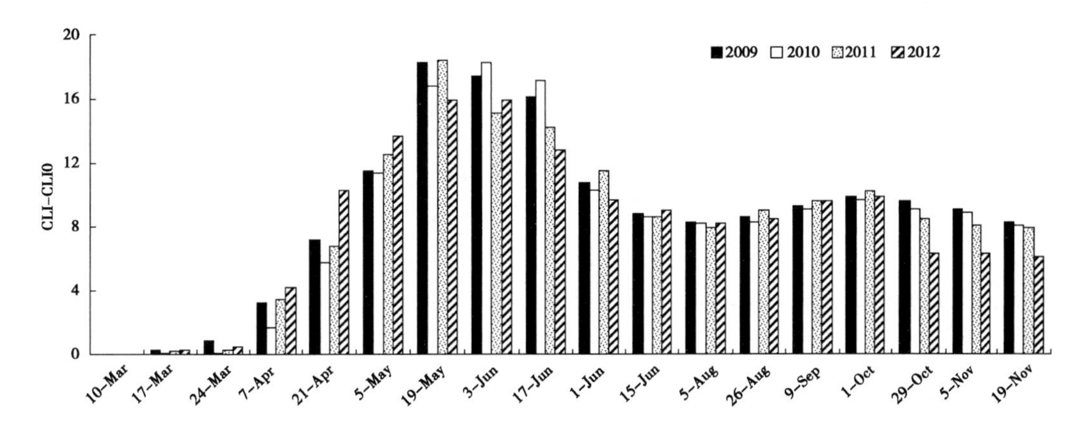

Fig. 2 Multiplication dynamics of lesions expresses by the increased canker lesion index (CLI). CLI (%) = (Σ (number of lesions of each class × each evaluation class) /total number of lesions × 7) × 100. The CLI value of 1 March was adopted as the

3.3 *Baseline sensitivity of B. dothidea population to trifloxystrobin*

A total of 10^7 single-conidium isolates of *B. dothidea* were tested for their sensitivities to trifloxystrobin. There was no evidence of geographical variation in the sensitivity, and the sensitivities of isolates remained unchanged during 2010—2013. In the presence of 100mg/L SHAM, trifloxystrobin had weak activity against the mycelial growth of *B. dothidea*, indicated by the EC_{50} values of > 100mg/L of 20.1% isolates (Fig. 3). During stage of spore germination, the frequency distribution of the EC_{50} values for 107 isolates was a unimodal curve (Fig. 4), ranging from 0.76mg/L to 36.7mg/L with an average of (9.73 ± 1.13) mg/L. The range-of-variation factor was 48.3. Thus, these sensitivity data could be used as a baseline for monitoring the shift of sensitivity in *B. dothidea* populations to trifloxystrobin.

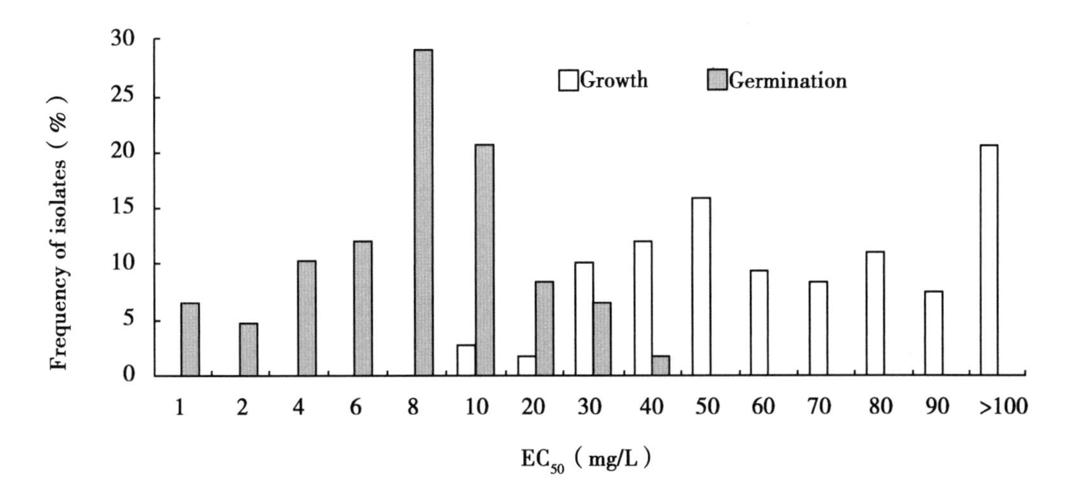

Fig. 3 *In vitro* baseline sensitivity of growth and spore germination to trifloxystrobin of *Botryosphaeria dothidea* population (n = 107)

3.4 *Control efficacy of canker disease of different fungicide programs*

In the two years and two sites experiments, three times of sprays were carried for each program. In 2014 (Table 1), if the first spray was done on 10 March (program A, C, E, F, G), a better efficacy (62.5% ~92.1%) than that on 20 March (program B, D) (≤59.7%) was found in both Tuankou and Changhua county. In 2015 (Table 2), if the first spray was done on 10 March (program A, C, E, F, G), a better efficacy (63.3% ~93.3%) than that on 20 March (program B, D) (≤59.8%) was also observed in both two sites. And, program E provided the highest control efficacy while program D provided the worst control efficacy in both years and sites (Table 1, 2).

Table 1　Efficacy of the canker disease control by different fungicide treatment programs (2014)

Treatments				Name	Site	Increased CLI value		Control efficacy (%)	
10 Mar	20 Mar	1 Apr	20 Apr			1 May	20 May	1 May	20 May
TRS[Y]	※	TRS	TRS	A	Tuankou	2.7	3.4	78.2b[X]	81.4b
※	TRS	TRS	TRS	B		6.7	8.2	45.9d	55.2d
DFC	※	DFC	DFC	C		2.6	5.7	79.0b	68.9c
※	DFC	DFC	DFC	D		8.3	9.2	33.1e	49.7e
DFC	※	TRS	TRS	E		1.9	2.6	84.7a	85.8a
TRS	※	DFC	DFC	F		4.5	6.3	63.7c	65.6c
DFC	※	DFC	TRS	G		2.2	3.9	82.3a	78.7b
control				CK		12.4	18.3	—	
TRS	※	TRS	TRS	A	Changhua	1.6	2.3	84.2b	86.9b
※	TRS	TRS	TRS			4.4	7.1	56.4d	59.7e
DFC	※	DFC	DFC			1.9	5.5	81.2bc	68.8d
※	DFC	DFC	DFC			6.6	8.4	34.7e	52.3f
DFC	※	TRS	TRS			0.8	1.7	92.1a	90.3a
TRS	※	DFC	DFC			2.2	6.6	78.2c	62.5e
DFC	※	DFC	TRS			1.3	3.1	87.1b	82.4c
control				CK		10.1	17.6	—	

[X]For each site, mean values with the same little letters within the same column were not significantly different with turkey tests at $P = 0.05$.

[Y]TRS = 50% trifloxystrobin WDG (Bayer) at 100 g a. i. per ha; ※ = no fungicide treatments at this day; DFC = 10% difenoconazole WDG (Syngenta) applied at 50 g a. i. per ha; control = no fungicide treatment control; Increased CLI value = CLI value or a survey-CLI_0. canker lesion index (CLI) = (Σ (number of lesions of each class × each evaluation class) /total number of lesions × 7) × 100. The CLI value of 1 March was adopted as the background value of this year (CLI_0).

Table 2 Efficacy of the canker disease control by different fungicide treatment programs (2015) Treatment

Treatments				Name	Site	Increased CLI value		Control efficacy (%)	
10 Mar	20 Mar	1 Apr	20 Apr			1 May	20 May	1 May	20 May
TRS[Y]	※	TRS	TRS	A	Tuankou	2. 3	3. 6	82. 0a[X]	80. 7b
※	TRS	TRS	TRS	B		6. 6	7. 8	48. 4c	58. 3d
DFC	※	DFC	DFC	C		2. 3	5. 6	82. 0a	70. 1c
※	DFC	DFC	DFC	D		8. 3	9. 2	35. 2d	50. 8e
DFC	※	TRS	TRS	E		1. 9	2. 7	85. 2a	85. 6a
TRS	※	DFC	DFC	F		4. 3	6	66. 4b	67. 9c
DFC	※	DFC	TRS	G		2. 1	4. 1	83. 6a	78. 1b
control				CK		12. 8	18. 7	—	
TRS	※	TRS	TRS	A	Changhua	1. 5	2. 4	85. 7b	85. 8b
※	TRS	TRS	TRS	B		4. 5	6. 8	57. 1d	59. 8f
DFC	※	DFC	DFC	C		1. 7	5. 2	83. 8bc	69. 2d
※	DFC	DFC	DFC	D		6. 3	8. 5	40. 0e	49. 7
DFC	※	TRS	TRS	E		0. 7	1. 7	93. 3a	89. 9a
TRS	※	DFC	DFC	F		2. 1	6. 2	80. 0c	63. 3e
DFC	※	DFC	TRS	G		1. 4	2. 9	86. 7b	82. 8c
control				CK		10. 5	16. 9	—	

4 Discussion

Little is known about the epidemiology and inoculum dispersal of *B. dothide* on Chinese hickory. Therefore, understanding the biology and epidemiology of *B. dothidea* is essential for developing appropriate management strategies to control it. Spores were captured continuously in the spore trap during the 4-year period.

To our knowledge, this is the first report of seasonal spore dispersal patterns by *B. dothidea* under the weather conditions of eastern China. Our related study showed that no other saprophytic or pathogenic species of *Botryosphaeriaceae* was trapped in the study area (Zhu et al., 2016). Therefore, the morphological traits (Phillips, 2002; Urbez-Torres et al., 2006) could be used to identification of trapped spores of *B. dothidea.*

Although the sexual stage of *B. dothidea* has been reported from *Sequoiadendron giganteum* and *Sequoia sempervirens* (Worral et al., 1986), perithecia of *B. dothidea* have not been observed and ascospores have not been trapped throughout the study period in 2009—2012. Conidia of *B. dothidea* are found to be the main source of inoculums on Chinese hickory coincided with that reported on pistachio (Ntahimpera et al., 2002). In California vineyards, *Botryosphaeriaceae* spp. spores were generally trapped from mid-fall to early spring. The maximum numbers of *Botryosphaeriaceae* spores were recorded during the winter months of December, January, and February (Urbez-Torres et al., 2010). Based on colony counts, the highest population of *Botryosphaeriaceous* fungi occurred with the winter months of December, January, and February with few to no spores trapped in the spring and summer seasons (Eskalen et al., 2013). Our study showed that the late spring to early summer (May to June) is the main spore dispersal period. The number of spores trapped during May 12 to 26 May was 43.9%, 52.8%, 51.4%, 49.2% for 2009, 2010, 2011, and 2012, respectively The results presented in this study indicated that the airborne inoculum of *B. dothide* on Chinese hickory, although present starting in end of march to early April, reaches peak in May and the spore deposition patterns were similar year after year. Spring (March to May) is the rainy season in Zhejiang (Shi et al., 2001) and most of the spores in Chinese hickory trapped were captured during May. In Chinese hickory forests at present, no artificial irrigation were done. These results suggested that the actual dispersal of these spores requires the impact of rain splash just as previous studies on pistachio and avocado showed *B. dothidea* to be a splash-dispersed pathogen (Ahimera et al., 2004; Eskalen et al., 2013). Meanwhile, results of the present study differed somewhat from those of previous research and the effect of temperatures on spore dispersal was indicated. High (July and August) and low temperatures (March) will limited the spore dispersal. This situation may be explained by the fact that both a lower and higher average temperature could have suppressed spore production during the winter to early spring cold months and during hot summer months. However, the effect of precipitation and temperatures on spore release requires further investigation.

Multiplication dynamics of canker lesions were also investigated during the same 4-year period. Results showed that the canker expansion and the weak re-expansion were observed in spring and after summer followed by the spore dispersal in each year. After comparisons of the patterns of canker lesion multiplication and spore dispersal, we speculated that *B. dothidea* overwintered on hickory trees, including the retained lesions and infections remained latent until suitable conditions for symptom development. Although inoculum sources can be diverse, asymptomatic infected (latent) trees is one of the most important sources of inoculum. However, this hypothesis need further research to be tested experimentally. This study revealed that the period of pycnidiospore dispersal of the canker pathogen was the susceptible period of Chinese hickory, thereby preliminary explaining how severe stem canker, one of the most important factors limiting Chinese hickory yield, may occur under favorable weather conditions, and providing a foundation for management of this disease.

Our previous study (Zhang et al., 2011) reported that demethylation-inhibitor fungicides (DMIs) such as difenoconazole showed excellent activity against the mycelial growth of *B. dothidea* with the mean EC_{50} value of 0.43 (± 0.11) mg/L for 150 isolates of *B. dothidea*. However, only about 60% of control efficacy was acquired when it applied three times from middle of April to early May. The present results hinted us that difenoconazole, a systematic chemical with excellent activity against the mycelial growth, might be applied on early March before the lesion multiplication to limit the growth of *B. dothidea* inside the trees according to our epidemiology and mode of inoculum dispersal results. The present study also showed that trifloxystrobin, a QoI fungicide (Margot et al.,

1998), had strong activity against the spore germination. QoIs constitute a new fungicide class developed from natural derivatives strobilurin A, oudemansin A, and myxothiazol A and inhibit mitochondrial respiration by binding at the Qo site of cytochrome b (Bartlett et al., 2002). And the average EC_{50} value of 9.73 ± 1.13mg/L for 10^7 isolates could be used as baselines for monitoring the shift of sensitivity of *B. dothidea* to trifloxystrobin (Russell, 2004). Control of canker disease on Chinese hickory by seven different fungicide programs is assessed. Results suggested that the time of the first spray is very important which should be done before the initiation of canker lesion expansion and far before the initiation of spore dispersal. The better choice for the first spray is a fungicide such as difenoconazole (Claire et al., 2015) which has strong inhibiting bioactivity against the mycelial growth of *B. dothidea* (Zhang et al., 2011). And, the following two sprays should adopt fungicides such as QoIs (Bartlett et al., 2002) or QoI + DMI to inhibit both the spore germination and the mycelial growth of *B. dothidea*. Moreover, other cultural management strategies should also be further assessed to integrated management of this disease.

Acknowledgments

This research was supported by the Special Fund for Forest-Scientific Research in the Public Interest (No. 201304403).

参考文献

[1] Ahimera N, Gisler S, Morgan D P, et al. Effects of single-drop impactions and natural and simulated rains on the dispersal of *Botryosphaeria dothidea* conidia [J]. Phytopathology, 2004, 94, 1 189 – 1 197.

[2] Avila-Adame C, Köller W. Impact of alternative respiration and target-site mutations on responses of germinating conidia of *Magnaporthe grisea* to Qo-inhibiting fungicides [J]. Pest Management Science, 2003, 59: 303 – 309.

[3] Bartlett DW, Clough J M, Godwin J R, et al. The strobilurin fungicides [J]. Pest Management Science, 2002, 58: 649 – 662.

[4] Claire L P, Josie E P, Andrew GS, et al. Azole fungicides-understanding resistance mechanisms in agricultural fungal pathogens [J]. Pest Management Science, 2015, 71: 1 054 – 1 058.

[5] Eskalen A, Faber B, Bianchi M. Spore trapping and pathogenicity of fungi in the *Botryosphaeriaceae* and *Diaporthaceae* associated with avocado branch canker in California [J]. Plant Disease, 2013, 97: 329 – 332.

[6] Jacobs K A, Rehner S A, Comparison of cultural and morphological characters and ITS sequences in anamorphs of *Botryosphaeria* and related taxa [J]. Mycologia, 1998, 90: 601 – 610.

[7] Ma L J, Lin J Y, Li Q, et al. Antifungal Constituents from the Husk of *Carya cathayensis*. Scientia Silvae Sinicae, 2009, 45: 90 – 94.

[8] Margot P, Huggenberger F, Amrein J, et al. CGA279202: a novel broad spectrum strobilurion fungicide [J]. Brighton Crop Protection Conference-Pests and Diseases, 1998, 2: 375 – 382.

[9] McDonald V, Eskalen A. Botryosphaeriaceae species associated with avocado branch cankers in California [J]. Plant Disease, 2011, 95: 1 465 – 1 473.

[10] Michailides T J. Pathogenicity, distribution, sources of inoculum, and infection courts of *Botryosphaeria dothidea* on pistachio [J]. Phytopathology, 1991, 81: 566 – 573.

[11] Michailides T J, Morgan D P. Spore release by *Botryosphaeria dothidea* in pistachio orchards and disease

control by altering the trajectory angle of sprinklers [J]. Phytopathology, 1993, 83: 145 - 152.

[12] Ntahimpera N, Driever G F, Felts D, et al. Dynamics and pattern of latent infection caused by *Botryosphaeria dothidea* on pistachio buds [J]. Plant Disease, 2002, 86: 282 - 287.

[13] Phillips A J L. *Botryosphaeria* species associated with diseases of grapevines in Portugal [J]. Phytopathology Mediterr, 2002, 41: 3 - 18.

[14] Russell PE. Sensitivity Baselines in Fungicide Resistance Research and Management. In: FRAC Monograph, vol. 3. CropLife International, Brussels. Available on line at: www. frac. info. 2004.

[15] Shi N, Ma L, Yuan X Y, et al. Climate features over Zhejiang Province in the last 50 years [J]. Journal of Nanjing Institute of Meteorology, 2001, 24: 207 - 213.

[16] Sinclair W A, Lyon H H. Diseases of trees and shrubs. 2nd edition [M]. NY: Cornell University Press, 2005.

[17] Slippers B, Smit W A, Crous P W, et al. Taxonomy, phylogeny and identification of Botryosphaeriaceae associated with pome and stone fruit trees in South Africa and other regions of the world. Plant Pathology, 2007, 56: 128 - 139.

[18] Urbez-Torres J R, Leavit G M, Voegel T M, et al. Identification and distribution of *Botryosphaeria* spp. associated with grapevine cankers in California. Plant Disease, 2006, 90: 1 490 - 1 503.

[19] Urbez-Torres J R, Battany M, Bettiga L J, et al. *Botryosphaeriaceae* species spore-trapping studies in California vineyards [J]. Plant Disease, 2010, 94: 717 - 724.

[20] Wang C L, Cao X H. Study on anti-tumor of *Juglans mandshurica* [J]. Food Science, 2004, 25: 285 - 287.

[21] White TJ, Bruns T, Lee S, et al. Amplification and direct sequencing of fungal ribosomal RNA genes for phylogenetics [C] //In 'PCR protocols: a guide to methods and applications'. San Diego, CA: Academic Press, 1990: 315 - 322.

[22] Worral J J, Correll J C, McCain A H. Pathogenicity and teleomorph-anamorph connection of *Botryosphaeria dothidea* on *Sequoiadendron giganteum* and *Sequoia sempervirens* [J]. Plant Disease, 1986, 70: 757 - 759.

[23] Zhang C Q, Xu B C. First report of canker on Chinese hickory (*Carya cathayensis*) caused by *Botryosphaeria dothidea* in China [J]. Plant Disease, 2011, 95: 1 319.

[24] Zhang C Q, Zhang Zhu G N. The mixture of kresoxim-methyl and boscalid, an excellent alternative controlling grey mould caused by *Botrytis cinerea* [J]. Annals of Applied Biology, 2008, 153: 205 - 213.

[25] Zhang C Q, Zhang Z P, Sun P L, et al. Comparison of sensitivity of *Botryosphaeria dothidea* to 7 fungicides and its baseline sensitivity to difenoconazole [J]. Chinese J Pesticide Science, 2011, 13: 84 - 86.

[26] Zhu C G, Deng X Y, Shi F. Evaluation of the antioxidant activity of Chinese hickory (*Carya cathayensis*) kernel ethanol extraction [J]. African Journal of Biotechnology, 2008, 7: 2 169 - 2 173.

[27] Zhu Z X, Shi H J, Lei F B, et al. Quantifying *Botryosphaeria dothidea* infection causing canker disease on Chinese hickory (*Carya cathayensis*) using real-time PCR [J]. Journal of Zhejiang A&F University, 2016, 33 (2): 364 - 368.

不同杀菌剂对油菜黑胫病菌的室内毒力测定*

宋培玲[1**]，吴　晶[1,2]，石　嵘[1,2]，燕孟娇[1]，皇甫海燕[1]，
郝丽芬[1]，皇甫九茹[1]，贾晓清[1]，李子钦[1***]
（1. 内蒙古农牧业科学院植物保护研究所，呼和浩特　010031；
2. 内蒙古大学生命科学学院，呼和浩特　010020）

摘要：采用生长速率法，在室内测定了7种杀菌剂对我国油菜黑胫病病原菌的毒力活性。结果表明：供试的7种杀菌剂在试验浓度下对油菜黑胫病菌菌丝的生长均有不同程度的抑制作用，相对抑制率与药剂浓度呈正相关，其中氟硅唑毒力活性最强，EC_{50}为0.11μg/mL。戊唑醇、咪鲜胺的次之，EC_{50}分别为0.326 8 μg/mL，0.670 8μg/mL，此研究筛选出了三种对油菜黑胫病菌有良好抑菌活性的杀菌剂。

关键词：*Leptosphaeria biglobosa*；杀菌剂；毒力测定

Toxicity of Different Fungicides against *Leptosphaeria biglobosa*

Song Peiling[1], Wu Jin[1,2], Shi Rong[1,2], Yan Mengjiao[1], HuangFu haiyan[1],
Hao Lifen[1], HuangFu jiuru[1], Jia Xiaoqing[1], Li Zinqin[1]
(1. *Plant Protection Institute*, *Inner Mongolia Academy of Agricultural & Animal Husbandry Sciences*, *Hohhot* 010031, *China*; 2. *School of Life Science*, *Inner Mongolia University*, *Hohhot* 010020, *China*)

Abstract: The toxic activity of seven fungicides on *Leptosphaeria biglobosa*, pathogen of canola blackleg was tested by the mycelial growth rate method *in vitro*. The results showed that all the fungicides could inhibit the mycelium growth at the different levels, and there was a positive correlation between the relative inhibitory rates and concentrations. The toxic activity of fluquinconazole was best with the EC_{50} value of 0.147 7μg/mL. Tebuconazole and prochloraz inhibition ranked the second with the EC_{50} values of 0.392 7μg/mL and 0.788 9μg/mL, respectively. In this research, the three fungicides showed powerful inhibition against mycelial growth of *L. biglobosa*.

Key words: *Leptosphaeria biglobosa*; Fungicide; Toxicity measurement

油菜黑胫病是世界各国油菜生产中的主要病害之一[1]，全球每个生产季的平均损失在5亿英镑以上[2]。其致病菌为小球腔菌属的真菌复合种，一般由 *Leptosphaeria maculans* 和 *Leptosphaeria biglobosa* 两种真菌混合感染引起。虽然后者在生产上不具有破坏性，仅在叶片和茎秆中上部形成有限病斑；前者能引起茎秆基部腐烂并造成严重的产量损失[3]。但近年来，

* 基金项目：内蒙古农牧业科学院创新基金（NM22032）；内蒙古农牧业科学院院青年创新基金（2014QNJJN03）
** 作者简介：宋培玲，女，内蒙古乌兰察布市人，博士，主要从事植物病理学研究
*** 通讯作者：李子钦，男，内蒙古鄂尔多斯人，研究员，博士，主要从事植物病理学研究；E-mail：ziqinli88@yahoo.com

诸多学者认为相对于*Leptosphaeria biglobosa*，*Leptosphaeria maculans* 是一个“年轻”的进化物种[4]。这一“年轻”物种可在病害流行地区快速蔓延，且有逐步取代弱侵染型病原菌 *Leptosphaeria biglobosa* 的趋势，这种趋势正严重威胁我国油菜的生产。

油菜黑胫病可以在多种气候条件下，不同的栽培措施和作物类型上暴发、流行，在冬、春性油菜上均可发生为害。随着国际间频繁的油菜贸易，目前 *L. maculans* 已经传入世界多数油菜生产国，并导致病害的发生[1,5-6]。目前，我国虽没有有关 *L. maculans* 发生为害的相关报道，但通过对我国多个省份的调查，发现中国多个油菜产区内都有 *L. biglobosa* 的存在。2008—2010 年，李国庆[7]在湖北省的 10 个主要冬油菜产区进行了黑胫病的调查，发现所调查的 10 个油菜产区都有黑胫病，发病率为 1% ~95%。在部分田块，病株率高于 30%，且大部分本地品种苗期就感病。2008—2012 年，李强生[8]等对我国 16 个省（市、自治区）60 个市（县）的冬、春油菜产区进行油菜茎基溃疡病/黑胫病的调查，发现 14 个省的 42 个市（县）存在黑胫病，发病田块约占调查田块的 10%，最重的田块发病率达 92%，整株死亡率达 5%。且从病症表现判断，我国冬、春性油菜茎部、基部病斑与欧洲、北美油菜黑胫病病斑相似。

目前，油菜黑胫病主要防控措施为轮作、改良土壤以及选用抗病品种等。而在生产实际中，实施这些措施又往往受到条件的限制。在没有高抗品种、且一些作物面临高危压力的情况下，化学防治是减少农作物产量损失最常用而且往往是最为有效地手段。药剂防治简单快捷，因此，随着油菜黑胫病害的不断加重，提前筛选防治药剂势在必行。目前，国内尚未有关于此病化学药剂防治的研究报道，笔者旨在应用几种杀菌剂对油菜黑胫病 *L. biglobosa* 进行毒力测定以筛选出最佳的防治药剂，为生产中油菜黑胫病的防治工作提供理论依据。

1 材料与方法

1.1 材料

1.1.1 供试菌株

油菜黑胫病菌株，2012 年采集分离自内蒙古额尔古纳拉不大林农场患病的油菜茎秆。采用常规组织分离法进行分离，切取小块油菜茎秆患病组织，自来水冲洗后，经 75% 的酒精及 1% NaClO 表面消毒，无菌水漂洗 3 次后接入水琼脂平板内，25℃恒温培养，待长出菌丝后，将菌丝转到 PDA 平板上，继续 25℃恒温培养。经纯化、鉴定及致病性测定后保存备用。

1.1.2 供试药剂（表 1）

表 1 供试药剂名称、剂型、厂家信息及浓度设置

编号	药剂	剂型	浓度（μg/mL）					生产厂家
			D1	D2	D3	D4	D5	
1	25% 多菌灵	可湿性粉剂 WP	4.17	3.125	2.5	1.25	0.625	四川国光农化股份有限公司
2	500g/L 异菌脲	悬浮剂 SC	100	50	33	25	20	拜耳作物科学（中国）有限公司
3	45% 咪鲜胺	乳油 EC	2.25	1.125	0.75	0.55	0.45	江苏宜兴市谷丰农业化学有限公司

（续表）

编号	药剂	剂型	浓度（μg/mL）					生产厂家
			D1	D2	D3	D4	D5	
4	10% 氟硅唑	乳油 EC	1	0.33	0.2	0.15	0.1	中国农科院植保所廊坊农药中试厂
5	430g/L 戊唑醇	悬浮剂 SC	0.86	0.43	0.28	0.21	0.17	拜耳作物科学（中国）有限公司
6	70% 甲基硫菌灵	可湿性粉剂 WP	466	311	207	138	92	浙江威尔达化工有限公司
7	250g/L 嘧菌酯	悬浮剂 SC	500	250	167	125	100	上虞颖泰精细化工有限公司

1.1.3　供试培养基

PDA 培养基：马铃薯 200g，琼脂 15g，葡萄糖 20g，水 1 000mL。

AEA 培养基和水杨肟酸。

1.2　方法

采用含毒介质法[9]。将所选药剂参照使用说明书提供的田间用药浓度，按有效成分从高浓度到低浓度用无菌水逐步稀释，根据预实验结果，每种药剂设定 5 个浓度梯度，进行测定，每个处理重复 3 次。按设置好的浓度梯度稀释药液，取 5mL 药液加入装有 45mL 温度为 40～50℃ PDA 培养基的三角瓶中，混均，倒入直径为 9cm 的培养皿中，制成含药平板。随后在平皿中央接种直径为 5mm 的菌饼，菌丝面向下接种在含药 PDA 培养基上，每皿一个菌饼，同时以加入等量无菌水的 PDA 平板为对照。25℃恒温培养，7d 后十字交叉法测量各浓度下菌落直径的大小，比较不同药剂对病原菌的抑制效果。

根据测定的菌落直径，以供试药剂浓度对数值为自变量，抑制率几率值为因变量建立毒力回归方程，算出各种药剂对供试病原菌的抑制中浓度 EC_{50}、相关系数 r。分析所需公式[10]：菌落净生长量 = 菌落直径 − 菌饼直径。

相对抑菌率（%）＝（对照菌落净生长量 − 处理菌落净生长量）/对照菌落净生长量 × 100

2　结果与分析

2.1　供试杀菌剂对油菜黑胫病病原菌的抑制作用

在各供试药剂所设定的浓度范围内，油菜黑胫病菌的菌丝生长均受到了不同程度的抑制，随着各药剂浓度的增加，药剂对病原菌的抑制作用随之增强，即相对抑菌率与药剂浓度成正相关，且药剂间抑制率差异极显著。由菌落直径及相对抑菌率可以看出，10% 氟硅唑抑菌效果最好，药剂浓度为 1μg/mL 时，菌落直径为 9.7mm，相对抑菌率 89.26%，当浓度低为 0.21μg/mL 时，菌落直径为 34.5mm，相对抑菌率 31.3%。45% 咪鲜胺抑菌效果较好，药剂浓度为 2.25μg/mL 时，菌落直径为 9.5mm，相对抑菌率 90%，当浓度低为 0.45μg/mL 时，菌落直径为 39.5mm，相对抑菌率 23.3%。而抑菌效果较差 500g/L 异菌脲，药剂浓度为 100μg/mL 时，菌落直径为 10.7mm，相对抑菌率 89.41%，当浓度低为 20μg/mL 时，菌落直径为 48.3mm，相对抑菌率仅为 19%。

2.2 供试杀菌剂对油菜黑胫病菌的毒力比较

由表2的抑制中浓度 EC_{50} 及相关系数r值可以看出，供试杀菌剂对油菜黑胫病菌表现出不同的毒力活性，其中三唑类的内吸性杀菌剂氟硅唑毒力最高，抑菌作用最强，EC_{50} 为0.11μg/mL。此外，三唑类的内吸性杀菌剂戊唑醇以及咪唑类广谱性杀菌剂咪鲜胺的毒力亦很高，EC_{50} 分别为 0.326 8μg/mL，0.670 8μg/mL。其次为苯并咪唑类杀菌剂多菌灵，其 EC_{50} 为1.472 6μg/mL。毒力活性一般的为二甲酰亚胺类高效广谱、触杀型杀菌剂异菌脲，其 EC_{50} 为34.939 9μg/mL。毒力最差的是甲氧基丙烯酸酯类杀菌剂嘧菌酯以及苯并咪唑类的具有内吸、预防和治疗作用的甲基硫菌灵，在试验中嘧菌酯与甲基硫菌灵对油菜黑胫病病原菌 *Leptosphaeria biglobosa* 菌丝生长所表现出的毒力较低，其 EC_{50} 分别为234.704 8μg/mL、165.904 5μg/mL（图）。

表2 7种杀菌剂对病原菌 *L. biglobosa* 的毒力测定

药剂	浓度（μg/mL）	菌落直径（mm）	相对抑菌率（%）	毒力回归方程	相关系数	EC_{50}
25%多菌灵	4.17	8.5 e E	92.22	$y=2.469x+4.585$	$R^2=0.904$	1.472 6
	3.125	14 d D	79.63			
	2.5	24.8 c C	55.93			
	1.25	31.2 b B	41.85			
	0.625	40.3 a A	21.48			
CK		50				
500g/L异菌脲	100	10.7 e E	89.41	$y=3.197x+0.066$	$R^2=0.921$	34.939 9
	50	15 d D	81.31			
	33	34.5 c C	44.86			
	25	38.8 b B	36.76			
	20	48.3 a A	19			
CK		58.5				
45%咪鲜胺	2.25	9.5 e E	90	$y=2.791x+5.484$	$R^2=0.934$	0.670 8
	1.125	14 d D	80			
	0.75	21.5 c C	63.33			
	0.55	32.2 b B	39.63			
	0.45	39.5a A	23.33			
CK		50				

（续表）

药剂	浓度（μg/mL）	菌落直径（mm）	相对抑菌率（%）	毒力回归方程	相关系数	EC_{50}
10%氟硅唑	1	9.7 e E	89.26	$y = 1.428x + 6.346$	$R^2 = 0.944$	0.110 0
	0.33	14.8 d D	78.15			
	0.2	18.9 c C	69.26			
	0.15	25 b B	55.56			
	0.1	31.3 a A	41.11			
CK		50				
430g/L戊唑醇	0.86	15.5 e E	77.41	$y = 2.059x + 6.000$	$R^2 = 0.927$	0.326 8
	0.43	21 d D	64.81			
	0.28	26.7 c C	51.85			
	0.21	34.5 b B	34.44			
	0.17	40 a A	22.22			
CK		50				
70%甲基硫菌灵	466	15.5 d D	78.82	$y = 1.974x + 0.618$	$R^2 = 0.923$	165.904 5
	311	16.2 d D	78.19			
	207	25.2 c C	50.16			
	138	30.8 b B	43.93			
	92	36.2 a A	31.78			
CK		58.5				
250g/L嘧菌酯	500	12.5 e E	60.75	$y = 0.977x + 2.684$	$R^2 = 0.956$	234.704 8
	250	13.8 d D	54.83			
	167	16.8 c C	43.93			
	125	18.4 b B	40.19			
	100	20.6 a A	34.27			
CK		58.5				

注：大、小写字母分别表示0.01和0.05水平上的差异显著性

3 讨论

为了筛选出可有效防治油菜黑胫病的安全、高效、低毒、低残留药剂，特别是具有不同作用机制的杀菌剂，以消除单一药剂对病原菌的选择压力，最大限度地减缓病原菌抗药性的发展蔓延，笔者在室内开展了7种国内外常见商品杀菌剂的毒力测定试验，氟硅唑、戊唑醇、咪鲜胺三种药剂对油菜黑胫病菌具有很好的抑制作用，且这三种药剂均具有保护和治疗作用，故初步确定氟硅唑、戊唑醇、咪鲜胺为可有效防治油菜黑胫病的药剂。

虽本试验所用7种药剂对油菜黑胫病病原菌均有不同程度的抑制作用，但其所表现的抑

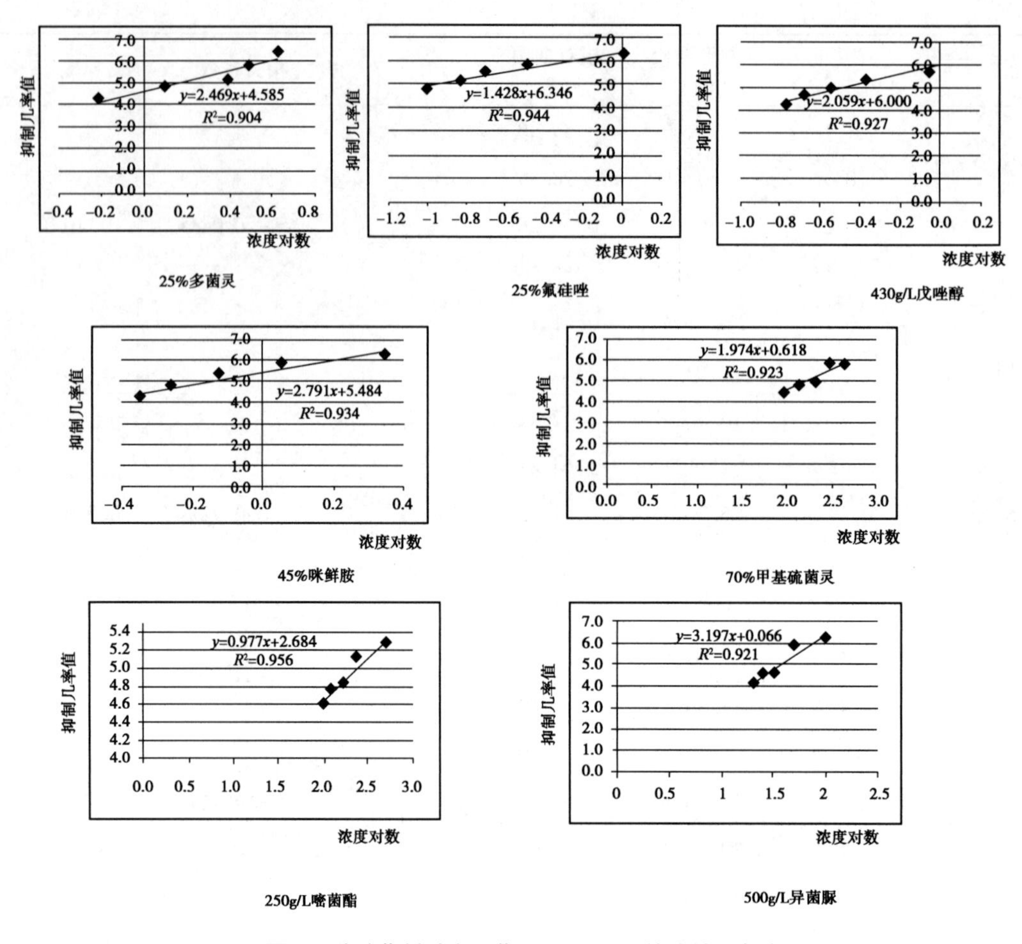

图　7 种杀菌剂对病原菌 *L. biglobosa* 的线性回归直线

制作用的强弱因药剂的化学结构、作用机制、浓度大小而异。试验结果表明，三唑类的内吸性杀菌剂氟硅唑对油菜黑胫病菌的抑制作用最好，毒力最强。其主要作用机理是破坏和阻止病菌的细胞膜重要组成成分麦角甾醇的生物合成，导致细胞膜不能形成，使病菌死亡。S. J. Marcroft[11]等亦发现，用氟硅唑对油菜种子进行包衣，可有效降低油菜黑胫病强毒型病原 *Leptosphaeia maculans* 的发生为害。其次，戊唑醇也表现出好的抑菌效果，其作用机理与所有的三唑类杀菌剂一样，同样是抑制真菌的麦角甾醇的生物合成，咪唑类广谱杀菌剂的咪鲜胺抑菌效果略低于三唑类杀菌剂，其作用机理虽与三唑类杀菌剂类似，通过抑制甾醇的生物合成而起作用，但它无内吸作用，因而其抑菌效果略差一些。早期亦有文献报道施用咪鲜胺可以很好的控制黑胫病。二甲酰亚胺类高效广谱、触杀型杀菌剂异菌脲抑菌效果一般，而据文献报道[12]，1983 年以后，英国许多冬油菜产区曾普遍施用异菌脲来防治油菜黑胫病。异菌脲的作用机制：能抑制蛋白激酶，控制许多细胞功能的细胞内信号，包括碳水化合物结合进入真菌细胞组分的干扰作用。因此，它即可抑制真菌孢子的萌发及产生，也可抑制菌丝生长。即对病原菌生活史中的各发育阶段均有影响，可有效防治对苯并咪唑类内吸杀菌剂有抗性的真菌。甲氧基丙烯酸酯类杀菌剂嘧菌酯以及苯并咪唑类的甲基硫菌灵对油菜黑胫病 *Leptosphaeria biglobosa* 毒力小，抑菌作用不明显，前者作用机理是通过抑制病菌的呼吸作用来破坏病菌的能量合成。后者的作用机理是当该药喷施于植物表面，并被植物体吸收后，在植物体内，经一系列生化反应，被分解为甲基苯并咪唑—乙—氨基甲酸酯（即多菌灵），进而干扰菌的有丝分裂中纺锤体的形成，使病菌孢子萌发长出的芽管扭曲异常，芽管细胞壁扭曲等，从而使病菌不能正常生长达到杀菌效果。

本研究在室内筛选出了几种对油菜黑胫病有良好抑菌效果的药剂，而病原、感病寄主和环境条件是植物病害发生发展的3个基本要素，病原和感病寄主之间的相互作用是在环境条件影响下进行的，因此，室内毒力测定所显示的抑菌活性可能与田间的抑菌活性并不一致，需要进一步开展田间防治试验，在不同年份、不同地区进行，以进一步掌握试验的几个杀菌剂的田间使用浓度、使用时期等，为油菜黑胫病的田间防治提供理论依据。

参考文献

[1] Fitt B D L, Brun H, Barbetti M J, et al. World-wide importance of phoma stem canker (*Leptosphaeria maculans* and *L. biglobosa*) on oilseed rape (*Brassica napus*) [J]. European Journal of Plant Pathology, 2006, 114 (1): 3-15.

[2] Fitt B D L, Hu B C, Li Z Q, et al. Strategies to prevent spread of *Leptosphaeria maculans* (phoma stem canker) onto oilseed rape crops in China; costs and benefits [J]. Plant pathology, 2008, 57 (4): 652-664.

[3] Williams R H, Fitt B D L. Differentiating A and B groups of *Leptosphaeria maculans*, causal agent of stem canker (blackleg) of oilseed rape [J]. Plant Pathology, 1999, 48 (2): 161-175.

[4] Mendes-Pereira E, Balesdent M H, Brun H, et al. Molecular phylogeny of *Leptosphaeria maculans-L. biglobosa* species complex [J]. Mycological Research, 2003, 107 (11): 1 287-1 304.

[5] Gabrielson R L. Blackleg disease of crucifers caused by *Leptosphaeria maculans* (*Phoma lingam*) and its control [J]. Seed Science and Technology, 1983, 11 (3): 749-780.

[6] West J S, Kharbanda P D, Barbetti M J, et al. Epidemiology and management of *Leptosphaeria maculans* (phoma stem canker) on oilseed rape in Australia, Canada and Europe [J]. Plant pathology, 2001, 50 (1): 10-27.

[7] 蔡翔，王转红，李国庆，等.湖北省油菜黑胫病的田间调查与病原鉴定 [C] //中国植物病理学会2011年学术年会论文集.北京：中国农业科学技术出版社，2011：71.

[8] 李强生，荣松柏，胡宝成，等.中国油菜黑胫病害分布及病原菌鉴定 [J]. 中国油料作物学报，2013，35 (4)：415-423.

[9] 张荣意，谭志琼.海南西瓜三叶枯病的病原鉴定及室内药剂筛选试验 [J]. 热带作物学报，1997，18 (1)：96-99.

[10] 陈年春.农药生物测定技术 [M]. 北京：中国农业大学出版社，1990：95-105.

[11] S. J. Marcroft and T. D. Potter. The fungicide fluquinconazole applied as a seed dressing to canola reduces *Leptosphaeria maculans* (blackleg) severity in south-eastern Australia [J]. Australasian Plant Pathology, 2008, 37: 396-401.

[12] Humpherson-Jones F M. The occurrence of virulent pathotypes of *Leptosphaeria maculans* in brassica seed crops in England [J]. Plant pathology, 2007, 35 (2): 224-231.

噻枯唑的光解特性：光解产物鉴定及生物学活性研究*

梁晓宇**，段亚冰，于晓玥，王建新，周明国***
（南京农业大学植物保护学院，南京 210095）

摘要：噻枯唑是我国防治水稻白叶枯病最常见的一种药剂。尽管之前有关于噻枯唑光解特性的研究，但其光解路径及其产物依然未知。我们进行了噻枯唑在光解箱中光解 4h 和 8h 的研究，发现噻枯唑光解后的溶液对白叶枯病菌的抑制活性显著强于未光解的溶液，证明光解能显著增强噻枯唑对白叶枯病菌的抑制活性。我们进一步利用液质联用技术分析鉴定出六种光解产物，并推导出噻枯唑的光解路径。噻枯唑和 2-氨基-1，3，4-噻二唑（AMT）对白叶枯病菌的抑制活性显著强于敌枯唑（ATDA），证明巯基结构在噻枯唑发挥抑菌活性中起着至关重要的作用。另外，我们推断噻枯唑和 AMT 对白叶枯病菌的活体和离体作用机制相似。

关键词：噻枯唑；光解；光解产物；抑制活性；水稻白叶枯病菌

Photochemical Degradation of Bismerthiazol: Structural Characterization of the Photoproducts and Their Inhibitory Activities against *Xanthomonas oryzae* pv. *oryzae*

Liang Xiaoyu, Duan Yabing, Yu Xiaoyue, Wang Jianxin, Zhou Mingguo
(*College of Plant Protection*, *Nanjing Agricultural University*, *Nanjing*, *Jiangsu Province* 210095, *China*)

Abstract: Bismerthiazol is a commonly used bactericide against rice bacterial leaf blight in China. Although previous research determined that bismerthiazol is susceptible to photolytic degradation, the photodegradation pathway and degradation products have remained unknown. In our study, the photodegradation of bismerthiazol was investigated after 4 and 8 hours of irradiation in a solar simulator. Inhibition of *Xanthomonas oryzae pv. oryzae* (*Xoo*) was greater with a photolyzed solution than with a non-photolyzed solution of bismerthiazol, indicating that Photodegradation increased the inhibitory activity of bismerthiazol against *Xoo*. Six photoproducts of bismerthiazol were characterized by LC-MS, and based on these products, a photodegradation pathway was inferred. Inhibition of *Xoo* was significantly greater with bismerthiazol and AMT than with 5-amino-1, 3, 4-thiadiazole (ATDA), suggesting that the sulfhydryl group was crucial for the inhibition of *Xoo* by bismerthiazol and its photoproducts. In addition, we infer that bismerthiazol and AMT might have a similar mode action *in vivo* and *in vitro*.

Key words: Bismerthiazol; Photodegradation; Photoproducts; Inhibitory activity; *Xanthomonas oryzae* pv. *oryzae*

---

* 基金项目：国家“973”计划（2012CB114000）；国家自然科学基金（31272065）；公益性行业（农业）科研专项（201303023）

** 作者简介：梁晓宇，男，博士研究生；E-mail：2012202020@ njau. edu. cn

*** 通讯作者：周明国，男，教授，博士生导师，主要从事杀菌剂药理学及病害防控研究；E-mail：mgzhou @ njau. edu. cn

噻唑、异噻唑、噻二唑和它们的衍生物具有良好的杀菌活性和诱导防卫反应能力，被广泛用于防治动植物上的病害[1-3]。噻枯唑（bismerthiazol），别名为叶青双或叶枯净，是由四川化工院自主研发的一种重要的噻二唑类杀细菌剂。噻枯唑主要被用于防治由水稻白叶枯病菌（*Xanthomonas oryzae* pv. *oryzae*，*Xoo*）引起的水稻白叶病（bacterial leaf blight，BLB）和水稻细菌性条斑病菌（*Xanthomonas oryzae* pv. *oryzicola*，*Xoc*）引起的水稻细菌性条斑病。自20世纪80年代起，噻枯唑一直作为我国防治水稻白叶枯病最常见的一种内吸性药剂[4-5]。但关于噻枯唑的作用机制研究甚少。

噻枯唑具有易光解的特性，陈锡岭利用薄层层析的方法发现噻枯唑的光解产物不多于四种，其中一种光解产物为2-氨基-1，3，4-噻二唑（AMT）[7]，而且AMT继续可以进行光解[8-9]。目前大多数的研究集中在测定噻枯唑和AMT在植物和环境中的残留情况[10-12]，还没有对噻枯唑其他的光解产物和光解路径进行研究。为了更好地安全使用噻枯唑，我们需要进一步测定噻枯唑光解产物的抑菌活性、毒理、安全性等方面的特性。

本实验采用了能够有效地鉴定未知化合物的液质联用技术（LC-MS）对噻枯唑的光解特性进行研究[13-15]。另外，我们评估了其中几种光解产物对噻枯唑的抑制活性，为噻枯唑的安全使用提供了更多的信息。

1　材料与方法

1.1　供试材料

供试菌株：水稻白叶枯病菌（*Xanthomonas oryzae* pv. *oryzae* ZJ173 和 *Xanthomonas oryzae* pv. *oryzae* 2－1－1）、水稻细菌性条斑病菌（*Xanthomonas oryzae* pv. *oryzicola* RS105），以上两种菌株均由南京农业大学杀菌剂实验室分离、鉴定并保存。

供试水稻：温室条件下培养的处于分蘖初期的IR24品种水稻植株。

供试药剂：98%噻枯唑原药由浙江龙湾化工有限公司提供，用N，N-二甲基亚酰胺和甲醇分别配成3.0×10^4μg/mL母液和10μg/mL母液。

培养基：牛肉膏蛋白胨培养基（NB）：牛肉浸膏3g，蛋白胨5g，酵母粉1g，蔗糖10g，加入蒸馏水溶解后定容至1L。固体培养基（NA）的配制是在每升液体培养基中加入20g琼脂粉。

1.2　光解条件

陈锡岭发现溶剂的类型不会影响噻枯唑光解特性[7]。为了方便实验操作和提高光解产物的回收效率，我们在实验室中利用甲醇作为噻枯唑的溶剂，制备了50mL的10μg/mL噻枯唑的工作液。50mL的不含噻枯唑的甲醇溶液作为空白对照。上述溶液黑暗储存于4℃冰箱中。

自然光照条件为夏季晴天中的太阳光，工作液暴露在实验室屋顶，光解时间从8：00—16：00。对照样品置于暗处相同时间，光解温度为25～30℃。样品每两天收集一次。光解箱条件为BL-GHX-V光化学反应器，装有工作液的玻璃管（110mm×90mm）置于氙灯（500W，10 500lx，λ＞300nm）下4小时或8小时，装有空白溶液的玻璃管置于暗处。图1为噻枯唑的光吸收波谱，噻枯唑在320、400和600nm的区域内有明显的吸收峰。一半的光解溶液用于液质联用技术分析，另一半用旋转蒸发仪50℃条件下浓缩并溶解于N，N-二甲基甲酰胺中，配制成10 000μg/mL的溶液用于抑菌实验测定。

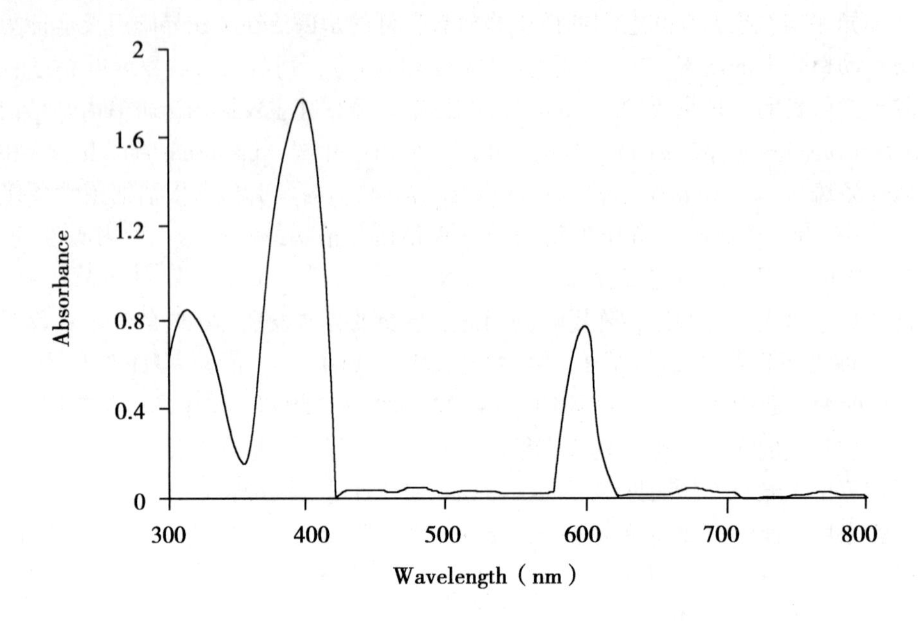

图1　噻枯唑在光谱300～800nm的原子吸收峰

Fig. 1　Absorption spectrum of bismerthiazol over 300～800nm

1.3　液质联用仪器与条件

液质联用仪器为带有C18柱的UHPLC（Dionex，Thermo，USA）和LTQ Orbitrap XL质谱仪（ThermoFisher Scientific）。液相色谱条件：流动相：A为含0.02%甲酸的水，B为含100%的乙腈，梯度洗脱：0～4.0min，10%流动相B；4.0～20.0min，10%～95%流动相B；20～24.5min，95%～10%流动相B；25～30min，10%流动相B。流速：0.20mL/min；柱温35℃；进样体积：5μL。质谱条件：离子源为ESI（+），质谱条件具体为：喷雾电压4 000V，离子源温度300℃，雾化气压力35psi，气帘气压力10psi。

1.4　噻枯唑光解溶液对*Xoo*的抑菌效果测定

噻枯唑光解溶液对*Xoo*的离体抑菌效果通过生长抑制法测定[16]。28℃培养*Xoo*于*NB*中至生长对数期，菌液稀释到浓度大约为10^7 CFU/mL，将100μL的稀释菌液加入含有12mg/L光解溶液的25mL培养基中后置于黑暗处培养（28℃，150r/min），当对照菌液浓度达到10^8 CFU/mL时，测定并计算各处理的抑制率。

在活体实验中，喷施50mL的水或200mg/L的光解溶液于水稻植株。作为空白对照，黑暗处喷施50mL的200mg/L的噻枯唑溶液于水稻植株。第二天用水稻植株剪叶接种浓度为10^8CFU/mL的*Xoo*菌液[18]，后置于25℃自然光条件下的温室。15d后测定叶片病斑长度并计算相对抑制率。每个处理三盆重复，实验重复三次。

1.5　噻枯唑、AMT和ATDA的抑制中浓度（Median effective concentrations，EC_{50}）测定

噻枯唑、AMT和ATDA对*Xoo* ZJ173、*Xoo* 2－1－1，和*Xoc* RS105的EC_{50}值通过生长抑制法测定。将100μL的稀释菌液加入含有梯度稀释药剂的25mL培养基中后置于黑暗处培养（28℃，150r/min）。噻枯唑的浓度梯度对*Xoo*和*Xoc*分别为0、1、2、4、8、16mg/L和0、2.5、5、10、20、40mg/L；AMT的浓度梯度对*Xoo*和*Xoc*为0、1、2、4、8、16mg/L；AMT的浓度梯度对*Xoo*和*Xoc*为0、25、50、100、125、150mg/L。当对照菌液浓度达到10^8 CFU/mL时，测定并计算各处理的抑制率及EC_{50}。每个浓度处理三个重复，实验重复三次。

1.6 噻枯唑、AMT 和 ATDA 的最小抑制浓度（Minimum inhibitory concentrations, MICs）测定

噻枯唑、AMT 和 ATDA 对 *Xoo* ZJ173、*Xoo* 2－1－1，和 *Xoc* RS105 的 MICs 值通过 NA 平板生长抑制法测定[17]。28℃培养 *Xoo* 于 *NB* 中至生长对数期，菌液稀释到浓度大约为 10^8 CFU/mL，将 100μL 的稀释菌液加入含有梯度稀释药剂的 NA 培养基中后置于黑暗处培养 72h，最小完全抑制菌体生长的药剂浓度即为 MICs 值。

1.7 噻枯唑和 AMT 对白叶枯病菌的活体抑制率测定

噻枯唑和 AMT 分别加水稀释至 0、25、50、100、150mg/L。喷施 200mL 的以上 5 种浓度的药剂溶液于水稻植株，第二天用水稻植株剪叶接种浓度为 10^8 CFU/mL 的 *Xoo* ZJ173 和 *Xoo* 2－1－1 菌液[18]，后置于 25℃自然光条件下的温室。15d 后测定叶片病斑长度并计算相对抑制率。每个处理三盆重复，实验重复三次。

1.8 数据处理

用 DPS 统计软件对试验数据进行统计分析。

2 结果与分析

2.1 噻枯唑的光解

含有 10μg/mL 噻枯唑的甲醇溶液暴露在自然光或光解箱下生成黄色颗粒状物质，使得光解溶液由无色变为黄色（图 2A）。这种物质可溶解于二硫化碳和硫化钠，但不溶于水、丙酮和甲醇。经氧瓶燃烧后用亚硝基铁氰化钠检验成深红色，证明是硫。该结果与陈锡岭报道的一致[7]。噻枯唑的光解溶液经旋转蒸发后浓缩于 N，N-二甲基亚酰胺中，并用于抑菌实验测定。

2.2 噻枯唑光解溶液对 *Xoo* 的抑制活性

无论自然光条件还是光解箱条件，噻枯唑的光解溶液对 *Xoo* 活体抑制活性显著强于未光解的溶液（图 2C），该结果与前人报道一致[7,9]。在离体实验中，噻枯唑自然光条件下光解 4 天和 6d 及光解箱条件下光解 4 小时和 8 小时的溶液对 *Xoo* 抑制活性显著强于未光解的溶液（图 2D）。综上所述，光解能够显著增强噻枯唑对 *Xoo* 的抑制活性。

2.3 光解产物鉴定

我们对光解箱条件下光解 4h 或 8h 的含有噻枯唑的甲醇溶液进行 LC-MS 分析，发现 6 种光解产物并标记为 PP1-PP6（图 3）。光解产物的保留时间和质谱结果列于表 1 中。PP1 和 PP6 的保留时间分别为 2.23min 和 3.62min，质谱结果表明它们相比于噻枯唑少 1～2 个巯基，可能是噻枯唑的巯基在光解条件下被移除（图 4）。PP5 的保留时间为 5.95min，分子量为 133，鉴定为 AMT。质谱结果表明，噻枯唑结构上的 C-N 键断裂后加氢形成了 AMT。PP3 的保留时间为 1.81min，可能由 AMT 被氧化得到的。PP2 来源于 AMT 的巯基移除，鉴定为 ATDA。PP4 的保留时间为 2.27min，可能是 PP2 和断裂游离的亚甲基结合后环化生成的产物。

PP1 和 PP2 被分别鉴定为敌枯双和敌枯唑。这两种药剂曾被广泛用于我国水稻白叶枯病害的防治中[19－20]，但是由于对动物和人体有毒副作用而被禁用多年[21]。所以噻枯唑在运输和储存中需要避光，在田间使用时需要减少噻枯唑在水稻收获时的用量。

PP3 和 PP4 之前未被研究，所以我们暂时无法测定它们对 *Xoo* 的抑制活性。但有报道发现 1，3，4-噻二唑的砜类衍生物对 *Xoo* 和 *Xoc* 的抑制活性显著强于噻枯唑。作为一种 1，3，

4-噻二唑的砜类衍生，PP3 可能也对这两种细菌有很好的抑制效果。这个也许可以解释为什么光解能增强噻枯唑的抑菌活性，当然也可能是光解产物协同防治效果所致。

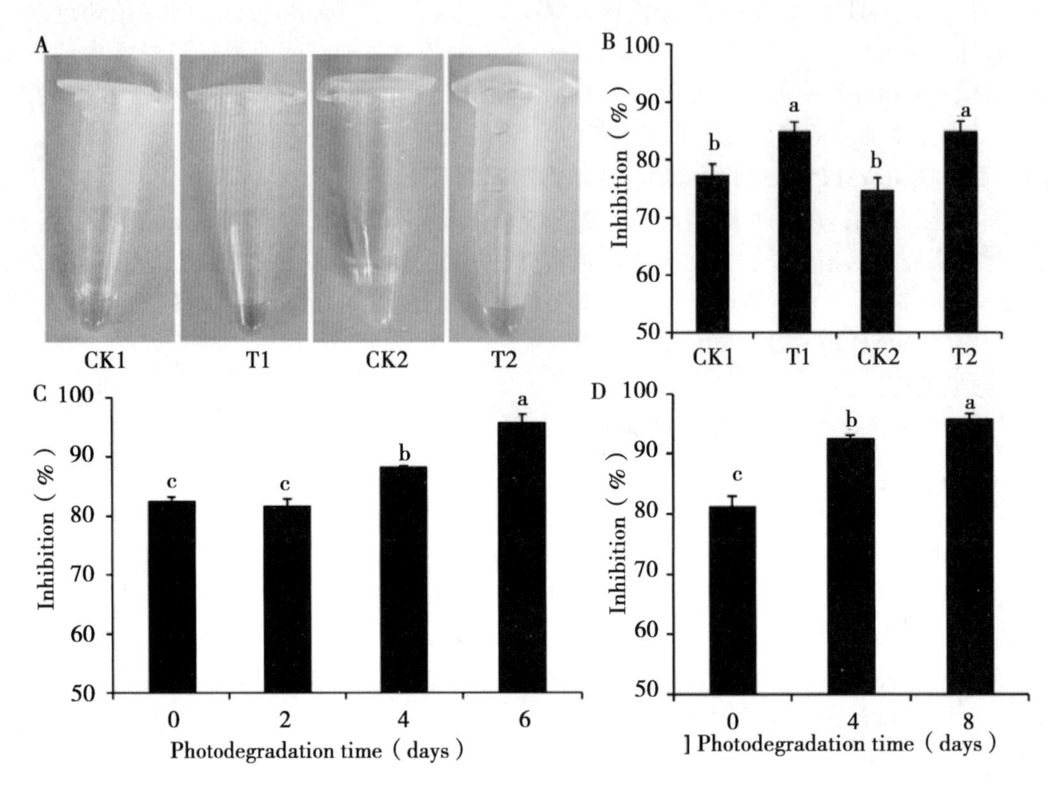

图 2　噻枯唑的光解溶液的颜色和对 *Xoo* 的抑制活性

Fig. 2　Effect of light on the color of a bismerthiazol solution and on its inhibition of *Xoo*.

注：（A）噻枯唑的光解溶液的颜色；（B）噻枯唑的光解溶液对 *Xoo* 的活体抑制活性；（C）自然光条件下噻枯唑光解溶液对 *Xoo* 的离体抑制活性；（D）光解箱条件下噻枯唑光解溶液对 *Xoo* 的离体抑制活性。A 和 B 图的实验处理如下：CK1，未自然光解；T1，自然光解 6d；CK2，未光解箱光解；T2，光解箱光解 8 小时。B、C、D 图中结果显著性分析利用 LSD 法分析（$P<0.05$）。

Note：（A） The color of the bismerthiazol solution after photodegradation. （B） *In vivo* inhibition of *Xoo* by a photolyzed bismerthiazol solution. （C） *In vitro* inhibition of *Xoo* by bismerthiazol solutions that had been exposed to natural sunlight for 0 to 6 days. （D） *In vitro* inhibition of *Xoo* by bismerthiazol solutions that had been exposed to lamplight for 0，4，or 8 hours. In A and B：CK1，not irradiated with natural sunlight；T1，irradiated with natural sunlight for 6 days；CK2，not irradiated with lamplight；T2，irradiated with lamplight for 8 hours. In B-D：inhibition is expressed relative to a control that was not exposed to bismerthiazol or its photodegradation products. For B，C，and D，means with different letters are significantly different according to Fisher's Least Significant Difference test（$P<0.05$）.

2.4　噻枯唑、PP2 和 PP5 对 *Xoo* 和 *Xoc* 的离体抑制活性

基于 EC_{50} 和 MIC 数值，噻枯唑和 PP5 对 *Xoo* 和 *Xoc* 的离体抑制活性显著强于 PP2（表 2）。尽管噻枯唑和 PP5 对 *Xoo* ZJ173 和 *Xoo* 2－1－1 的 EC_{50} 值相差不大，但 PP5 对 *Xoc* 的离体抑制活性显著强于噻枯唑。由此可见巯基结构在噻枯唑发挥抑菌活性中起着至关重要的作用。这些结果表明结构减半并不会影响噻枯唑的离体抑制活性。

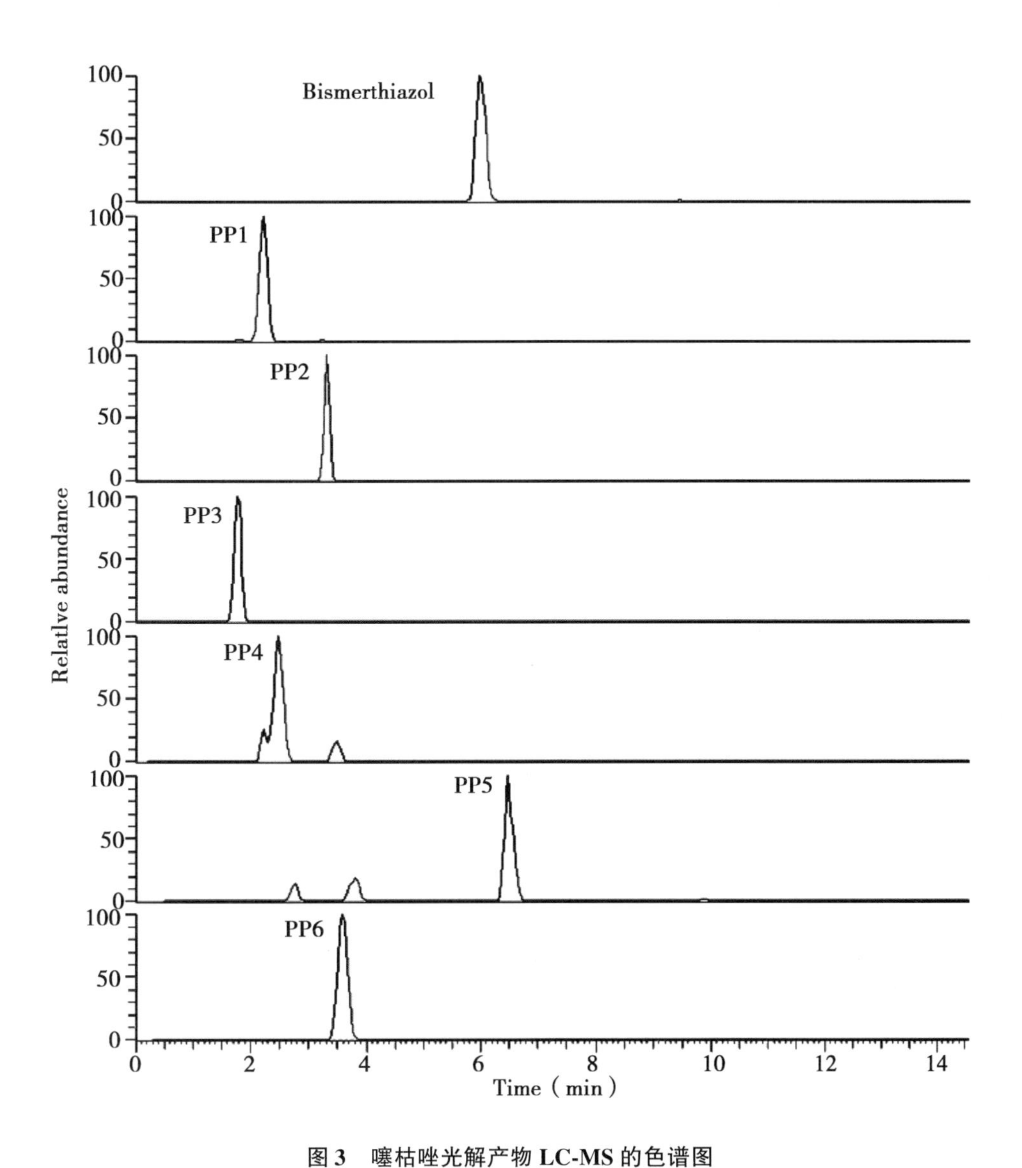

图3　噻枯唑光解产物 LC-MS 的色谱图

Fig. 3　LC-MS chromatograms of a bismerthiazol solution that had been irradiated in a solar simulator. Six photoproducts of bismerthiazol（PP1 to PP6）were characterised by mass spectra.

2.5　噻枯唑和 PP5 对 *Xoo* ZJ173 和 *Xoo* 2－1－1 的的活体抑制活性

噻枯唑对 *Xoo* ZJ173 的活体抑制活性显著强于 PP5（表4）。尽管 *Xoo* 2－1－1 对噻枯唑和 PP5 表现为活体抗性，但离体条件下却表现为敏感。我们推断噻枯唑和 2-氨基-1，3，4-噻二唑对白叶枯病菌的活体和离体作用机制相似。

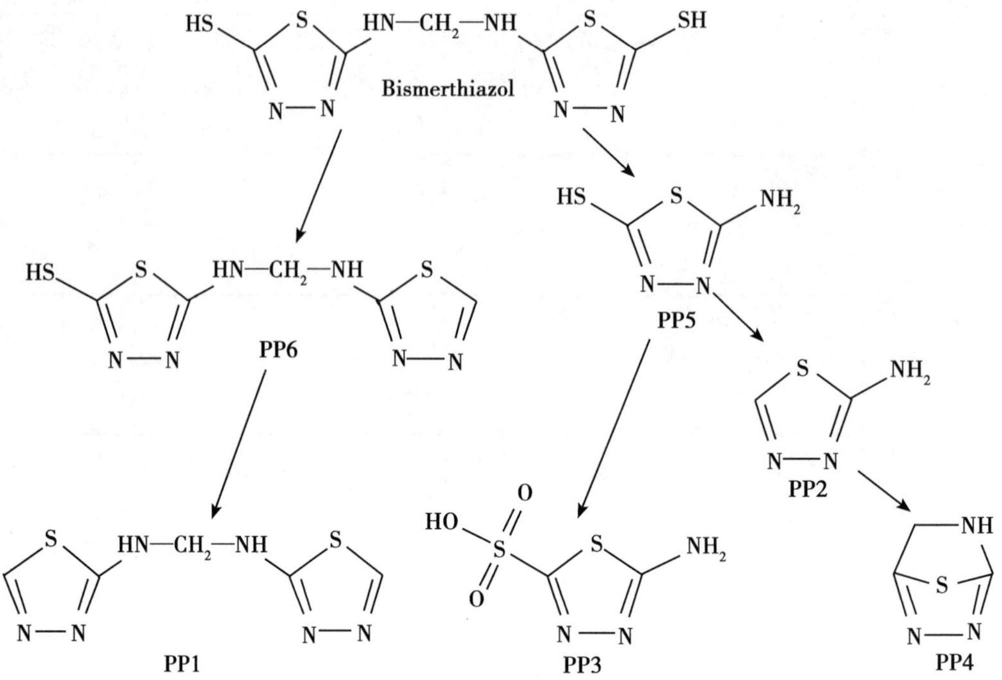

图 4　噻枯唑可能的光解路径

Fig. 4　Possible photodegradation pathway for bismerthiazol.

表 1　噻枯唑光解产物 LC-MS 鉴定结果

Table 1　Mass measurements for the [M + H]$^{+}$ ions of bismerthiazol and degradation products by LC-MS.

化合物 Compound	实际值 Input m/z [M + H]$^{+}$	理论值 Calculated [M + H]$^{+}$	保留时间 Retention time (min)	误差 Error (ppm)	结构式 Formula	4h 光解 Irradiation for 4 h	8h 光解 Irradiation for 8 h
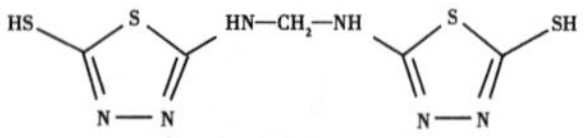 HS S HN—CH$_2$—NH S SH N—N N—N Bismerthiazol	278.960 51	278.690 65	5.98	-1.590	$C_5H_6N_6S_4$	Found	ND*
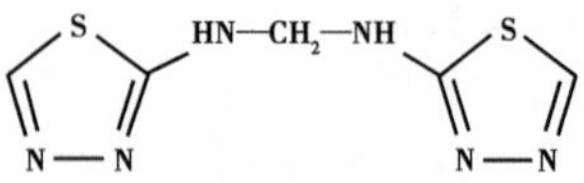 S HN—CH$_2$—NH S N—N N—N PP1	215.016 49	215.016 81	2.23	-1.499	$C_5H_6N_6S_2$	Found	Found
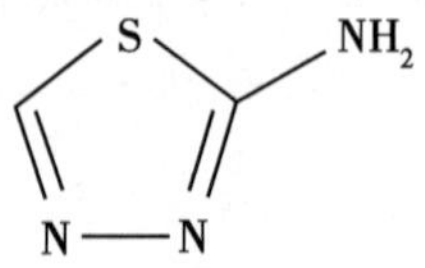 S NH$_2$ N—N PP2	102.011 97	102.012 04	3.32	-0.730	$C_2H_3N_3S$	Found	ND

（续表）

化合物 Compound	实际值 Input m/z [M+H]$^+$	理论值 Calculated [M+H]$^+$	保留时间 Retention time（min）	误差 Error （ppm）	结构式 Formula	4 小时光解 Irradiation for 4 h	8 小时光解 Irradiation for 8 h
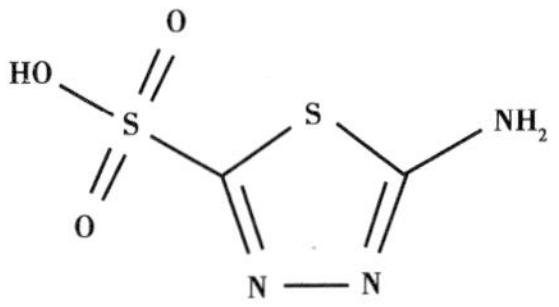 PP3	181.968 46	181.968 86	1.81	-2.192	$C_2H_3N_3O_3S_2$	Found	Found
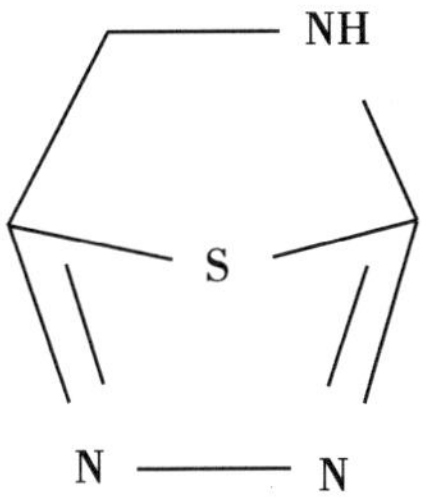 PP4	114.011 79	114.012 4	2.27	-2.232	$C_3H_3N_3S$	Found	Found
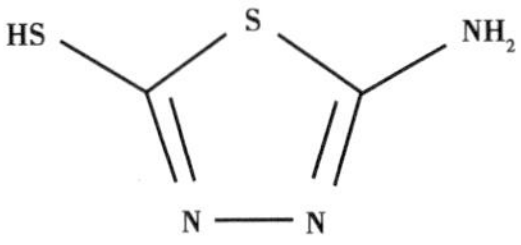 PP5	133.983 88	133.984 11	5.95	-1.008	$C_2H_3N_3S_2$	Found	ND
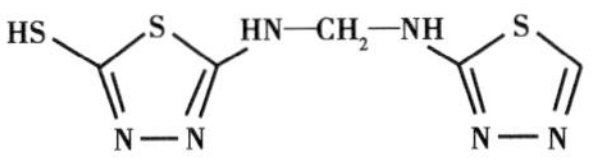 PP6	246.988 63	246.988 88	3.62	-1.024	$C_5H_6N_6S_3$	Found	ND

注：ND 为未发现

ND: not found.

表 2 噻枯唑、PP2 和 PP5 对 *Xoo* 和 *Xoc* 的离体 EC_{50} 值

Table 2 *In vitro* inhibition of *Xoo* and *Xoc* by bismerthiazol, PP2, and PP5 as indicated by EC_{50} values based on growth on NA medium.

化合物 Compound	EC_{50}（mg/L）		
	Xoo ZJ173	*Xoo* 2-1-1	*Xoc* RS105
Bismerthiazol	11.68 ±0.94 b	5.54 ±0.20 b	17.58 ±0.82 b
PP2	101.43 ±0.89 a	50.23 ±0.94 a	125.59 ±5.63 a
PP5	10.27 ±1.76 b	6.91 ±0.37 b	6.39 ±0.46 c

注：EC_{50}值为平均值 ± 标准偏差值，利用 LSD 法分析平均值。

* Values are means ± standard deviations (SD). Means in a column followed by the same letter are not different according to Fisher's LSD test (P=0.05).

表3 噻枯唑、PP2 和 PP5 对 *Xoo* 和 *Xoc* 的 MIC 值

Table 3 The MICs of bismerthiazol, ATDA, and AMT against *Xoo* and *Xoc* based on growth on NA medium.

化合物 Compound	MIC（mg/L）		
	Xoo ZJ173	*Xoo* 2－1－1	*Xoc* RS105
Bismerthiazol	60	50	70
PP2	>200	>200	>200
PP5	60	50	60

表4 噻枯唑和 PP5 对 *Xoo* ZJ173 和 *Xoo* 2－1－1 的的活体抑制活性

Table 4 *In vivo* inhibition of *Xoo*（strains ZJ173 and 2－1－1）by bismerthiazol and PP5

菌株 *Xoo* strain	抑制率 Reduction in lesion length（%）							
	噻枯唑浓度 Bismerthiazol concentration（mg/L）				PP5 浓度 PP5 concentration（mg/L）			
	25	50	100	150	25	50	100	150
ZJ173	42.31 ± 5.26	74.60 ± 4.00	96.08 ± 1.06	97.92 ± 0.98	22.86 ± 5.33	43.55 ± 7.64	47.36 ± 2.07	55.76 ± 7.06
2－1－1	17.41 ± 4.51	19.83 ± 3.19	23.65 ± 3.29	26.00 ± 6.89	10.95 ± 4.42	17.76 ± 6.38	29.74 ± 2.84	41.79 ± 3.38

3 结论

本实验中，利用 LC-MS 鉴定出噻枯唑六种光解产物，其中两种化合物为第一次报道，三种化合物是噻枯唑结构上的 1～2 个巯基移除形成的产物，其他的是噻枯唑氧化或重组得到的产物。基于以上产物的结构，我们推测出可能的噻枯唑光解途径。活体离体抑菌实验证明光解能够显著增强噻枯唑对 *Xoo* 的抑制活性。噻枯唑和 AMT 对白叶枯病菌的抑制活性显著强于 ATDA，证明巯基结构在噻枯唑发挥抑菌活性中起着至关重要的作用。另外，我们推断噻枯唑和 AMT 对白叶枯病菌的活体和离体作用机制相似。

参考文献

[1] Turan-Zitouni G, Demirayak Ş, Özdemir A, et al. Synthesis of some 2-［（benzazole-2-yl）thioacetyl-amino］thiazole derivatives and their antimicrobial activity and toxicity［J］. European Journal of Medicinal Chemistry, 2004, 39: 267－272.

[2] Leoni A, Locatelli A, Morigi R, et al. Novel thiazole derivatives: a patent review（2008—2012. Part 2）［J］. Expert Opin. Ther. Patents, 2014, 24: 759－777.

[3] Oostendorp M, Kunz W, Dietrich B, et al. Induced disease resistance in plants by chemicals［J］. European Journal of Plant Pathology, 2001, 107: 19－28.

[4] 马忠华，周明国，叶钟音. 噻枯唑对水稻白叶枯病菌作用机制研究初报［J］. 植物病理学报，1997，3：46－50.

[5] 沈光斌，周明国. 水稻白叶枯病菌对噻枯唑的抗药性监测［J］. 植物保护，2002，1：9－11.

[6] Zhu X F, Xu Y, Peng D, et al., Detection and characterization of bismerthiazol-resistance of *Xanthomonas oryzae* pv. *oryzae* [J]. Crop Protection, 2013, 47: 24 - 29.

[7] 陈锡岭. 叶青双光解产物的初步鉴定 [J]. 河南职技师院学报, 1993, 2: 28 - 32.

[8] 陈锡岭, 石明旺. 新型杀菌剂叶青双光解动态研究 [J]. 农业环境保护, 2000, 4: 239 - 241.

[9] 高扬帆, 薛凤珍, 李广领, 等. 叶青双在水中的降解特性研究 [J]. 西北农业学报, 2006, 6: 233 - 235.

[10] 陈锡岭, 周红. 稻米中杀菌剂叶青双及代谢产物的快速高效液相色谱测定法 [J]. 河南职技师院学报, 2000, 1: 27 - 34.

[11] 高扬帆, 陈锡岭, 孔凡彬, 等. 水中叶青双及其降解物的反相液相色谱测定研究 [J]. 河北农业大学学报, 2006, 2: 83 - 99.

[12] Wu J, Zhang H, Wang K, et al. Determination and study on dissipation and residue of bismerthiazol and its metabolite in Chinese cabbage and soil [J]. Environmental Monitoring and Assessment, 2014, 186: 1 195 - 1 202.

[13] García-Galán M J, Díaz-Cruz M S, Barceló D. Identification and determination of metabolites and degradation products of sulfonamide antibiotics [J]. Trends in Analytical Chemistry, 2008, 27: 1 008 - 1 022.

[14] Fernández-Alba AR, García-Reyes JF. Large-scale multi-residue methods for pesticides and their degradation products in food by advanced LC-MS [J]. Trends in Analytical Chemistry, 2008, 27: 973 - 990.

[15] Hisaindee S, Meetani M, Rauf M. Application of LC-MS to the analysis of advanced oxidation process (AOP) degradation of dye products and reaction mechanisms [J]. Trends in Analytical Chemistry, 2013, 49: 31 - 44.

[16] Li J, Zhou M, Li H, et al. A study on the molecular mechanism of resistance to amicarthiazol in *Xanthomonas campestris* pv. *citri* [J]. Pest Management Science, 2006, 62: 440 - 445.

[17] Kauffman H, Reddy A, Hsieh S, et al. An improved technique for evaluating resistance of rice varieties to Xanthomonas oryzae [J]. Plant Dis Rep, 1973, 57, 537 - 541.

[18] McManus P, Jones A. Epidemiology and genetic analysis of streptomycin-resistant *Erwinia amylovora* from Michigan and evaluation of oxytetracycline for control [J]. Phytopathology, 1994, 84: 627 - 632.

[19] 孙爱民, 杨卫东, 傅丽, 等. 地面水中叶枯灵的高效液相色谱测定 [J]. 四川环境, 1995, 2: 55 - 56.

[20] Xu C Y, Luo S Y. Preliminary study of the synthesis of N, N'-methylene-bis (2-amino-1, 3, 4-taiadiazole) and control of rice bacterial leaf blight [J]. Agrochemicals, 1973, 7: 14 - 17.

[21] Hill D L. Aminothiadiazoles [J]. Cancer Chemotherapy and Pharmacology, 1980, 4: 215 - 220.

[22] Li P, Shi L, Gao M N, et al. Antibacterial activities against rice bacterial leaf blight and tomato bacterial wilt of 2-mercapto-5-substituted-1, 3, 4-oxadiazole/thiadiazole derivatives [J]. Bioorganic & Medicinal Chemistry Letters, 2015, 25: 481 - 484.

7种杀菌剂对安徽省烟草根黑腐病菌的毒力作用

王文凤，江　寒，叶　磊，檀根甲*
（安徽农业大学植物保护学院，合肥　230036）

摘要：本文采用菌丝生长速率法和孢子萌发法测定了7种药剂对安徽省烟草根黑腐病菌［*Thielaviopsis basicola*（Berk. and Br.）Ferraris］的毒力，并在此基础上进行药剂的复配组合，以期筛选出高效、低毒、低残留复配组合。结果表明：7种单剂中，多菌灵、咪鲜胺和吡唑嘧菌酯对烟草根黑腐病菌的菌丝生长和孢子萌发均有较好的抑制作用，在复配组合中，多菌灵与苯醚甲环唑的5种配比组合均起到增效作用，多菌灵与申嗪霉素、吡唑嘧菌酯、咪鲜胺的部分组合起增效作用，吡唑嘧菌酯与申嗪霉素的3种配比组合表现出较强的增效作用。

关键词：烟草根黑腐病菌；杀菌剂；毒力测定；菌丝生长；分生孢子

Toxicity of Seven Fungicides to *Thielaviopsis basicola* in Tobacco

Wang Wenfeng，Jiang Han，Ye Lei，Tan Genjia*
（*School of Plant Protection*，*AnhuiAgricultural University*，*Hefei*，230036）

Abstract：The toxicity of seven fungicides to *Thielaviopsis basicola*（Berk. and Br.）Ferraris were tested by the mycelium growth rate and spore germination method，and compose difference fungicide to select fungicide and mixture with high efficiency，low toxicity and low residue. Results show that the 7 kinds of single dose，Carbendazim and prochloraz，Pyraclostrobin have better inhibitory effect on pathogen mycelial growth and spore germination. In combination，five ratio combinations of carbendazim and difenoconazole both have synergistic effect，part ratio combinations of carbendazim and phenazine-1-carboxylic acid，prochloraz，Pyraclostrobin also have synergistic effect，part of ratio combination of phenazine-1-carboxylic acid and Pyraclostrobin showed a greater synergy.

Key words：Thielaviopsisbasicola；Fungicide；Toxicity test；Mycelium；Conidiospore

烟草根黑腐病是一种世界性的真菌病害，与黑胫病一起被认为是为害烟草根茎部的两种最重要真菌病害。在我国多个省份均有发病，发病分布较广，烟草幼苗和较大烟株均可染病，毁坏烟苗，致使病株叶片变黄、变薄，严重影响烟草产量和质量[1]。

目前，烟草根黑腐病在安徽省南方烟区时有发生，且有逐年加重之趋势，具有潜在的危险性[2]。本文开展了7种杀菌剂对安徽省烟草根黑腐病菌的室内毒力测定，以期为安徽省烟草根黑腐病菌的药剂防治提供理论基础，减少烟草根黑腐病菌给烟草生产带来的经济损失。

1　材料与方法

1.1　供试菌株

烟草根黑腐病菌［*Thielaviopsis basicola*（Berk. et Br.）Fer.］，由安徽农业大学植物保护

* 通讯作者：檀根甲，博士，教授，博士生导师，主要从事植病流行与绿色防控技术研究；E-mail：tgj63@163.com

学院病理研究室分离纯化获得。

1.2 供试药剂

10%苯醚甲环唑水分散粒剂（瑞士先正达作物保护有限公司），250g/L 嘧菌酯悬浮剂（瑞士先正达作物保护有限公司），80%多菌灵可湿性粉剂（江苏泰仓农化有限公司），50%咪鲜胺锰盐可湿性粉剂（江苏辉丰农化有限公司），戊唑醇原药（上海农乐生物制品有限公司），申嗪霉素原药（上海农乐生物制品有限公司），250g/L 吡唑嘧菌酯乳油（巴斯夫有限公司）。

1.3 供试培养基

马铃薯蔗糖琼脂培养基（PDA）培养基：马铃薯 200g，蔗糖 15g，琼脂 15g，蒸馏水 1 000mL[3]。

1.4 杀菌剂单剂毒力测定

1.4.1 单剂对烟草根黑腐病菌菌丝生长的影响

采用陈年春的菌丝生长速率法测定 7 种杀菌剂对烟草根黑腐病菌菌丝生长的抑制作用[4-5]。将上述 7 种杀菌剂分别配制成 6 个质量浓度梯度母液，无菌水为对照，分别吸取 6mL 配制好的药剂加入 45mL 的融化后的 PDA 培养基中，摇匀后分别倒成平板，配制成相应浓度的含药培养基平板，用直径为 4mm 的打孔器在烟草根黑腐病菌培养 9d 后的 PDA 平板上取菌碟，分别置于含药 PDA 平板中央，每处理 3 个重复，置于 25℃恒温培养箱中培养，9d 后用十字交叉法测量菌株在不同药剂不同浓度梯度处理下的菌丝生长直径，根据公示计算菌丝生长抑制率，并利用 DPS 软件，以浓度对数为自变量，相对抑制率的几率值为纵坐标，绘制标准曲线，计算出 7 种药剂对烟草根黑腐病菌的毒力回归方程和有效抑制中浓度（EC_{50}）、EC_{90}的值，根据 EC_{50}、EC_{90}的数值比较 7 种杀菌剂的毒力大小[6]。

菌丝生长抑制率(%)=(对照组菌落直径－处理组菌落直接)/(对照菌丝生长直径－4)×100

1.4.2 单剂对烟草根黑腐病菌孢子萌发及芽管伸长的影响

将烟草根黑腐病菌在 PDA 平板上培养 9d 后，用无菌水水洗菌丝并过滤配制孢子悬浮液（低倍镜视野下每视野 30～50 个分生孢子），加入不同浓度药剂配制成不同浓度梯度的含药孢悬液，取 150μL 含药孢悬液置于凹玻片中，25℃恒温培养箱中培养，6h 后观察孢子萌发情况并测量芽管长度。每处理观察统计 300 个孢子，利用公式计算孢子萌发抑制率以及芽管伸长抑制率。

孢子萌发抑制率（%）＝（对照组孢子萌发率－处理组孢子萌发率）/对照组孢子萌发率×100

1.5 复配组合的配方筛选

根据上述 7 种杀菌剂的毒力测定试验结果，将苯醚甲环唑、申嗪霉素分别与多菌灵、咪鲜胺和吡唑嘧菌酯按照有效成分为 D：A、D：B、A：A、A：B、C：A 5 种不同比例进行复配，并将多菌灵、咪鲜胺、吡唑醚菌酯三种药剂按照上述比例进行两两复配。实验方法同 1.4.1。

参照 1.4.1 数据处理方法，利用 DPS 软件，绘制标准曲线，计算出每组复配药剂对烟草根黑腐病菌的毒力回归方程，相关系数和 EC_{50}，并根据孙云沛法计算出每组复配药剂的共毒系数（CTC），比较复配杀菌剂对烟草根黑腐病菌的毒力效果。其中，CTC 明显大于 100 时，表示复配药剂对烟草根黑腐病菌的毒力测定起到增效作用，明显小于 100 时表示拮抗作用[7-8]。共毒系数（CTC）的计算公式如下：

$$共毒系数（CTC）= A的EC_{50} \times B的EC_{50}/M的EC_{50} \times (PA \times B的EC_{50} + PB \times A的EC_{50})$$

其中：A和B分别表示两种单剂，M表示这两种单剂的混合药剂，PA和PB分别表示A和B在混剂中的比例。

2 结果与分析

2.1 单剂对烟草根黑腐病菌菌丝生长的影响

7种不同杀菌剂对烟草根黑腐病菌菌丝生长的影响表明（表1），多菌灵和咪鲜胺的抑菌效果最好，明显优于其他药剂，EC_{50}分别为0.045 9μg/mL和0.087 6μg/mL。其次，抑菌效果较好的是吡唑嘧菌酯和苯醚甲环唑，EC_{50}分别为0.435μg/mL和0.518 3μg/mL；其中，申嗪霉素和戊唑醇对烟草根黑腐病菌的菌丝生长有一定的抑制作用，EC_{50}分别为4.764 7μg/mL和4.052 1μg/mL；而嘧菌酯对烟草根黑腐病菌的菌丝生长基本无抑制作用。

表1　不同杀菌剂对烟草根黑腐病菌菌丝生长的影响

杀菌剂 Fungicide	毒力回归方程 Virulence equation	相关系数（*r*） Coefficient（*r*）	EC_{50}（95%置信区间）（μg/mL）	EC_{90}（95%置信区间）（μg/mL）
申嗪霉素	$y=4.671\ 7+0.484\ 2x$	0.982 2	4.764 7（2.272 7～9.989 3）	2 112.495 9（569.038 5～8 576.291 4）
戊唑醇	$y=4.084\ 5+1.506\ 5x$	0.906 5	4.052 1（0.682 4～24.060 6）	28.729 1（4.728 0～174.569 5）
咪鲜胺	$y=7.192\ 4+2.072\ 8x$	0.960 4	0.087 6（0.066 0～8.306 0）	0.363 6（0.128 9～1.025 2）
多菌灵	$y=9.180\ 6+3.123\ 8x$	0.910 4	0.045 9（0.007 5～0.281 0）	0.118 0（0.022 0～0.633 6）
苯醚甲环唑	$y=5.277\ 2+1.021\ 6x$	0.901 5	0.518 3（0.067 7～3.970 4）	8.117 7（1.211 5～54.392 1）
吡唑嘧菌酯	$y=5.290\ 8+0.802\ 0x$	0.997 0	0.435 0（0.328 0～0.574 2）	17.196 5（10.293 7～28.728 2）
嘧菌酯	$y=3.678\ 0+0.360\ 4x$	0.982 7	4 655.854 8（1 005.391 5～21 560.739 0）	39 887.312 0（72 278.505 4～544 995.722 4）

2.2 单剂对烟草根黑腐病菌孢子萌发及芽管伸长的影响

7种不同杀菌剂对烟草根黑腐病菌孢子萌发影响的试验中表明（表2），吡唑嘧菌酯和多菌灵两种单剂对烟草根黑腐病菌分手孢子萌发的抑制效果最好，EC_{50}分别为0.024 7和0.567μg/mL；苯醚甲环唑、咪鲜胺、戊唑醇的抑制效果依次降低，EC_{50}分别为1.858 4、2.422 2和9.021 3μg/mL；申嗪霉素和嘧菌酯的效果最差，EC_{50}分别为17.153 1和23.375 6μg/mL。

表 2 不同杀菌剂对烟草根黑腐病菌孢子萌发的影响

杀菌剂 Fungicide	毒力回归方程 Virulence equation	相关系数 (r) Coefficient (r)	EC_{50}（95% 置信区间）(μg/mL)	EC_{90}（95% 置信区间）(μg/mL)
申嗪霉素	y = 3.508 6 + 1.208 3x	0.909 1	17.153 1（1.824 8 ~ 161.237 6）	197.245 7（7.897 7 ~ 4 926.215 4）
戊唑醇	y = 3.725 8 + 1.333 8x	0.924 1	9.021 3（1.434 4 ~ 56.735 7）	82.429 6（6.334 5 ~ 1 072.692 6）
咪鲜胺	y = 4.776 4 + 0.582 0x	0.980 0	2.422 2（1.115 9 ~ 5.268 6）	385.667 3（81.137 8 ~ 1 833.169 1）
多菌灵	y = 5.139 7 + 0.567 1x	0.980 7	0.567 0（0.308 3 ~ 1.042 6）	103.154 1（36.312 2 ~ 293.035 2）
苯醚甲环唑	y = 4.871 5 + 0.477 5x	0.983 2	1.858 4（0.962 3 ~ 3.728 1）	897.482 2（184.263 3 ~ 4 372.200 1）
吡唑嘧菌酯	y = 7.374 5 + 1.477 2x	0.942 5	0.024 7（0.003 4 ~ 0.178 5）	0.182 0（0.041 7 ~ 0.795 4）
嘧菌酯	y = 4.166 0 + 1.899 9x	0.892 3	23.375 6（2.656 6 ~ 205.666 6）	211.453 2（10.577 7 ~ 4 222.725 1）

2.3 复配组合的配方筛选

由表 3 可知，多菌灵与苯醚甲环唑的组合均表现出明显的增效作用，5 种配比组合的共毒系数（CTC）均大于 100；多菌灵与申嗪霉素配比为 A : D 时，CTC 为 199.62，表现出明显的增效作用，而其他 4 种配比表现出拮抗作用；多菌灵与吡唑嘧菌酯的组合 A : B 和 A : D 的 CTC 分别为 118.31、179.31，两种药剂之间有增效作用，其余配比则表现出拮抗作用。咪鲜胺与申嗪霉素、苯醚甲环唑、吡唑醚菌酯三种药剂进行组合时，以及吡唑嘧菌酯与苯醚甲环唑进行组合，其 CTC 均小于 100，均表现出拮抗作用。其中，咪鲜胺与多菌灵的组合 C : B、A : A、A : B 的 CTC 均大于 100，具有一定的增效作用；吡唑嘧菌酯与申嗪霉素的组合 C : B、A : A、A : D 的 CTC 均大于 200，表现出较强的增效作用。

表 3 几种杀菌剂复配对烟草根黑腐菌的毒力测定

混剂	配比	毒力回归方程	相关系数 (r)	EC_{50}	95% 置信区间	共毒系数 (CTC)
多菌灵：苯醚甲环唑	C : B	y = 7.377 8 + 1.876 7x	0.944 3	0.054 1	0.012 0 ~ 0.244 2	133.45
	A : A	y = 7.475 2 + 1.765 0x	0.942 9	0.039 6	0.007 8 ~ 0.200 1	212.80
	B : A	y = 8.361 0 + 2.269 3x	0.912 8	0.033 0	0.004 0 ~ 0.276 1	166.44
	A : B	y = 6.945 5 + 1.816 4x	0.883 8	0.084 9	0.009 1 ~ 0.789 5	137.74
	A : D	y = 7.261 2 + 2.034 3x	0.959 6	0.077 3	0.023 5 ~ 0.254 8	231.61

（续表）

混剂	配比	毒力回归方程	相关系数（r）	EC_{50}	95%置信区间	共毒系数（CTC）
多菌灵：申嗪霉素	C：B	$y=7.0710+2.0334x$	0.9463	0.0856	0.0218～0.3359	88.79
	A：A	$y=6.6465+2.2809x$	0.9323	0.1897	0.0463～0.7778	47.93
	B：A	$y=7.9954+2.8437x$	0.9227	0.0884	0.0168～0.4665	77.59
	A：B	$y=6.5656+2.1647x$	0.9004	0.1891	0.0326～1.0971	71.51
	A：D	$y=6.9045+1.9925x$	0.9110	0.1107	0.0193～0.6336	199.62
多菌灵：吡唑嘧菌酯	C：B	$y=6.9661+1.8684x$	0.8911	0.0887	0.0117～0.6728	80.58
	A：A	$y=6.9337+1.8940x$	0.8923	0.0953	0.0130～0.6983	87.13
	B：A	$y=7.0197+1.8827x$	0.9028	0.0846	0.0125～0.5707	77.38
	A：B	$y=6.9585+1.9263x$	0.9020	0.0962	0.0147～0.6315	118.31
	A：D	$y=6.9836+1.8970x$	0.9009	0.0900	0.0133～0.6097	179.31
咪鲜胺：多菌灵	C：B	$y=8.3741+2.3100x$	0.9233	0.0346	0.0049～0.2450	185.69
	A：A	$y=7.1627+1.8788x$	0.9197	0.0706	0.0121～0.4114	111.08
	B：A	$y=7.1370+2.0211x$	0.9464	0.0856	0.0255～0.3417	76.83
	A：B	$y=8.2964+2.4170x$	0.9228	0.0433	0.0066～0.2844	126.12
	A：D	$y=7.2351+1.9867x$	0.9482	0.0750	0.0191～0.2948	67.64
咪鲜胺：申嗪霉素	C：B	$y=5.3506+0.7768x$	0.9976	0.3537	0.2765～0.4526	48.64
	A：A	$y=5.3829+0.7783x$	0.9930	0.3221	0.2110～0.4918	44.78
	B：A	$y=5.0630+0.5661x$	0.9980	0.7739	0.6101～0.9817	7.55
	A：B	$y=4.8706+0.5457x$	0.9412	1.7263	0.3721～8.0083	14.70
	A：D	$y=5.3108+0.7982x$	0.9547	0.4080	0.1335～1.2467	99.99
咪鲜胺：苯醚甲环唑	C：B	$y=5.4642+0.6698x$	0.9883	0.2027	0.1159～0.3546	65.46
	A：A	$y=4.8823+0.6581x$	0.9942	1.5039	0.9681～2.3864	9.95
	B：A	$y=5.0528+0.5679x$	0.9976	0.7897	0.6067～1.0280	15.34
	A：B	$y=4.9867+0.7800x$	0.9774	1.0401	0.4443～2.4350	18.86
	A：D	$y=4.9488+0.6237x$	0.9717	1.2080	0.4542～3.2128	21.55
咪鲜胺：吡唑醚菌酯	C：B	$y=5.1369+0.5970x$	0.9744	0.5897	0.2530～1.3746	21.82
	A：A	$y=5.0428+0.6187x$	0.9918	0.8526	0.5204～1.3968	17.10
	B：A	$y=4.9876+0.6749x$	0.9951	1.0433	0.7054～1.5432	11.45
	A：B	$y=5.0158+0.7330x$	0.9974	0.9515	0.7173～1.2622	19.71
	A：D	$y=4.9773+0.6911x$	0.9965	1.0787	0.7757～1.5000	22.49

（续表）

混剂	配比	毒力回归方程	相关系数（r）	EC_{50}	95%置信区间	共毒系数（CTC）
吡唑嘧菌酯：申嗪霉素	C：B	$y=5.0883+0.5694x$	0.990 9	0.699 6	0.421 2～1.161 9	225.40
	A：A	$y=4.8776+0.7232x$	0.991 0	1.476 3	0.842 7～2.586 3	212.17
	B：A	$y=5.2142+0.8057x$	0.978 1	0.542 1	0.249 4～1.178 3	36.18
	A：B	$y=5.2484+0.6134x$	0.995 7	0.393 6	0.282 1～0.546 0	63.98
	A：D	$y=4.8776+0.7232x$	0.991 0	1.476 3	0.842 7～2.586 3	390.48
吡唑嘧菌酯：苯醚甲环唑	C：B	$y=4.8592+0.6372x$	0.994 0	1.663 2	1.040 8～2.657 8	22.77
	A：A	$y=4.8624+0.6859x$	0.991 0	2.068 9	0.825 9～3.311 9	22.39
	B：A	$y=4.9138+0.4763x$	0.993 3	1.516 8	1.299 7～1.770 1	25.41
	A：B	$y=4.8221+0.6932x$	0.972 3	1.805 7	0.640 4～5.091 1	26.86
	A：D	$y=4.9237+0.5674x$	0.990 0	2.387 0	1.775 9～2.998 1	20.77

3 结论与讨论

在所测 7 种单剂对烟草根黑腐病菌的毒力结果表明，多菌灵和咪鲜胺对菌丝生长抑制效果最好，明显优于其他药剂，其次抑菌效果较好的是吡唑嘧菌酯和苯醚甲环唑；在药剂对分生孢子萌发的抑制试验中，吡唑嘧菌酯和多菌灵两种单剂抑制效果最好，咪鲜胺与苯醚甲环唑效果其次。这与不同药剂本身的作用机理相关，多菌灵的抑制机理为干扰真菌有丝分裂中纺锤体的形成，继而影响细胞的分裂；咪鲜胺作用机理是通过抑制麦角甾醇的生物合成，从而使菌体细胞膜功能受破坏而起作用[9]；吡唑醚菌酯则通过抑制线粒体呼吸作用，使线粒体不能产生和提供细胞正常代谢所需要的能量，从而阻止病菌菌丝生长和孢子萌发。7 种单剂中，多菌灵和吡唑嘧菌酯对菌丝生长和孢子萌发都具有较好的抑制效果，但多菌灵连续单一使用容易引致病菌产生抗药性[10]，吡唑醚菌酯相比多菌灵更具有广谱、高效，利于作物生长的特点，增强作物对环境影响的耐受力，提高作物产量的效果。

前人研究报道表明，将不同作用机理的杀菌剂进行合理的混合使用能够提高防效的同时减少用药量[11]，复配剂的研发可以解决由于长期单一使用某一种杀菌剂引起的诸多问题，如防效降低、持效期缩短、病原菌抗药性产生等。本研究中，由不同单剂组合对烟草根黑腐病菌抑制试验结果表明，最优的复配组合分别为多菌灵与苯醚甲环唑的配比组合 A：A 和 A：D；吡唑嘧菌酯与申嗪霉素的配比组合 A：B 和 C：B。

通过室内毒力测定可以得出不同单剂和复配组合对病菌的抑制效果，为安徽省烟草根黑腐病防治提供理论基础，有效单剂和复配剂的防治效果还需进一步通过田间试验验证。

参考文献

［1］ 西南大学植保学院. 烟草根黑腐病［EB/OL］.［2007－3－29］.

［2］ 朱贤朝，王彦亭，王智发. 中国烟草病害［M］. 北京：中国农业出版社，2001：36－451.

［3］ 方中达. 植病研究方法（第 3 版）［M］. 北京：中国农业出版社，1998：46－47.

［4］ 陈年春. 农药生物测定技术［M］. 北京：北京农业大学出版社，1991：95－112.

［5］ 李广领，高扬帆. 8 种新型杀菌剂对 2 种玉米致病菌的室内毒力测定［J］. 安徽农业科学，2005，33（12）：2 265 – 2 266.

［6］ 张志祥. EXCELE 在毒理回归计算中的应用［J］. 昆虫知识，2002，30（1）：33.

［7］ Benton J M，Cobb A H. The plant growth regulator activity of the fungicide，epoxiconazole，on *Galiumaparine* L.［J］Plant Quarantine 1995，17（4），243 – 245.

［8］ Luo Jinyan. Biological characteristic and its quarantine significance of rice Burkholderiaglumae［J］. Plant Quarantine，2003，17（4）：243 – 245.

［9］ 陈平，柳训. 咪鲜胺的应用概况及其残留检测研究［J］. 湖北农业科学，2007（5）：53.

［10］ 赵善欢. 植物化学保护［M］. 北京：中国农业出版社，2000：141 – 142.

［11］ 刘学敏. 杀菌剂混剂的增效作用［J］. 农药科学与管理，2002，23（5）：12 – 15.

山东链霉菌所产抗生素效价测定研究*

马井玉[1**]，李德舜[2]，王付彬[1***]
（1. 山东省济宁市农业科学研究院，济宁 272031；
2. 山东大学微生物技术国家重点实验室，济南 250100）

摘要：对山东链霉菌所产三烯抗生素效价测定方法进行了探索，通过进行生物量法和浊度法实验，最终确定了浊度法作为山东链霉菌所产抗生素效价的测定方法。

关键词：山东链霉菌；效价测定；生物量法；浊度法

Potency Bioassay of Antibiotics Produced by *Streptomyces shandongensis* *

Mang Jingyu[1**], Li Deshun[2], Wang Fubin[1***]
(1. *Jining Academy of Agricultural Sciences*, *Jining* 272031, *China*; 2. *State Key Lab of Microbial Technology*, *Shandong Univ.*, *Jinan* 250100, *China*)

Abstract: According the experiment of *Streptomyces shandongensis* titer determination of exploration, we contrast the biomass method and turbidimetry and found that turbidimetry is superior. Final, we just make turbidimetry as a measurement method of the determination titer of antibiotics.

Key words: *Streptomyces shandongensis*; Potency test; Biomass method; Turbidimetry

农用抗真菌抗生素是微生物代谢过程中的产物，在一定浓度下具有防治植物病虫害，调节植物抗病能力和生长的作用[1-3]。随着化学农药大量使用所带来的一系列问题，农用抗生素的研究一直成为人们所关注的话题。我国在农业抗真菌抗生素方面起步较晚，但是发展迅速。经过几十年的研究和发展，取得了一定的成果，公主岭霉素、井冈霉素、农抗120、武夷菌素、春雷霉素等农用抗生素相继被研发。近年来，随着技术的不断成熟与发展又研究出了天柱菌素、中生菌素等农用抗生素。山东链霉菌是实验室分离的一种淡黄色链霉菌，其产的抗生素具有很好的抗真菌和潜在抗肿瘤效果[4-6]，并且市场上关于抗真菌农药及其医用药物非常稀少，且具有一定的生理毒性。因此对于山东链霉菌所产抗生素的研究具有重大意义。抗生素效价是抗生素生产环节一个重要的指标，对于抗生素生产具有重要的指示作用，须根据抗生素自身情况，选择合适、有效的鉴定方法[7-8]。本文研究的主要目的是对提纯的产品进行抗生素标定，找出一种简单有效的抗生素效价测定方法。

---

* 基金项目：山东省现代农业产业技术体系项目（SDAIT－05－12）

** 第一作者：马井玉，男，山东汶上人，研究员，主要从事作物病虫害综合防控技术研究与应用；E-mail：mjy309@163. com

*** 通讯作者：王付彬，农艺师，主要从事植物病虫害防控技术研究；E-mail：fbw2007@163. com

1 实验材料和方法

1.1 材料

1.1.1 菌种

山东链霉菌（*Streptomyces shandongensis*），白地霉（*Geotrichum candidum*），小麦根霉病菌（*Helminthosporium sorokinum*），棉花枯萎病菌（*Fusrium oxysporum* f. sp. *vasinfectum*），苹果腐烂菌（*Valsa mali*），玉米青枯病菌（*Pythium graminicola* Subram），马铃薯环腐菌（*Corynebacterium sepednicum*），枯草芽孢杆菌（*Bacillus subtilis*），金黄色葡萄球菌（*Staphylococcus aureus*），毛霉（*Mucor* sp.），黑曲霉（*Aspergillus niger*），根霉（*Rhizopus*），酿酒酵母（*Saccharomyces cerevisiae*），热带假丝酵母（*Candida tropicalis*），白色念珠菌（*Monilia albican*），产朊假丝酵母（*Candida utilis*），汉逊酵母（*Hansenula*），以上菌种均有山东大学生命科学学院菌种室提供。

1.1.2 培养基

土豆汁液体培养基：土豆200g，蔗糖20g，水1 000mL 。土豆沸水煮半个小时，过滤加入蔗糖，定容至1 000mL。

土豆汁半固体培养基：土豆200g，蔗糖20g，KH_2PO_4 3g，$MgSO_4$ 1.5g，琼脂10g，水1 000mL。

高氏一号固体培养基：可溶性淀粉20g，KNO_3 1g，NaCl 0.5g，$MgSO_4$ 0.5g，K_2HPO_4 0.5g，$FeSO_4$ 0.01g，琼脂15g，水1 000mL，pH值为7.2～7.4配制时先用少量冷水将淀粉调成糊状，倒入煮沸的水中，边加入边搅拌，然后加入其他成分，121℃高压灭菌20min。

高氏一号液体培养基：KNO_3 1g，NaCl 0.5g，$MgSO_4$ 0.5g，可溶性淀粉20g，K_2HPO_4 0.5g，$FeSO_4$ 0.01g，水1 000mL，pH值为7.2～7.4，121℃高压灭菌20min。

酵母膏麦芽汁培养基：酵母膏10g，葡萄糖4g，麦芽膏10g，水1 000mL，pH值为7.0（如果半固体加10g琼脂，固体加入15～20g琼脂粉）115℃高压灭菌25min。

LB培养基：酵母提取物5g，氯化钠10g，胰蛋白胨10g，水1 000mL，pH值为7.2～7.5（如果半固体加10g琼脂，固体加入15～20g琼脂粉）121℃高压灭菌20min。

水琼脂：琼脂粉15g，1 000mL水，pH值自然，121℃高压灭菌20min。

1.1.3 试剂及耗材

无水乙醇，鼎国；大孔树脂x、y，南开大学化工厂；色谱甲醇，天津四友；微孔滤膜，上海新亚净化器件厂；牛津杯，本实验室提供；氯仿、乙酸乙酯、正丁醇，分析纯。

1.1.4 主要仪器

SHZ-D（Ⅲ）型循环水真空泵，河南省予华仪器有限公司；303-4A数显电热保温箱，宁波自动化仪表研究所；Bio-2000发酵罐（5L），中国镇江东方；WXJ-9388核酸蛋白检测仪，康特高科生物仪器有限公司；RE-52C旋转蒸发仪，河南省予华仪器有限公司；闪式浓缩仪，北京金鼐科技发展有限公司；LC-10Atvp型高效液相色谱，日本Shimadzu；LC-6AD型半制备高效液相色谱，日本Shimadzu；ALPHA 1-2 LD plus冷冻干燥机，德国Christ；UV3001紫外可见分光光度计，岛津；HZQ-C空气浴振荡器，哈尔滨东联电子技术开发有限公司；千分尺，实验室原有。

1.2 研究方法

1.2.1 抗生素纯品的提取

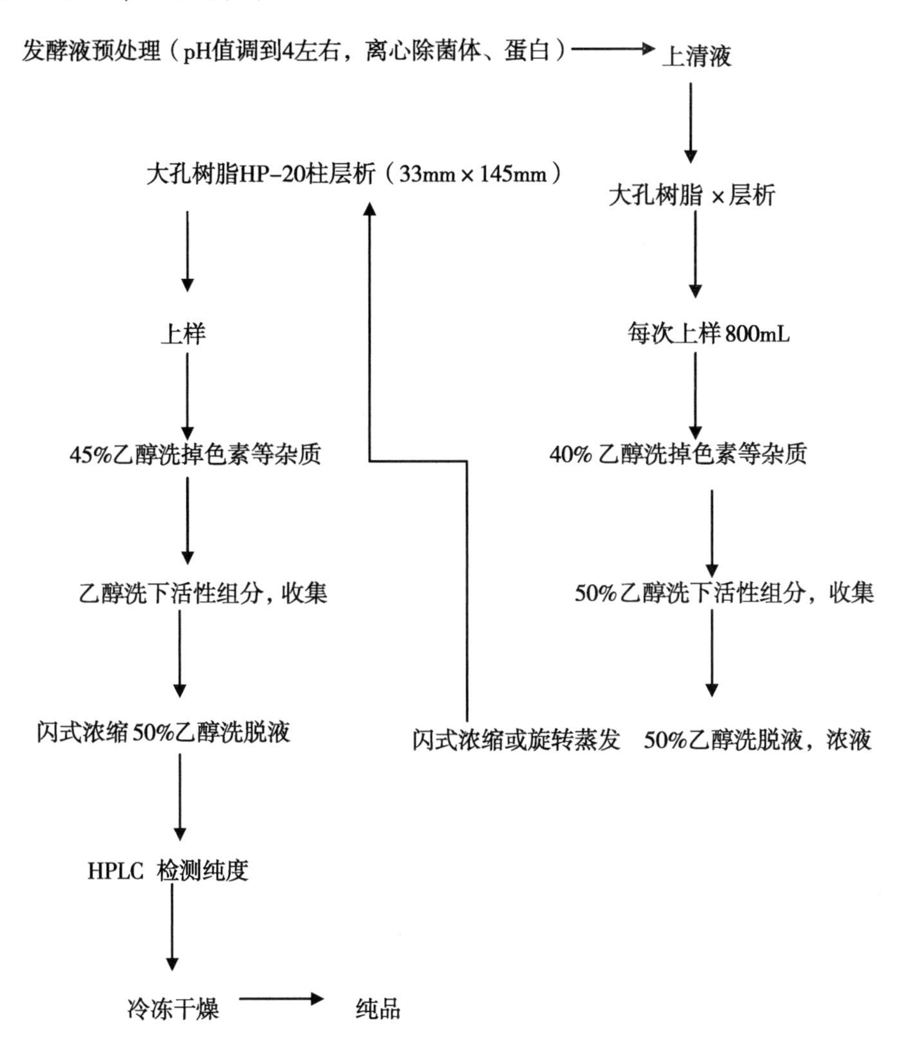

1.2.2 抗生素发酵液的制备

种子液制备：从山东链霉菌茄子瓶斜面中刮取孢子于含玻璃珠的无菌水中，振荡20min，吸取5mL于含50mL发酵液的300mL三角瓶中，28℃培养48h。

发酵液制备：吸取10mL种子液，接种到含有90mL发酵液的500mL三角瓶中，28℃培养6d，最后调pH值，离心除菌体。

发酵罐中发酵液制备：5L发酵罐，按照10%接种量，接种种子液，280r/min发酵6～7d，最后离心除菌体。

1.2.3 指示菌的选择

由于生物量法测定抗生素效价对指示菌的要求较高，不同的指示菌边缘整齐程度及指示菌浓度的控制有所区别，因此，不同的指示菌对于抗生素效价的测定有不同的影响。三烯霉素具有强烈抗真菌作用，因此，本实验选取了具有代表性的真菌和一些细菌，对指示菌进行筛选，从而选择合适的指示菌。所选指示菌包括白地霉、小麦根霉病菌、棉花枯萎病菌、苹果腐烂菌、玉米青枯病菌、土豆环腐菌、枯草芽孢杆菌、金黄色葡萄球菌、毛霉、黑曲霉、根霉、酿酒酵母、汉逊酵母、白色念珠菌、产朊假丝酵母、热带假丝酵母。采用管碟法观察不同指示菌抑菌圈的大小以及抑菌圈边缘的整齐程度。

1.2.4 指示菌的培养及抑菌圈测定效价方法

指示菌的培养：取指示菌 1mL，加入到含 50mL 发酵液的 300mL 三角瓶中，30℃培养至 OD_{650} 为 0.8 即可。

指示菌抑菌圈测定方法：采用双层平板法，现在平板底部倒一层水琼脂培养基（15mL），待培养基冷却后再倒一层混有指示菌的 PDA 培养基（5mL），待冷却后加入牛津杯，放置大约 10min 后，加入发酵液约 200μL，放入 4℃冰箱 8～9h，以便发酵液能够充分扩散。最后将培养基放入 30℃恒温培养箱，培养 36h，测量抑菌圈大小。根据浓度对数与抑菌圈直径的关系，对效价测定准确度进行评估。

1.2.5 比浊法测定效价

把发酵液稀释成不同的浓度，进行紫外扫描，测其在 271nm 下的吸收值，进行线性回归分析，对比浊法测定效价效果进行预测。

1.2.6 纯品标准曲线的制备和高效液相纯品的标定

纯品标准液制备：取液相制备出来的纯品，配制成 500mg/mL 的母液，然后进行稀释，测定不同浓度的纯品在 271nm 下的吸收值。

1.2.7 HPLC 外标法测定抗生素含量（对效价测定方法的准确性进行标定）

HPLC 纯品标定：取一定量液相制备出来的纯品，分别配置成 1 000、750、500、250、100 以及 50mg/L 的溶液，在液相条件下（流动相为甲醇和 20mm 的乙酸铵），测定其峰面积，制作标准曲线。

效价测定方法的准确性测定：取一定量发酵液，根据现有效价测定方法估测其抗生素含量，然后 HPLC 外标法测定其准确含量，两者进行比较，对效价测定方法的准确性进行评定。

2 结果与分析

2.1 指示菌的选择

由三烯霉素对于丝状真菌具有较强的抗性，对于非丝状真菌并没有表现出抗菌活性，对于某些细菌具有微弱的抗性（表 1）。因此，指示菌的选择应选择丝状真菌。但是由于小麦根霉病菌、棉花枯萎病菌、苹果腐烂菌等丝状真菌产孢子能力弱，且抑菌圈边缘不整齐，对于效价测定误差较大，因此不适合做指示菌。而白地霉虽然属于酵母，但是性质居于酵母和霉菌之间，用其做指示菌，不仅具有较大的抑菌圈，而且抑菌圈边缘整齐，菌体生长速度快，前期数量可以测定，因此，选择白地霉作为效价测定的指示菌能够有效地降低效价测定中的误差。

表 1 不同指示菌的性质及抑菌圈直径

指示菌	抑菌圈直径（mm）	性质
小麦根霉病菌	55.3	抑菌圈边缘不光滑
棉花枯萎病菌	47.6	抑菌圈边缘不光滑

（续表）

指示菌	抑菌圈直径（mm）	性质
苹果腐烂菌	53.3	抑菌圈边缘不光滑
玉米青枯病菌	51.4	抑菌圈边缘不光滑
马铃薯环腐菌	49.2	抑菌圈边缘不光滑
毛霉	42.6	抑菌圈边缘不光滑
黑曲霉	47.5	抑菌圈前期光滑，后期具有一定不规整性
根霉	46.4	抑菌圈边缘不光滑
白地霉	36.8	边缘光滑，抑菌圈直径规整
酿酒酵母	无	
汉逊酵母	无	
白色念珠菌	无	
产朊假丝酵母	无	
热带假丝酵母	无	
枯草芽孢杆菌	无	
金黄色葡萄球菌	9.2	边缘光滑

2.2 管碟法测定抗生素效价

2.2.1 管碟法发酵液抗生素效价预测（表2，图1）

表2 不同浓度发酵液抑菌圈直径

发酵液相对浓度	发酵液相对浓度*120	浓度取对数	抑菌圈直径（mm）			抑菌圈平均直径（mm）
1	120	2.079 181	31.2	28.3	30.2	29.9
2/3	80	1.903 09	28.4	31.1	26.3	28.6
1/2	60	1.778 151	25.6	24.7	26.4	25.57
1/3	40	1.602 06	22.4	23.6	21.4	22.47
1/6	20	1.301 03	13.2	14.8	13.7	13.9

进行回归分析，数据如下：

SUMMARY OUTPUT

回归统计	
Multiple R	0.981 988
R Square	0.964 301
Adjusted R Square	0.952 401

（续表）

回归统计	
标准误差	0.064 954
观测值	5

方差分析

	df	*SS*	*MS*	*F*	Significance *F*
回归分析	1	0.341 896	0.341 896	81.036 07	0.002 894
残差	3	0.012 657	0.004 219		
总计	4	0.354 553			

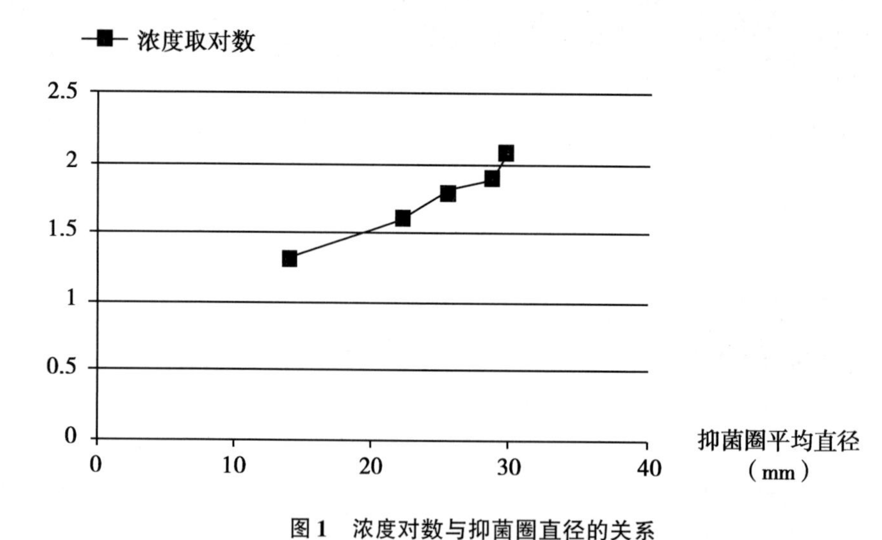

图 1　浓度对数与抑菌圈直径的关系

Fig. 1　Log concentration with inhibition zone diameter

由回归分析可见相关性系数 $R=0.982$，因此具有相当高的相关性。

2.2.2　管碟法标准曲线建立（表 3，图 2）

表 3　抑菌圈直径与浓度的关系

浓度（mg/L）	500	350	250	100	50	25
浓度对数	2.698 9	2.544 06	2.397 9	2.00	1.698 9	1.397 9
抑菌圈直径（mm）	32.4	27.8	25.8	20.6	17.4	15.6

2.3　比浊法测定抗生素效价测定

2.3.1　比浊法发酵液抗生素效价预测

由图 3 和表 4 可以看出相关系数 R 很高，因此具有很高的相关性。发酵液 3 为发酵罐发酵，可能是发酵罐发酵剪切力大导致部分菌体自溶，包内物质如蛋白、核酸等对于吸光度具有一定的影响。

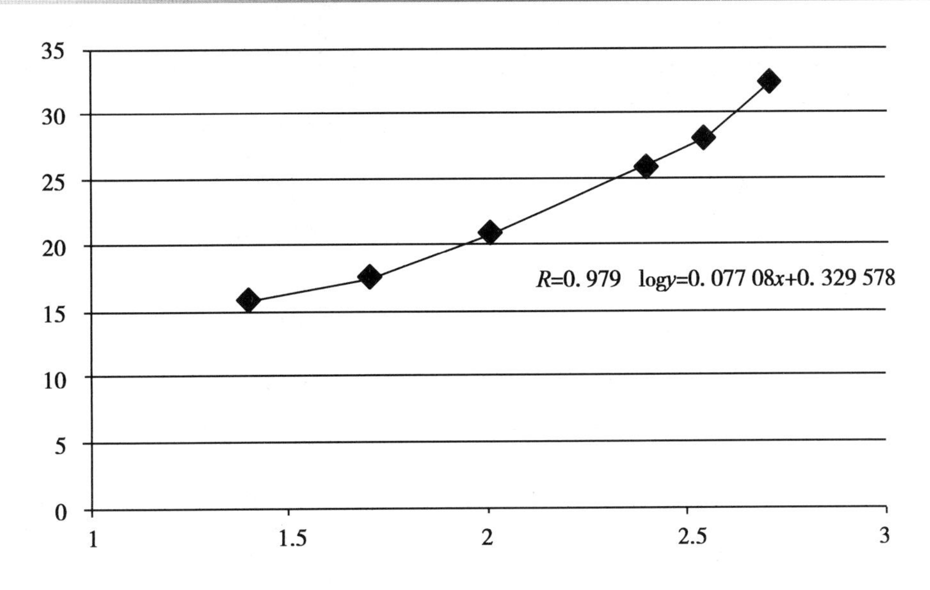

图 2　浓度对数与抑菌圈直径关系

Fig. 2　Log concentration with inhibition zone diameter

表 4　三种发酵液紫外吸收与浓度的关系

发酵液 1			发酵液 2			发酵液 3		
浓度	浓度×120	紫外吸收	浓度	浓度×120	紫外吸收	浓度	浓度×120	紫外吸收
1/3	40	2.834	1/3	40	3.286	1/3	40	3.127
1/4	30	2.147	1/4	30	2.413	1/4	30	2.513
1/5	24	1.67	1/5	24	1.938	1/5	24	1.865
1/6	20	1.403	1/6	20	1.649	1/6	20	1.57
1/10	12	0.873	1/8	15	1.211	1/8	15	1.171

2.3.2　比浊法标准曲线建立（表 5、图 4，图 5）

表 5　纯品紫外吸收与浓度的关系

浓度（mg）	500	250	100	50	25
紫外吸收	18.8	9.21	3.818	1.921	0.985

2.4　HPLC 对两种效价测定方法准确性检验

2.4.1　HPLC 标准曲线的绘制

高效液相下峰面积与浓度的关系（表 6）；由图 6 可见峰面积与纯品的浓度线性关系的相关度很高，符合实验要求。

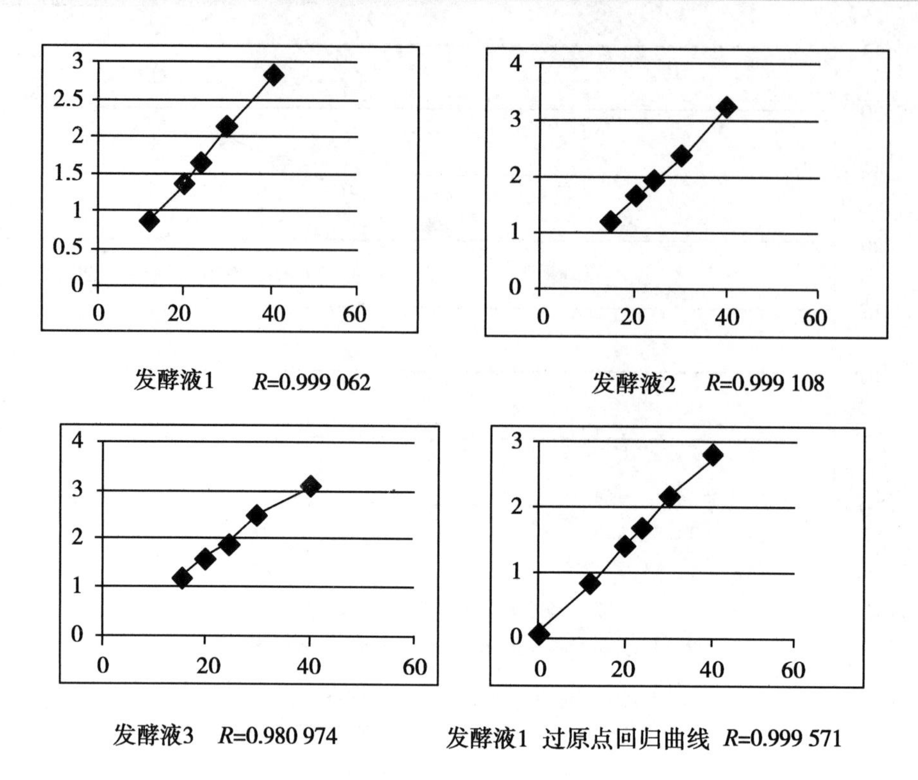

图3　3种发酵液紫外吸收与浓度关系（Y轴为紫外吸收，X轴为相对浓度）

Fig. 3　Relationship of fermented liquid ultraviolet absorption and concentration

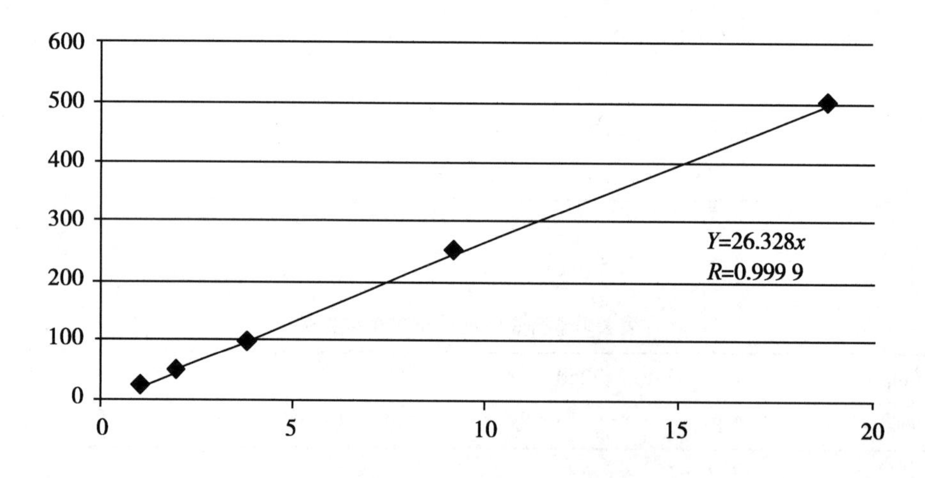

图4　纯品浓度与紫外吸收的线性分析

Fig. 4　Linear analysis of concentration and ultraviolet absorption

表6　高效液相下峰面积与浓度的关系

浓度（mg/L）	1 000	750	500	250	100
峰面积	13 611 272	9 882 341	6 226 752	3 154 728	1 134 613

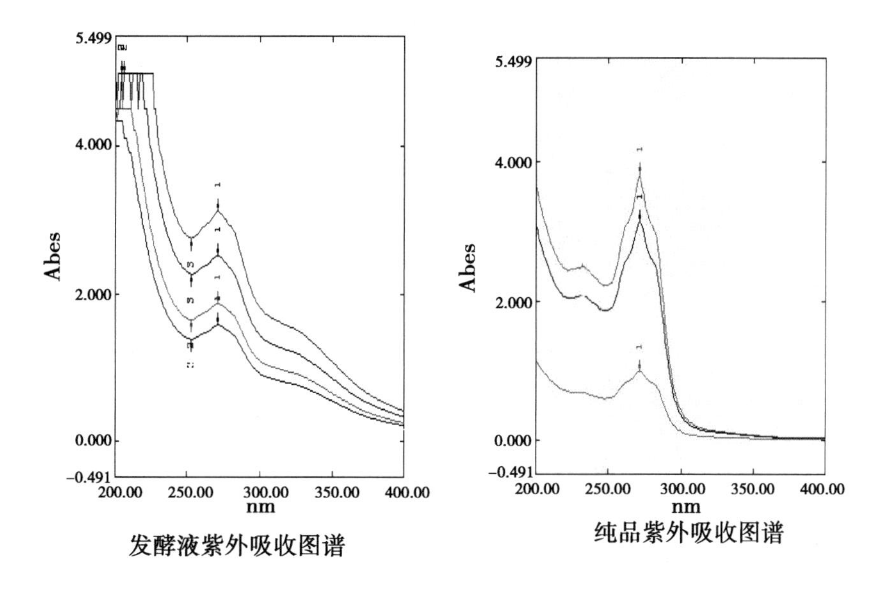

图 5　发酵液紫外吸收图谱和纯品紫外吸收图谱

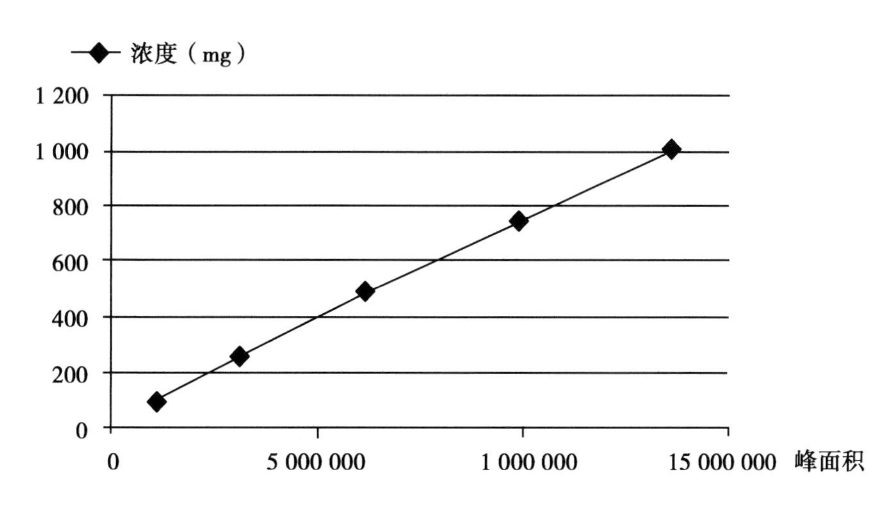

图 6　浓度与峰面积的线性关系（R = 0.998 471）

Fig. 6　Linear relationship of the concentration and peak area

2.4.2　比浊法准确性验证

表 7　估测值与实测值的关系

紫外吸收	9.03	4.18	3.25	2.32
估测值（mg）	237.489	109.934	85.475	61.016
HPLC 实测值（mg）	224.85	112.647	86.47	59.624

由表 7 和表 8 可以看出，F = 0.002 327远小于 $F_{0.01(1,7)}$ 和 $F_{0.05(1,7)}$，因此不具有显著性差异，准确性很高。

表 8　单因素方差分析

SUMMARY

组	观测数	求和	平均	方差
行 1	4	493.914	123.478 5	6 175.893
行 2	4	483.591	120.897 8	5 271.296

方差分析

差异源	*SS*	*df*	*MS*	*F*	*P*-value	*F* crit
组间	13.320 54	1	13.320 54	0.002 327	0.963 089	5.987 378
组内	34 341.57	6	5 723.594			
总计	34 354.89	7				

2.4.3　管碟法准确性验证

由表 9 和表 10 可以看出，$F=0.011\,094$小于 $F_{0.05(1,7)}$，但是大于 $F_{0.01(1,7)}$，因此具有一定的准确性。

表 9　管碟法估测值与实测值之间的关系

抑菌圈直径	21.6	15.4	30	26.7	23
估测值	98.7	32.85	438.5	244.1	126.6
实测值	105	30.3	406.4	232.6	115.4

表 10　管碟法单因素方差分析

组	观测数	求和	平均	方差
行 1	5	940.75	188.15	25 428.33
行 2	5	889.7	177.94	21 552.69

方差分析

差异源	*SS*	*df*	*MS*	*F*	*P*-value
组间	260.610 25	1	260.610 25	0.011 094	0.918 708
组内	187 924.072	8	23 490.509		

3　小结与讨论

通过以上微生物测定法（管碟法）和浊度法测定三烯霉素效价，我们可以看出这两种方法都具有一定的可行性和准确性，但是相比来说浊度法的准确性要高于管碟法，这可能是管碟法存在的误差较大，如指示菌菌浓度和活性每次测定都具有一定的偏差，培养基的 pH、牛津杯摆放方式、操作误差以及培养箱所放位置不同造成的温度偏差等，都对实验结果造成一定的影响。而浊度法，在测定发酵罐发酵的发酵液时，准确性稍低，造成这样的原因可能是发酵罐剪切力大，造成部分菌体内物质外泄，从而对结果造成一定的影响，因此，在测定采用发酵罐发酵后的发酵液需要进行预处理，如对发酵液进行除蛋白和核酸等。但从上述实验结果看出，采用浊度法测定三烯霉素的效价具有准确性高且操作简单的特点。因此，本实验采用比浊法作为抗生素效价的测定方法。

参考文献

[1] Xu F Y, Li Z H, Zeng H C. Progress on the research in antifungal agricultural antibiotics [J]. Chinese Journal of Tropical Agriculture（热带农业科学）, 2005, 25 (1): 60-65.

[2] Cui Z J, Zhang K C, She G M, et al. Progress on the Research of Active Components in Antifungal Agricultural Antibiotics [J]. Chinese Agricultural Science Bulletin（中国农学通报）, 2010, 26 (5): 213-218.

[3] Shen Y C. Recent progress on the research and development in Agricultural Antibiotics [J]. Plant Protection Technology and Extension（植保技术与推广）, 1997, 17 (6): 35-37.

[4] Li D S, Su Z R, Yuan Z G, et al. Primary studies on antibiotics produced by *Streptomyces shandongensis* sp. nov. [J]. Journal of Shandong University（山东大学学报（理学版））, 2004, 39 (6): 121-124.

[5] Ma J Y, Li D S. The Study of Mechanism of Antibiotic from Streptomyces shandongensis Effect to Fungi [J]. Edible Fungi of China（中国食用菌）, 2008, 27 (suppl.): 61-65.

[6] Feng Y P, Gao F R, Ma J Y. Control and yield increasing of tomato and cucumber diseases in Greenhouse of *Streptomyces shandongensis* [J]. Shandong Agricultural Sciences（山东农业科学）, 2010, (8): 80-82.

[7] Chang R H. Study on the factors influencing the experimental results in the determination of antibiotics titer [J]. Jiangsu Pharmaceutical and Clinical Research（江苏药学与临床研究）, 1999, 7 (4): 60-61.

[8] Jia C F, Wang X M, Liu M, et al. Some improvements on the experiment of antibiotic titer determination [J]. Bulletin of Biology（生物学通报）, 2008, 43 (10): 48-49.

黄瓜霜霉病菌对不同药剂敏感性及相应药剂田间防效验证*

孟润杰**，王文桥***，赵建江，马志强，韩秀英

（河北省农林科学院植物保护研究所，河北省农业有害生物综合防治工程技术研究中心，农业部华北北部作物有害生物综合治理重点实验室，保定 071000）

摘要：为明确黄瓜霜霉病菌对不同药剂敏感性及相应药剂对黄瓜霜霉病的田间防效，采用叶盘漂浮法测定了2014年和2015年从河北省定兴县贤寓镇龙华村采集的黄瓜霜霉病菌对不同药剂的敏感性，并进行田间药效试验验证。结果表明，所采集的黄瓜霜霉病菌对甲霜灵和嘧菌酯的抗性频率均为100%，平均抗性水平分别为583.4倍和312.9倍，抗性菌株均为高抗菌株；对烯酰吗啉、双炔酰菌胺和氟吡菌胺的抗性频率分别为50%、20%和30%，平均抗性水平分别为3.9倍、0.8倍和1.1倍，抗性菌株均为低抗菌株；田间按照推荐剂量第三次施药后7d，58%甲霜灵·代森锰锌WP及68%精甲霜灵·代森锰锌WG对黄瓜霜霉病的防效为67.0%~71.2%，与80%代森锰锌WP的防效（68.0%~70.6%）相当，250g/L嘧菌酯SC对霜霉病的防效为75.6%~78.5%，显著低于50%烯酰吗啉WP、250g/L双炔酰菌胺SC和687.5g/L氟吡菌胺·霜霉威盐酸盐SC的防效（>85%）。黄瓜霜霉病菌田间抗性发生情况为相应药剂的田间药效所验证：黄瓜霜霉病菌对甲霜灵和嘧菌酯产生抗性而导致58%甲霜灵·代森锰锌WP、68%精甲霜灵·代森锰锌WG和250g/L嘧菌酯SC防效丧失或较差，而霜霉病菌对烯酰吗啉、双炔酰菌胺和氟吡菌胺仍保持敏感，50%烯酰吗啉WP、250g/L双炔酰菌胺SC和687.5g/L氟吡菌胺·霜霉威盐酸盐SC可用于黄瓜霜霉病的防治。

关键词：黄瓜霜霉病菌；杀菌剂；敏感性；抗性；田间防治效果

Sensitivity to Fungicides of *Pseudoperonospora cubensis* and Controlling Efficacy of Corresponding Fungicides against Cucumber Downy Mildew*

Meng Runjie**, Wang Wenqiao***, Zhao Jianjian, Ma Zhiqiang, Han Xiuying

(*Institute of Plant Protection, Hebei Academy of Agricultural and Forestry Sciences, IPM Center of Hebei Province, Key Laboratory of Integrated Pest Management on Crops in Northern Region of North China, Ministry of Agriculture, Baoding* 071000, *China*)

Abstract: To confirm the sensitivity to different fungicides of *P. cubensis* and validate the efficacies by field trails. The sensitivity to different fungicides of *Pseudoperonospora cubensis* isolates collected from Longhua village, Xianyu town, Dingxing county in Hebei province in 2014 and 2015 was determined by leaf disk floating test. The control efficacies of different fungicides against cucumber downy mildew were tested in 2014—2015. The results showed that the resistance frequency ofmetalaxyl and azoxystrobin of *P. cubensis*were 100% and the average resistance level

* 基金项目：公益性行业（农业）科研专项（201303023）；国家科技支撑计划（2012BAD19B06）

** 第一作者：孟润杰，男，助理研究员，研究方向为杀菌剂应用研究；E-mail：runjiem@163.com

*** 通讯作者：王文桥，研究员；E-mail：wenqiaow@163.com

were 583.4 times and 312.9 times, the isolates were all high resistance. The resistance frequency of dimethomorph, mandipropamid and fluopicolide of *P. cubensis* were 50%, 20% and 30%, but the isolates were all low resistance and the average resistance level were 3.9 times, 0.8 times and 1.1 times. Seven days after 3 sprays at the recommended rates, the control efficacies of metalaxyl-mancozeb 58WP and mefenoxam-mancozeb 68WP were 67.0% ~ 71.2%, were not significantly different from the controlling efficacy of mancozeb 80WP (68.0% ~70.6%) against cucumber downy mildew, the control efficacy of azoxystrobin 250g/L SC was 75.6% ~78.5%, significantly less effective than dimethomorph 50WP, mandipropamid 250g/L SC and fluopicolide-propamocab hydrochloride 687.5g/L SC (>85%) against cucumber downy mildew. Severe resistance occurrence in *P. cubensis* to metalaxyl and azoxystrobin led to loss or decrease of efficacy of metalaxyl-mancozeb 58WP, mefenoxam-mancozeb 68WP and azoxystrobin 250g/L SC, but no severe resistance occurred to dimethomorph, mandipropamid and fluopicolide, therefore, dimethomorph 50WP, azoxystrobin 250g/L SC and fluopicolide-propamocab hydrochloride 687.5g/L SC can be used for control of cucumber downy mildew.

Key words: *Pseudoperonospora cubensis*; Fungicides; Sensitivity; Resistance; Field control efficacy

黄瓜霜霉病是由古巴假霜霉（*Pseudoperonospora cubensis*）引起的一种毁灭性病害[1]，该病流行性强、传播速度快、发病重，给黄瓜生产造成经济损失严重，生产上主要依靠化学药剂进行防治[2]。目前，防治霜霉病的内吸杀菌剂主要有苯基酰胺类杀菌剂甲霜灵、精甲霜灵，羧酸酰胺类杀菌剂烯酰吗啉、氟吗啉、双炔酰菌胺，乙酰胺类杀菌剂霜脲氰，氨基甲酸酯类杀菌剂霜霉威，甲氧基丙烯酸酯类杀菌剂嘧菌酯，苯甲酰胺类杀菌剂氟吡菌胺等[3-4]。由于生产中化学药剂的不合理使用以及不正确的使用技术等问题，导致黄瓜霜霉病菌对部分化学药剂产生抗性从而导致防治失败[5]。甲霜灵、烯酰吗啉、嘧菌酯、双炔酰菌胺和氟吡菌胺已在我国广泛用于黄瓜霜霉病等卵菌病害多年，当前的研究多为单方面的研究黄瓜霜霉病菌对药剂的敏感性或药剂对黄瓜霜霉病的田间防效，缺少将黄瓜霜霉病菌对这些药剂敏感性与相应药剂田间药效反应相互验证的研究报道，而这又直接制约着生产上选药、用药及如何制定黄瓜霜霉病高效、可持续化学控制策略。因此，本研究测定了同一地区不同年份霜霉病菌对甲霜灵、嘧菌酯、烯酰吗啉、双炔酰菌胺及氟吡菌胺的敏感性，同时在该地区进行相应药剂的田间药效验，以明确黄瓜霜霉病菌对各药剂敏感性与其田间防效的对应关系，得出更加准确的评价，为黄瓜霜霉病菌抗药性治理和杀菌剂的合理使用提供依据。

1 材料与方法

1.1 供试菌株

黄瓜霜霉病菌（*Pseudoperonospora cubensis*）。2014 年和 2015 年从河北省定兴县贤寓镇龙华村黄瓜温室采集新鲜的霜霉病叶，装入低温保温箱带回实验室，先用清水冲去病斑上的孢子囊，然后在 18 ~20℃条件下保湿培养至产生大量新生孢子囊，用去离子水冲洗下新鲜孢子囊，2 500r/min 离心 3 次，加去离子水配制成 1×10^5 个孢子囊/mL 的悬浮液用于接种。

1.2 供试药剂

98%甲霜灵原药，浙江禾本科技有限公司；95%嘧菌酯原药，南京金土地化工有限公司；97.6%烯酰吗啉原药，河北冠龙农化有限公司；93%双炔酰菌胺原药，先正达作物保护有限公司；96%氟吡菌胺原药，拜耳作物科学（中国）有限公司。将原药用丙酮溶解，加入无菌水稀释，配制成 1 000μg/mL 母液，保存于 4℃冰箱中，用于测定黄瓜霜霉病菌对不同药剂的敏感性。

50%烯酰吗啉WP（阿克白），巴斯夫（中国）有限公司；250g/L嘧菌酯SC（阿米西达）、68%精甲霜灵·代森锰锌WG（金雷），先正达（苏州）作物保护有限公司；58%甲霜灵·代森锰锌WP（雷佳米），深圳诺普信农化股份有限公司；687.5g/L氟吡菌胺·霜霉威盐酸盐SC（银法利），拜耳作物科学（中国）有限公司；80%代森锰锌WP（大生），美国陶氏益农化工有限公司。用于田间药效试验。

1.3 供试作物

黄瓜品种为“新泰密刺”，种植于河北省农科院植保所日光温室中，待植株长至6～7片真叶时，采集叶片并打取直径1.5cm的叶盘供试。

1.4 敏感性测定

采用叶盘漂浮法[6]测定黄瓜霜霉病菌对不同杀菌剂的敏感性。将1.2中配制好的甲霜灵和嘧菌酯母液稀释为200μg/mL、100μg/mL、50μg/mL、10μg/mL、5μg/mL的药液，烯酰吗啉母液稀释为100μg/mL、50μg/mL、10μg/mL、5μg/mL、1μg/mL的药液，双炔酰菌胺和氟吡菌胺母液稀释为5μg/mL、1μg/mL、0.5μg/mL、0.1μg/mL、0.05μg/mL的药液。将配好的药液倒入9 cm的培养皿中（20mL/皿），空白对照倒去离子水。将1.3中打取的叶盘叶背朝上漂浮于培养皿中的药液上，每皿15个叶盘，设3次重复。每叶盘中心点1滴（10μL）1.1中配制好的霜霉病菌孢子囊悬浮液，盖上皿盖，置于19℃、16h光照/8h黑暗的生长室中培养7～10d，对照充分发病后调查叶盘上的发病情况。根据产孢子囊面积占整个叶盘面积的百分率划分病级：0级，无病；1级，1%～5%；3级，6%～10%；5级，11%～25%；7级，26%～50%；9级，>50%。计算病情指数和相对防效，通过7.05版的DPS数据分析软件进行药剂浓度与防效之间的线性回归分析，求出药剂的毒力回归方程$Y=bx+a$、相关系数（r）以及有效抑制中浓度（EC_{50}）。

$$病情指数=\sum（叶盘数\times相对级数）/（叶盘总数\times最高级数）\times100$$

$$相对防效（\%）=［（对照病情指数-处理病情指数）/对照病情指数］\times100$$

1.5 敏感性划分

王文桥[7]、刘晓宇[8]、闫磊[9]、崔继敏[10]等采用叶盘漂浮法建立的黄瓜霜霉病菌对甲霜灵、烯酰吗啉、嘧菌酯、氟吡菌胺和双炔酰菌胺的敏感基线分别为0.047μg/mL、0.65μg/mL、0.0062μg/mL、0.15μg/mL和0.358μg/mL。参照上述敏感基线，将黄瓜霜霉病菌对供试药剂的抗性类型划分为5个级别：敏感菌株（S），抗性水平≤2倍；低抗菌株（LR），2倍<抗性水平≤10倍；中抗菌株（MR），10倍<抗性水平≤100倍；高抗菌株（HR），100倍<抗性水平≤1 000倍；特高抗菌株（EHR），抗性水平>1 000倍。

$$抗性水平=敏感性（EC_{50}）/敏感基线（EC_{50}）$$

$$抗性频率（\%）=抗性菌株数/全部供试菌株数\times100$$

1.6 田间药效试验

分别于2014年和2015年在河北省定兴县贤寓镇龙华村日光温室进行不同杀菌剂防治黄瓜霜霉病田间药效试验。黄瓜品种为“津优303”，垄距1m，株距0.3m，一垄双行。试验地土质为壤土，偏碱，土壤肥力较高，定植前亩施鸡粪$5m^3$，栽培管理条件一致。黄瓜霜霉病零星发生时开始用药，间隔期7d，共用药3次。采用Jacto-40型手动喷雾器叶面喷雾，用药液量$900L/hm^2$。

试验处理：250g/L嘧菌酯SC 150 g a.i./hm^2；68%精甲霜灵·代森锰锌WG 1 224g a.i./hm^2；58%甲霜灵·代森锰锌WP 1 566g a.i./hm^2；50%烯酰吗啉WP 300g

a. i. /hm^2；687. 5g/L 氟吡菌胺·霜霉威盐酸盐 SC 1 031g a. i. /hm^2；80%代森锰锌 WP 2 400g a. i. /hm^2。设清水对照。每处理4次重复，小区面积21m^2，采用随机区组排列。

施药前调查病情基数，因零星发病，病情基数视为零；最后一次用药后7天再次调查发病情况。每小区随机5点取样，每点调查2株，每株调查全部叶片，按各叶片上病斑面积占整个叶片面积的百分率划分病级：0级，无病；1级，<5%；3级，6%~10%；5级，11%~25%；7级，26%~50%；9级，>51%。记录各处理病情，计算病情指数及防治效果。

病情指数=［Σ（病叶数×相对级值数）/（调查总叶数×最高级值数）］×100

防治效果（%）=［（对照区病情指数-处理区病情指数）/对照区病情指数］×100

1.7 数据统计分析

试验数据采用7.05版的DPS数据分析软件进行统计分析。

2 结果与分析

2.1 黄瓜霜霉病菌对不同杀菌剂敏感性

2014年和2015年从定兴县贤寓镇龙华村采集的40株黄瓜霜霉病菌对甲霜灵和嘧菌酯的抗性频率均为100%，所测菌株对甲霜灵和嘧菌酯的平均抗性水平为583.4倍和312.9倍，均为高抗菌株，表明所采集的黄瓜霜霉病菌对甲霜灵和嘧菌酯普遍产生较高水平的抗性；对烯酰吗啉、双炔酰菌胺和氟吡菌胺的抗性频率分别为50%、20%和30%，平均抗性水平分别为3.9倍、0.8倍和1.1倍，所检测到的抗性菌株均为低抗菌株，虽然检测到霜霉病菌对烯酰吗啉、双炔酰菌胺和氟吡菌胺的低抗菌株，但抗性水平较低，这3种药剂对黄瓜霜霉病菌毒力较高（表1）。

表1 黄瓜霜霉病菌对不同药剂的敏感性

Table 1 Sensitivity of *Pseudoperonospora cubensis* to different fungicides

杀菌剂	菌株数	平均抗性水平	不同类型菌株百分率（%）					抗性频率（%）
			S	LR	MR	HR	EHR	
甲霜灵	40	583.4	0.0	0.0	0.0	100.0	0.0	100.0
嘧菌酯	40	312.9	0.0	0.0	0.0	100.0	0.0	100.0
烯酰吗啉	40	3.9	50.0	50.0	0.0	0.0	0.0	50.0
双炔酰菌胺	40	0.8	80.0	20.0	0.0	0.0	0.0	20.0
氟吡菌胺	40	1.1	70.0	30.0	0.0	0.0	0.0	30.0

注：S代表敏感，LR代表低抗，MR代表中抗，HR代表高抗，EHR代表特高抗

Note: S stands for sensitive, LR stands for low resistance, MR stands for intermediate resistance, HR stands for high resistance, EHR stands for extremely high resistance.

2.2 不同杀菌剂对设施黄瓜霜霉病的防治效果

2014年和2015年田间药效试验表明，按照推荐剂量进行3次喷施，68%精甲霜灵·代森锰锌 WG 1 224g a. i. /hm^2 和58%甲霜灵·代森锰锌 WP 1 566 a. i. /hm^2 的防效为67.0%~71.2%，与80%代森锰锌 WP 2 400g a. i. /hm^2 的防效相当，表明混剂68%精甲霜灵·代森锰锌 WG 和58%甲霜灵·代森锰锌 WP 中的甲霜灵和精甲霜灵对黄瓜霜霉病菌的

防效几乎丧失，起防治作用的应该为代森锰锌；喷施250g/L嘧菌酯SC 150g a. i. /hm^2 的处理对黄瓜霜霉病的防效为75.6% ~78.5%，对黄瓜霜霉病防效较差的原因可能是由于霜霉病菌对嘧菌酯的敏感性下降造成的；喷施687.5g/L氟吡菌胺·霜霉威盐酸盐SC 1 031g a. i. /hm^2、50%烯酰吗啉WP 300g a. i. /hm^2 和250g/L双炔酰菌胺SC对黄瓜霜霉病的防效均高于85%，防效良好且稳定（表2）。

表2 不同杀菌剂对黄瓜霜霉病田间防效

Table 2 Efficacy of different fungicides in controlling cucumber downy mildew in the field

杀菌剂	剂量（g a. i. /hm^2）	2014年		2015年	
		病情指数	防效（%）	病情指数	防效（%）
50%烯酰吗啉WP	300	5.12	86.0 a	5.65	86.6 a
250g/L双炔酰菌胺SC	150	4.08	88.8 a	4.93	88.3 a
250g/L嘧菌酯SC	150	7.86	78.5 b	10.26	75.6 b
687.5g/L氟吡菌胺·霜霉威盐酸盐SC	1 031	3.86	89.4 a	4.86	88.4 a
68%精甲霜灵·代森锰锌WG	1 224	10.95	70.0 c	12.11	71.2 c
58%甲霜灵·代森锰锌可湿性粉剂	1 556	12.06	67.0 c	13.62	67.6 c
80%代森锰锌可湿性粉剂	2 400	11.68	68.0 c	12.35	70.6 c
空白对照		36.52		42.05	

注：根据LSD分析，表中同组、同列数据后相同小写字母表示在0.05的水平下差异不显著。

Note: Data in a column followed by the same letters are not significantly different at $P = 0.05$ by Fisher's LSD test

3 讨论

由于受田间种植环境、药剂的加工技术及药剂特性等因素的影响，病原菌对药剂的敏感性并不完全等同于药剂的田间防效，单方面的研究病菌对药剂的敏感性或药剂的田间防效都是片面的，不能得到全面准确的结论。本研究结合敏感性检测和田间防效试验推断：试验地区黄瓜霜霉病菌对甲霜灵和嘧菌酯敏感性下降，产生严重抗性，由于苯基酰胺类化合物之间存在交互抗性，混剂58%甲霜灵·代森锰锌WP和68%精甲霜灵·代森锰锌WG中甲霜灵和精甲霜灵和对黄瓜霜霉病的防治作用几乎丧失。250g/L嘧菌酯SC在我国最初投入使用时对黄瓜霜霉病的防效达90%以上[12]，但本试验中250g/L嘧菌酯SC对黄瓜霜霉病田间防效为75.6% ~78.5%，防治效果的降低可能是由于黄瓜霜霉病菌对嘧菌酯产生抗性而造成的。黄瓜霜霉病菌对甲霜灵和嘧菌酯普遍产生抗性的原因可能在于霜霉病菌抗性菌株适合度好，设施栽培的黄瓜温室大棚中，气候条件非常适合霜霉病发生，苯基酰胺类及甲氧基丙烯酸酯类药剂长期普遍使用，频繁施药，对病菌群体持续施加选择压力有关。因此，在黄瓜霜霉病菌对甲霜灵和嘧菌酯普遍产生抗性的黄瓜产区应停止或慎用以这些药剂为主要成分的制剂。

黄瓜霜霉病菌对烯酰吗啉、双炔酰菌胺和氟吡菌胺较为敏感，含有这3种有效成分的制剂687.5g/L氟吡菌胺·霜霉威盐酸盐SC、50%烯酰吗啉WP和250g/L双炔酰菌胺SC对黄瓜霜霉病具有良好的防治效果。但由于黄瓜霜霉病菌为高抗药性风险病原菌[3]，烯酰吗啉、双炔酰菌胺等羧酸酰胺类杀菌剂被认为是具有中等抗性风险的药剂[5]，黄瓜霜霉病菌对氟吡菌胺存在中等抗性风险[10]，因此，使用烯酰吗啉、双炔酰菌胺和氟吡菌胺防治霜霉病时不应忽视其抗药性风险管理，应加强田间抗药性监测，制定合理的抗药性治理策略，同一生长季节不要重复、单一使用同一种类型的药剂，而应将不同作用机理的药剂混合或轮换使用，以延缓或避免抗药性的产生。提倡在霜霉病预测预报或发病中心监测的基础上，采用不同作用机理的杀菌剂交替、精准施药，限制每个生长季节每种药剂的使用次数不超过 2 次，在发病前或发病初期按照推荐剂量和间隔期施药，避免在发病较重时铲除性施药等。另外，防控策略上提倡高效环保型化学防控技术与利用抗病品种、使用无滴膜、地膜覆盖、膜下滴灌、通风降湿、及时摘除病残叶等防控措施相结合的综合防控措施，减少化学农药的使用，降低杀菌剂对整个病菌群体的抗药性选择压，以获得延缓抗药性发生或发展、安全有效防控霜霉病的效果，加强黄瓜霜霉病菌对各类药剂的敏感性监测，及时了解田间病菌对各类药剂的抗性发展动态，以便于及时制定合理的抗性治理策略。

参考文献

[1] AlešLebeda, Yigal Cohen. Cucurbit downy mildew (*Pseudoperonospora cubensis*) —biology, ecology, epidemiology, host-pathogen interaction and control [J]. Plant Pathol, 2010, 129: 157 - 192.

[2] Savory E A, Granke L L, Quesada-Ocampo L M, et al. The cucurbit downy mildew pathogen *Pseudoperonospora cubensis* [J]. Molecular Plant Pathology, 2011, 12 (3): 217 - 226.

[3] Gisi U. Chemical control on downy mildews [A]. In: PTN Spencer-Phillips, U Gisi, A Lebeda (eds.). Advances in downy mildew research vol. 1 [C]. Dordrecht: Kluwer Academic Publishers, 2002: 119 - 159.

[4] Urban J, Lebeda A. Fungicide resistance in cucurbit downy mildew-methodological, biological and population aspects [J]. Annals of Applied Biology, 2006, 149: 63 - 75.

[5] Keith J, Brent. Fungicide resistance in crop pathogens: how can it be managed (second, revised edition) [M]. Brussels: FRAC, Newline Graphics Reprinted, 2007. 1 - 57.

[6] Schwinn F, Sozzi D. Recommended methods for the detection and measurement of resistance of plant pathogens to fungicides: Method for fungicide resistance in late blight of potato [J]. FAO Method NO. 30 FAO Plant Protection Bulletin, 1982, 30: 69 - 71.

[7] 王文桥，刘国容，严乐恩，等. 黄瓜和葡萄霜霉病菌的抗药性监测 [J]. 南京农业大学学报，1996，19（增刊）：127 - 131.

[8] 刘晓宇. 黄瓜霜霉病菌和番茄早疫病菌对嘧菌酯的敏感性基线及抗药性风险评估 [D]. 南京：南京农业大学，2004.

[9] Wang Wenqiao, Yan Lei, Meng Runjie, et al. Sensitivity to fluopicolide of wild type isolates and biological characteristics of fluopicolide-resistant mutants in *Pseudoperonospora cubensis* [J]. Crop Protection, 2014, 55: 119 - 126.

[10] 崔继敏，杨晓津，赵建江，等. 黄瓜霜霉病菌对双炔酰菌胺的敏感基线及其抗性突变体生物学性状研究 [J]. 农药学学报，2013，15（5）：496 - 503.

[11] 吴新平，顾宝根，刘乃炽，等. 农药田间药效试验准则（一）杀菌剂防治黄瓜霜霉病 [M]. 北京：中国标准出版社，2005.

[12] 康丽娟，韩秀英，马志强. 嘧菌酯对三种蔬菜病害的毒力、防效及安全性研究 [J]. 农药学学报，2004，6（1）：85 - 88.

西瓜蔓枯病菌对啶酰菌胺敏感基线的建立及抗性监测*

王少秋**，李　雨，谭　蕊，黄萌雨，余　洋，杨宇衡，毕朝位***
（西南大学植物保护学院，重庆　400715）

摘要：2014—2015 年从渝贵不同地区采集分离了 368 株西瓜蔓枯病菌，利用菌丝生长速率法分别测定随机筛选的 100 个西瓜蔓枯病菌菌株对啶酰菌胺的敏感性及 MIC 值，并依据敏感基线及 MIC 值对各地的菌株进行田间抗性监测，研究结果表明，测定的 100 株菌株对啶酰菌胺的 EC_{50} 分布在 0.033 2～0.180 3μg/mL 间，平均 EC_{50} 值为（0.085 1±0.036 77）μg/mL，其敏感性频率分布呈连续性单峰曲线的近正态分布，因此平均 EC_{50} 值可作为西瓜蔓枯病菌对啶酰菌胺的敏感性基线。以所测得的 MIC 值 8μg/mL 结合敏感基线确定以 10μg/mL 作为西瓜蔓枯病菌对啶酰菌胺的抗性监测标准，对 368 株西瓜蔓枯病菌进行抗药性监测结果发现所采集的这些菌株均未对啶酰菌胺产生抗性，因此啶酰菌胺可以作为多菌灵的替代药剂用于西瓜蔓枯病的防治。

关键词：西瓜蔓枯病菌；啶酰菌胺；敏感基线；MIC 值；抗性监测

西瓜蔓枯病（gummy stem blight）是影响瓜类蔬菜生产的重要病害之一，它可以造成减产 15%～30%，重病田块可减产 80% 以上，果实品质也会受到极大影响，严重威胁西瓜生产[1]。我国最早于 1930 年有所记载，至今在我国主要西瓜产区都有发生，尤其在西南地区是露地西瓜的主要病害之一。目前生产上对西瓜蔓枯病以化学防治为主，常用药剂有多菌灵、百菌清、甲基托布津、代森锰锌等。由于杀菌剂的长期大量使用使得抗性普遍严重，在生产上常常会有药剂防效下降或失效的现象，本实验室对重庆地区西瓜蔓枯病菌的抗性监测发现田间多菌灵抗性菌株比例达 80% 以上，并且均为高抗菌株。

啶酰菌胺，是由巴斯夫公司 1992 年开发的琥珀酸脱氢酶抑制剂类杀菌剂，作用机制主要是抑制病原菌琥珀酸脱氢酶活性，从而干扰其呼吸作用[2]。啶酰菌胺 2005 年在我国正式登记上市，现已在 50 多个国家获得登记，用于防治 100 多种作物上的 80 多种病害，其对灰霉病、菌核病、白粉病及各种腐烂病、根腐病等均有良好的防治效果[3-4]，是目前使用范围最广、用量最大的 SDHIs 类杀菌剂。啶酰菌胺上市后仅 5 年就有了抗性报道，被国际杀菌剂抗性行动委员会归类为中等抗性风险杀菌剂[5]。Avenot 等[6] 在美国加利福尼亚州监测到该药剂的田间首例抗性菌－马铃薯早疫病菌（*Alternaria alternata*）。2007 年，Stevenson 等[7] 在田间检测到因 SdhB 亚单位上的氨基酸由组氨酸突变为酪氨酸而引起西瓜蔓枯病菌（*Didymella bryoniae*）对啶酰菌胺的抗性菌株。该药现已逐渐在我国西瓜蔓枯病防治上投入使用，但有关该病原菌对啶酰菌胺抗药性方面的研究在我国未见报道。因此，及时建立西瓜蔓枯病菌对啶酰菌胺的敏感基线等相关研究，对该杀菌剂的田间科学使用和延缓药剂的使用寿命具

* 基金项目：公益性行业（农业）科研专项（201303023）

** 作者简介：王少秋，女，甘肃省陇西县人，硕士研究生，主要从事植物病院真菌抗药性研究；E-mail：627917384@qq.com

*** 通讯作者：毕朝位，男，副教授，主要从事植物真菌病害及病原菌抗药性研究；E-mail：chwbi@swu.edu.cn

有重要指导意义。

1 材料与方法

1.1 供试材料

2014—2015 年对重庆、贵州地区采集西瓜蔓枯病的发病组织，然后进行分离保存。

啶酰菌胺（98%）原药，由西南大学植物病理真菌研究室提供。啶酰菌胺原药用丙酮溶解，配制成 $1\times10^4\mu g/mL$ 母液于 4℃冰箱保存备用。

马铃薯葡萄糖琼脂培养基（PDA）：马铃薯 200g、葡萄糖 20g、琼脂粉 20g，加水定容至 1 L，高压蒸汽灭菌（121℃）30min 备用。

1.2 试验方法

采用菌丝生长速率法[8]测定西瓜蔓枯病菌对啶酰菌胺的敏感性：在含 0.006、0.061、0.197、0.629、2.013μg/mL 啶酰菌胺的 PDA 培养基上接种直径 7mm 的西瓜蔓枯病菌菌饼，以不含药的 PDA 培养基为对照，每个浓度设置 3 个重复，25℃恒温培养，约 4d 后用十字交叉法测量菌落直径。按（1）式计算菌丝生长的抑制率。

菌丝生长抑制率（%）＝（1－处理菌落生长直/对照菌落生长直径）×100　　（1）

本研究采用区分剂量法对西瓜蔓枯病菌对啶酰菌胺的田间抗药性进行抗性监测[9]。根据预实验，设定终浓度介于 5～10μg/mL 等差为 1μg/mL 的啶酰菌胺的浓度梯度，随机选取 100 株对啶酰菌胺敏感的野生型西瓜蔓枯病菌菌株，测定其 MIC 值。根据所测的 MIC 值并结合对啶酰菌胺的敏感性基线，确定敏感和抗性西瓜蔓枯病菌对啶酰菌胺的区分剂量值，以监测菌株菌丝能否在含区分剂量药剂的 PDA 培养基上生长为标准，建立西瓜蔓枯病菌对啶酰菌胺的田间抗药性监测标准。

2 结果与分析

2.1 西瓜蔓枯病菌对啶酰菌胺的野生敏感基线

采用菌丝生长速率法测定了 100 株西瓜蔓枯病菌对啶酰菌胺的敏感性，得到了西瓜蔓枯病菌对啶酰菌胺的敏感性分布（图）。其中 EC_{50} 值最小为 0.033 2μg/mL，最大为 0.180 3

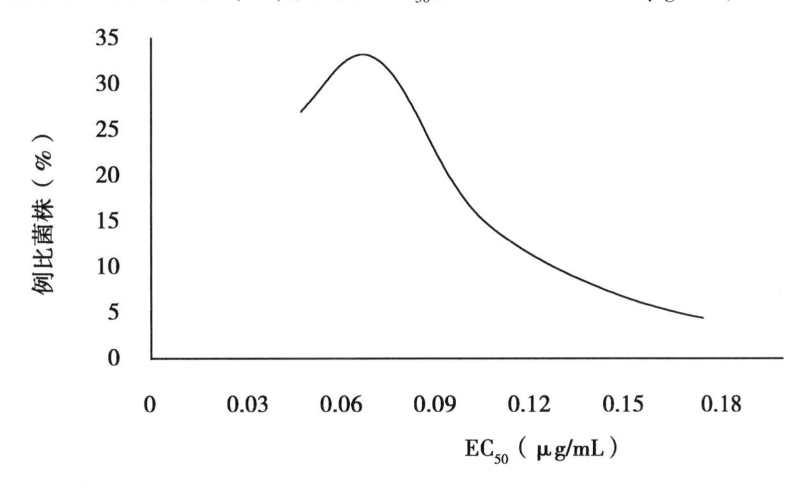

图　西瓜蔓枯病菌对多菌灵的敏感性频率分布

μg/mL，相差 5.4 倍，平均 EC_{50}值为（0.085 1 ±0.036 77）μg/mL。100 株西瓜蔓枯病菌对啶酰菌胺的敏感性频率分布呈连续单峰曲线，没有出现敏感性下降的抗药性群体。因此，可以采用这些菌株的平均 EC_{50}值作为将来进行田间西瓜蔓枯病菌对啶酰菌胺抗药性检测的敏感基线。

2.2 不同地区西瓜蔓枯病菌对啶酰菌胺的敏感性比较

采自渝贵 5 个地区的西瓜蔓枯病菌群体对啶酰菌胺的敏感性无显著差异（表 1），其 EC_{50}平均值为 0.065 416 306 ~ 0.100 876 456μg/mL。其中武隆地区的菌株最敏感，其平均 EC_{50}值为 0.065 416 306μg/mL，合川地区的菌株最不敏感，其平均 EC_{50}值为 0.100 876 456 μg/mL。地区之间西瓜蔓枯病菌群体对啶酰菌胺的敏感性相差 1.5 倍。

表 1 渝贵不同地区西瓜蔓枯病菌对啶酰菌胺的敏感性

地区	菌株数	EC_{50}范围（μg/mL）	EC_{50}平均值（μg/mL）
息烽	21	0.039 7623 3 ~ 0.166 897 305	0.090 617 62a
北碚	33	0.041 437 23 ~ 0.157 171 271	0.083 475 49a
武隆	22	0.034 852 12 ~ 0.122 788 561	0.065 416 31a
永川	7	0.040 836 65 ~ 0.175 833 139	0.099 997 35a
合川	17	0.033 241 06 ~ 0.180 224 102	0.100 876 45a

注：表中相同字母表示采用 Duncan 氏新复极差法检验数值间差异不显著（$P \geqslant 0.05$）

2.3 啶酰菌胺对西瓜蔓枯病菌的抗性监测结果

通过对 100 株田间敏感型西瓜蔓枯病菌株对啶酰菌胺的 MIC 测定。所测菌株对啶酰菌胺的 MIC 值在分布在 4 ~ 8μg/mL，最大 MIC 值为 8μg/mL。根据所测结果与对啶酰菌胺敏感性基线综合分析，最终确定以能在含 10μg/mL 啶酰菌胺 PDA 培养基上生长的为抗药性菌株，而不能在含 10μg/mL 啶酰菌胺 PDA 培养基上生长为敏感性菌株作为西瓜蔓枯病菌对啶酰菌胺的抗性监测标准。

以所确定的对吡唑醚菌酯抗性监测标准，对 2014—2015 年渝贵地区所分离获得的 368 株西瓜蔓枯病菌进行抗药性监测（表 2）。测定结果表明，所有西瓜蔓枯病菌菌株均不能在含 10μg/mL 啶酰菌胺 PDA 培养基上生长，田间尚无抗性菌株。

表 2 西南地区西瓜蔓枯病菌对啶酰菌胺的田间抗药性监测

	菌株来源	监测菌株数	抗啶酰菌胺菌株数
2014 年	息烽	74	0
	北碚	70	0
	武隆	38	0
	合川	27	0
2015 年	息烽	15	0
	北碚	23	0
	武隆	42	0
	永川	43	0
	合川	36	0
总计	5 个地区	368	0

3 讨论

从西瓜蔓枯病菌对啶酰菌胺的敏感性分布可以看出，敏感性变化范围小，敏感性最低和最高的菌株 EC_{50}值相差 1.5 倍。但西瓜蔓枯病菌对啶酰菌胺仍很敏感，敏感性最低菌株的 EC_{50}值为 0.100 876 456μg/mL，表明自然界中西瓜蔓枯病菌对啶酰菌胺尚处于非常敏感的水平。但啶酰菌胺对不同西瓜蔓枯病菌菌株的 EC_{50}值还是存在一定的差异，这与自然界病原菌本身存在的生理差异有关，也可能与生产中大量使用的其他类型药剂间存在着微弱的交互抗性有关，此问题尚需进一步研究。

根据所测 100 株田间敏感型西瓜蔓枯病菌株对啶酰菌胺的 MIC 值与对啶酰菌胺敏感性基线综合分析，最终确定以 10μg/mL 作为西瓜蔓枯病菌对啶酰菌胺的抗性监测标准。对 368 株西瓜蔓枯病菌进行抗药性监测发现所采集的样本中无抗性菌株，说明田间还未产生抗啶酰菌胺的西瓜蔓枯病菌株。啶酰菌胺还有着不易与其他类型的杀菌剂产能生交互抗性的特点，与同类型的杀菌剂中只与萎锈灵和吡噻菌胺有交互抗性，且吡噻菌胺尚未在我国登记上市，因此啶酰菌胺可以代替一些生产中常用的已产生抗药性的药剂用于西瓜蔓枯病的防治与抗药性治理。

啶酰菌胺上市后仅 5 年就有了抗性报道，被国际杀菌剂抗性行动委员会归类为中等抗性风险杀菌剂，国外已有西瓜蔓枯病菌对啶酰菌胺抗药性的报道。但其在中国尚属新药剂，未发现其有西瓜蔓枯病抗药菌株出现的报道。生产中单一、过量地用药，将使病菌抗药性的产生成为可能。因此，啶酰菌胺在西瓜蔓枯病上替代其他药剂大量使用时一定要注意采取必要的抗性风险预测及治理策略研究。

参考文献

[1] 刘书林，顾兴芳. 瓜类蔬菜蔓枯病研究概况 [J]. 中国蔬菜，2013 (18)：1 - 10.

[2] RUSSELL P E. Fungicide Resistance Action Committee (FRAC): A Resistance ActivityUpdate [J]. Outlooks on Pest Management, 2009, 20 (3): 122 - 125.

[3] 亦冰. 新颖杀菌剂 - 啶酰菌胺 [J]. 世界农药，2006，28 (5)：51 - 53.

[4] 颜范勇，刘冬青，司马利锋，等. 新型烟酰胺类杀菌剂 - 啶酰菌胺 [J]. 农药，2008，47 (2)：132 - 135.

[5] 李良孔，袁善查，潘洪玉，等. 琥珀酸脱氢酶抑制剂类 (SDHI) 杀菌剂及其抗性研究进展 [J]. 农药，2011，50 (3)：165 - 169，172.

[6] Avenot H F, Michailides T J. Resistance to boscalid in *Alternarla alternate* isolates from Pistachio in Califomia [J]. Plant Disease, 2007, 91 (10): 1 345 - 1 350.

[7] Stevenson K L, Langston D B, Sanders F. Baseline Sensitivity and Evidence of Resistance to Boscalid in Didymellabryoniae [J]. Phytopathology, 2007, 98: S151.

[8] 纪明山，祁之秋，王英姿，等. 番茄灰霉病菌对嘧霉胺的抗药性 [J]. 植物保护，2003，30 (4)：396 - 400.

[9] Taga M, Waki T, Tsuda M, Ueyama A. Fungicide sensitivity and genetics of IBP - resistant mutants of Pyriculariaoryzae [J]. Phytopathology, 1982, 72: 905 - 908.

4种药剂对尖孢炭疽病菌的室内毒力及田间防效*

高杨杨，禾丽菲，李北兴，刘　峰**
（山东农业大学植物保护学院，泰安　271018）

摘要：采用菌丝生长速率法、孢子萌发法和芽管伸长法测定了四氟醚唑、氟吡菌酰胺、肟菌酯及对照药剂代森锰锌对山东地区辣椒炭疽病菌——尖孢炭疽病菌（*Colletotrichum acutatum*）的室内抑制毒力。结果表明四氟醚唑可抑制尖孢炭疽病菌菌丝的生长，其 EC_{50} 为2.70μg/mL；氟吡菌酰胺可有效抑制该病菌的孢子萌发和芽管伸长，其 EC_{50} 为12.51和2.00μg/mL，但是对菌丝生长基本无抑制作用；肟菌酯抑制菌丝生长、孢子萌发和芽管伸长的毒力均较高，其 $EC_{50} \leq$ 2.29μg/mL；代森锰锌对该病菌孢子萌发和芽管伸长的抑制效果明显，EC_{50} 为0.56和1.97μg/mL，显著高于对菌丝生长的抑制毒力。田间药效试验结果表明，4%四氟醚唑水乳剂32g a. i. /hm^2 对辣椒炭疽病的防效为66.54%，42.8%氟菌·肟菌酯悬浮剂（露娜森）160g a. i. /hm^2 和80%代森锰锌可湿性粉剂2 000g a. i. /hm^2 的防效分别为58.63%和52.02%。表明四氟醚唑和氟菌·肟菌酯具有作为辣椒炭疽病田间防治的替代药剂的潜力。

关键词：四氟醚唑；氟吡菌酰胺；肟菌酯；代森锰锌；辣椒炭疽病；田间药效

Indoor Toxicity and Field Control Efficacies of Four Fungicides Against *Colletotrichum acutatum*

Gao Yangyang, He Lifei, Li Beixing, Liu Feng
(*College of Plant Protection*, *Shandong Agricultural University*, *Tai'an* 271018, *China*)

Abstract: To evaluate the control efficacies of tetraconazole, fluopyram, trifloxystrobin and mancozed on pepper anthracnose caused by *Colletotrichum acutatum*, the inhibitory effects of four fungicides on its mycelia growth, spore germination and germ tube elongation were determined. The results indicated that tetraconazole could strongly inhibit the mycelial growth of *C. acutatum*, with EC_{50} value of 2.70μg/mL. Fluopyram could inhibit the spore germination and germ tube elongation with EC_{50} value of 12.51 and 2.00μg/mL, but the inhibition on mycelial growth was low. Trifloxystrobin could strongly inhibit the mycelial growth, spore germination and germ tube elongation of *C. acutatum*, with EC_{50} lower than 2.29μg/mL. Mancozed had higher activity in inhibiting spore germination and germ tube elongation of *C. acutatum* than of mycelial growth, with EC_{50} values of 0.56 and 1.97μg/mL. In field experiments, tetraconazole at the dose of 32 g a. i. /hm^2, fluopyram + trifloxystrobin at the dose of 160 g a. i. /hm^2 and mancozed at dose of 2 000g a. i. /hm^2 all showed favorable efficacy in controlling pepper anthracnose, the control efficacies were 66.54%, 52.02% and 54.63%, respectively. Overall, tetraconazole and fluopyram + trifloxystrobin can effectively control pepper anthracnose and they could be used as the alternative fungicides.

Key words: Tetraconazole; Fluopyram; Trifloxystrobin; Mancozed; Pepper anthracnose; Control efficacy

---

* 基金项目：国家重点研发计划项目（2016YFD0200500）
** 通讯作者：刘峰；E-mail：fliu@ sdau. edu. cn

近年来，随着辣椒经济效益的提高，山东地区辣椒种植面积逐年扩大，已成为山东省重要的经济作物之一。炭疽病是辣椒产区的常见病害，具有分布广、蔓延快、危害重的特点，可造成产量损失 30% ~40%[1]。据报道，我国辣椒炭疽病菌主要包括 4 种：红色炭疽病菌（*C. gloeosporioides*）、黑点炭疽病菌（*C. capsici*）、黑色炭疽病菌（*C. coccodes*）和尖孢炭疽病菌（*C. acutatum*）[2]。笔者调查发现，山东济宁、菏泽、德州等地露地栽培条件下，辣椒炭疽病发生极为普遍。主要原因可能是辣椒结果期正值雨季，雨量大，温度适宜，有利于病害发生。目前，尖孢炭疽菌被认为是造成辣椒果实发生炭疽病的主要病原菌之一，该菌为害重、寄主范围广，除侵染辣椒外，还可侵染草莓、苹果、茄子、杧果和银杏[3-5]等作物，对农业生产危害极大。

由于目前农业防治和生物防治效果有限，生产中推广难度大，化学防治是目前国内防治辣椒炭疽病的主要手段[6]。但是，有报道常规药剂如多菌灵、甲基硫菌灵、代森锰锌和百菌清对辣椒炭疽病的防治效果不理想[2]，并且辣椒炭疽病菌已经对苯菌特产生抗性[7-8]，因此需要筛选新的高效药剂。本研究比较了四氟醚唑、氟吡菌酰胺、肟菌酯和代森锰锌对尖孢炭疽菌的室内毒力，并在田间条件下评价了四氟醚唑、氟吡菌酰胺·肟菌酯和代森锰锌对辣椒炭疽病的控制效果，以期为山东地区辣椒炭疽病的有效控制提供参考。

1 材料与方法

1.1 材料

1.1.1 菌株和培养基

辣椒炭疽病病果采自山东省济宁和菏泽等地，采用常规的方法进行分离[2]，剪取病健交界处，在 75% 乙醇溶液中浸泡 30s 后，用无菌蒸馏水冲洗 3 次，接种在 PDA 培养基上，所得到的纯培养物保存在 4℃ 冰箱中备用。通过菌落形态、孢子形态和转录间隔区（ITS）的序列对菌种进行鉴定。

1.1.2 药剂及器械

95% 四氟醚唑原药（tetraconazole），4% 四氟醚唑水乳剂，浙江省杭州宇龙化工有限公司。97% 肟菌酯原药（trifloxystrobin），江苏省南通泰禾化工有限公司。96% 氟吡菌酰胺原药（fluopyram），42.8% 氟菌·肟菌酯悬浮剂（露娜森），德国拜耳作物科学公司。90% 代森锰锌原药（mancozeb），山东潍坊润丰化工股份有限公司。80% 代森锰锌可湿性粉剂，山东东信生物农药有限公司。98% 水杨肟酸（salicylhydroxamic Acid，SHAM），日本东京化成工业株式会社。

水杨肟酸用甲醇溶解，四氟醚唑、肟菌酯和氟吡菌酰胺原药用丙酮溶解，制成 1×10^4 μg/mL 的母液；代森锰锌母液制备方式为：1% 原药，2% T－80，97% 水在研磨机中研磨制成悬浮剂母液。于 4℃ 中保存，进行室内毒力测定试验。

田间施药器械为 MATABI－16 型背负式手动喷雾器。

1.1.3 供试品种和试验田

辣椒品种为朝天椒（三英八号）。试验地点为山东省济宁市金乡县十里铺村。

1.2 室内毒力测定

1.2.1 药剂对尖孢炭疽菌菌丝生长速率的影响[9]

将所采集的尖孢炭疽菌菌株纯化，进行室内毒力试验。将适当体积的四氟醚唑、氟吡菌酰胺、肟菌酯和代森锰锌母液加入到提前冷却至 50℃ 的 PDA 培养基中，混匀后倒入培养皿中，每皿 20mL，其中肟菌酯的 PDA 平板上需要加入 100μg/mL 的水杨肟酸。每个浓度 4 个培养皿，每个处理 3 个重复，以不加药剂的 PDA 培养基为空白对照。用打孔器在 7d 生长旺盛的菌落边缘打取菌柄（8mm），将菌柄倒置放在培养皿的中央，置于 25℃ 黑暗培养 7d 后，采用十字交叉法测量菌落直径。

菌丝生长抑制率（%）=［（对照菌落生长直径－处理菌落生长直径）/对照菌落生长直径］×100

1.2.2　药剂对尖孢炭疽菌孢子萌发和芽管伸长的影响[10]

先利用无菌水冲洗培养 7d 的尖孢炭疽菌平板，在无菌条件下过滤，滤液于 10 000r/min 离心 5min，倒掉上清液，将分生孢子重新悬浮于无菌水中，至 1×10^5/mL，制备得孢子悬浮液。再将适当体积的药液加入到融化好的水琼脂培养基中制成试验浓度，每个浓度 4 个板每个处理 3 个重复，用移液器吸取 100μL 孢子悬浮液打在凝固好的水琼脂平板上，利用涂布器涂匀后，置于 25℃黑暗培养 8h 观察每个处理的孢子萌发率（当芽管长度大于孢子长度的一半时视为萌发）和芽管长度（每个视野至少调查 20～30 个芽管）。

孢子萌发抑制率（%）=（对照孢子萌发率－处理孢子萌发率）/对照孢子萌发率 ×100

芽管伸长抑制率（%）=（对照芽管长度－处理芽管长度）/对照芽管长度 ×100

1.3　田间药效试验

试验共 4 个处理，包括 4% 四氟醚唑水乳剂 32g a. i. /hm^2、42. 8% 氟菌·肟菌酯悬浮剂（露娜森）160g a. i. /hm^2、80% 代森锰锌可湿性粉剂 2 000g a. i. /hm^2 和清水对照（CK）。每个处理 3 次重复，小区面积为 $30m^2$，小区设计采用随机区组设计。每公顷用水量为 675L，计算各处理所需要的用水量。2015 年 8 月 22 日辣椒炭疽病发生时开始施第一次药，采用手动喷雾器在辣椒表面均匀施药，每隔 6d 施药一次，共施药 2 次。

病害等级分别在第一次施药之前，第一次施药后 6d 和最后一次施药后 7d 开始调查，每个小区采取 5 点取样的方式，每个点调查 2 株的全部果实。病害等级划分：0 级：无病斑；1 级：病斑面积占整个果实的 2% 一下；3 级：病斑面积占整个果实的 3% ～8%；5 级：病斑面积占整个果实的 9% ～15%；7 级：病斑面积占整个果实的 16% ～25%；9 级：病斑面积占整个果实的 25% 以上。根据 GB/T17980. 33—2000 规定的辣椒炭疽病田间防治试验标准[11]计算防治效果。

$$病情指数=\frac{\sum(各级病果数\times相对病级数)}{调查总果数\times9}\times100$$

$$防治效果（\%）=\left[1-\frac{空白对照区施药前病情指数\times药剂处理区施药后病情指数}{空白对照区施药后病情指数\times药剂处理区施药前病情指数}\right]\times100$$

2　结果与分析

2.1　药剂对尖孢炭疽菌的室内毒力

2.1.1　药剂对尖孢炭疽菌菌丝生长的影响

通过菌丝生长速率法测定了 4 种药剂对尖孢炭疽菌的抑制毒力，结果表明（表 1）除氟吡菌酰胺以外，四氟醚唑、肟菌酯、对照药剂代森锰锌对病原菌的菌丝生长都具有一定的抑制效果，其中四氟醚唑和肟菌酯的毒力最高，EC_{50}值为 2. 70 和 2. 29μg/mL。

表 1　4 种药剂对尖孢炭疽菌菌丝生长的毒力

Table 1　Inhibitory rates of 4 fungicides on mycelia growth of *Colletotrichum acutatum*

药剂	毒力回归方程	相关系数	EC_{50}（μg/mL）	EC_{90}（μg/mL）
四氟醚唑	$y = 4.2918 + 1.6401x$	0.989 7	2.70	16.34
氟吡菌酰胺	—	—	>100	—
肟菌酯	$y = 4.7887 + 0.5862x$	0.958 4	2.29	351.94
代森锰锌	$y = 4.0137 + 0.8225x$	0.962 9	15.82	571.85

2.1.2　药剂对尖孢炭疽菌孢子萌发和芽管伸长的影响

4 种杀菌剂抑制尖孢炭疽菌孢子萌发的毒力大小依次为肟菌酯 > 代森锰锌 > 氟吡菌酰胺 > 四氟醚唑，肟菌酯和代森锰锌的毒力最高，EC_{50}分别为 0.16 和 0.56μg/mL，其次为氟吡菌酰胺，EC_{50}为 12.51μg/mL，而四氟醚唑基本不抑制该病菌的菌丝生长（表 2）。除四氟醚唑外，肟菌酯、代森锰锌和氟吡菌酰胺对芽管伸长均具有很高的抑制效果，其 EC_{50}分别为 0.19、1.97 和 2.00μg/mL（表 3）。

表 2　4 种药剂对尖孢炭疽菌孢子萌发的毒力

Table 2　Inhibitory rates of 4 fungicides on spore germination of *Colletotrichum acutatum*

药剂	毒力回归方程	相关系数	EC_{50}（μg/mL）	EC_{90}（μg/mL）
四氟醚唑	—	—	>50	—
氟吡菌酰胺	$y = 3.6219 + 1.2561x$	0.973 4	12.51	131.04
肟菌酯	$y = 5.2577 + 0.3227x$	0.994 6	0.16	>100
代森锰锌	$y = 5.4885 + 1.9589x$	0.983 2	0.56	2.54

表 3　4 种药剂对尖孢炭疽菌芽管伸长的毒力

Table 3　Inhibitory rates of 4 fungicides on germ tube elongation of *Colletotrichum acutatum*

药剂	毒力回归方程	相关系数	EC_{50}（μg/mL）	EC_{90}（μg/mL）
四氟醚唑	—	—	>100	—
氟吡菌酰胺	$y = 4.7645 + 0.7846x$	0.997 3	2.00	85.80
肟菌酯	$y = 5.2110 + 0.2896x$	0.958 4	0.19	>100
代森锰锌	$y = 4.6626 + 1.1426x$	0.976 7	1.97	26.11

2.2　田间药效试验

2.2.1　药剂对辣椒炭疽病的田间防治效果

由于氟吡菌酰胺对尖孢炭疽菌的室内毒力较低，因此在田间药效试验中选择氟菌·肟菌酯混剂（露娜森），由表 4 可知，3 种药剂对辣椒炭疽病具有一定的防治效果，其防效均高于 50%，其中四氟醚唑的防治效果略高于对照药剂代森锰锌，其防效为 66.54%。此外，最后一次施药后防治效果明显高于第一次施药后的防效，可以证明试验药剂对辣椒炭疽病在田间的蔓延具有一定的控制效果。

表4 药剂对辣椒炭疽病的田间防治效果

Table 4 Control effects of fungicides on pepper anthracnose disease in the field

药剂	药剂剂量（g a. i. / hm²）	病情指数			防效（%）	
		药前	第一次药后6d	第二次药后7d	第一次药后6d	第二次药后7d
4%四氟醚唑水乳剂	32	5.12	6.57	7.76	35.32 ± 1.76 a	66.54 ± 2.61 a
42.8%氟菌·肟菌酯悬浮剂	160	4.36	5.67	7.92	28.44 ± 3.54 a	58.63 ± 1.30 ab
80%代森锰锌可湿性粉剂	2 000	4.84	6.02	9.05	29.36 ± 2.64 a	54.63 ± 1.57 b
CK		4.67	8.43	20.74		

注：表中数据为平均值±标准误。同列数据不同字母表示经Tukey检验在 $P < 0.05$ 水平差异显著。Data are mean ± SE. Different letters in the same column indicate significant difference at $P < 0.05$ level by Tukey's test

2.2.2 安全性

田间施药2次后观察，各处理小区的辣椒均未发现药害症状，表明在施用剂量下各药剂对该辣椒安全。

3 讨论

本研究表明，肟菌酯对尖孢炭疽的每个发育阶段均具有抑制作用，尤其对孢子萌发的毒力最高。病原菌孢子萌发阶段需要的能量最多，肟菌酯属于甲氧基丙烯酸酯类杀菌剂，该类杀菌剂抑制病原菌能量的产生[12]，从而抑制孢子呼吸。Karadimous等研究表明甲氧基丙烯酸酯类杀菌剂对病原菌的孢子萌发毒力最高[13]。四氟醚唑属于三唑类杀菌剂，主要通过抑制甾醇的合成从而起到抑菌的效果。有研究表明，三唑类药剂氟菌唑抑制灰霉菌菌丝生长的毒力最高[14]。代森锰锌属于保护性杀菌剂，治疗和铲除的作用较差。王文桥等研究表明代森锰锌对马铃薯晚疫病菌的孢子萌发的毒力高于对菌丝生长的抑制效果[15]，与本研究结果一致。

四氟醚唑对尖孢炭疽菌的菌丝生长具有很高的抑制活性，这与田间其较好的治疗效果相一致。肟菌酯对炭疽病菌的各个发育阶段均具有很好的抑制活性，但是氟吡菌酰胺只对芽管伸长的毒力较高。商品化的氟菌·肟菌酯（露娜森）为后两种药剂的混剂，可以看出两者混用一方面可以弥补氟吡菌酰胺抑制效果不佳的现象，另一方面药剂复配在一定程度上还可以延缓病原菌抗药性的产生。此外，代森锰锌虽然对菌丝、孢子和芽管的都有较强的抑制作用，但是田间施药剂量为2 000g a. i. /hm² 时其防效仍然较低，一方面可能与其没有内吸活性；另一方面可能是由于可湿性粉剂喷施后润湿展着性能不佳，辣椒果实上着药量低，不耐雨水冲刷有关[16]。

四氟醚唑和氟菌·肟菌酯（露娜森）虽然对辣椒炭疽病的防治效果略高于对照药剂代森锰锌，但是田间防治效果均不突出，性价比优势不明显。另外，试验田为露天，在此条件下各种因素（孢子侵染高峰期、植株种植密度、施药质量、风、雨等）对病情的扩展和药剂药效的发挥都有较大影响。此外，本研究施药时偏晚，辣椒炭疽病的病情指数较高，也不利于以保护性为主的杀菌剂药效的发挥。不同施药时机及不同地区和场景下，两种新药剂对辣椒炭疽病的田间防效仍需要进一步评价。

参考文献

[1] 杨青，易图永. 辣椒炭疽病及其防治研究进展［J］. 江西农业学报，2009，21：107 - 109.

[2] 夏花，朱宏建，周倩，等. 湖南芷江辣椒上一种新炭疽病的病原鉴定［J］. 植物病理学报，2012，42：120 - 125.

[3] Peres N A，Timmer L W，Adaskaveg J E，et al. Life styles of *Colletotrichum acutatum*［J］. Plant Disease，2005，89：784 - 796.

[4] Liao C Y，Chen M Y，Chen Y K，et al. Formation of highly branched hyphae by *Colletotrichum acutatum* within the fruitcuticles of *Capsicum* spp.［J］. Plant Pathology，2012a，61：262 - 270.

[5] Liao C Y，Chen M Y，Chen Y K，et al. Characterization of three *Colletotrichum acutatum* isolates from *Capsicum* spp［J］. European Journal Plant Pathology，2012b，133：599 - 608.

[6] 曾庆华，消仲久，向金玉，等. 3 种杀菌剂对黑点型辣椒炭疽病菌的室内毒力测定［J］. 贵州农业科学，2010，38：93 - 94.

[7] Peres N A R，Souza N L，Zitko S E，et al. Activity of benomyl of control of postbloom fruit drop of citrus caused by *Colletotrichum acutatum*［J］. Plant Disease，2002，86：620 - 624.

[8] Talhinhas P，Sreenivasaprasad S，Neves - Martins J，et al. Molecular and phenotypic analyses reveal association of diverse *Colletotrichum acutatum* groups and a low level of *C. gloeosporioides* with olive anthracnose［J］. Applied Environment Microbiology，2005，71：2 987 - 2 998.

[9] 孙广宇，宗兆锋. 植物病理学实验技术［M］. 北京：中国农业出版社，2002：139 - 146.

[10] 陈雨，张文芝，周明国. 氰烯菌酯对禾谷镰孢菌分生孢子萌发及菌丝生长的影响［J］. 农药学学报，2007，9：235 - 239.

[11] 农药田间药效准则（一）：杀菌剂防治辣椒炭疽病［S］. 2000，GB/T 17980. 33—2000.

[12] Allen，P. J，Metabolic aspects of spore germination in fungi［J］. Annu Rev Phytopathol，2003，3：313 - 342.

[13] Karadimos D A，Karaoglanidis G S and Tzavella-Klonari K. Biological activity and physical modes of action of the Qo inhibitor fungicides trifloxystrobin and pyraclostrobin against *Cercospora beticola*［J］. Crop protection，2005，24：23 - 29.

[14] Song Y，Xu D，Lu H，et al. Baseline sensitivity and efficacy of the sterol biosynthesis inhibitor triflumizole against *Botrytis cinerea*［J］. Australasian Plant Pathology，2015，45：1 - 8.

[15] 王文桥，马志强，韩秀英，等. 霜脲氰和代森锰锌对马铃薯晚疫病菌的离体活性及混合增效作用［J］. 农药学学报，2002，1：29 - 33.

[16] 杨睿，张秋利，饶全武. 农药可湿性粉剂的技术控制［J］. 河南化工，2009，26：19 - 21.

灰葡萄孢菌对咯菌腈的抗药性监测及抗药性机制研究

任维超，武东霞，段亚冰，侯毅平，王建新，周明国，陈长军
（南京农业大学植物保护学院，南京 210095）

摘要：本研究监测了2012—2014年江苏和山东两省不同地区不同寄主的灰葡萄孢菌对咯菌腈的敏感性，于2013年首次在山东寿光黄瓜灰霉菌株中筛选到2株高抗菌株，高抗频率为2.67%，2014年在山东潍坊番茄灰霉菌株中筛选到4株高抗菌株，高抗频率为1.30%，而在江苏地区未发现咯菌腈高抗菌株。通过室内药剂驯化获得6株咯菌腈抗药性突变体，抗性稳定，抗性倍数从30到10 000以上不等。与野生敏感菌株相比，田间、室内咯菌腈抗性突变体的生物适合度均显著降低，而且都对渗透胁迫高度敏感。当用1μg/mL咯菌腈处理4h后，敏感菌株的胞内甘油含量及*Bchog*1基因表达量均显著上升，而田间抗性菌株及室内诱抗突变体的上升幅度远低于敏感菌株。田间、室内咯菌腈高抗突变体均在*Bos*1基因检测到突变，但存在差异：田间高抗菌株的突变位点均位于近N末端的HAMP结构域内，而室内药剂驯化获得的高抗突变体的突变类型分为两种：一类位于HAMP结构域，另一类位于近C末端的HATpase_ c结构域内。本文从生理生化和分子水平上研究了咯菌腈田间、室内抗药性突变体的特性，为明确灰葡萄孢菌对咯菌腈的抗性机理提供了理论基础。

关键词：咯菌腈；灰葡萄孢菌；抗药性；生物学特性；基因突变

Study on Resistance of *Botrytis cinerea* to Fludioxonil

Ren Weichao, Wu Dongxia, Duan Yabing, Hou Yiping,
Wang Jianxin, Zhou Mingguo, Chen Changjun
(*College of Plant Protection*, *Nanjing Agricultural University*, *Nanjing* 210095)

Abstract: In this study, the sensitivity of *B. cinerea* isolates which were collected from different hosts in different regions of Jiangsu Province and Shandong Province were tested *in vitro* from 2012 to 2014. Among the 75 isolates from infected cucumbers in Shouguang City, Sahndong Province in 2013, there were 2 fludioxonil-HR isolates with a resistance frequency of 2.67%. Of the 308 isolates from infected tomatoes in Weifang City, Shandong Province in 2014, there were 4 HR isolates with a resistance frequency of 1.30%, while all the isolates collected from Jiang Province were sensitive to fludioxonil. Six fludioxonil-resistant (FR) mutants were obtained from 4 sensitive isolates by repeating exposure them to fludioxonil in the laboratory. Fludioxonil-resistance of these mutants was stable and the level of fludioxonil-resistance, as indicated by the RF values, ranged from 34.38 to above 10 000. All these field and laboratory mutants showed reduced fitness in mycelial growth, sporulation and pathgenicity, and increased sensitivity to osmotic stress. When treated with 1μg/mL of fludioxonil for 4 h, glycerol concentrations and expression levels of *Bchog*1 in the sensitive isolates increased significantly, while those in the field and laboratory FR mutants increased slightly. Mutations in the *Bos*1 gene were detected in the field and laboratory fludioxonil-HR mutants, but there are some differences. Mutations in the field mutants were all located in the HAMP domains of N-terminal region. Mutations in the laboratory mutants were classified into 2 types: mutations located in HAMP domains (Type I) and mutations located in the HATpase_ c domain of C-terminal region (Type II). These results will increase our understanding of resistance mechanism of *B. cinerea* to fludioxonil.

Key words: Fludioxonil; *Botrytis cinerea*; Fungicide resistance; Biological characterization; Gene mutation

由灰葡萄孢菌（*Botrytis cinerea* Pers. ex Fr.）感染引起的灰霉病，可以危害蔬菜、果树、花卉作物等200多种植物，造成严重的经济损失[1]。由于国内尚未培育出优良的抗病品种，所以生产中对灰霉病的控制仍采以化学防治为主[2]。灰葡萄孢菌具有繁殖速度快、易产生遗传变异和适合度高等特性[3]，随着单一作用靶标位点杀菌剂的长期频繁使用，灰葡萄孢菌对生产中常用的杀菌剂如苯并咪唑类多菌灵、二甲酰亚胺类腐霉利和异菌脲、N-苯氨基甲酸酯类乙霉威、琥珀酸脱氢酶抑制剂类啶酰菌胺和苯胺基嘧啶类嘧霉胺等杀菌剂均产生了抗药性，并且抗性群体数量逐年上升致使对该病害的防治效果大幅下降，甚至失效[4]。本课题组监测了2012—2013年采自镇江和南京的灰霉菌株对多菌灵、腐霉利、乙霉威、嘧霉胺、醚菌酯5种常用杀菌剂的抗药性情况，高抗频率高达78%以上，因此上述药剂在检测地区已失去使用价值。

咯菌腈（fludioxonil）属于苯基吡咯类杀菌剂，具有杀菌作用机理独特、杀菌谱广、毒性等优点，是目前防治灰霉病的高效杀菌剂，但在大面积推广应用之前应对其进行抗药性风险评估，并且对投入生产使用的地区监测灰霉病菌对其的抗性动态，研究抗性机理，从而制订科学合理的施药方案。

本研究监测了2012—2014年江苏、山东部分使用咯菌腈地区的灰霉菌对咯菌腈的敏感性动态，并从生理生化和分子水平上研究了田间、室内抗咯菌腈突变体的特性，这对评估灰葡萄孢菌对咯菌腈的抗性风险及探索灰葡萄孢菌对咯菌腈的抗性机理具有重要的参考意义。

1 材料与方法

1.1 供试菌株

供试灰葡萄孢菌是2012—2014年从江苏南京、镇江、盐城和山东寿光、潍坊地区的不同寄主（草莓、番茄、黄瓜、芹菜）采集分离得到的单孢菌株。菌株编号及数量如表1所示。

表1 供试灰葡萄孢菌的采集地、采集时间、寄主、编号及数量

Table 1 Origin of *Botrytis cinerea* isolates from different geographical regions and different hosts in Jiangsu and Shandong Province of China during 2012—2014

Location	Year	Host	Number of isolates	Code
Nanjing, Jiangsu	2012	strawberry	62	Nj1-X ~ Nj6-X
Zhenjiang, Jiangsu	2012	strawberry	228	Bt1-X ~ Bt22-X
Nanjing, Jiangsu	2013	strawberry	213	Nj7-X ~ Nj29-X
Yancheng, Jiangsu	2013	celery	16	Yc-1 ~ Yc-16
Shouguang, Shandong	2013	cucumber	75	Sg-1 ~ Sg-75
Weifang, Shandong	2014	tomato	308	Wf-1 ~ Wf-308

1.2 供试药剂与培养基

97.9%咯菌腈原药（fludioxonil，由江苏省扬农化工有限公司提供）预溶于甲醇配制成

50mg/mL 的储备母液；98% 腐霉利原药（procymidone，由住友化学有限公司提供）和 96.2% 异菌脲原药（iprodione，由江苏快达农化股份有限公司提供）预溶于丙酮配制成 50mg/mL 的储备母液。

马铃薯葡萄糖琼脂（PDA）培养基：马铃薯 20g，葡萄糖 20g，琼脂粉 16g，用蒸馏水定容至 1L。

马铃薯葡萄糖培养液（PDB）：配方中除不加琼脂粉外，其他与 PDA 培养基相同。

酵母蛋白胨葡萄糖培养液（YEPD）：酵母提取物 3g，蛋白胨 10g，葡萄糖 20g，用蒸馏水定容至 1L。

以上培养基均经高压高温灭菌后，常温保存。

1.3 室内药剂驯化获得抗咯菌腈突变体

随机选取 10 株野生型咯菌腈敏感菌株（Nj1-2、Nj3-2、Nj4-6、Nj5-10、Bt2-4、Bt3-4、Bt6-4、Bt10-3、Yc-2 和 Yc-6），将这些菌株在 PDA 平板上预培养 3 天后，分别用打孔器打制 5mm 的菌碟接种到含 0.1μg/mL 咯菌腈的 PDA 平板上，每皿接种 5 个菌碟，每个菌株接种 30 皿，置于 25℃的恒温培养箱中培养 14d。挑取扇形菌落边缘的菌丝转接到含 5μg/mL 咯菌腈的 PDA 平板上，按照此方法依次提高咯菌腈浓度（10、50 和 100μg/mL）进行药剂驯化，将能快速生长的菌丝转接到不含药的 PDA 斜面保存待用。

1.4 田间筛选获得抗咯菌腈突变体

为了获得田间抗咯菌腈菌株，采用区分剂量法监测了 2012—2014 年采自江苏省和山东省不同地区不同寄主的共 902 株灰葡萄孢菌对咯菌腈的敏感性。根据 Fillinger 等的研究，采用的区分剂量为 0.1μg/mL、1μg/mL 和 5μg/mL 的咯菌腈。敏感（S）菌株：在含 0.1μg/mL 咯菌腈的 PDA 平板上不能生长；低抗（HR）菌株：在含 0.1μg/mL 咯菌腈的 PDA 平板上能正常生长，但是在含 1μg/mL 咯菌腈的 PDA 平板上不能生长；高抗（HR）菌株：在含 5μg/mL 咯菌腈的 PDA 平板上能正常生长。每个菌株每药剂浓度接 3 皿，本试验重复 2 次。抗性频率（%）=（抗性菌株数/测定菌株数）× 100。将田间咯菌腈高抗菌株保存至 PDA 斜面，留做进一步研究。

1.5 抗性遗传稳定性及抗性倍数测定

为了测定咯菌腈抗药性突变体的遗传稳定性及抗性倍数，将室内药剂驯化得到的抗药性突变体及其亲本菌株和田间筛选得到的咯菌腈高抗菌株及敏感菌株在不含药剂的 PDA 平板上连续转接 10 代，分别在第一代和第十代测定这些抗、感菌株对咯菌腈的 EC_{50} 值。对于敏感菌株，所设咯菌腈浓度梯度为 0、0.003 12、0.006 25、0.012 5、0.025 和 0.05μg/mL；对于抗药性突变体，所设咯菌腈浓度梯度为 0、0.390 6、1.562 5、6.25、25 和 100μg/mL。每个处理重复 4 皿，试验重复 2 次。抗性倍数（RF，resistance facto）计算公式如下：RF = 抗药性突变体的 EC_{50}/亲本菌株或者敏感菌株的 EC_{50}。抗性遗传稳定性用敏感性变化（factor of sensitivity change，RFC）来衡量，RFC = 转接第一代的抗性倍数/转接 10 代后的抗性倍数。

1.6 交互抗性测定

采用菌丝生长速率法，分别测定了田间、室内咯菌腈抗药性突变体及敏感菌株对二甲酰亚胺类杀菌剂腐霉利和异菌脲的敏感性，采用线性相关分析咯菌腈与腐霉利和异菌脲是否存在交互抗性。所用药剂浓度梯度设置如表 2，每药剂各处理浓度设置 3 个重复，本试验重复 3 次。

表 2　交互抗性测定各药剂设置的浓度梯度

Table 2　Concentrations of fungicides used in the cross-resistance assay.

Fungicides	Fungicide concentrations（μg/mL）	
	fludioxonil-sensitive isolates	fludioxonil-resistant mutants
fludioxonil	0，0.003 125，0.006 25，0.012 5，0.025，0.05	0，0.390 6，1.562 5，6.25，25，100
procymidone	0，0.039 1，0.156 2，0.625，2.5，10	0，0.390 6，1.562 5，6.25，25，100
iprodione	0，0.039 1，0.156 2，0.625，2.5，10	0，0.390 6，1.562 5，6.25，25，100

1.7　灰葡萄孢菌抗咯菌腈突变体的生物适合度测定

供试菌株：室内药剂驯化获得的抗药性突变体及其亲本菌株，田间筛选获得的高抗菌株及当地的敏感菌株。

1.7.1　菌丝生长速率测定

将供试田间、室内咯菌腈抗药性突变体及敏感菌株打取 5mm 菌碟，分别接种于不含药剂的 PDA 平板上，每个菌株重复 6 皿。25℃培养 3d 后，采用十字交叉法测量菌落直径。该试验重复 3 次。

1.7.2　产孢能力测定

将供试田间、室内咯菌腈抗药性突变体及敏感菌株的菌碟分别接种于不含药剂的 PDA 平板上，每个菌株设置 6 个重复。在 12h 光周期光条件下，于 25℃培养两周后，收集孢子。将孢子悬浮液 5 000r/min 离心 1min 后，倒去上清，将孢子沉淀物重新溶于 1mL 的灭菌蒸馏水中，使用血球计数板计算各菌株的产孢量。本试验重复 3 次。

1.7.3　致病力测定

致病力测定试验是在草莓植株叶片及番茄果实上进行的。将预培养的田间、室内咯菌腈抗药性突变体及敏感菌株分别沿菌落边缘打取 5mm 的菌碟，菌丝面朝下接种于长势基本一致的草莓叶片上，每个菌株接种 3 盆植株，每盆 6 张叶片，套袋保湿，并保持温室温度 20～28℃、湿度 75%～90%，5 天后测量病斑直径，计算病斑面积。本试验重复 3 次。

另外，将从供试田间、室内咯菌腈抗药性突变体及敏感菌株菌落边缘打制的菌碟菌丝面朝上接种于大小基本一致的番茄果实上，每个菌株接种 4 个番茄，套袋保湿，置于 25℃恒温光照生长箱中培养 3 d 后测量病斑直径，计算病斑面积。本试验重复 3 次。

1.8　抗药性突变体及敏感菌株对渗透压、离子胁迫的敏感性测定

将预培养的田间、室内咯菌腈抗药性突变体及敏感菌株分别沿菌落边缘打取 5mm 的菌碟，接种于含有 0.5mol/L NaCl、0.5mol/L KCl、0.8mol/L 葡萄糖（glucose）、0.5mol/L 山梨醇（sorbitol）、0.2mol/L $CaCl_2$ 和 0.3mol/L $MgCl_2$ 的 PDA 平板上，以接种于空白 PDA 平板作为对照。每菌株各处理重复 4 皿，25℃培养 3d 后采用十字交叉法测量菌落直径，计算出菌丝生长抑制率。本试验重复 3 次。

$$\text{抵制率（\%）}=\frac{\text{对照菌落直径（mm）}-\text{各处理菌落直径（mm）}}{\text{对照菌落直径（mm）}-5}\times 100$$

1.9　抗药性突变体及敏感菌株的菌丝内甘油含量测定

将预培养的田间、室内咯菌腈抗药性突变体及敏感菌株分别沿菌落边缘打取 5mm 的菌碟，接入含有 100mL YEPD 培养液的锥形瓶中，每瓶接种 5 个菌碟，每菌株重复 6 瓶。25℃

175r/min 振荡培养 60h 后，每个菌株取出 3 瓶加入终浓度为 1μg/mL 的咯菌腈，另外 3 瓶不加药剂作为空白对照。继续摇培 4h 后，过滤收集菌丝，剔除菌碟，用超纯水冲洗 3 次，然后用滤纸吸干水分风干待用。将上述制备的菌丝样品用液氮研磨成粉末，称取 100mg 样品置于 2mL 离心管中，加入 1mL 灭菌蒸馏水，在涡旋仪上涡旋 30min 中后，5 000r/min 离心 20min，取上清液即得到样品胞内甘油溶液。用购自北京普利莱基因技术有限公司的液体样品甘油含量测定试剂盒（GPO-POD 酶法）测定菌丝胞内甘油含量，实验重复 3 次。

1.10　抗药性突变体及敏感菌株的双组分组氨酸激酶基因 *Bos*1 的克隆分析

根据灰葡萄孢菌 *B. cinerea* 核基因组测序菌株 B05.10 的双组分组氨酸激酶基因 *Bos*1（Broad Institue 登录号：BC1G_ 00374），来扩增抗药性突变体及敏感菌株的完整 *Bos*1 基因序列，通过基因序列比对软件进行分析。

1.11　灰葡萄孢菌抗咯菌腈突变体对 *Bchog*1 和 *Bos*1 基因表达的影响

将预培养的田间、室内咯菌腈抗药性突变体及敏感菌株分别沿菌落边缘打取 5mm 的菌碟，接入含有 100mL YEPD 培养液的锥形瓶中，每瓶接种 5 个菌碟，每菌株重复 6 瓶。25℃ 175r/min 振荡培养 60h 后，每个菌株取出 3 瓶加入终浓度为 1μg/mL 的咯菌腈，另外 3 瓶不加药剂作为空白对照。继续摇培 4h 后，过滤收集菌丝，提取菌体总 RNA。根据灰葡萄孢菌 *Bos*1 基因（BC1G_ 00374）及其调控的下游基因 *Bchog*1（BC1G_ 03001）的编码区序列，分别设计引物 os1-QF/QR 和 hog1-QF/QR（os1-QF/QR：5′-TACTACTGCGATCCTGCAAAC-3′/5′-ATCTGCACCTGGTAACCTAATC-3′； hog1-QF/QR： 5′-CGGCACCACCTTTGAGATTA-3′/5′-AACCGGTGAGGTTATC TTTGG-3′）；以灰葡萄孢菌肌动蛋白基因 *Actin*（BC1G_ 08198）作为内参，设计引物 actin-QF/QR（5′-TGTCACCAACTGGGATGATATG-3′/5′-CTGTTGGACTTT-GGGTT GATTG-3′）。数据分析采用 7 500 system 软件和 2-△△CT 的方法，且本试验进行 3 次生物学重复。

2　结果与分析

2.1　灰葡萄孢菌抗咯菌腈突变体的筛选

通过室内药剂驯化，共获得 6 株能在含有 100μg/mL 咯菌腈的 PDA 平板上稳定生长的抗药突变体。由亲本菌株 Nj5-10 和 Bt3-4 各得到 1 株抗药性突变体，编号分别为 Nj5-10R 和 Bt3-4R；由亲本菌株 Bt6-4 得到 2 株抗药性突变体，编号为 Bt6-4R1 和 Bt6-4R2；由亲本菌株 Yc-6 得到 2 株抗药性突变体，编号为 Yc-6R1 和 Yc-6R2。

田间监测了 2012—2014 年采自江苏省和山东省不同地区不同寄主的共 902 株灰葡萄孢菌对咯菌腈的敏感性，结果如表 3。2012—2013 年分离自江苏省南京和镇江不同草莓大棚中的 503 株灰葡萄孢菌以及 2013 年从采集自江苏省盐城地区的感病芹菜叶片上分离得到的 16 株灰葡萄孢菌均对咯菌腈高度敏感，不存在咯菌腈抗性菌株。然而 2013—2014 年，在山东省寿光和潍坊地区的黄瓜和番茄田块检测到了咯菌腈低抗（LR）和高抗（HR）菌株。在 2013 年采自山东省寿光黄瓜田块的 75 株灰葡萄孢菌中，检测到 9 株低抗菌株和 2 株高抗菌株，抗性频率分别为 10.67% ^{LR}和 2.67% ^{HR}；在 2014 年采自山东省潍坊番茄田块的 308 株灰葡萄孢菌中，检测到 82 株低抗菌株和 4 株高抗菌株，抗性频率分别为 17.21% ^{LR}和 1.30% ^{HR}。

表 3　2012—2014 年田间咯菌腈抗药性突变体的筛选结果

Table 3　Sensitivity to fludioxonil of *Botrytis cinerea* isolates from different geographical regions in Jiangsu and Shandong Province of China during 2012—2014

Location	Year	Host	Number of isolates	Number			Frequency [d] (%)	
				S [a]	LR [b]	HR [c]	LR	HR
Nanjing, Jiangsu	2012	strawberry	62	62	0	0	0	0
Zhenjiang, Jiangsu	2012	strawberry	228	228	0	0	0	0
Nanjing, Jiangsu	2013	strawberry	213	213	0	0	0	0
Yancheng, Jiangsu	2013	celery	16	16	0	0	0	0
Shouguang, Shandong	2013	cucumber	75	64	9	2	10.67	2.67
Weifang, Shandong	2014	tomato	308	222	53	4	17.21	1.30

[a] S = sensitive, [b] LR = lowly resistant, [c] HR = highly resistant, [d] Frequency of resistance isolates.

2.2　抗性遗传稳定性及抗性倍数测定

室内药剂驯化获得的 6 株咯菌腈抗药性突变体均可在含有 100μg/mL 咯菌腈的 PDA 平板上稳定生长，但是抗性水平不等，Nj5-10R、Bt6-4R1 和 Yc-6R1 的 EC_{50}值大于 100μg/mL，Bt3-4R 的 EC_{50}值大于 4μg/mL，Bt6-4R2 和 Yc-6R 的 EC_{50}值大于 0.2μg/mL 而小于 0.5μg/mL，相应的抗性倍数分别为大于 10 000、400 和 30～60（表 4）。6 株田间高抗菌株的 EC_{50}值均大于 100μg/mL，抗性倍数高达 3 000以上（表 4）。将田间、室内抗药性突变体及敏感菌株在不含药 PDA 平板上连续转代培养后，测定对咯菌腈抗药性水平变化，结果表明田间和室内抗药性突变体在无药剂下连续培养 10 代后抗性水平基本不变，这说明抗药性表型能够通过无性繁殖稳定遗传（表 4）。

表 4　咯菌腈抗药性突变体的抗性遗传稳定性及抗性倍数

Table 4　Stability and level of fludioxonil-resistance for the laboratory and field mutants of *Botrytis cinerea*

Isolate or mutant	Sensitivity to fludioxonil [a]	Origin [b]	EC_{50} (μg/mL) [c]		RF [d]		FSC [e]
			1st	10th	1st	10th	
Nj5-10	S	field isolate	0.005 6	0.006 9	—	—	—
Nj5-10R	HR	laboratory mutant	>100	>100	>10 000	>10 000	—
Bt3-4	S	field isolate	0.008 6	0.008 3	—	—	—
Bt3-4R	HR	laboratory mutant	4.013 5	3.344 6	466.69	402.96	0.86
Bt6-4	S	field isolate	0.007 3	0.007 8	—	—	—
Bt6-4R1	HR	laboratory mutant	>100	100	>10 000	>10 000	—
Bt6-4R2	LR	laboratory mutant	0.421 5	0.369 2	57.74	47.33	0.82
Yc-6	S	field isolate	0.007 5	0.006 8	—	—	—
Yc-6R1	HR	laboratory mutant	>100	>100	>10 000	>10 000	—

（续表）

Isolate or mutant	Sensitivity to fludioxonil [a]	Origin [b]	EC_{50}(μg/mL) [c]		RF [d]		FSC [e]
			1st	10th	1st	10th	
Yc-6R2	LR	laboratory mutant	0.286 5	0.233 8	38.2	34.38	0.90
Sg-2	S	field isolate	0.037 7	0.035 6	—	—	—
Sg-28	S	field isolate	0.015 6	0.016 5	—	—	—
Sg-17	HR	field mutant	>100	>100	>3 000	>3 000	—
Sg-38	HR	field mutant	>100	>100	>3 000	>3 000	—
Wf-20	S	field mutant	0.009 1	0.010 2	—	—	—
Wf-142	S	field isolate	0.013 4	0.012 8	—	—	—
Wf-55	HR	field mutant	>100	>100	>8 000	>8 000	—
Wf-161	HR	field mutant	>100	>100	>8 000	>8 000	—
Wf-192	HR	field mutant	>100	>100	>8 000	>8 000	—
Wf-202	HR	field strain	>100	>100	>8 000	>8 000	—

[a] S: sensitive; LR: lowly resistant; HR: highly resistant.

[b] Laboratory mutants were obtained by mass selection on fludioxonil-amended medium; Field mutant were collected from the field locations.

[c] EC_{50}: effective concentration for 50% inhibition of mycelial growth at the 1st transfer and the 10th transfer.

[d] RF: resistance factor, a ratio of EC_{50} for a fludioxonil-resistant mutant relative to the EC_{50} for the sensitive isolate.

[e] FSC: the ratio of RF values at the 1st and 10th transfer.

2.3 交互抗性测定

交互抗性测定结果显示，田间筛选和室内药剂驯化获得的抗咯菌腈灰葡萄孢菌均对二甲酰亚胺类杀菌剂腐霉利和异菌脲表现出抗性，这说明咯菌腈与腐霉利、异菌脲存在正交互抗性（表5）。

表5　抗咯菌腈灰葡萄孢菌的交互抗性测定结果

Table 5　Sensitivity of fludioxonil-resistant mutant and fludioxonil-sensitive isolates of *Botrytis cinerea* to procymidone and iprodione

Isolate or mutant	Sensitivity to fludioxonil [a]	Origin [b]	EC_{50} (μg/mL) [c]		
			fludioxonil	procymidone	iprodione
Nj5-10	S	field isolate	0.006 2 e	0.378 6 i	0.051 5 f
Nj5-10R	HR	laboratory mutant	>100 a	>100 ao5	>100 a
Bt3-4	S	field isolate	0.008 7 e	1.952 4 h	2.596 1 cde
Bt3-4R	HR	laboratory mutant	3.737 9 b	>100 a	>100 a
Bt6-4	S	field isolate	0.007 8 e	2.214 fg	2.928 cd
Bt6-4R1	HR	laboratory mutant	>100 a	>100 a	>100 a

（续表）

Isolate or mutant	Sensitivity to fludioxonil [a]	Origin [b]	EC_{50} （μg/mL）[c]		
			fludioxonil	procymidone	iprodione
Bt6-4R2	LR	laboratory mutant	0. 397 4 c	3. 628 3 e	3. 533 9 c
Yc-6	S	field isolate	0. 007 2 e	0. 188 7 j	0. 225 f
Yc-6R1	HR	laboratory mutant	>100 a	>100 a	>100 a
Yc-6R2	LR	laboratory mutant	0. 255 5 d	15. 575 b	3. 708 c
SG-2	S	field isolate	0. 349 de	2. 110 5 g	1. 250 7 def
Sg-28	S	field isolate	0. 016 e	3. 935 2 c	1. 203 6 ef
Sg-17	HR	field mutant	>100 a	>100 a	52. 197 b
Sg-38	HR	field mutant	>100 a	>100 a	>100 a
Wf-20	S	field mutant	0. 009 7 e	2. 279 4 f	1. 231 1 ef
Wf-142	S	field isolate	0. 013 3 e	3. 806 9 d	2. 305 2 cde
Wf-55	HR	field mutant	>100 a	>100 a	>100 a
Wf-161	HR	field mutant	>100 a	>100 a	>100 a
Wf-192	HR	field mutant	>100 a	>100 a	>100 a
WF-202	HR	field mutant	>100 a	>100 a	>100 a

[a] S：sensitive；LR：lowly resistant；HR：highly resistant.

[b] Laboratory mutants were obtained by mass selection on fludioxonil-amended medium；Field mutant were collected from the field locations.

[c] EC_{50}：effective concentration for 50% inhibition of mycelial growth；means in a column followed by the same letter are not different according to Fisher's least significant difference（LSD）（$P = 0.05$）.

2.4 灰葡萄孢菌抗咯菌腈突变体的生物适合度测定

2.4.1 菌丝生长速率测定

除了 YC-6R2 的生长速率快于亲本菌株，室内药剂驯化的获得的其余 5 株抗药性突变体的菌丝生长速率与亲本菌株相比均有所下降；同样，田间筛选获得的咯菌腈高抗菌株与敏感菌株相比，菌丝生长速率也明显下降（表 6）。田间抗性种群与室内抗性种群相比，它们的菌丝生长速率无显著差异（P t-test >0. 05）。

2.4.2 产孢能力测定

除了 Bt6-4R2 的产孢量与亲本菌株相当，室内药剂驯化的获得的其余 5 株抗药性突变体的产孢量均显著低于亲本菌株；同样，田间筛选获得的咯菌腈高抗菌株与敏感菌株相比，产孢量也显著下降（表 6）。田间抗性种群与室内抗性种群相比，它们的产孢量无显著差异（P t-test >0. 05）。

2.4.3 致病力测定

与亲本菌株相比，室内药剂驯化的获得的 6 株抗药性突变体在草莓叶片和番茄果实上的致病力均显著降低；同样，与田间敏感菌株相比，田间高抗菌株在草莓叶片和番茄果实上的致病力也显著降低（表 6）。田间抗性种群与室内抗性种群相比，它们的致病力无显著差异

（P t-test >0.05）。

2.5 抗药性突变体及敏感菌株对渗透压、离子胁迫的敏感性测定

试验结果显示：与亲本菌株或敏感菌株相比，室内诱抗突变体和田间高抗菌株对0.5mol/L NaCl 和0.5mol/L KCl 更加敏感（图1A，B）；0.8mol/L 葡萄糖、0.5mol/L 山梨醇、0.2mol/L $CaCl_2$ 和0.3mol/L $MgCl_2$ 促进咯菌腈敏感菌株的生长，但室内诱抗突变体和田间高抗菌株的菌丝生长均受到显著抑制（图1C ~ F）。田间抗性菌株与室内诱抗突变体比较，对山梨醇更敏感（P t-test < 0.05），而对其他测定的胁迫因子的敏感性无显著差异（P t-test >0.05）。以上结果表明咯菌腈抗性影响菌体对高糖、高盐渗透压及离子胁迫的响应调控。

表6 咯菌腈抗、感菌株的生物适合度比较[a]

Table 6 Fitness of *B. cinerea* strains that were resistant or sensitive to fludioxonil[a]

Isolate or mutant	Origin[b]	Mycelial growth (cm)	Sporulation ($\times 10^6$ spores/ Petri plate)	Pathogenicity (lesion area) (cm^2)	
				Strawberry leaves	Tomato fruits
Nj5-10	field-S	6.12 d	43.8 f	3.90 f	6.99 e
Nj5-10R	laboratory-HR	5.53 def	0.1j	0.08 i	0.5 ij
Bt3-4	field-S	7.39 ab	100 a	14.41 a	14.79 a
Bt3-4R	laboratory-HR	4.26 g	4.2 j	0.22 i	0.26 ij
Bt6-4	field-S	6.22 d	63.7 d	7.30 cd	8.64 d
Bt6-4R1	laboratory-HR	5.68 de	7.9 h	0.96 hi	0.21 ij
Bt6-4R2	laboratory-LR	6.35 cd	60.8 de	4.01 f	2.36 gh
Yc-6	field-S	7.9 ab	43.3 f	4.12 ef	10.65 c
Yc-6R1	laboratory-HR	6.38 cd	0.2 j	2.15 g	3.32 fg
Yc-6R2	laboratory-LR	8.01 ab	0 j	0 i	0 j
SG-2	field-S	8.20 a	77.7 b	6.61 d	12.54 b
Sg-28	field-S	7.50 ab	67.7 c	5.09 e	11.80 bc
Sg-17	field-HR	5.53 def	0.5 j	0.63 hi	1.33 hij
Sg-38	field-HR	4.53 g	0.4 j	2.28 g	2.12 gh
Wf-20	field-S	7.17 bc	57.9 e	7.72 bc	10.72 c
Wf-142	field-S	7.73 ab	67.4 c	8.75 b	13.14 b
Wf-55	field-HR	4.87 efg	20.2 g	2.47 g	3.94 f
Wf-161	field-HR	4.67 fg	0.3 j	0.56 i	1.55 hi
Wf-192	field-HR	4.40 g	0.4 j	1.63 gh	1.73 hi
WF-202	field-HR	4.93 efg	2.9 ij	1.64 gh	1.02 hij

[a] Means in a column followed by the same letter are not different according to Fisher's least significant difference (LSD) (P = 0.05).

[b] field-S: field sensitive isolates; laboratory-HR/LR: highly resistant mutants/lowly resistant mutants which were obtained by mass selection on fludioxonil-amended medium in the laboratory; field-HR/LR: highly resistant mutants/lowly resistant mutants which were collected from the field locations.

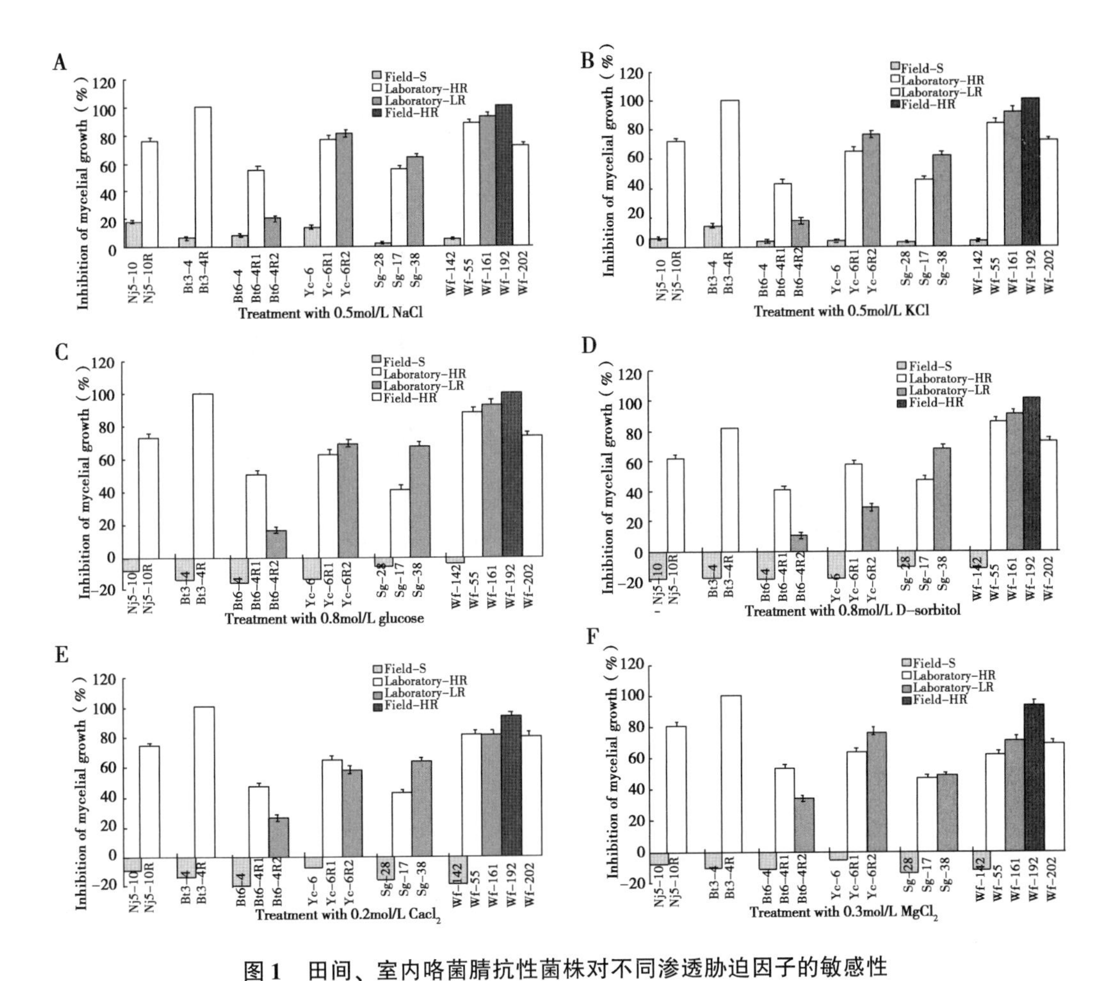

图1　田间、室内咯菌腈抗性菌株对不同渗透胁迫因子的敏感性

Fig. 1　Sensitivity of the laboratory and field fludioxonil-resistant mutants and sensitive isolates to osmotic stress, which were generated by NaCl (A), KCl (B), glucose (C), D-sorbitol (D), $CaCl_2$ (E) or $MgCl_2$ (F). Bars denote the standard errors of three repeated experiments.

2.6　抗药性突变体及敏感菌株的菌丝甘油含量测定

在正常条件下，室内诱抗突变体与亲本菌株、田间高抗菌株与敏感菌株的胞内甘油含量无明显差异。当用1μg/mL咯菌腈处理4 h后，亲本菌株或敏感菌株的胞内甘油含量显著上升，而室内、田间高抗菌株的胞内甘油含量变化不显著；2株抗性水平较低的室内诱抗突变体Bt6-4R2和Yc-6R2的甘油含量也显著上升，但是增加幅度要远低于亲本菌株（图2）。以上结果表明，咯菌腈抗性抑制了菌体在药剂胁迫下甘油合成的调控，并且与抗性水平相关，抗性水平越高，抑制作用越显著。

2.7　抗药性突变体及敏感菌株的双组分组氨酸激酶基因*Bos*1的克隆分析

编码灰葡萄孢菌双组分组氨酸蛋白激酶Bos1的基因长度为4 347bp，氨基酸序列长度为1 310aa，将氨基酸序列提交KEGG网站Motif预测Bos1蛋白的结构和功能。灰葡萄孢菌的双组分组氨酸蛋白激酶Bos1具有Ⅲ型组氨酸激酶的特征，包括6个重复的HAMP结构域、组氨酸激酶结构域（HisKA）、ATPase结构域（HATPase_ c）和响应调节蛋白结构域

（Rec）（图3）。

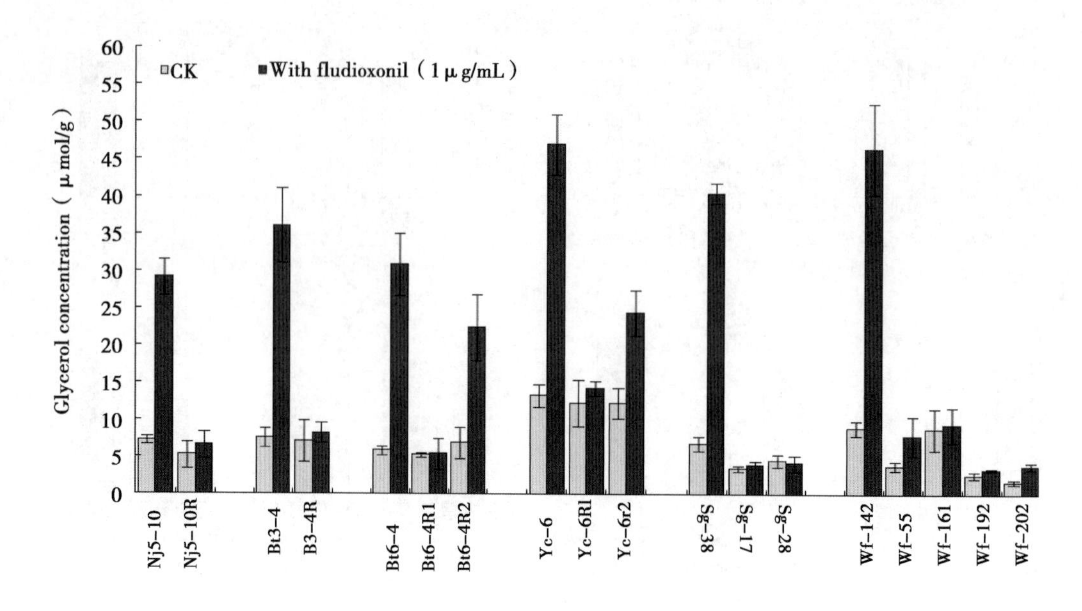

图2 菌丝甘油含量测定

Fig. 2 Comparison in intracellular glycerol concentration of field and laboratory fludioxonil-resistant mutants and sensitive isolates

测序结果表明：供试的咯菌腈敏感菌株 *Bos*1 基因序列与麻省理工大学灰葡萄孢菌基因组数据库上登录的标准菌株 B05.10 的序列完全一致；2 株抗性水平较低的室内诱抗突变体 Bt6-4R2 和 Yc-6R2 的序列也标准菌株 B05.10 的序列完全一致；其余室内、田间咯菌腈高抗菌株均在 *Bos*1 基因发生突变，突变位点多样（表7）。

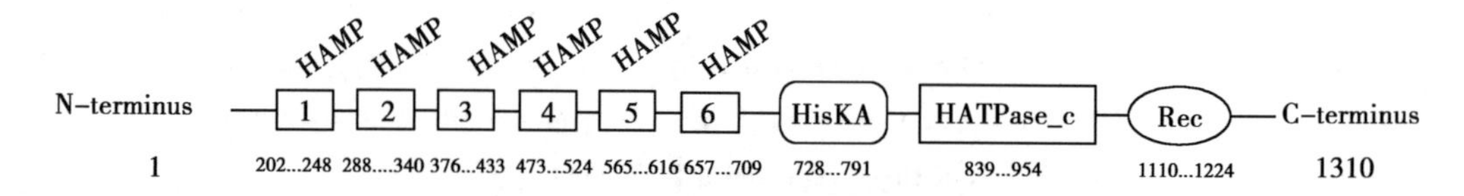

图3 灰葡萄孢菌双组分组氨酸激酶 Bos1 结构示意图

Fig. 3 Schematic representation of the domain architecture of two-component histidine kinase *Bos*1 from *Botrytis cinerea*

The characteristic domains of *Bos*1 include six HAMP domain repeats, the His Kinase A（phospho-acceptor）domain（HisKA）, the HK-like ATPase domain（HATPase_ c）and the response regulator domain（Rec）.

根据 *Bos*1 发生突变的位置分为两种类型：一种是突变发生在近 N 末端的 6 个 HAMP 重复结构域内，另一种突变是发生在近 C 末端的 HATPase_ c 结构域。其中，田间筛选获得的 6 株高抗菌株均只在 HAMP 结构域内检测到点突变，属于第一种突变类型：分离自黄瓜的 2 株田间高抗菌株 Sg-17 和 Sg-38 在 HAMP 结构域上发生多个点突变，且二者有 2 个相同的点突变；分离自番茄的 4 株田间高抗菌株均发生单点突变，具体突变情况如表 6。室内药剂驯化获得的 4 株高抗突变体：Bt3-4R 和 Yc-6R1 在 HAMP 结构域发生单点突变，属于第一种突变类型，具体点突变如表7；Nj5-10R和Bt6-4R1的突变均发生在HAT-Pase_ c结构域，属于第二种突变类型。Nj5-10R在846位由谷氨酰胺突变为终止密码子（CAA→TAA）导致提前终止，Bt6-4R1在835位插入一段55 bp的重复序列，导致移码突变提前在838位终止。

表7 田间、室内抗咯菌腈突变体 *Bos*1 突变类型

Table 7 Mutation in *Bos*1 peptide sequence in laboratory and field mutants of *Botrytis cinerea* resistant to fludioxonil

Isolate or mutant	Origin [a]	Bos1 peptide sequence [b]	Structural domain [c]
Nj5-10	field-S	wt	—
Nj5-10R	laboratory-HR	Q846STOP	HATPase_ c
Bt3-4	field-S	wt	—
Bt3-4R	laboratory-HR	E253D	Between HAMP2 and HAMP3
Bt6-4	field-S	wt	—
Bt6-4R1	laboratory-HR	a 55-bp duplication at amino acid position 835 and resulting in 838STOP	HATPase_ c
Bt6-4R2	laboratory-LR	no mutation	—
Yc-6	field-S	wt	—
Yc-6R1	laboratory-HR	G415D	HAMP3
Yc-6R2	laboratory-LR	no mutation	—
SG-2	field-S	wt	—
Sg-28	field-S	wt	—
Sg-17	field-HR	R319K, V336M, D337N, V346I, A350S, Q369P, N373S	HAMP2, HAMP3, and between them
Sg-38	field-HR	G262S, Q369P, N373S	HAMP1, between HAMP2 and HAMP3
Wf-20	field-S	wt	—
Wf-142	field-S	wt	—
Wf-55	field-HR	G311R	HAMP2
Wf-161	field-HR	G265D	Between HAMP 1 and HAMP 2
Wf-192	field-HR	N609T	HAMP5
WF-202	field-HR	G545E	Between HAMP4 and HAMP5

2.8 灰葡萄孢菌抗咯菌腈突变体对 *Bchog*1 和 *Bos*1 基因表达的影响

灰葡萄孢菌 *Bchog*1 基因（BC1G_ 03001）与酿酒酵母中 MAPK 途径中的 *Hog*1 基因同源，是位于 HOG-MAPK 途径中的下游基因，受上游基因 *Bos*1 调控。由于田间和室内筛选获得的咯菌腈高抗突变体，在 *Bos*1 均发生突变，因此本试验通过测定 *Bchog*1 的相对表达量来研究咯菌腈抗药性突变体在药剂胁迫下是否对 *Bchog*1 基因的表达有影响。结果表明：在正常条件下，室内诱抗突变体与亲本菌株、田间高抗菌株与敏感菌株的 *Bchog*1 基因表达量无显著差异。当用 1μg/mL 咯菌腈处理 4 h 后，亲本菌株或敏感菌株的 *Bchog*1 基因表达量显著上调，而室内、田间高抗菌株变化不显著；2 株抗性水平较低的室内诱抗突变体 Bt6-4R2 和 Yc-6R2 的 *Bchog*1 基因表达量也显著上调，但是上调幅度要远低于亲本菌株（图 4）。以上结果表明，咯菌腈抗性抑制了菌体在药剂胁迫下 HOG-MAPK 途径的激活，并且与抗性水平相关，抗性水平越高，抑制作用越显著。同时，我们也测定了田间、室内咯菌腈抗药性突变体在药剂胁迫下对 *Bos*1 基因的表达是否有影响，但未发现明显规律（图 5）。

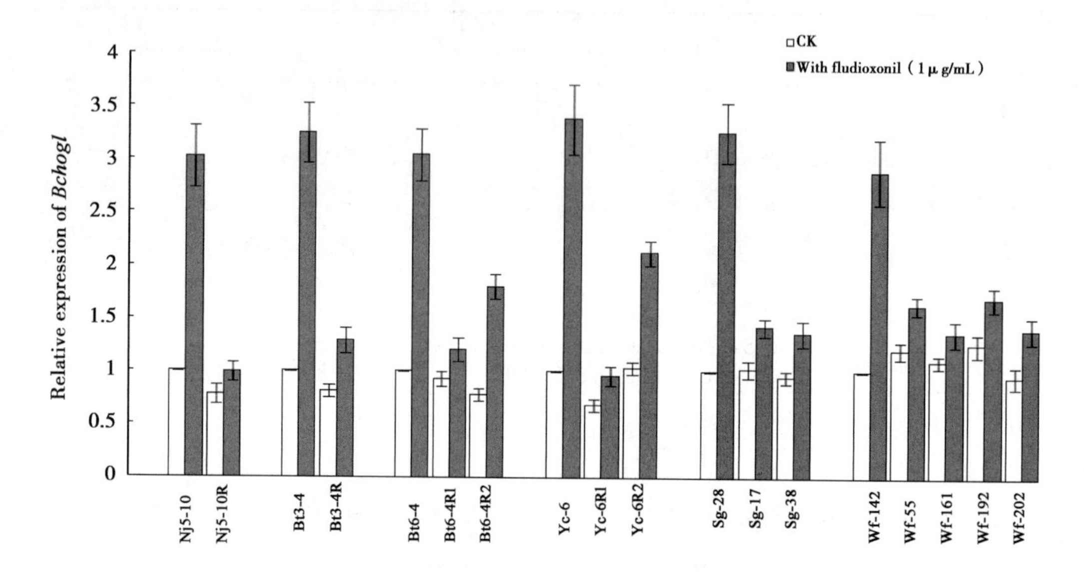

图 4　田间、室内咯菌腈抗药性突变体对 *Bchog*1 基因表达水平的影响

Fig. 4　Relative expression level of *Bchog*1 gene in the wild-type isolates, field and laboratory fludioxonil-resistant mutants of *Botrytis cinerea*

3　结果与讨论

苯基吡咯类杀菌剂咯菌腈抗菌谱广，对子囊菌、担子菌、半知菌等很多病原菌都有很好的抑菌活性，对灰葡萄孢菌有特效，既能够抑制孢子的萌发，又能够抑制菌丝体的生长。在国内，咯菌腈近年来才开始应用于蔬菜、水果等灰霉病的防治。目前国内尚未有咯菌腈田间抗性的报道，但是在国外许多国家已陆续发现田间低抗或中抗的灰霉菌种群[5-7]。2012—2014 年，我们监测了部分已经开始使用咯菌腈防治灰霉病的地区其灰霉菌对咯菌腈的敏感性。在江苏地区未检测到咯菌腈田间抗性菌株，而 2013 年和 2014 年分别在山东寿光黄瓜和山东潍坊番茄灰霉菌中发现了对咯菌腈敏感性下降的菌株，抗性频率为百分之十几，且多数表现为低抗。同时，我们也筛选到 6 株高抗菌株，这是国内外首次在田间发现咯菌腈高抗菌株，因此对田间高抗菌株的生物学及生理生化特性进行研究尤为重要。

由于此前尚未发现咯菌腈田间高抗菌株，所以现有与咯菌腈抗性相关的报道多以室内诱抗突变体为材料进行研究的。本文通过室内药剂驯化获得 6 株抗性稳定、抗性水平不等的咯菌腈抗性突变体，以此作为参照，研究比较了田间、室内咯菌腈抗药性突变体的特性。我们发现，田间、室内抗咯菌腈突变体的在菌丝生长、产孢能力、致病力方面均显著降低，结果与现有咯菌腈室内抗性在灰霉菌及其他真菌中的报道相似[8-9]，这预示着咯菌腈抗性菌株在

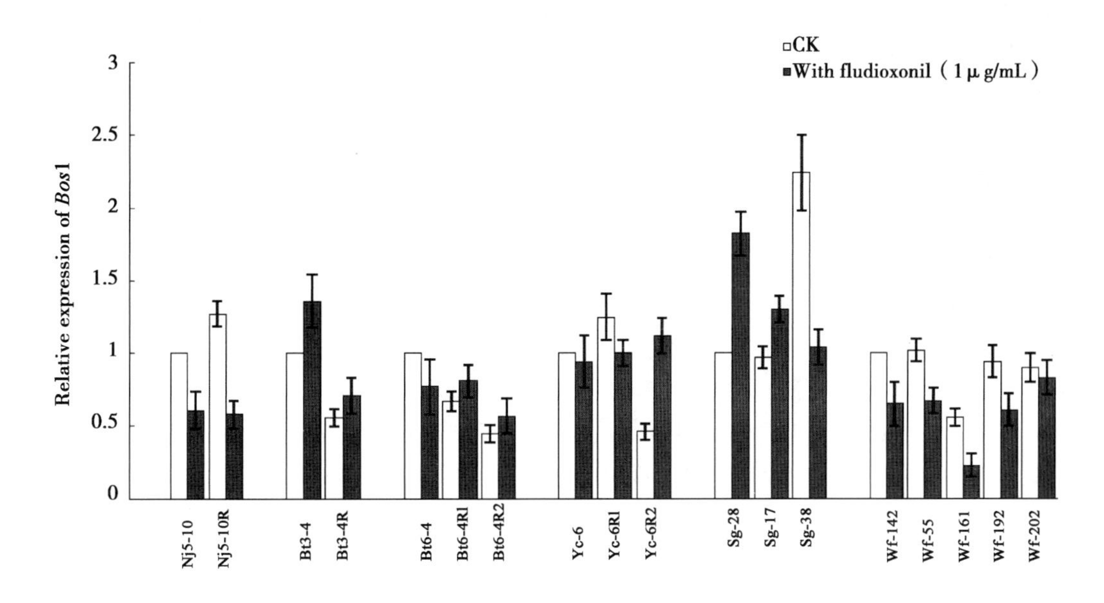

图5 田间、室内咯菌腈抗药性突变体对 *Bos*1 基因表达水平的影响

Fig. 5 Relative expression level of *Bos*1 gene in the wild-type isolates, field and laboratory fludioxonil-resistant mutants of *Botrytis cinerea*

自然环境的生存和竞争能力要远远低于野生敏感菌株，如果降低选择压力，如减少或停止用药，抗性菌株在自然群体中的比例就会下降，不易形成抗药性群体。所以在使用咯菌腈防治病害时，建议与其他不同作用机制的杀菌剂轮换或混配使用。

交互抗性分析结果表明田间、室内灰葡萄孢菌抗咯菌腈菌株对二甲酰亚胺类杀菌剂腐霉利和异菌脲呈正交互抗性。在许多其他植物病原真菌中，这两类药剂也存在正交互抗性[10-12]。目前有大量的研究资料表明灰葡萄孢菌对二羧酰亚胺类和苯基吡咯类杀菌剂产生抗性与Ⅲ型双组分组氨酸激酶（HK）相关[13-14]。有趣的是，Fillinger 等发现苯基吡咯类杀菌剂诱导的高抗菌株对二甲酰亚胺类杀菌剂也是高抗的，而二甲酰亚胺类药剂诱导的高抗菌株对苯基吡咯类杀菌剂却多数为敏感性，少数表现为低抗[15]，这表明这两类杀菌剂的抗性机制是存在差异的。HK 同时涉及对不利环境的适应性调控，例如渗透压胁迫与氧胁迫。Viaud 等研究表明 HK 对灰霉病菌生长和致病性发挥了重要作用[16]，这也解释为什么对这咯菌腈产生高抗的菌株在田间很难发现。

丝状真菌的Ⅲ型双组分组氨酸激酶在功能上与酿酒酵母的 Sln1 同源，调控 HOG MAPK 途径。酿酒酵母菌的 HOG 途径是研究最多的 MAPK 系统，它由两条上游支路（Sln1 和 Sho1）和下游的 MAPK 级联组成的[17]。在正常条件下，渗透感应蛋白组氨酸激酶 Sln1 通过将磷酸传递到 Ypd1 和响应调控蛋白 Ssk1 从而对 HOG 途径起负调控作用；高渗刺激后 Sln1 的活性暂时受抑制引发 HOG 途径的激活，并且表达渗透响应基因[18]。多项研究表明Ⅲ型组氨酸激酶是咯菌腈的作用靶标[19-21]，野生敏感菌在咯菌腈处理下，会持续激活 HOG-MAPK 途径合成大量的甘油致使细胞死亡。我们推测抗性菌株因其信号转导途径中关键酶基因突变，在药剂胁迫下抑制 HOG 途径的激活，不会积累大量甘油，因而可以在药剂处理下存活。因此，我们测定了田间、室内抗药性突变体在药剂胁迫下菌体内甘油含量的积累以及 *Bchog*1 基因的表达量，试验结果与设想一致：咯菌腈处理后，敏感菌株体内甘油含量及

*Bchog*1 表达量显著上升，而田间、室内咯菌腈抗性突变体的上升幅度远低于敏感菌株，这表明咯菌腈抗性抑制了菌体在药剂胁迫下 HOG-MAPK 途径的激活，影响了对甘油合成的调控，并且实验结果显示这种抑制作用与抗性水平相关，抗性水平越高，抑制作用就越显著。而菌体正是通过激活 HOG-MAPK 途径合成甘油来响应对外界胁迫的调控，这也解释了田间、室内咯菌腈抗性突变体为何对高渗胁迫敏感。

本文对田间、室内抗咯菌腈突变体的 *Bos*1 基因进行克隆分析，除了 2 株室内较低水平抗性菌株的 *Bos*1 基因没有发生突变，其他田间、室内高抗菌株均在 Bos1 发生突变，但是发生突变类型的有差异，有单点突变，有多点突变，还有插入重复序列等多种情况，另外突变的位置也呈多样性，田间高抗菌株的突变均发生在 HAMP 区域内，但是室内诱抗突变体有的发生在 HAMP 区域内，而是发生在近 C 末端的 HATpase_ c 结构域内。这种突变类型在之前从未报道过，打破了认为咯菌腈高水平抗性只与Ⅲ型双组分组氨酸激酶 HAMP 重复结构的突变有关的局限性，对更加深入地了解咯菌腈的抗性机制具有重要的意义。

参考文献

[1] Williamson B, Tudzynski B, Tudzynski P, et al. *Botrytis cinerea*: the cause of grey mould disease [J]. Molecular Plant Pathology, 2007, 8 (5): 561 – 580.

[2] Rosslenbroich H J, Stuebler D. *Botrytis cinerea*—history of chemical control and novel fungicides for its management [J]. Crop protection, 2000, 19 (8): 557 – 561.

[3] Leroux P, Fritz R, Debieu D, et al. Mechanisms of resistance to fungicides in field strains of *Botrytis cinerea* [J]. Pest Management Science, 2002, 58 (9): 876 – 888.

[4] Elad Y, Yunis H, Katan T. Multiple fungicide resistance to benzimidazoles, dicarboximides and diethofencarb in field isolates of *Botrytis cinerea* in Israel [J]. Plant Pathology, 1992, 41 (1): 41 – 46.

[5] Leroch M, Plesken C, Weber R, et al. Gray mold populations in German strawberry fields show multiple fungicide resistance and are dominated by a novel clade close to *Botrytis cinerea* [J]. Applied and environmental microbiology, 2012.

[6] Kretschmer M, Leroch M, Mosbach A, et al. Fungicide-driven evolution and molecular basis of multidrug resistance in field populations of the grey mould fungus *Botrytis cinerea* [J]. PLoS Pathogens, 2009, 5 (12): e1000696.

[7] Fernández-Ortuño D, Bryson P, Grabke A, et al. First report of fludioxonil resistance in *Botrytis cinerea* from a strawberry field in Virginia [J]. European Journal of Plant Pathology, 2014.

[8] Ziogas B, Kalamarakis A. Phenylpyrrole fungicides: mitotic instability in *Aspergillus nidulans* and resistance in *Botrytis cinerea* [J]. Journal of Phytopathology, 2001, 149 (6): 301 – 308.

[9] Ochiai N, Fujimura M, Motoyama T, et al. Characterization of mutations in the two-component histidine kinase gene that confer fludioxonil resistance and osmotic sensitivity in the os-1 mutants of *Neurospora crassa* [J]. Pest Management Science, 2001, 57 (5): 437 – 442.

[10] Avenot H, Simoneau P, Iacomi-Vasilescu B, et al. Characterization of mutations in the two-component histidine kinase gene AbNIK1 from *Alternaria brassicicola* that confer high dicarboximide and phenylpyrrole resistance [J]. Current Genetics, 2005, 47 (4): 234 – 243.

[11] Iacomi-Vasilescu B, Avenot H, Bataillé-Simoneau N, et al. *In vitro* fungicide sensitivity of *Alternaria* species pathogenic to crucifers and identification of *Alternaria brassicicola* field isolates highly resistant to both dicarboximides and phenylpyrroles [J]. Crop protection, 2004, 23 (6): 481 – 488.

[12] Kanetis L, Förster H, Jones C, et al. Characterization of genetic and biochemical mechanisms of fludioxonil and pyrimethanil resistance in field isolates of *Penicillium digitatum* [J]. Phytopathology, 2008, 98 (2): 205 - 214.

[13] Cui W, Beever R E, Parkes S L, et al. Evolution of an osmosensing histidine kinase in field strains of *Botryotinia fuckeliana* (*Botrytis cinerea*) in response to dicarboximide fungicide usage [J]. Phytopathology, 2004, 94 (10): 1 129 - 1 135.

[14] Cui W, Beever R E, Parkes S L, et al. An osmosensing histidine kinase mediates dicarboximide fungicide resistance in *Botryotinia fuckeliana* (*Botrytis cinerea*) [J]. Fungal Genetics and Biology, 2002, 36 (3): 187 - 198.

[15] Fillinger S, Ajouz S, Nicot P C, et al. Functional and structural comparison of pyrrolnitrin-and iprodione-induced modifications in the class III histidine-kinase Bos1 of *Botrytis cinerea* [J]. PloS ONE, 2012, 7 (8): e42 520.

[16] Viaud M, Fillinger S, Liu W, et al. A class III histidine kinase acts as a novel virulence factor in *Botrytis cinerea* [J]. Molecular Plant-Microbe Interactions, 2006, 19 (9): 1 042 - 1 050.

[17] Hohmann S. Osmotic stress signaling and osmoadaptation in yeasts [J]. Microbiology and Molecular Biology Reviews, 2002, 66 (2): 300 - 372.

[18] Chen R E, Thorner J. Function and regulation in MAPK signaling pathways: lessons learned from the yeast *Saccharomyces cerevisiae* [J]. Biochimica et Biophysica Acta (BBA) -Molecular Cell Research, 2007, 1773 (8): 1 311 - 1 340.

[19] Motoyama T, Ohira T, Kadokura K, et al. An Os-1 family histidine kinase from a filamentous fungus confers fungicide-sensitivity to yeast [J]. Current Genetics, 2005, 47 (5): 298 - 306.

[20] Bahn Y-S, Kojima K, Cox G M, et al. A unique fungal two-component system regulates stress responses, drug sensitivity, sexual development, and virulence of *Cryptococcus neoformans* [J]. Molecular Biology of the Cell, 2006, 17 (7): 3 122 - 3 135.

[21] Hagiwara D, Matsubayashi Y, Marui J, et al. Characterization of the NikA histidine kinase implicated in the phosphorelay signal transduction of *Aspergillus nidulans*, with special reference to fungicide responses [J]. Bioscience, Biotechnology, and Biochemistry, 2007, 71 (3): 844 - 847.

灰葡萄孢菌（*Botrytis cinerea*）对啶菌噁唑的敏感性基线及啶菌噁唑防效

朱 赫*，黄成田，纪明山**
（沈阳农业大学植物保护学院，沈阳 100161）

摘要：灰葡萄孢菌（*Botrytis cinerea*）是一种重要的空气传播的植物病原菌，有着广阔的寄主范围。由该病菌引起的灰霉病常导致产量减少、质量降低，引起严重的损失。在国内，由灰葡萄孢菌引起的灰霉病可以导致蔬菜生产中 10% ~20% 的损失，在一些发病严重的地区甚至可以高达 60% 以上。新型杀菌剂啶菌噁唑已在国内注册用于防治番茄灰霉病，目前暂未广泛使用。本研究基于 2012 年采自辽宁省的 165 株番茄灰霉病菌的 EC_{50}值，建立了番茄灰霉病菌对啶菌噁唑的敏感基线。结果表明啶菌噁唑 EC_{50}值的分布频率是一个单峰曲线，带有右侧拖尾。EC_{50}值的平均值为（0.067 6 ± 0.040 9）（SD）μg/mL，分布范围为 0.012 8 ~ 0.198 7μg/mL。啶菌噁唑与多菌灵和腐霉利之间没有交互抗药性。采用离体叶片法和盆栽法测定了啶菌噁唑的防效。在离体叶片试验中，啶菌噁唑在 100，200，400μg/mL 的浓度下可以达到 80% 以上的保护性防效和 50% 以上的治疗性防效。在盆栽试验中，啶菌噁唑在 100，200，400μg/mL 的浓度下可以达到 80% 以上的保护性防效和 60% 以上的治疗性防效。这些结果说明啶菌噁唑在防治番茄灰霉病方面是一个兼具保护性和治疗性作用的高效杀菌剂。

关键词：敏感性基线；灰葡萄孢菌；防效；交互抗药性；啶菌噁唑

Baseline Sensitivity and Control Efficacy of Pyrisoxazole Against *Botrytis cinerea*

Zhu He，Huang Chengtian，Ji Mingshan
(*College of Plant Protection*，*Shenyang Agricultural University*，*Shenyang*，100161，*China*)

Abstract：*Botrytis cinerea* is an important airborne plant pathogen with a wide host range. Gray mold caused by *B. cinerea* usually leads to both yield and quality reduction and causes severe losses. In China，the percentage of yield loss caused by *B. cinerea* gray mold in vegetable production is generally 10% to 20%，sometimes up to over 60% in some severely infected regions. The novel fungicide pyrisoxazole was registered for the control of tomato gray mold in China and has not been put into widespread use. In this study，baseline sensitivity of *B. cinerea* to pyrisoxazole was established based on EC_{50} values of 165 isolates sampled from tomato from Liaoning Province of China in 2012. Results showed that the frequency distribution of pyrisoxazole EC_{50} values was a unimodal curve with a right－hand tail. The mean EC_{50} value was 0.067 6 ± 0.040 9（SD）μg/mL and the range of individual EC_{50} values was from 0.012 8 to 0.198 7μg/mL. There was no cross－resistance between pyrisoxazole，carbendazim and procymidone. The control efficacy of pyrisoxazole was assessed on detached leaves and in pot experiments. Pyrisoxazole provided over 80% preventive control efficacy and over 50% curative control efficacy at 100，200，400μg/mL on detached leaves. It also provided over 80% preventive control efficacy and over 60% curative control efficacy at 100，200，400μg/mL in pot experiments. These results indicate that pyrisoxazole is a highly effective fungicide with both preven-

* 第一作者：朱赫，女，博士学位，主要研究方向为农药毒理学；E-mail：zhuhev5@126.com
** 通讯作者：纪明山，教授，主要研究方向为生物农药与农药毒理学；E-mail：jimingshan@163.com

tive and curative effect for the control of *B. cinerea*.

Key words: Baseline sensitivity; *Botrytis cinerea*; Control efficacy; Cross - resistance; Pyrisoxazole

灰葡萄孢菌（*Botrytis cinerea* Pers. : Fr. ; 有性型：*Botryotinia fuckeliana*（de Bary）Whetzel）是一种重要的空气传播的植物病原菌，有着广阔的寄主范围。在世界范围内，有超过200个植物种可被其感染，包括水果、蔬菜、观赏花卉以及其他大量的作物[1-2]。温室培育可以增加灰葡萄孢菌对作物的侵染风险，因为温室中较高的湿度尤其适合灰葡萄孢菌的生长[1,3]。由灰葡萄孢菌引起的灰霉病是一种高破坏性的病害，尤其在国内的温室蔬菜上广泛为害。它常导致产量减少、质量降低，引起蔬菜生产中的严重的损失，其中包括番茄、黄瓜、韭菜、茄子、辣椒、生菜、洋葱以及菜豆等。在国内，灰霉病可以导致蔬菜生产中10% ~20%的损失，在一些发病严重的地区甚至可以高达60%以上[4]。

一直以来，灰霉病都很难防治，主要因其具有多样的入侵模式、快速的侵染速度、广泛的寄主、以及较强的生存能力[1-2]。目前来说，化学防治仍然是灰霉病综合治理中最有效的方法。不幸的是，灰葡萄孢菌对杀菌剂有着高抗性发展风险[5-6]。在20世纪，灰葡萄孢菌已经对苯并咪唑类杀菌剂（包括苯菌灵、多菌灵），N－苯胺基甲酸酯类（乙霉威），二甲酰亚胺类（包括异菌脲、腐霉利）以及苯胺基嘧啶类（包括嘧菌环胺、嘧菌胺、嘧霉胺）产生抗性[7-8]。近年来，陆续有报道其田间菌株对各种杀菌剂的抗性发生，其中包括羟基苯胺类杀菌剂环酰菌胺[9-11]，三唑类杀菌剂戊菌唑、戊唑醇、三唑酮[11]，醌外部抑制剂类（QoI）杀菌剂醚菌酯[12]、嘧菌酯[12-13]、唑菌胺酯[14-16]，琥珀酸脱氢酶抑制剂类（SDHI）杀菌剂啶酰菌胺[14-16]、氟吡菌酰胺、氟唑菌酰胺、噻菌胺[17]，苯基吡咯类杀菌剂咯菌腈[18]。随着灰霉病菌对常用杀菌剂的抗性不断发展，选用具有不同作用模式的杀菌剂作为传统杀菌剂的替代品可以降低选择性压力并达到灰霉病菌的有效且可持续的治理。

新型杀菌剂啶菌噁唑（研究代码 SYP－Z048）是由沈阳化工研究院自主研发的一种兼具保护性和治疗性的内吸性杀菌剂[4,19]。啶菌噁唑按化学结构划分属于吡啶类杀菌剂，按作用方式划分属于脱甲基抑制剂类（DMI）杀菌剂，其作用靶标为甾醇生物合成中的14－α甾醇脱甲基酶[20]。它的化学结构为3－［5－（4－氯苯基）－2，3－二甲基－3－异噁唑烷基］吡啶[21]。啶菌噁唑对子囊菌、担子菌、半知菌引起的多种植物病害都有良好的防效[19,22]。已有报道啶菌噁唑可以很好地防治黄瓜、番茄、韭菜、草莓、辣椒、生菜、洋葱、菜豆以及许多其他蔬菜上的灰霉病[4]。前人研究表明啶菌噁唑对番茄和黄瓜灰霉病菌[4,22]以及番茄叶霉病菌有着很好的抑制作用[19]。在田间试验中，啶菌噁唑对番茄和黄瓜灰霉病菌的防效明显高于嘧霉胺、菌核净、腐霉利、多菌灵以及甲基托布津与乙霉威的混剂[4,22]；对番茄叶霉病菌的防效明显高于腈菌唑、代森锰锌、多菌灵以及甲基硫菌灵[19]。

啶菌噁唑于2008年在国内正式注册（90%原药注册号 PD20080773；25%乳油注册号 PD20080774），用于防治灰葡萄孢菌引起的番茄灰霉病（中国农药信息网，http://www.chinapesticide.gov.cn）。目前国内尚未大范围使用啶菌噁唑进行灰霉病的防治。在其广泛应用前，有必要建立灰葡萄孢菌对啶菌噁唑的敏感基线，以便于未来抗性监测。本次的研究主要目的：①建立灰葡萄孢菌对啶菌噁唑的敏感基线；②检验啶菌噁唑与两种广泛使用的杀菌剂多菌灵和腐霉利的交互抗药性；③测定离体与活体条件下啶菌噁唑对灰葡萄孢菌的保护性和治疗性防效。

1 材料与方法

1.1 供试材料

供试菌株：165 株单孢分离的番茄灰霉病菌菌株，PDA 斜面 4℃黑暗保存。这些菌株于 2012 年采自辽宁省各地温室番茄，其中沈阳、抚顺各 10 株；大连、本溪各 19 株；鞍山 12 株；朝阳、丹东、葫芦岛、盘锦各 20 株；阜新 15 株。采集温室未施用过啶菌噁唑或其他 DMI 类杀菌剂，各温室间相隔 20km 以上。

供试药剂：93% 啶菌噁唑原药由沈阳化工研究院提供，以丙酮溶解得到 1 000μg a. i. /mL 的母液。93% 腐霉利原药购自连云港优士化学品有限公司，以丙酮溶解得到 1 000μg a. i. /mL 的母液。98% 多菌灵原药购自长春长双化学品有限公司，以 0.1 mol/L 盐酸（HCl）溶解得到 1 000μg a. i. /mL 的母液。以上母液放置于 4℃冰箱黑暗保存，5 日内使用。

1.2 番茄灰霉病菌对啶菌噁唑的敏感基线的建立

EC_{50}值的测定采用菌丝生长抑制法。从预培养 3 d 的番茄灰霉病菌菌落边缘打取菌饼（直径 5mm），菌饼以菌丝面向下接种于含有 0，0.006 25，0.012 5，0.025，0.05，0.1 以及 0.2μg a. i. /mL 系列浓度啶菌噁唑的含药平板中央，每处理重复 3 次。不含药的平板作为对照。25℃恒温培养箱中黑暗培养 72 h 后，用十字交叉法两次量取菌落直径（减去菌饼直径 5mm），并以其平均值计算防效。

$$防效(\%) = \frac{对照组菌落直径平均值 - 处理组菌落直径平均值}{对照组菌落直径平均值} \times 100$$

根据啶菌噁唑的浓度及其对应的防效，利用 DPS 软件求出线性回归方程、EC_{50}值以及相关系数。基于 165 个菌株的 EC_{50}值分布频率建立番茄灰霉病菌对啶菌噁唑的敏感基线。

1.3 啶菌噁唑与多菌灵、腐霉利之间的交互抗药性

根据 Li 等[23]的方法，对 3 个多菌灵抗性菌株、3 个腐霉利抗性菌株以及 3 个敏感菌株进行测定，用于检验啶菌噁唑与多菌灵、腐霉利之间的交互抗药性。这 9 个菌株对 3 种杀菌剂的 EC_{50}值测定方法同前。对于敏感菌株，平板中多菌灵的系列浓度为 0，0.062 5，0.125，0.25，0.5，1 以及 2μg/mL；平板中腐霉利的系列浓度为 0，037 5，0.075，0.15，0.3，0.6 以及 1.2μg/mL。对于抗性菌株，平板中多菌灵的系列浓度为 0，12.5，25，50，100，200 以及 400μg/mL；平板中腐霉利的系列浓度为 0，0.25，0.5，1，2，4 以及 8μg/mL。根据 EC_{50}值的对数值，利用 SPSS 软件中的双变量相关性分析计算灰霉病菌对啶菌噁唑敏感性与其对多菌灵、腐霉利敏感性之间的相关性系数（r）。

1.4 啶菌噁唑对灰霉病菌离体的保护性和治疗性活性

啶菌噁唑对灰霉病菌离体的保护性和治疗性活性采用离体叶片法测定。在测定其保护性活性时，在 3～4 叶期的番茄植株上摘取大小相当（直径 6～7cm）且生长方向一致的叶片，无菌水冲洗 3 次并晾干，放入铺有湿润滤纸的 90mm 培养皿中。啶菌噁唑母液用 0.1%（v/v）吐温 80 水溶液连续稀释成系列浓度。离体叶片喷洒 0，25，50，100，200 以及 400μg a. i. /mL 的啶菌噁唑，每个处理 4 次重复。喷洒了等量不含啶菌噁唑的 0.1%（v/v）吐温 80 水溶液的叶片用作对照。腐霉利用作对照药剂。从预培养 3 d 的番茄灰霉病菌的菌落上用打孔器打出菌饼（直径 5mm），菌饼以菌丝面向下接种在表面晾干后的叶片正面中间。接种后的叶片转入 25℃ 生长室中培养。接种后的 24 h 和 48 h，十字交叉法两次测量病斑直径（减去菌饼直径 5mm）并取平均值。防效计算公式参见抑制百分数的计算公式，病斑直径取

代菌落直径。在测定其治疗性活性时，测定步骤与上述相同，只是杀菌剂喷洒时间点为接种后 24 h。接种后的叶片在喷药前后均在 25℃生长室中培养。接种后的 24 h 和 48 h，用前述方法记录病斑直径并计算防效。

1.5 啶菌噁唑对灰霉病菌活体的保护性和治疗性活性

啶菌噁唑对灰霉病菌活体的保护性和治疗性活性采用盆栽法测定。测定的步骤与离体叶片法相似，试验材料由离体叶片替换为 3 ~4 叶期的番茄植株。番茄植株在 25℃生长室（光周期：16 ~8h = 光/暗；空气相对湿度 85%）中培养。在活体盆栽试验中，每个处理 4 次重复。在测定其保护性活性时，接种后 48 h 和 72 h 记录病斑直径；在测定其保护性活性时，杀菌剂喷施后 48 h 和 72 h 记录病斑直径。

1.6 数据分析

线性回归方程与 EC_{50}值由 DPS 软件中的生物测定程序进行计算；辽宁省 10 个地区灰葡萄孢种群的平均 EC_{50}值之间是否有显著性差异由 SPSS 软件中的 Tukey 检验（$\alpha = 0.05$ 水平）确定；165 株菌株的 EC_{50}值是否符合正态分布由 SPSS 软件中的 Kolmogorov – Smirnov 检验确定；啶菌噁唑与多菌灵和腐霉利的交互抗药性由 SPSS 软件中的双变量相关性分析检验；啶菌噁唑与腐霉利的防效之间是否存在显著性差异由 SPSS 软件中的 t 检验确定，其中，防效数据分析前进行了反正弦平方根转换。

2 结果与分析

2.1 灰霉病菌对啶菌噁唑的敏感基线

165 株灰葡萄孢菌的平均 EC_{50}值为（0.067 6 ±0.040 9）（SD）μg/mL，其 EC_{50}值范围为 0.012 8 ~0.198 7μg/mL（表 1）。最高和最低 EC_{50}的比值为 15.5。

表 1　辽宁省 10 个地区番茄灰霉病菌种群对啶菌噁唑的敏感性

地区	菌株数量	EC_{50}（μg/mL）[a]	
		平均值 ± 标准差（SD）	范围
沈阳	10	0.076 8 ±0.034 3a[b]	0.012 8 ~0.198 7
大连	19	0.070 3 ±0.036 6a	0.019 4 ~0.150 7
鞍山	12	0.078 1 ±0.045 4a	0.016 2 ~0.176 1
抚顺	10	0.078 8 ±0.048 8a	0.014 8 ~0.171 7
本溪	19	0.068 1 ±0.041 9a	0.019 4 ~0.171 7
丹东	20	0.066 6 ±0.044 6a	0.013 1 ~0.177 2
阜新	15	0.050 1 ±0.017 9a	0.024 9 ~0.131 5
朝阳	20	0.067 1 ±0.036 5a	0.014 6 ~0.161 3
葫芦岛	20	0.057 0 ±0.029 4a	0.020 1 ~0.194 0
盘锦	20	0.063 6 ±0.047 2a	0.018 4 ~0.145 0
总计	165	0.067 6 ±0.040 9	0.012 8 ~0.198 7

[a] EC_{50}值的测定采用菌丝生长速率法，在 PDA 培养基上测得

[b] 本栏中的相同字母表明 10 个地区间的番茄灰霉病菌平均 EC_{50}值并没有显著性差异。该分析使用 SPSS 软件中的 Tukey 检验（$\alpha = 0.05$ 水平）确定

辽宁省 10 个地区灰葡萄孢种群的平均 EC_{50} 值之间没有显著性差异（$P = 0.911$）（表

1），表明敏感性并没有地区间差异。经 Kolmogorov - Smirnov 检验，165 株灰葡萄孢菌的的 EC_{50}值的频率分布符合正态分布（$P=0.105$）。因此，其平均值可以作为灰葡萄孢菌对啶菌噁唑的敏感基线。从 EC_{50}值的频率分布图可以看出其分布为带有右侧长拖尾的单峰曲线（图 1）。

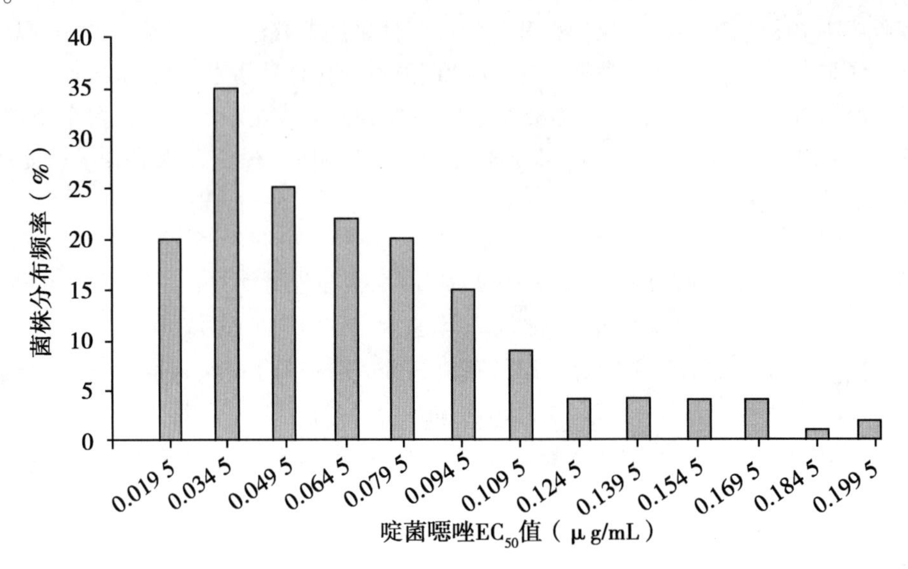

图 1　辽宁省不同地区 165 株番茄灰霉病菌对啶菌噁唑 EC_{50}值的频率分布

（EC_{50}值的分组区间为 0.015μg/mL。X 轴上的值表示各区间中值）

2.2　啶菌噁唑与多菌灵、腐霉利之间的交互抗药性

灰葡萄孢菌对啶菌噁唑的敏感性和多菌灵的敏感性之间的相关性（$P=0.774$，$r=-0.112$，图 2A）以及和腐霉利的敏感性之间的相关性（$P=0.208$，$r=-0.464$，图 3）并不显著，表明啶菌噁唑与多菌灵和腐霉利之间没有交互抗药性。

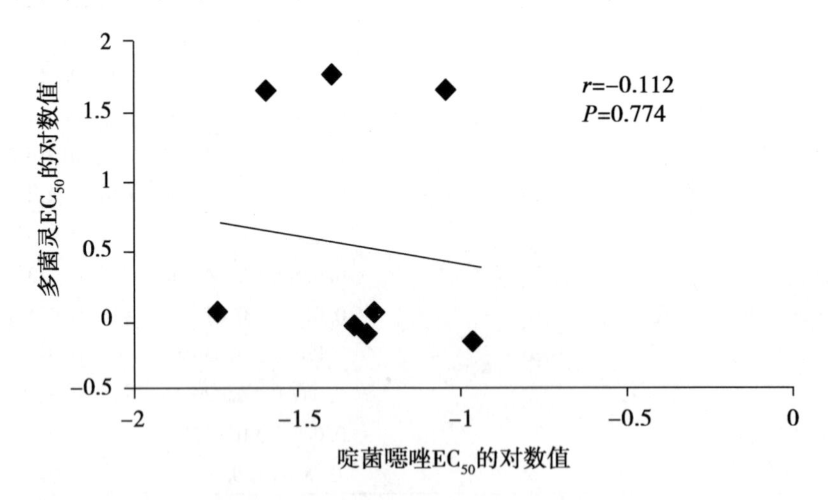

图 2　啶菌噁唑与多菌灵敏感性的皮尔森相关系数（r）

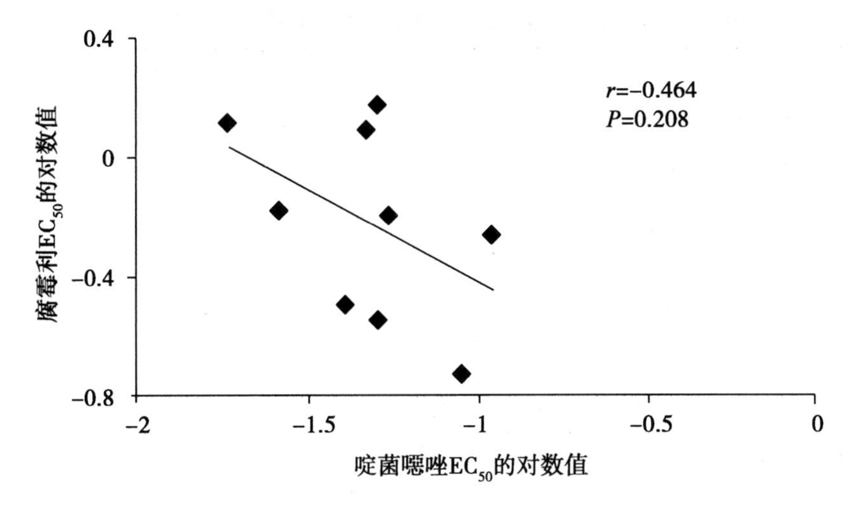

图3　啶菌噁唑与腐霉利敏感性的皮尔森相关系数（r）

2.3　啶菌噁唑对灰霉病菌的保护性和治疗性活性

离体叶片法测定的结果表明，啶菌噁唑的保护性和治疗性防效在测试的各浓度和时间点均高于对照药剂腐霉利。接种后 24 h 和 48 h 后，400μg/mL 啶菌噁唑对灰葡萄孢菌的保护性防效均为 100%；200μg/mL 啶菌噁唑对灰葡萄孢菌的保护性防效均高于 80%。接种后 24 h 和 48 h，400μg/mL 腐霉利对灰葡萄孢菌的保护性防效分别为 54.84% 和 65.91%（表2）。接种后 24 h，100，200，400μg/mL 三个浓度的啶菌噁唑的治疗性防效均高于 50%；接种后 48 h，3 个浓度的啶菌噁唑的治疗性防效均高于 70%。相比之下，接种 24 h 和 48 h 后，腐霉利最高的治疗性防效分别为 47.73% 和 63.04%，且均在最高浓度 400μg/mL。盆栽法测定也得到相似的结果，48 h 和 72 h 后，100，200，400μg/mL 三个浓度的啶菌噁唑的保护性防效高于 80%，保护性防效高于 60%（表3）。25% 啶菌噁唑乳油（商品名菌思奇）的田间推荐剂量为 200～400 g a.i./hm^2。

表2　离体叶片法测定啶菌噁唑对灰葡萄孢菌的保护性与治疗性活性

浓度（μg/mL）	保护性防效（%）				治疗性防效（%）[a]			
	接种后 24 h		接种后 48 h		施药后 24 h		施药后 48 h	
	啶菌噁唑	腐霉利	啶菌噁唑	腐霉利	啶菌噁唑	腐霉利	啶菌噁唑	腐霉利
25	32.26	19.35*	59.09	21.59*	13.64	9.09*	44.57	8.70*
50	51.61	22.58*	72.73	27.27*	31.83	18.18*	60.33	18.48*
100	80.65	29.03*	90.91	40.91*	50.00	27.27*	72.83	34.78*
200	93.55	45.16*	97.73	54.55*	59.09	34.09*	78.26	54.35*
400	100.00	54.84*	100.00	65.91*	63.64	47.73*	82.61	63.04*

[a]测定治疗性防效时，喷药时间为接种后 24 h

*说明经 t 检验（α = 0.05 水平），同等条件下的腐霉利防效与啶菌噁唑防效有显著性差异

表 3 盆栽法测定啶菌噁唑对灰葡萄孢菌的保护性与治疗性活性

浓度（μg/mL）	保护性防效（%）				治疗性防效（%）[a]			
	接种后 48 h		接种后 72 h		施药后 48 h		施药后 72 h	
	啶菌噁唑	腐霉利	啶菌噁唑	腐霉利	啶菌噁唑	腐霉利	啶菌噁唑	腐霉利
25	56.04	21.98*	58.33	22.92*	40.63	8.33*	54.11	15.01*
50	69.23	27.47*	70.83	32.29*	53.13	19.79*	66.01	25.21*
100	84.62	32.97*	85.42	47.92*	68.75	35.42*	79.04	43.34*
200	94.51	51.65*	95.83	65.63*	77.08	52.08*	85.84	59.21*
400	100.00	58.24*	100.00	72.92*	83.33	60.42*	90.93	72.80*

[a] 测定治疗性防效时，喷药时间为接种后 24 h

* 说明经 t 检验（$\alpha=0.05$ 水平），同等条件下的腐霉利防效与啶菌噁唑防效有显著性差异

3 结论与讨论

考虑到番茄灰霉病菌的严重为害及其对市面已有杀菌剂的抗性的逐步增加，引入与现有杀菌剂有着不同作用模式的杀菌剂势在必行。吡啶类杀菌剂啶菌噁唑则是这样一种新型杀菌剂，已在国内注册用于防治番茄灰霉病。

本文首次研究了啶菌噁唑对辽宁省温室番茄灰霉病菌的敏感基线。啶菌噁唑对 165 株番茄灰霉病菌菌丝的平均 EC_{50} 值为 0.067 6 ± 0.040 9（SD）μg/mL 可作为番茄灰霉病菌对啶菌噁唑的敏感基线。这一敏感基线呈偏斜分布并带有右侧长拖尾，暗示其潜在的抗性风险并警示我们应采取严格的预防抗性措施。这一抗性风险或许不可避免，毕竟番茄灰霉病菌属于高抗性风险的病原菌[5-6]；同时，也有可能因为啶菌噁唑作为 DMI 杀菌剂归属于甾醇生物合成抑制剂（SBI）一类之中，而 SBI 类的杀菌剂已被杀菌剂抗性行动委员会（FRAC）根据杀菌剂的抗性发展程度划分为中度抗性[24]。

已有报道称啶菌噁唑是内吸性杀菌剂，兼具保护性与治疗性。作为内吸性杀菌剂，啶菌噁唑在喷施后可以穿透番茄小叶并在组织内移动，进而在植株体内向顶、向基或横向移动[21]。本次研究评估了啶菌噁唑对番茄灰霉病菌离体与活体的保护性和治疗性防效。离体与活体生物测定的结果均表明啶菌噁唑有着显著高于腐霉利的保护性与治疗性防效。啶菌噁唑优异的杀菌活性应与其化学结构相关。它结构中同时具有一个吡啶环和一个异噁唑烷环，而二者都具有杀菌作用[21]。也有报道称啶菌噁唑对使用者和环境安全[22]。综合其保护性与治疗性防效、高杀菌毒力、安全并对环境友好的特性，啶菌噁唑有着极大潜力，可作为防治番茄灰霉病菌的替代药剂。

在菌株采集时得知，多菌灵与腐霉利广泛应用于辽宁省温室灰霉病防治。交互抗性检测的结果表明啶菌噁唑与多菌灵和腐霉利之间并没有交互抗药性。因此，啶菌噁唑可以与多菌灵和腐霉利轮用或者混用，以更好治理灰霉病的。杀菌剂混用，尤其是不同作用模式的杀菌剂混用，可以降低选择性压力从而减缓抗性发展，是更持续的防治真菌病害的方法。已有报道称啶菌噁唑与保护性杀菌剂百菌清或者福美双混用对防治灰葡萄孢菌非常有效[25]。事实上，啶菌噁唑与 40% 福美双悬乳剂的混剂已在国内注册（注册号 PD20093355）用于防治灰霉病（中国农药信息网，http：//www.chinapesticide.gov.cn）。除此之外，啶菌噁唑与其他

杀菌剂（包括百菌清、啶酰菌胺、唑菌胺酯、唑菌酯、氟吗啉、氟吡菌酰胺、噻菌胺、乙菌定、粉唑醇、甲基代森锌以及丙硫菌唑等）的多种不同剂型的混剂已申请获得国内专利，用于防治灰葡萄孢菌及其他植物病原菌（国家知识产权局专利检索与分析网页，http：//www. pss – system. gov. cn）。

鉴于当前灰葡萄孢菌对现有广泛使用的杀菌剂有着严重的抗性问题，并对近年新进市场的杀菌剂逐步产生抗性，啶菌噁唑作为一种有着高杀菌活性、新作用模式的新型杀菌剂，对未来灰葡萄孢菌的抗性防控有着极大潜力。虽然如此，仍需谨慎使用，最好可以与其他杀菌剂轮用或混用以延缓抗性的发生。

参考文献

[1] Elad Y, Williamson B, Tudzynski P, et al. *Botrytis* spp. and diseases they cause in agricultural systems-an introduction. In: Y. Elad et al. (Ed.), *Botrytis*: Biology, Pathology and Control (pp. 1 – 8) [M]. Dordrecht, the Netherlands: Springer, 2007.

[2] Williamson B, Tudzynski B, Tudzynski P, et al. *Botrytis cinerea*: the cause of gray mould disease [J]. Molecular Plant Pathology, 2007, 8 (5): 561 – 580.

[3] Rosslenbroich H J, Stuebler D. *Botrytis cinerea*-history of chemical control and novel fungicides for its management [J]. Crop Protection, 2000, 19: 557 – 561.

[4] Si N G, Zhang Z J, Liu J L, et al. Biological activity and application of a novel fungicide, SYP – Z048 (II) [J]. Chinese Journal of Pesticides, 2004, 43 (2): 61 – 63.

[5] FRAC (Fungicide Resistance Action Committee). Pathogen risk list [Z]. 2013. http://www.frac.info/docs/default – source/publications/pathogen – risk/pathogen – risk – list. pdf? sfvrsn = 8.

[6] Russell P E. Sensitivity baselines in fungicide resistance research and management [Z]. FRAC Monograph No. 3. Brussels, Belgium. 2004. http://frac. sw. aa – g. de/docs/default – source/publications/monographs/monograph – 3. pdf? sfvrsn = 8.

[7] Leroux, P. Chemical control of *Botrytis* and its resistance to chemical fungicides. In: Y. Elad et al. (Ed.), Botrytis: Biology, Pathology and Control (pp. 195 – 222) [M]. Dordrecht, the Netherlands: Springer. 2007.

[8] Topolovec – Pintarić S. Resistance to botryticides. In: Nooruddin Thajuddin (Ed.), Fungicides – Beneficial and Harmful Aspects. ISBN: 978 – 953 – 307 – 451 – 1, (pp. 19 – 44) [M]. InTech. 2011. http://www. intechopen. com/books/fungicides – beneficial – and – harmful – aspects/resistance – to – botryticides.

[9] Esterio M, Ramos C, Walker A S, et al. Phenotypic and genetic characterization of Chilean isolates of *Botrytis cinerea* with different levels of sensitivity to fenhexamid [J]. Phytopathologia Mediterranea, 2011, 50: 414 – 420.

[10] Myresiotis C K, Karaoglanidis G S, Tzavella – Klonari K. Resistance of *Botrytis cinerea* isolates from vegetable crops to anilinopyrimidine, phenylpyrrole, hydroxyanilide, benzimidazole, and dicarboximide fungicides [J]. Plant Disease, 2007, 91 (4): 407 – 413.

[11] Zhang C Q, Zhu J W, Wei F L, et al. Sensitivity of *Botrytis cinerea* from greenhouse vegetables to DMIs and fenhexamid [J]. Phytoparasitica, 2007, 35 (3): 300 – 313.

[12] Ishii H, Fountaine J, Chung W H, et al. Characterisation of QoI – resistant field isolates of *Botrytis cinerea* from citrus and strawberry [J]. Pest Management Scicence, 2009, 65: 916 – 92.

[13] Huangfu Y H, Dai D J, Shi H J, et al. Study on resistance of *Botrytis cinerea* to azoxystrobin collected from fruits and vegetables in Zhejiang Province [J]. Chinese Journal of Pesticide Science, 2013, 15

(5): 504 -510.

[14] Bardas G A, Veloukas T, Koutita O, et al. Multiple resistance of *Botrytis cinerea* from kiwifruit to SDHIs, QoIs and fungicides of other chemical groups [J]. Pest Management Scicence, 2010, 66: 967 -973.

[15] Fernández - Ortuño D, Chen F P, Schnabel G. Resistance to Pyraclostrobin and Boscalid in *Botrytis cinerea* Isolates from Strawberry Fields in the Carolinas [J]. Plant Disease, 2012, 96: 1 198 -1 203.

[16] Kim Y K, Xiao C L. Resistance topyraclostrobin and boscalid in populations of *Botrytis cinerea* from stored apples in Washington State [J]. Plant Disease, 2010, 94: 604 -612.

[17] Amiri A., Heath S M, Peres N A. Resistance to fluopyram, fluxapyroxad, and penthiopyrad in *Botrytis cinerea* from strawberry [J]. Plant Disease, 2014, 98: 532 -539.

[18] Li X P, Fernández - Ortuño D, Grabke A, et al. Resistance to fludioxonil in *Botrytis cinerea* isolates from blackberry and strawberry [J]. Phytopathology, 2014, 104: 724 -732.

[19] Liu J L, Si N G, Chen L, et al. Biological activity against tomato leaf mold and application of a novel fungicide, SYP - Z048 (III) [J]. Chinese Journal of Pesticides, 2004, 43 (3): 103 -105.

[20] FRAC (Fungicide Resistance Action Committee). FRAC Code List 2015: Fungicides sorted by mode of action (including FRAC Code numbering) [Z]. 2015. http: //www. frac. info/docs/default - source/publications/frac - code - list/frac - code - list - 2015 - finalC2AD7AA36764. pdf? sfvrsn = 4. Accessed 24 June 2015.

[21] Chen F P, Han P, Liu P F, et al. Activity of the novel fungicide SYP - Z048 against plant pathogens [J]. Scientific Reports, 2014, 4: 6473, doi: 10. 1038/srep06473.

[22] Si N G, Zhang Z J, Liu J L, et al. Biological Activity and Application of a Novel Fungicide: SYP - Z048 (I) [J]. Chinese Journal of Pesticides, 2004, 43 (1): 16 -18.

[23] Li J L, Liu X Y, Di Y L, et al.. Baseline sensitivity and control efficacy of DMI fungicide epoxiconazole against *Sclerotinia sclerotiorum* [J]. European Journal of Plant Pathology, 2015, 141: 237 -246.

[24] Brent K J, Hollomon D W. Fungicide resistance: the assessment of risk. FRAC Monograph No. 2, second (revised) edition [Z]. FRAC, Brussels, Belgium. 2007. http: //www. frac. info/docs/default-source/publications/monographs/monograph -2. pdf? sfvrsn = 8.

[25] Shao J X, Zhou X F, Zhang J X. Field experimental results in controlling tomato gray mold with SYP - Z048 25% EC [J]. Agrochemicals, 2006, 45 (7): 488 -490.

辽宁省番茄灰霉病菌抗药性研究

杜　颖*，王　智，刘　妍，纪明山**
（沈阳农业大学植物保护学院，沈阳　100161）

摘要：灰霉病是温室番茄生产上的重要病害之一，对温室番茄的生产造成严重减产。目前，在生产上主要应用化学杀菌剂来防治番茄灰霉病，但随着杀菌剂长久而广泛的应用，抗药性问题已经越来越严重。本实验利用菌丝生长速率法测定了辽宁省番茄灰霉病菌对多菌灵、啶酰菌胺、咯菌腈、嘧霉胺和腐霉利的敏感性。对多菌灵的 EC_{50} 值为 3.349 0 ~ 1 062.763 8μg/mL，平均 EC_{50} 值为 169.885 8μg/mL，最大值和最小值相差 317.337 7倍。对啶酰菌胺的 EC_{50} 值为 0.080 0 ~ 7.787 2μg/mL，平均值为 2.194 9μg/mL；对咯菌腈的 EC_{50} 值分布在 0.000 7 ~ 0.024 4μg/mL，平均值为 0.007 3μg/mL，最大值和最小值相差 34.86 倍；对嘧霉胺的 EC_{50} 值在 2.679 3 ~ 79.383 2μg/mL，平均值为 29.383 2μg/mL；对腐霉利的 EC_{50} 值在 0.187 7 ~ 7.983 3μg/mL，平均 EC_{50} 值为 2.686 7μg/mL。采用 SPSS 软件对所测得的数据进行 K－S 法非参数性检验，所测 117 株番茄灰霉病菌对啶酰菌胺和咯菌腈的敏感性频率分布符合正态分布。辽宁番茄灰霉病菌除对咯菌腈全部表现敏感以外，对其他的杀菌剂均有抗性菌株的出现，并得出番茄灰霉病菌对啶酰菌胺、咯菌腈、腐霉利、多菌灵、嘧霉胺抗药性频率分别为 0、0、35.04%、100%、96.58%。

关键词：番茄灰霉病菌；啶酰菌胺；咯菌腈；抗药性

The Study on Resistance of *Botrytis Cinerea* in Tomato from Different Areas of Liaoning Province

Du Ying，Wang Zhi，Liu Yan，Ji Mingshan*
（*College of Plant Protection*，*Shenyang Agricultural University*，*Shenyang*，100161，*China*）

Abstract：Grad，causing serious reduction of production，is an important disease on tomato. At present，chemical synthesized fungicides are main to control of the gray mold，and the resistance problem had been more and more serious with the widely use of fungicide. The sensitivity of *B. cinerea* from Liaoning province to fludioxonil，procymidone，carbendazim and pyrimethanil were tested by measuring mycelial. The EC_{50} value ranged from 3.349 0 to 1 062.763 8μg/mL for carbendazim，and the average EC_{50} value was 169.885 8μg/mL，the ratio of maximum value to the minimum value was 317.337 7 times. The EC_{50} values ranged from 0.080 0 to 7.787 2μg/mL for boscalid，and the average EC_{50} value was 2.194 9μg/mL. The EC_{50} values ranged from 0.000 7 to 0.024 4μg/mL for fludioxonil，and the average EC_{50} value was 0.007 3μg/mL，the ratio of maximum value to the minimum value was 34.68 times. The EC_{50} value ranged from 0.090 0 to 11.499 6μg/mL for procymidone，and the average EC_{50} value was 1.527 5μg/mL，the ratio of maximum value to the minimum value was 127.773 3 times. The EC_{50} values ranged from 0.154 2 to 73.193 2μg/mL for pyrimethanil，and the average EC_{50} value was 3.639 5μg/mL，the ratio of maximum value to the minimum value was 474.664 1 times. The data was analyzed by using to K－S method of

---

* 第一作者：杜颖，女，硕士，主要研究方向为农药毒理学；E-mail：duying92m@163.com
** 通讯作者：纪明山，教授，主要研究方向为生物农药与农药毒理学；E-mail：jimingshan@163.com

nonparametric test from SPSS software. The frequency distribution of the sensitivities of 117 strains to boscalid represented a curve of continuous single peak which approached a unimodal curve. In addition to all of sensitive strains of *B. cinerea* to fludioxonil, it has the appearance of resistant strains of *B. cinerea* to other fungicides, and *B. cinerea* on fludioxonil, procymidone, carbendazim, pyrimethanil resistance frequency was 0, 35.04%, 100%, 96.58%.

Key words: *Botrytis cinerea*; Boscalid; Fludioxonil, Resistance

由灰葡萄孢菌（*Botrytis cinerea*）引起的番茄灰霉病是一种重要的世界性病害，菌在植物发病部位产生大量分生孢子梗和分生孢子，形成肉眼可见的灰色霉层，因而这种病害叫做灰霉病[1]。灰葡萄孢属能侵害的植物多达数百种，其中包括20余种重要蔬菜。包括番茄、黄瓜、葡萄等。中国是番茄制品生产国和出口国的主要国家之一，番茄的种植面积在145.5万hm^2左右，其中15%左右的种植面积是保护地番茄，每年番茄的产量约为837.6 t，且每年都在逐渐增长[2-3]。现今灰霉病已经使番茄栽培生产受到了严重的限制[4]。随着保护地番茄种植比例的不断增大，番茄灰霉病的危害也随之日益加重，逐步成为番茄保护地上的主要病害之一，可使番茄的产量下降20%～30%，严重时番茄产量可下降60%左右，甚至无产量[5-6]。目前生产上的主要方法是化学防治，常用的杀菌剂有苯并咪唑类的多菌灵、二甲酰亚胺类的腐霉利、N－苯氨基甲酸酯类的乙霉威、苯胺基嘧啶类的嘧霉胺、甲氧基丙烯酸酯类的嘧菌酯等。在番茄灰霉病菌防治中，由于药剂选择压作用等因素的影响，番茄灰霉病菌对这些常用的化学杀菌剂的抗药性问题表现得十分突出，但对于新型杀菌剂啶酰菌胺和咯菌腈的抗药性还没有普遍发生，仅个别地区个别菌株出现抗药性[7]。啶酰菌胺是一种新型烟酰胺类杀菌剂，由德国巴斯夫公司研制，在美国、欧洲等50多个国家注册登记使用[8]。其杀菌活性高，杀菌谱广泛，主要用于葡萄、蔬菜和果树上灰霉病、菌核病、白粉病和各种腐烂病的防治[9]。Kretschmer等对德国灰葡萄孢菌对咯菌腈的抗药性进行研究测定，没有发现抗药性菌株的产生[10]。Bardas等对希腊灰霉病菌对咯菌腈的敏感性进行了研究测定，未发现对咯菌腈具有抗药性的菌株产生[11]。赵建江等对河北省和山东省番茄灰霉病菌对咯菌腈的抗药性情况进行了研究测定，未发现对咯菌腈具有抗药性的菌株产生[12]。因此对新型杀菌剂咯菌腈和啶酰菌胺的抗药性研究具有十分重要的意义，为新型杀菌剂咯菌腈和啶酰菌胺在今后的生产推广应用实践提供重要的依据。

1 材料与方法

1.1 供试菌株

辽宁省的沈阳市、鞍山市、大连市、盘锦市、抚顺市、朝阳市、阜新市、本溪市、丹东市和葫芦岛市的不同地区采集长有番茄灰霉病菌的病果和病叶，每个大棚采集3～5个病果或病叶于自封袋中装好带回实验室。在无菌条件下经过分离纯化共得到165株番茄灰霉病菌菌株，移接于PDA斜面上，放置于4℃冰箱内保存备用。

1.2 供试药剂

50%多菌灵（carbendazim）可湿性粉剂，安徽华星化工股份有限公司提供；87.4%啶酰菌胺（boscalid）原药，中国农业科学院蔬菜花卉研究所提供；98%咯菌腈（fludioxonil）原药，中国农业科学院蔬菜花卉研究所提供；98.2%嘧霉胺（pyrimethanil）原药，昆山瑞泽农药股份有限公司提供；93%腐霉利（procymidone）原药，连云港优士化学品有限公司提供。

1.3 试验方法

采用菌丝生长速率法[13]测定番茄灰霉病菌对多多菌灵、啶酰菌胺、咯菌腈、腐霉利和嘧霉胺的敏感性。在无菌条件下，将被测菌株移植于PDA平板上，于25℃恒温培养箱中培养3d，使用打孔器沿培养好的菌株菌落边缘打取直径为5mm的菌饼，将正面朝下，移植于含系列浓度的多菌灵（0、2.5、10、40、160和640μg/mL）、啶酰菌胺（0、0.062 5、0.25、1、4和16μg/mL）、咯菌腈（0、0.000 8、0.004、0.02、0.1和0.5μg/mL）、腐霉利（0、0.062 5、0.25、1、4和16μg/mL）和嘧霉胺（0、0.5、2、8、32和128μg/mL）的PDA含药平板（其中啶酰菌胺、腐霉利和咯菌腈使用PDA培养基，嘧霉胺使用FGA培养基）中央。于25℃恒温培养箱中培养3d，每个处理3次重复，用十字交叉法测量各菌落直径，按照下列公式计算各浓度药剂对菌丝生长的抑制率。

$$\text{菌丝生长抑制率}(\%) = \frac{\text{对照组菌落直径平均值} - \text{处理组菌落直径平均值}}{\text{对照组菌落直径平均值} - 0.5} \times 100$$

利用SPSS数据处理系统统计不同浓度杀菌剂对番茄灰霉病菌菌丝生长的抑制率，以抑制率几率值为纵坐标（y）、药剂浓度对数值为横坐标（x），求出毒力回归方程 $y = a + bx$，有效中浓度（EC_{50}）值和相关系数（r）[14]。

2 结果与分析

2.1 番茄灰霉病菌对不同药剂敏感性的测定

采用菌丝生长速率法测定辽宁省不同地区的117株番茄灰霉病菌菌株对多菌灵、啶酰菌胺、咯菌腈、嘧霉胺、腐霉利的敏感性，测定结果如表1和图1所示。结果显示番茄灰霉病菌对多菌灵的EC_{50}值为3.349 0～1 062.763 8μg/mL，平均EC_{50}值为169.885 8μg/mL。对啶酰菌胺的EC_{50}值为0.080 0～7.787 2μg/mL，平均值为2.194 9μg/mL。对咯菌腈的EC_{50}值为0.000 7～0.024 4 μg/mL，平均值为0.007 3 μg/mL。对杀菌剂嘧霉胺的EC_{50}值为0.154 2～73.193 2μg/mL，平均EC_{50}值为3.639 5μg/mL。对腐霉利的EC_{50}值为0.090 0～11.499 6μg/mL，平均值为1.527 5μg/mL。

表1 辽宁省不同地区番茄灰霉病菌对5种EC_{50}范围的比较

采集地	菌株数	EC_{50}范围（μg/mL）				
		多菌灵	啶酰菌胺	咯菌腈	嘧霉胺	腐霉利
沈阳	10	3.349 0～268.300 3	0.668 1～2.301 0	0.002 7～0.019 2	2.202 1～4.719 9	0.842 0～1.963 5
抚顺	15	4.165 9～432.768 5	0.849 8～5.956 9	0.001 9～0.016 7	0.315 1～4.294 3	0.090 0～4.312 8
普兰店	14	5.643 1～417.650 6	2.584 2～7.787 2	0.001 8～0.016 3	1.420 8～5.831 4	0.091 5～2.223 7
辽中	18	5.246 6～756.480 1	0.080 0～3.880 2	0.002 6～0.015 5	0.669 1～5.037 2	0.900 3～6.848 6
海城	20	4.113 0～869.786 3	0.393 4～1.807 9	0.001 6～0.024 4	0.154 2～1.865 8	0.092 8～5.041 2

（续表）

采集地	菌株数	EC_{50}范围（μg/mL）				
		多菌灵	啶酰菌胺	咯菌腈	嘧霉胺	腐霉利
北票	40	3.710 7 ~ 1 062.763 8	0.940 7 ~ 3.502 6	0.000 7 ~ 0.015 3	0.191 8 ~ 17.859 1	0.135 2 ~ 2.581 1

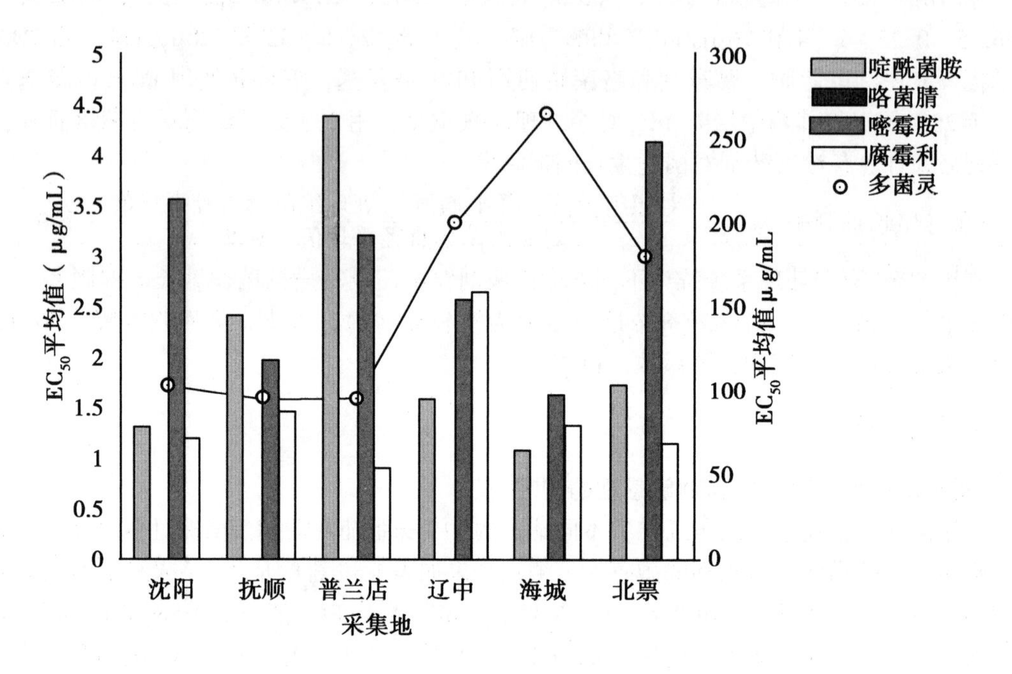

图1　辽宁省不同地区番茄灰霉病菌对5种杀菌剂EC_{50}平均值的比较

2.2　番茄灰霉病菌对不同杀菌剂抗性的发生频率和抗性水平

当菌株的EC_{50}值小于敏感基线的5倍时，为敏感菌株（S）；当EC_{50}值处于敏感基线的5倍和10倍之间时，为低抗菌株（LR）；当EC_{50}值处于敏感基线的10倍和40倍之间时为中抗菌株（MR）；当EC_{50}值大于敏感基线的40倍时，为高抗菌株（HR）[15]（表2）。

表2　番茄灰霉病菌对不同杀菌剂的抗药性参考标准

杀菌剂	敏感基线	参考文献
啶酰菌胺	（1.973 1 ±1.001 1）mg/L	辽宁省番茄灰霉病菌对啶酰菌胺敏感性基线的建立
咯菌腈	（0.024 1 ±0.009 4）mg/L	辽宁省番茄灰霉病菌对咯菌腈敏感性基线的建立
腐霉利	（0.31 ±0.08）mg/L	番茄灰霉病菌对腐霉利的抗药性检测及生物学性状研究
嘧霉胺	0.091 1mg/L	番茄灰霉病菌对嘧霉胺抗药性的试验

采用菌丝生长速率法对辽宁省不同地区采集的117株番茄灰霉病菌进行了抗药性检测。结果如表3所示，番茄灰霉病菌对不同药剂的抗性表现型不同，对啶酰菌胺和咯菌腈均为敏感菌株，无抗药性菌株，抗药性频率为0；对腐霉利敏感的菌株为76个，低抗的菌株为31个，中抗的菌株为10个，抗药性频率为35.04%；对嘧霉胺敏感的菌株为4个，低抗的菌

株为6个，中抗的菌株为74个，高抗的菌株为33个，抗药性频率为96.58%；对多菌灵敏感的菌株为0个，低抗的菌株为49个，中抗的菌株为9个，高抗的菌株为59个，抗药性频率为100%。

表3　番茄灰霉病菌对不同杀菌剂的抗药性菌株个数

杀菌剂	敏感（个）	抵抗（个）	中抗（个）	高抗（个）	抗性频率（%）
啶酰菌胺	117	0	0	0	0
咯菌腈	117	0	0	0	0
腐霉利	76	31	10	0	35.04
嘧霉胺	4	6	74	33	96.58
多菌灵	0	49	9	59	100

3　结果与讨论

采用菌丝生长速率法测定了辽宁省不同地区117株番茄灰霉病菌菌株对咯菌腈、腐霉利、嘧霉胺和多菌灵的敏感性。结果显示（图1，表1）：番茄灰霉病菌对多菌灵的EC_{50}值为3.349 0～1 062.763 8 μg/mL，平均值为169.885 8 μg/mL，最大值与最小值相差317.337 7倍；对啶酰菌胺的EC_{50}值为0.080 0～7.787 2μg/mL，平均值为2.194 9μg/mL，最大值与最小值相差97.34倍；对咯菌腈的EC_{50}值为0.000 7～0.024 4μg/mL，平均值为0.007 3μg/mL，最大值和最小值相差34.86倍；对杀菌剂嘧霉胺的EC_{50}值为0.154 2～73.193 2μg/mL，平均值为3.639 5μg/mL，最大值与最小值相差474.664 1倍；对腐霉利的EC_{50}值为0.090 0～11.499 6 μg/mL，平均值为1.527 5 μg/mL，最大值与最小值相差127.773 3倍。试验结果显示，番茄灰霉病菌除了对啶酰菌胺和咯菌腈全部表现为敏感，而对其他3种杀菌剂都有抗性菌株的出现。番茄灰霉病菌对啶酰菌胺、咯菌腈、腐霉利、多菌灵、嘧霉胺抗药性频率分别为0、35.04%、100%、96.58%。

试验得出辽宁省不同地区的番茄灰霉病菌菌株对啶酰菌胺和咯菌腈的敏感性差异不显著，每个地区番茄灰霉病菌菌株对咯菌腈的敏感性都很强，没有抗药性菌株的出现。这可能与这些地区对新型药剂啶酰菌胺和咯菌腈的施药频率和使用年数等因素有关。该结果可为今后咯菌腈在辽宁省的使用提供重要依据，说明新型杀菌剂啶酰菌胺和咯菌腈对辽宁省番茄灰霉病菌的防治具有良好的效果，可以被广泛使用。

参考文献

[1] 李保聚，朱国仁，赵奎华，等.番茄灰霉病在果实上的侵染部位及防治新技术［J］. 植物病理学报，1999，1：64－68.

[2] 常法平，张雪江，石振飞.日光温室番茄几种主要病害的综合防治［J］. 农业工程技术温室园艺，2006，12：42.

[3] 李保聚，朱国仁.番茄灰霉病发展症状诊断及综合防治［J］. 植物保护，1998，24（6）：18－20.

[4] 康立娟，张小风，王文桥，等.灰霉菌的抗药性与适合度测定［J］. 农药学学报，2006，2（3）：39－42.

[5] 纪明山，祁之秋，王英姿，等.番茄灰霉病菌对嘧霉胺的抗药性［J］.植物保护学报，2003，30（4）：396－400.
[6] 纪明山，程根武，张益先，等.灰霉病菌对多菌灵和乙霉威抗性研究［J］.沈阳农业大学学报，2006，3：17－20.
[7] 祁之秋，王建新，陈长军，等.现代杀菌剂抗性研究进展［J］.农药，2006，10：655－659.
[8] 熊晓姝.杀菌剂新品种——啶酰菌胺制剂的研究［J］.农药研究与应用，2006，10（4）：21－23.
[9] 颜范勇，刘冬青，司马利锋，等.新型烟酰胺类杀菌剂——啶酰菌胺［J］.农药，2008，47（2）：132－135.
[10] 杨玉柱，焦必宁.新型杀菌剂咯菌腈研究进展［J］.现代农药，2007，6（5）：35－39.
[11] Kretschmer M，Hahn M. Fungicide resistance and genetic diversity of *Botrytis cinerea* from a vineyard in Germany［J］. Journal of Plan Diseases and Protection，2008，148（7－8）：214－219.
[12] Bardas G A，Veloukas T，Koutita O，et al. Multiple resistance of *Botrytis cinerea* from kiwifruit to SDHIs，QoIs and fungicides of other chemical groups［J］. Pest Management Science，2010，66（9）：967－973.
[13] 赵建江，张小风，马志强，等.番茄灰霉病菌对咯菌腈的敏感性基线及其与不同杀菌剂的交互抗性［J］.农药，2013，52（9）：684－686.
[14] 慕立义.植物化学保护研究方法：1版［M］.北京：中国农业出版社，2013：142.
[15] 李兴红，乔广行，黄金宝，等.北京地区番茄灰霉病菌对嘧霉胺的抗药性检测［J］.植物保护，2012，38（4）：141－143.

Oxathiapiprolincan Effectively Control Downy Mildew of Cucumber, Grape and Chinese Cabbage

Li Beixing[1,2 *], Zhang Wenjuan[3 *], Zhang Daxia[1,2], Mu Wei[2,3], Liu Feng[1,3 **]

(1. *Shandong Provincial Key Laboratory for Biology of Vegetable Diseases and Insect Pests, College of Plant Protection, Shandong Agricultural University, Tai'an, Shandong* 271018, *P. R. China*; 2. *Research Center of Pesticide Environmental Toxicology, Shandong Agricultural University, Tai'an, Shandong* 271018, *China*; 3. *Key Laboratory of Pesticide Toxicology& Application Technique, Shandong Agricultural University, Tai'an, Shandong* 271018, *P. R. China*)

Abstract: The efficacy of oxathiapiprolin in controlling downy mildew of cucumber, grape and Chinese cabbage were investigated under field conditions. The results showed that oxathiapiprolin can effectively control downy mildew of cucumber and Chinese cabbage at doses of 10, 20 and 30 g/hm^2 under field conditions. This fungicide also exhibited favorable control efficacy on downy mildew of grape at concentrations of 33.3, 40 and 50mg/L. Overall, oxathiapiprolin possessed excellent preventive and curative activity against cucumber downy mildew. Moreover, oxathiapiprolin treatments were capable of significantly promoting cucumber yield over a long harvest period compared to the dimethomorph 250 g/hm^2 treatment and the control group.

Key words: Oxathiapiprolin; Downy mildew; Efficacy; Cucumber; Grape; Chinese cabbage

1 Introduction

Downy mildews, caused by the well - known pathogenic oomycetes[1-2], are often the most devastating diseases of vegetables and fruits. These pathogens can cause frequent re - infection when the incubation period is short[3] and subsequently cause prevalent epidemics under suitable conditions. Downy mildew is a major problem in the production of cucumber, where it causes great losses in both yield and quality[4-5]. In China, in recent years, many high - standard greenhouses have emerged and have rapidly increased in number as a result of the 'Vegetable Basket Project'[6]. The suitable temperatures and high humidity of greenhouses provide favorable environmental conditions for the development of downy mildew, which can then result in infections and epidemics[7]. Downy mildew also dramatically threatens the production of Chinese cabbage and grape, which are planted worldwide[8-9].

In terms of the integrated management of downy mildew, cultural practices, resistant cultivars and plant quarantine often provide little assistance[10]. Although genetically modified crops and bio-

* Li Beixing and Zhang Wenjuan share joint first authorship.

** Corresponding author: Liu Feng, Professor; E-mail: fliu@ sdau. edu. cn logical control are current subjects of much research, economical options seem unlikely to be available within the next few years. Chemical control is therefore still the dominant strategy for managing downy mildew. Farmers rely on synthetic chemicals to ensure high yields and a decent quality of their agricultural products[11-12]. Despite the application of fungicides, epidemics of downy mildew remain prevalent. We postulate that three factors primarily contribute to the failure of chemical management. First, only a limited number of fungicides are currently available for controlling downy mildew. Outdated fungicides such as Bordeaux mixture, mancozeb and ziram are still commonly used for managing this disease in China. Second, most of the available fungicides have favorable protective activities, however, few commercial fungicides provide ideal curative activities. Despite the complexity of local climatic conditions, the spraying of fungicides in a timely manner, especially prior to infection or at the early stages of pathogen infection, has a high probability of suppressing epidemics. However, this procedure may fail so long as disease spots are observed. Third, the occurrence and rapid development of chemical resistance of downy mildew is another factor that jeopardizes disease management[11,13]. Given the severe effects of downy mildew infections, there is an urgent need to develop and make rational use of novel, high-efficient and safe fungicides for combating downy mildew.

Oxathiapiprolin was the first of the piperidinyl thiazole isoxazoline fungicides discovered and developed by the DuPont company in 2007[14]. Several publications demonstrate that oxathiapiprolin has high activity against various oomycetes, including *Phytophthora capsici*, *Phytophthora infestans*, *Peronospora belbahrii*, *Peronophythora litchi* and *Pythium ultimum*[10,15-16]. However, the efficacy of oxathiapiprolin in managing plant-pathogenic oomycetes under field conditions has been rarely reported. As a proof, we conducted the current study to evaluate the effectiveness of oxathiapiprolin in controlling downy mildew in cucumber, grape and Chinese cabbage in field experiments.

2 Materials and methods

2.1 Tested fungicides

Oxathiapiprolin 10% dispersible oil suspension (OD) was provided by DuPont Company; dimethomorph 50% wettable powder (WP) was purchased from BASF SE; azoxystrobin 250 g/L suspension concentrate (SC) was purchased from Syngenta Company; propineb 70% WP was purchased from Bayer Crop Science.

2.2 Field experiments for controlling downy mildew of cucumber

Field experiments were conducted in greenhouses where cucumber had been produced for more than 5 years. The experimental sites had a history of heavy natural infestation by cucumber downy mildew. In 2012, the trial site was located in Beiteng village, Tai'an, P. R. China (117.14 E, 35.96 N, at an altitude of 104 m above mean sea level). Cucumber seedlings with the brand name of Dongguan No. 3 were transplanted to the greenhouse on July 23. Fungicides were first sprayed at the appearance of initial symptoms of cucumber downy mildew using an MATABI-16 sprayer (spray pressure: 0.3 MPa, cone nozzle, flow rate: 650mL min^{-1}) with a spray volume of 750 L/hm^2. Oxathiapiprolin 10% OD was sprayed at doses of 10, 20 and 30 g active ingredient (a. i.) per hectare. Dimethomorph 50% WP was used as the reference fungicide, whereas water was sprayed as blank control. Treatments were placed using a randomized block design with four replications. Individual plots consisted of 5 rows, with plot size of 15 m^2, and there were approximately 150 cucumber plants per plot. Fungicides were applied on August 25, September 5 and September 15. Disease severity was investigated before the fungicide application, and 7 and 14 d after the last application. Verification experiments were conducted in 2013 and 2014, and both trials were located in Zhaozhuang village, Tai'an, P. R. China (117. 16 E, 36. 13 N, at an altitude of 119 m above mean sea level). In 2013, cucumber seedlings (cultivar Shandong Mici) were transplanted to the greenhouse on February 5. Fungicides (the same as the above mentioned) were sprayed every 14 d from April 18, for a total of three times. Disease severity was evaluated before fungicide application, 14 d after the first application, and 7 and 14 d after the last application. In 2014, cucumber seedlings (cultivar Shandong Mici) were transplanted to the greenhouse on January 23. Fungicides (the same as the above mentioned) were sprayed every 14 d from April 15, for a total of three times. Disease severity was investigated before fungicide application, and 7 and 14 d after the last application. Cucumber yield was evaluated on May 18 and every following two or three days (according to conventional farming operations) for approximately one month. In each plot, ten randomly selected plants were labeled and re – evaluated each time for the quantification of cucumber yield. The disease severity of cucumber downy mildew was ranked using a 0 – 9 index according to GB/T 17980. 26—2000 (Pesticide guidelines for field efficacy trials – Fungicides against downy mildew of cucumber) in China. In detail, five points were randomly removed from each plot for disease severity evaluation. At each point, all leaves of two plants were investigated. The disease index was graded as follows: 0, no colony; 1, diseased area was less than 5%; 3, 5% ≤ diseased area < 10%; 5, 10% ≤ diseased area < 25%; 7, 25% ≤ diseased area < 50%; and 9, more than 50% of the leaf surface was covered with mildew. The equations for the calculation of disease severity and control efficacy are as follows:

$$\text{Disease severity (DS)} = \frac{\sum \text{(number of leaves allocated to an individual index} \times \text{disease index)}}{\text{total number of leaves investigated} \times 9} \times 100$$

$$\text{Control efficacy (\%)} = \left(1 - \frac{\text{DS of blank control before treatment} \times \text{DS after fungicide treatment}}{\text{DS of blank control after treatment} \times \text{DS before fungicide treatment}}\right) \times 100$$

2. 3 Field experiments for controlling downy mildew of grape

Field experiments were carried out in 2012 and 2013. Both trial sites were located in Wangzhuang, Tai'an, P. R. China (117. 14 E, 35. 96 N, at an altitude of 104 m above mean sea level). The grapes (cultivar Jufeng) were planted in an open field with a plant density of 6 600 per hectare. Fungicides were first sprayed before the appearance of the initial symptoms of downy mildew on grape using an MATABI – 16 sprayer with a spray volume of 0. 5 L for every plant. Oxathiapiprolin 10% OD was sprayed at concentrations of 33. 3, 40 and 50mg/L (a. i.). Azoxystrobin 250 g/L SC was used as the reference fungicide, whereas water was sprayed as blank control. Treatments were placed using a randomized block design with four replications. There were 10 plants in each plot, with a plot size of approximately 15 m^2. In 2012, fungicides were applied on August 2 and August 16. Disease severity was evaluated 14 d after the first fungicide application (August 16), and 14 d after the last application (August 30). In 2013, fungicides (the same as above) were sprayed on July 27 and August 10. Disease severity was evaluated 14 d after the first fungicide application (August 10), and 7 d and 14 d after the last application. The disease severity of grapevine downy mildew was ranked using a 0 ~ 9 index according to GB/T 17980. 122—2004 (Pesticide guidelines for field efficacy trials – Fungicides against downy mildew of grape) in China. In detail, 10 newly emerged vines were randomly removed from each plot for disease severity evaluation. Ten leaves of each vine were investigated from the top down. The disease index was graded as follows: 0, no colony; 1, diseased area was less than 5%; 3, 5% ≤ diseased area < 25%; 5, 25% ≤ diseased area < 50%; 7, 50% ≤ diseased area < 75%; and 9, more than 75% of the leaf surface was covered with mildew. The equation for the calculation of disease severity is as follows:

$$\text{Disease severity (DS)} = \frac{\sum (\text{number of leaves allocated to an individual index} \times \text{disease index})}{\text{total number of leaves investigated} \times 9} \times 100$$

Because the first evaluation was conducted before the appearance of the initial symptoms of downy mildew of grape, the disease severity was 0 prior to fungicide application. Thus, the equation for the calculation of control efficacy is as follows:

$$\text{Control efficacy(\%)} = \left(\frac{\text{DS of blank control} - \text{DS of fungicide treatment group}}{\text{DS of blank control}}\right) \times 100$$

2.4 Field experiments for controlling downy mildew of Chinese cabbage

Field experiments were carried out in 2013 and 2014. In 2013, the trial site was located in Zhaizi village, Tai'an, P. R. China (117. 15 E, 36. 15 N, at an altitude of 126 m above mean sea level). Chinese cabbage with the brand name of Weibai No. 7 was planted in an open field on August 17 (spacing: 60 cm × 40 cm). Fungicides were first sprayed at the appearance of initial symptoms of downy mildew using an MATABI – 16 sprayer with a spray volume of 675 L/hm^2. Oxathiapiprolin 10% OD was sprayed at doses of 10, 20 and 30 g a. i. per hectare. Propineb 70% WP (2 250 g/hm^2) was used as the reference fungicide, whereas water was sprayed as a blank control. Treatments were placed using a randomized block design with four replications. The individual plot size was 20 m^2. Fungicides were applied on September 20 and September 30. Disease severity was evaluated before fungicide application, 10 d after the first application and 14 d after the last application. In 2014, the trial site was located at the experimental field of Shandong Agricultural University, Tai'an, P. R. China (117. 16 E, 36. 17 N, at an altitude of 131 m above mean sea level). Chinese cabbage (cultivar Weibai No. 7) was planted in an open field on August 20. Fungicides were sprayed on September 25 and October 5. Disease severity was evaluated before fungicide application, 10 d after the first application and 10 d after the last application. Five samples were taken from each plot for disease severity evaluation, and each point had 4 individual plants. The disease severity of Chinese cabbage downy mildew was ranked using a 0 ~ 9 index according to GB/T 17980. 115—2000 (Pesticide guidelines for field efficacy trials – Fungicides against downy mildew of Chinese cabbage) in China. In detail, five points were removed from each plot for disease severity evaluation, with each point having 4 individual plants. All external leaves (except the head) of each plant were investigated. The disease index was graded as follows: 0, no disease spots; 1, diseased area was less than 5%; 3, 5% ≤ diseased area < 10%; 5, 10% ≤

diseased area < 20% ; 7, 20% ≤ diseased area < 50% ; and 9, more than 50% of the leaf surface was covered with mildew. The equations for the calculation of disease severity and control efficacy are as follows:

$$\text{Disease severity (DS)} = \frac{\sum \text{(number of leaves allocated to an individual index} \times \text{disease index)}}{\text{total number of leaves investigated} \times 9} \times 100$$

$$\text{Control efficacy (\%)} = \left(1 - \frac{\text{DS of blank control before treatment} \times \text{DS after fungicide treatment}}{\text{DS of blank control after treatment} \times \text{DS before fungicide treatment}}\right) \times 100$$

2.5 Data analysis

The data analysis was performed with the DPS software (version 7.05). Control efficacy is presented as the average value ± standard deviation (n = 4). Differences among the treatments were tested using Duncan's multiple range test (a = 0.05).

3 Results

3.1 The efficacy of oxathiapiprolin in controlling downy mildew of cucumber

Table 1 Control efficacy of oxathiapiprolin on cucumber downy mildew in 2012

Treatment (g/hm^2)	DS before application	7 d after 3rd application		14 d after 3rd application	
		DS	Control efficacy	DS	Control efficacy
Oxathiapiprolin - 10	3.50	0.46	90.56 ± 3.65 b	0.59	89.02 ± 2.44 b
Oxathiapiprolin - 20	3.44	0.24	95.16 ± 3.14 ab	0.33	93.72 ± 1.84 a
Oxathiapiprolin - 30	3.56	0.15	97.19 ± 3.10 a	0.18	96.68 ± 0.83 a
Dimethomorph - 250	3.44	1.08	78.90 ± 4.09 c	1.18	77.58 ± 2.38 c
Control	3.70	5.27		5.62	

Note: DS indicates disease severity. Control efficacy is presented as the average value ± standard deviation (n = 4). The values with different lower case letters are significantly different at the $P < 0.05$ level, according to Duncan's multiple range test (the same as below)

In 2012, a significant reduction in the disease severity of cucumber downy mildew was observed after the fungicide applications in the field experiments (Table 1). Oxathiapiprolin showed favorable control efficacy of cucumber downy mildew 7d after the third application, even at the lowest dose. When oxathiapiprolin was sprayed at doses of 10, 20 and 30 g/hm^2, control efficacies were 90.56% , 95.16% and 97.19% , respectively, whereas that of the reference fungicide (dimethomorph 250g/hm^2) was 78.90%. Oxathiapiprolin also exhibited excellent lasting activity in its control of cucumber downy mildew. All three treatments of oxathiapiprolin had control efficacies higher than 89% , even 14 d after the third application. All three oxathiapiprolin treatments were significantly higher than that of the dimethomorph 250g/hm^2 treatment.

Table 2 Control efficacy of oxathiapiprolin on cucumber downy mildew in 2013

Treatment (g/hm^2)	DS before application	14 d after 1st application		7 d after 3rd application		14d after 3rd application	
		DS	Control efficacy	DS	Control efficacy	DS	Control efficacy
Oxathiapiprolin - 10	2. 64	0. 54	89. 32 ± 4. 68 a	0. 78	86. 09 ± 3. 80 b	1. 03	82. 37 ± 8. 79 a
Oxathiapiprolin - 20	2. 59	0. 38	92. 07 ± 5. 50 a	0. 59	89. 22 ± 3. 25 ab	0. 77	86. 53 ± 6. 46 a
Oxathiapiprolin - 30	2. 67	0. 24	95. 13 ± 2. 25 a	0. 34	93. 35 ± 5. 11 a	0. 69	89. 04 ± 1. 25 a
Dimethomorph - 250	2. 62	1. 15	77. 57 ± 3. 15 b	1. 52	73. 67 ± 5. 47 c	2. 08	67. 23 ± 3. 60 b
Control	2. 66	5. 23	—	5. 82	—	6. 27	—

The field experiments carried out in 2013 revealed a similar trend as that of 2012. A slight difference was that when disease severity was evaluated 14 d after the first application, the control efficacy of oxathiapiprolin was also excellent (Table 2). This indicates that oxathiapiprolin is capable of controlling cucumber downy mildew even when only sprayed once. The experiments carried out in 2012 and 2013 effectively demonstrated the quick - acting and lasting activities of oxathiapiprolin in controlling downy mildew of cucumber. However, whether oxathiapiprolin has effective therapeutic activity is unknown, as the fungicides were first sprayed at the appearance of the initial symptoms of cucumber downy mildew in 2012 and 2013. Fungicides were first sprayed when disease severity was much higher (at a rank of approximately 5. 5, as shown in Table 3) in the field experiments conducted in 2014. Disease severity in the reference fungicide treatment increased dramatically 7 d after the third application, and even more so at 14 d, resulting in an unfavorable control efficacy. Although the control efficacy of oxathiapiprolin was not as high in 2014 as in 2012 and 2013, oxathiapiprolin also exhibited high therapeutic activity against cucumber downy mildew. When oxathiapiprolin was sprayed at doses of 10, 20 and 30g/hm^2, the control efficacies were 79. 56%, 82. 74% and 84. 89%, respectively, even 14 d after the third application. Non - significant differences were observed among the different experimental doses in terms of control efficacy (14 d after the third application); however, they were all significantly higher than that of the dimethomorph 250g/hm^2 treatment, of which the control efficacy was only 42. 96%.

Table 3 Control efficacy of oxathiapiprolin on cucumber downy mildew in 2014

Treatment	Application rate (g/hm^2)	DS before application	7 d after 3rd application		14 d after 3rd application	
			DS	Control efficacy	DS	Control efficacy
Oxathiapiprolin	10	5. 18	3. 67	80. 46 ± 3. 48 b	6. 14	79. 56 ± 2. 45 a
	20	5. 39	3. 36	82. 69 ± 3. 41 ab	5. 31	82. 74 ± 3. 65 a
	30	5. 41	2. 35	87. 93 ± 2. 49 a	4. 56	84. 89 ± 4. 82 a
Dimethomorph	250	5. 56	9. 33	53. 92 ± 4. 81 c	18. 09	42. 96 ± 13. 80 b
Control		5. 35	19. 62		31. 24	

Tosystematically evaluate the application performance of oxathiapiprolin on cucumber, cucumber yield was also measured. As shown in Figure 1, no significant differences in cucumber yield were observed among the oxathiapiprolin treatments and the dimethomorph 250g/hm^2 treatment over the first three consecutive harvests. In the next period, however, cucumber yield for all of the oxathiapiprolin treatments dramatically increased compared with those of the reference fungicide treatment and the untreated control. In the following three periods, a significant increase in cucumber yield was observed in the oxathiapiprolin treatments, whereas those of the reference fungicide treatment and the untreated control continued decreasing. For the last harvest period, the cucumber yields of the oxathiapiprolin treatments were approximately 4 times higher than that of the dimethomorph 250g/hm^2 treatment and 8 times higher compared to the untreated control. These experimental data indicated the favorable ability of oxathiapiprolin in increasing cucumber yield over a long harvest period.

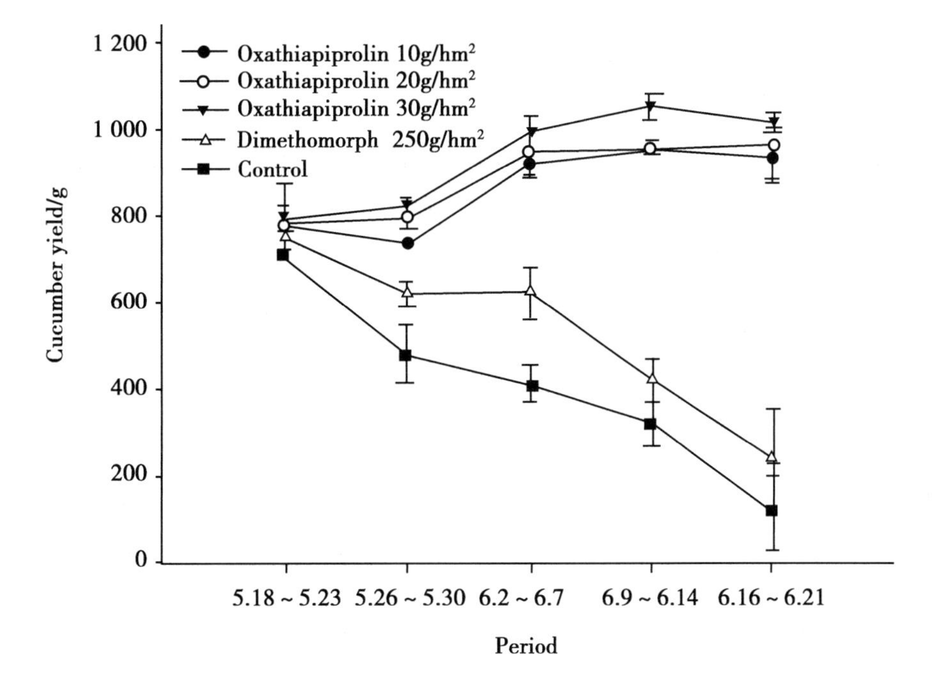

Figure 1 Effect of oxathiapiprolin treatments on cucumber yield. Cucumber yield is displayed as the average value ± standard deviation of three consecutive harvests for each treatment

3. 2 The efficacy of oxathiapiprolin in controlling downy mildew of grape

Oxathiapiprolin was also capable of controlling downy mildew of grape. Disease severity was first investigated 14 d after the first application of the fungicide. As shown in Table 4, control efficacies were 85. 84%, 87. 79% and 90. 77%, when oxathiapiprolin was sprayed at concentrations of 33. 3, 40 and 50mg/L, respectively. All of these were significant higher than the control efficacy of the azoxystrobin 166. 7mg/L treatment. When oxathiapiprolin was sprayed twice, its control efficacy was much higher. The control efficacy of the treatment with the highest concentration reached 97. 83%, which was significantly higher than that of the reference fungicide. Field experiments carried out in 2013 revealed a similar trend as that of 2012. All of the control efficacies of the oxathiapiprolin treatments were significantly higher than azoxystrobin 166. 7mg/L treatment, regardless of when disease severities were evaluated.

Table 4 Control efficacy of oxathiapiprolin on downy mildew of grape

Treatment (mg/L)	Control efficacy in 2012		Control efficacy in 2013		
	14 d after the first application	14 d after the last application	14 d after the first application	7 d after the last application	14d afterthe last application
Oxathiapiprolin – 33. 3	85. 84 ± 5. 23 b	90. 88 ± 5. 08 bc	87. 52 ± 5. 96 a	92. 11 ± 3. 47 b	89. 01 ± 1. 11 a
Oxathiapiprolin – 40	87. 79 ± 3. 17 ab	95. 51 ± 2. 72 ab	89. 80 ± 1. 71 a	97. 35 ± 1. 00 a	94. 00 ± 0. 81 a
Oxathiapiprolin – 50	90. 77 ± 2. 00 a	97. 83 ± 0. 73 a	92. 47 ± 3. 23 a	99. 68 ± 0. 65 a	95. 29 ± 2. 08 a
Azoxystrobin – 166. 7	81. 03 ± 3. 03 c	85. 87 ± 3. 55 c	79. 26 ± 11. 83 b	84. 42 ± 3. 16 c	76. 43 ± 8. 88 b

3. 3 The efficacy of oxathiapiprolin in controlling downy mildew of Chinese cabbage

As shown in Figure 2, downy mildew of Chinese cabbage developed slowly in both 2013 and 2014, with the disease severity increasing from 3. 60 to 4. 15 and 3. 31 to 3. 96, respectively. All fungicide treatments showed favorable activity in controlling downy mildew of Chinese cabbage, including oxathiapiprolin at doses of 10, 20 and 30g/hm^2 and propineb 2 250g/hm^2. In 2013, disease severities for all fungicide treatments were approximately 1. 10 on average ten days after the first application. After the second fungicide application was administered, disease severities continued to decrease significantly. The disease severity in the oxathiapiprolin 10g/hm^2 treatment was not significantly different from that of the propineb 2 250g/hm^2 treatment, whereas the disease severities in the oxathiapiprolin 20 and 30g/hm^2 treatments were both significantly lower than that of the propineb 2 250g/hm^2 treatment. In 2014, the trend was similar. There were no significant differences among the four fungicide treatments ten days after the first application in terms of disease severity. Disease severities for all fungicide treatments continued to decrease dramatically after the second application. Similarly, the disease severities in the oxathiapiprolin 20 and 30g/hm^2 treatments were both significantly lower than those of oxathiapiprolin 10g/hm^2 and the reference fungicide propineb 2 250g/hm^2.

4 Discussion and conclusions

The current study demonstrated thatoxathiapiprolin can effectively control downy mildew of cucumber and Chinese cabbage at doses of 10, 20 and 30g/hm^2 under field conditions. It also exhibited favorable control efficacy for downy mildew of grape at concentrations of 33. 3, 40 and 50mg/L. In addition, oxathiapiprolin treatments were capable of significantly promoting cucumber yield over a long harvest period compared to the dimethomorph 250g/hm^2 treatment and the control group.

Miao et al. reported that oxathiapiprolin can produce EC_{90} values of 1. 21 $\times 10^{-2}$ and 2. 64 $\times 10^{-3}$mg/L against *Pseudoperonospora cubensis* in terms of average diameters of lesions and sporangial

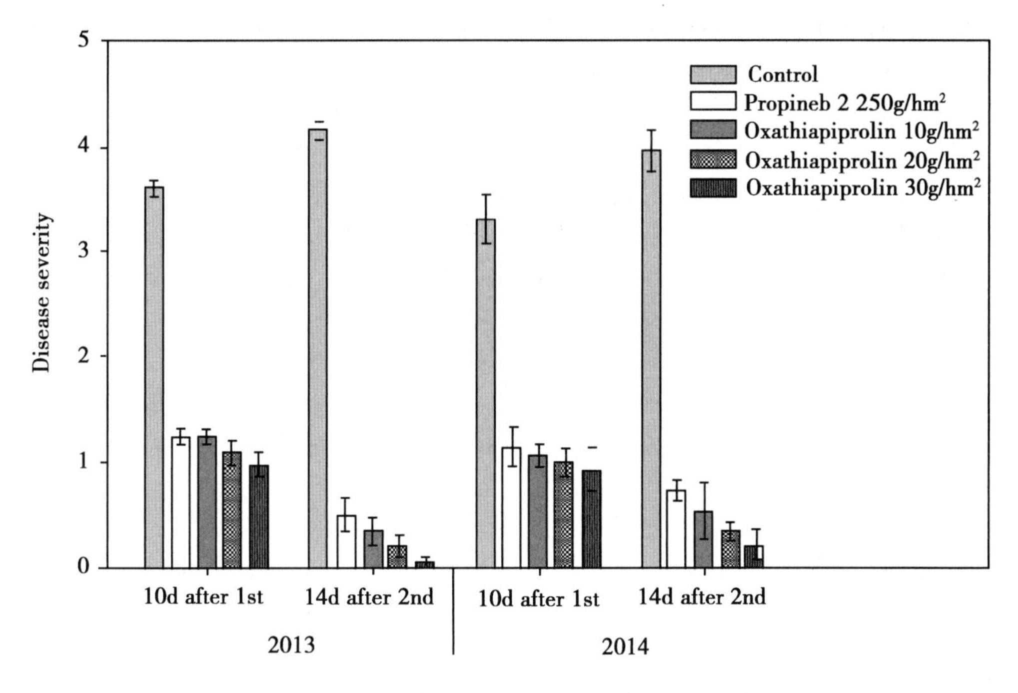

Figure 2 Disease severities for the oxathiapiprolin and propineb treatments on downy mildew of Chinese cabbage

Disease severity is displayed as the average value ± standard deviation ($n = 4$).

production *in vitro*. Our experiments carried out under field conditions for three years were consistent with their findings in laboratory tests. We also demonstrated that oxathiapiprolin can provide favorable protective and curative activity compared with the dimethomorph 250g/hm^2 treatment, although the control efficacies of the oxathiapiprolin treatments decreased when it was sprayed at high disease severity. The ability of oxathiapiprolin to greatly increase cucumber yield over a long harvest period was an interesting phenomenon. When cucumber yield was investigated between May 18 and May 23, approximately 7 d after the last fungicide application, disease severities in all treatments were relatively sustainable. Thus, cucumber yield varied from 710.9 to 794.6 g (Figure 1). However, when the evaluation was conducted between May 26 to May 30, cucumber yields for the control and dimethomorph 250g/hm^2 treatments dropped to 481.8 and 621.2 g, respectively, whereas those of the oxathiapiprolin treatments slightly increased. This time period was approximately 14 d after the last fungicide application. Indeed, disease severities of the control and dimethomorph 250g/hm^2 treatments (31.24 and 18.09, respectively) were much greater than those of the oxathiapiprolin treatments (6.14, 5.31 and 4.56). The number of functional leaves was another possible key factor influencing cucumber yield, as more functional leaves lead to a better capacity for photosynthesis. The average number of leaves for ten investigated cucumber plants in each treatment were 130.8, 148.5, 148.2, 126.8 and 126.0 for the oxathiapiprolin 10, 20 and 30g/hm^2, dimethomorph 250g/hm^2 and control treatments, respectively.

Publicationsreporting on the efficacy of oxathiapiprolin have increased gradually in recent years. In addition to favorable activity against *P. cubensis*[17], oxathiapiprolin has also been shown to be highly efficient in inhibiting other oomycetes. Ji and Csinos[16] found that the asexual life stages of *Phytophthora capsici* were more sensitive to oxathiapiprolin than other compounds used in the control of oomycete pathogens. Bittner and Mila[18] demonstrated the highly efficacious activity of oxathiapiprolin against *Phytophthora nicotianae*, whereas Ji *et al.* [15] reported that it was effective in managing black shank on tobacco (caused by *P. nicotianae*) at a dose of 140g/hm^2 in transplant water, followed by directed sprays at the time of first cultivation and lay - by at 70g/hm^2. It also has the potential to manage downy mildew of basil caused by *Peronospora belbahrii*[19].

Oxathiapiprolin was first registered in China on 19 December 2015 for controlling grape downy mildew (caused by *Plasmopara viticola*, 33.3 ~ 50mg/L), cucumber downy mildew (caused by *Pseudoperonospora cubensis*, 15 ~ 30g/hm^2), potato late blight and tomato late blight (caused by *Phytophthora infestans*, 22.5 ~ 30g/hm^2 and 15 ~ 30g/hm^2, respectively) and pepper Phytophthora blight (caused by *P. capsici*, 22.5 ~ 37.5g/hm^2) with the certificate number of LS20150355[20]. Our experiments lead to a recommendation of almost the same doses as the registration, except we found that a slightly lower dose of oxathiapiprolin was also capable of controlling cucumber downy mildew. Miao's experimental data proved favorable preventive and curative activity of oxathiapiprolin against pepper Phytophthora blight, with control efficacies of 85.66% and 69.50%, respectively, when applied at 30g/hm^2[10]. Our results demonstrated excellent preventive and curative activity of oxathiapiprolin against cucumber downy mildew. In addition, we are the first to report the favorable control efficacy of oxathiapiprolin against downy mildew of Chinese cabbage under field conditions. Spraying oxathiapiprolin at doses of 10, 20 and 30g/hm^2 twice can effectively control downy mildew of Chinese cabbage (caused by *Peronospora parasitica*). All of these results demonstrate the effectiveness of oxathiapiprolin in managing diseases caused by plant - pathogenic oomycetes.

Althoughoxathiapiprolin acts on a novel target (oxysterol binding protein, OSBP)[21], the site-specific mode of action also makes it prone to cause resistance. According to the FRAC (Fungicide Resistance Action Committee) Code List 2016, its resistance risk is assumed to be medium to high, resistance management is therefore required. In terms of the application of oxathiapiprolin, alternation or combination with other fungicides (with different modes of action) is recommended, including CAA - fungicides (carboxylic acid amides) and QoI - fungicides (quinone outside inhibitors). In addition, appropriate physical measures should be implemented during the crop season. For instance, to regulate the temperature, ventilation should occur in time to lower the humidity in the greenhouse, and the amount of nitrogen fertilizer should also be managed. Breeding resistant varieties is another effective strategy to control downy mildew because different species have been reported to vary in their susceptibility to downy mildew[22-24].

Acknowledgement

This work was supported by a grant fromNational Natural Science Foundation of China (31572040) and the Special Fund to reduce use of chemical fertilizers from the Ministry of Science and Technology of China.

Declaration of interest statement

The authors declare no competing financial interest.

REFERENCES

[1] Thines M, Kamoun S. Oomycete-plant coevolution: recent advances and future prospects [J]. Current Opinion in Plant Biology, 2010, 13: 427 – 433.

[2] Runge F, Thines, M. Host matrix has major impact on the morphology of Pseudoperonospora cubensis [J]. European Journal of Plant Pathology, 2010, 129: 147 – 156.

[3] Buloviene V, Surviliene E. Effect of environmental conditions and inocolum concentration on sporulation of Peronospora destructor [J]. Agronomy Research, 2006: 147 – 150.

[4] Shetty N V, Wehner T C, Thomas C E, et al. Evidence for downy mildew races in cucumber tested in Asia, Europe, and North America [J]. Scientia Horticulturae, 2002, 94: 231 – 239.

[5] Yang X, Li M, Zhao C. Early warning model for cucumber downy mildew in unheated greenhouses [J]. New Zealand Journal of Agricultural Research, 2007, 50: 1 261 – 1 268.

[6] Yang Y, Liang Y, Junyi F U, et al. Consideration on the Construction of Jiangsu Vegetable Basket Project Base [J]. Journal of Changjiang Vegetables, 2014: 72 – 75.

[7] Hong C, Xiao Y, Zeng S. Modelling of the cucumber downy mildew under plastic shelter conditions [J]. Journal of Plant Protection, 1989, 16: 217 – 220.

[8] Dick M W. Binomials in the Peronosporales, Sclerosporales and Pythiales [A]. In Advances in Downy Mildew Research, Spencer – Phillips P T N, Gisi U, Lebeda A. Eds. Springer Netherlands: Dordrecht, 2002: 225 – 265.

[9] Gessler C, Pertot I, Perazzolli M. Plasmopara viticola: A review of knowledge on downy mildew of grapevine and effective disease management [J]. Phytopathologia Mediterranea, 2011, 50: 3 – 44.

[10] Miao J, Dong X, Lin D, et al. Activity of the novel fungicide oxathiapiprolin against plant – pathogenic oomycetes [J]. Pest Management Science, 2015, DOI: 10.1002/ps. 4 189.

[11] Gisi U, Sierotzki H. Fungicide modes of action and resistance in downy mildews [J]. European Journal of Plant Pathology, 2008, 122: 157 – 167.

[12] Lebeda A, Cohen Y. Cucurbit downy mildew (Pseudoperonospora cubensis) —biology, ecology, epidemiology, host – pathogen interaction and control [J]. European Journal of Plant Pathology, 2010, 129: 157 – 192.

[13] Judelson H S, Roberts S. Multiple Loci Determining Insensitivity to Phenylamide Fungicides in Phytophthora infestans [J]. Phytopathology, 1999, 89: 754 – 760.

[14] Pasteris Robert J, Hanagan Mary A N N, Shapiro R. Fungicidal azocyclic amides. 2007/06/22/Application date, 2007.

[15] Ji P, Csinos A S, Hickman L L, et al. Efficacy and application methods of oxathiapiprolin for management of black shank on tobacco [J]. Plant Disease, 2014, 98: 1 551 – 1 554.

[16] Ji P, Csinos A S. Effect of oxathiapiprolin on asexual life stages of Phytophthora capsici and disease development on vegetables [J]. Annals of Applied Biology, 2015, 166: 229 – 235.

[17] Cohen Y. The novel oomycide oxathiapiprolin inhibits all stages in the asexual life cycle of Pseudoperonospora cubensis – causal agent of cucurbit downy mildew [J]. PLoS ONE, 2015, 10: 14 – 15.

[18] Bittner R J, Mila A L. Effects of oxathiapiprolin on Phytophthora nicotianae, the causal agent of black shank of tobacco [J]. Crop Protection, 2016, 81: 57 – 64.

[19] Patel J S, Costa de Novaes M I, Zhang S. Evaluation of the new compound oxathiapiprolin for control of downy mildew in Basil [J]. Plant Health Progress, 2015, 16: 165 – 172.

[20] China I o, Newly approved registration (1) new temporary registration [CP]. Pesticide Registration Public Announcement, 2016, 9.

[21] Pasteris R J, Hanagan M A, Bisaha J J, et al. Discovery of oxathiapiprolin, a new oomycete fungicide that targets an oxysterol binding protein [J]. Bioorganic & Medicinal Chemistry, 2016, 24: 354 - 361.

[22] Boso S, Alonso - Villaverde V, Gago P, et al. Susceptibility to downy mildew (Plasmopara viticola) of different Vitis varieties [J]. Crop Protection, 2014, 63: 26 - 35.

[23] Cadle - Davidson L. Variation within and between *Vitis* spp. for foliar resistance to the downy mildew pathogen Plasmopara viticola [J]. Plant disease, 2008, 92: 1 577 - 1 584.

[24] Boso S, Alonso-Villaverde V, Gago P, et al. Susceptibility of 44 grapevine (*Vitis vinifera* L.) varieties to downy mildew in the field. Australian Journal of Grape and Wine Research, 2011, 17, 394 - 400.

氟啶胺与氰霜唑不同混配组合对马铃薯晚疫病菌增效作用试验研究

时春喜，胡文渊，曲子瑞，张　腾

（西北农林科技大学植物保护学院，植保资源与病虫害治理教育部重点实验室，杨凌　712100）

摘要：为了明确氟啶胺与氰霜唑混配的增效作用，本研究采用菌丝生长速率法测定了97%氟啶胺原药、96.8%氰霜唑原药以及5个氟啶胺与氰霜唑混配组合对马铃薯晚疫病菌的室内毒力。研究结果表明：97%氟啶胺原药、96.8%氰霜唑原药的抑菌活性较强，EC_{50}分别为0.70、0.45μg/mL。5个混配组合均表现出较强的抑菌活性，且25%氟啶胺+5%氰霜唑混配组合的增效作用显著，增效系数（SR）为1.52。结果表明，二者混配后不仅有较强的抑菌效果，而且还可以减少化学农药的使用量，是开发环保，高效，安全杀菌剂的合理混配组合。

关键词：氟啶胺；氰霜唑；马铃薯晚疫病；混配；增效研究

Research on Synergistim Activity of Fluazinam and Cyazofamid Different Mixtures Against *Phytophthora infestans*（Mont.） de Bary

Shi Chunxi，Hu Wenyuan，Qu Zirui，Zhang Teng

Abstract： In order to clear the synergistic activity of Fluazinam and Cyazofamid Different mixtures，in this study，the mycelial groth rate was used to evaluate the activity of 97% fluazinam TC，96.8% Cyazofamid TC and five mixtures against Phytophthora infestans（Mont.） de Bary. The result showed that the bactericidal activity of 97% fluazinam TC，96.8% Cyazofamid TC were very high，and their EC_{50} were 0.70，0.45μg/mL。five mixtures all showed high bactericidal activity，and the synergistic activity of 25% fluazinam + 5% Cyazofamid mixture was most significantly that its SR reached 1.52. The results showed that the mixtures nont only have a strong antibacterial effect，but also can be effective in reducing the amount of pesticide exposure.

Key words： Fluazinam；Cyazofamid；*Phytophthora infestans*（Mont.） de Bary；fungicide mixture；the research of synergism

马铃薯是世界第四大粮食作物，也是我国高海拔，高纬度地区重要的粮食作物，我国已成为世界上第一大马铃薯生产国，由于它有较高的经济效益，马铃薯的种植遍及全国各个省[1]。马铃薯晚疫病是中国马铃薯产区的主要病害之一，是一种导致马铃薯茎叶死亡和块茎腐烂的毁灭性病害[2]。在多雨，冷凉，适于晚疫病流行的地区和年份，植株提前枯死，病情严重时可造成20%～40%的减产，因此一直是马铃薯产业病害防治工作的重点[3]。

化学药剂防治仍是当前防治马铃薯晚疫病的主要手段。目前化学药剂防治晚疫病的方法主要有种薯处理，根施颗粒剂和生长期喷药等三种。生产上防治马铃薯晚疫病多采用氟啶胺、代森锰锌、烯酰吗啉和代森锌等，而在已登记的防治马铃薯晚疫病的产品中，氟啶胺所占的比例较大。由于长期单一使用一些内吸性杀菌剂导致马铃薯晚疫病菌对其产生抗性，从而加大了防治难度，因此，筛选新的马铃薯晚疫病的防治药剂，成为防治马铃薯晚疫病，控制其发生与危害的关键。

氟啶胺氟啶胺是日本石原产业株式会社发现的新杀菌剂，氟啶胺杀菌谱广、活性高、具有预防及治疗作用与苯基酰胺类杀菌剂无交互抗性，对由鞭毛菌、接合菌引起的多种植物病害有良好的防治效果[4]。

氰霜唑[5]是日本石原株式会社研制与 BASF 共同开发的新一代咪唑类杀菌剂。用于防治以霜霉病、疫病为代表的卵菌纲病害，具有很高的杀菌活性，对疫霉菌生活史的各个成长阶段包括孢子囊的形成、萌发、卵孢子的形成、游动孢子的释放、移动以及菌丝的生长等都具有很高的抑制作用[6]。氰霜唑使用剂量低，对人畜低毒，是安全性非常高的药剂。对有益生物、花粉媒介昆虫等的天敌类几乎没有影响，适合于有害生物综合治理[7]。氰霜唑无交互抗性作用位点与其他杀菌剂不同，因此能有效防治对常用杀菌剂霜脲·锰锌、恶霜灵、甲霜灵等已产生抗性的病原菌，可与其他杀虫、杀菌剂等混用[8]。

近年来农药新品种研制困难，市场开发周期长再加上有害生物的抗药性日趋严重使得防治易产生抗性病害的有效药剂越来越稀缺。随着人们对各种农药的结构活性研究的不断深入，农药合理复配成为研究热点。一方面，农药的合理复配能够提高防效，延缓抗药性；另一方面能够降低成本，减少污染。氟啶胺与氰霜唑混配在马铃薯晚疫病防治上是否有增效作用此前没有报道过，因此作者进行了不同混配比例的氟啶胺与氰霜唑对马铃薯晚疫病的增效性研究，以明确其合理的混配比例及增效作用，为化学农药的混级加工以及在蔬菜等病害防治上的应用提供科学依据。

1　材料与方法

1.1　实验材料

1.1.1　供试病菌来源

马铃薯晚疫病［*Phytophthora infestans*（Mont.）de Bary］（病菌菌种经西北农林科技大学植物保护学院杀菌剂生测室分离纯化而得）。

1.1.2　供试病原菌及其培养

马铃薯晚疫病菌［*P. infestans*（Mont.）de Bary］，从陕西省咸阳市乾县北索村果园发病的新红星病果上分离纯化所得，在 PDA 斜面上于4℃冰箱中培养保存。

1.1.3　室内毒力测定方法

1.1.4　供试药剂

97%氟啶胺原药由陕西先农生物科技有限公司提供；96.8%氰霜唑原药由陕西先农生物科技有限公司提供。

1.1.5　试验仪器设备

电子天平（灵敏度0.001g）；生物培养箱；培养皿；移液枪；接种针；打孔器；游标卡尺等。

1.2　试验方法

1.2.1　生长速率

参照黄彰欣[9]等的方法采用生长速率法测定杀菌剂对病菌的抑制效果。在预备试验的基础上将啶酰菌胺和吡唑醚菌酯设5个混配组合，每个处理设4个重复。将供试药剂配制成所需浓度的 PDA 含药平板（表1）。在平板中央置一块直径为5mm 的马铃薯晚疫病菌菌饼，带有菌丝一面接触培养基。20℃培养箱中，黑暗条件下倒置培养 10d，用十字交叉法测量病菌菌落直径，计算抑制率[10]，用 DPS 软件求出各药剂对病菌的毒力回归方程和抑制中浓度 EC_{50}。

1.2.2 药剂配制

将 97% 氟啶胺原药称量好后加入到预先加好 10mL 无菌水的灭菌三角瓶中，充分振荡使其溶解，然后再加无菌水定容至 100mL，制成浓度为 800μg/mL 的 97% 氟啶胺原药母液；将 96.8% 氰霜唑原药和 5 个氟啶胺原药 + 氰霜唑原药混配组合共 6 个处理，称好后，依次加入预先加好 10mL 无菌水的灭菌三角瓶中，充分振荡使其溶解，然后再加无菌水定容至 100mL，制成浓度为 400、800μg/mL 的试验母液各 100mL 备用。

并将配制好的试验母液分别按照浓度由高到低稀释成 5 个浓度梯度（表 1）。

表 1 供试药剂配比、用药量及使用浓度

处理编号	药剂	原药量（g）	有效成分（g）	试验浓度（μg/mL）
①	97% 氟啶胺原药	0.082	0.080	80、40、20、10、5
②	96.8% 氰霜唑原药	0.041	0.040	40、20、10、5、2.5
③	21% 氟啶胺 +9% 氰霜唑	0.058 +0.025	0.056 +0.024	80、40、20、10、5
④	23% 氟啶胺 +7% 氰霜唑	0.063 +0.019	0.061 +0.019	80、40、20、10、5
⑤	25% 氟啶胺 +5% 氰霜唑	0.069 +0.014	0.067 +0.013	80、40、20、10、5
⑥	27% 氟啶胺 +3% 氰霜唑	0.074 +0.008	0.072 +0.008	80、40、20、10、5
⑦	29% 氟啶胺 +1% 氰霜唑	0.080 +0.003	0.077 +0.003	80、40、20、10、5

1.2.3 药剂处理

在无菌操作条件下，分别量取预先融化的灭菌培养基 18mL 与预先配制好的药液 2mL 加入无菌锥形瓶中（此时药液浓度将稀释 10 倍），从低浓度到高浓度依次定量吸取药液，分别加入上述锥形瓶中，充分摇匀。然后立即倒入直径 9cm 的培养皿中，制成相应浓度的含药培养基，另设不含药液的处理作空白对照，重复 4 次。

1.2.4 接种

将培养好的病原菌，在无菌条件下用直径为 5mm 的打孔器，自菌落边缘切取菌饼，用接种针将菌饼接种于含药平板培养基中央，菌丝面靠培养基，盖上皿盖，置于 20℃、黑暗条件下培养 10d。

1.2.5 调查

采用游标卡尺测量菌落直径，单位为毫米（mm）。在供试菌落培养 10d 后，每个菌落采用十字交叉法垂直测量增长直径各一次，取其平均值（差异过大的则舍去）。

1.2.6 计算方法

根据调查数据，计算各处理浓度对供试靶标菌的菌丝生长抑制率，单位为百分率（%），计算结果保留小数点后两位：

$$D = D_1 - D_2 \tag{1}$$

式中：

D——菌落增长直径；

D_1——菌落直径；

D_2——菌饼直径。

$$I = \frac{D_0 - D_t}{D_0} \times 100 \qquad (2)$$

式中：

I——菌丝生长抑制率；

D_0——空白对照菌落增长直径；

D_t——药剂处理菌落增长直径。

1.2.7 统计分析

进行药剂联合毒力测定时，本试验采用 Wadley 法计算混剂的增效系数（SR），评价混剂的联合作用类型。

Wadley 法：根据增效系数（SR）来评价药剂混用的增效作用，即 SR＜0.5 为拮抗作用，0.5≤SR≤1.5 为相加作用，SR＞1.5 为协同增效作用。增效系数（SR）按公式（3）、（4）计算：

$$X_1 = \frac{P_A + P_B}{P_A/A + P_B/B} \times 100 \qquad (3)$$

式中：

X_1——混剂 EC_{50} 理论值，单位为毫克每升（mg/L）；

P_A——混剂中 A 的百分含量，单位为百分率（%）；

P_B——混剂中 B 的百分含量，单位为百分率（%）；

A——混剂中 A 的 EC_{50} 值，单位为毫克每升（mg/L）；

B——混剂中 B 的 EC_{50} 值，单位为毫克每升（mg/L）。

$$SR = \frac{X_1}{X_2} \qquad (4)$$

式中：

SR——混剂的增效系数；

X_1——混剂的 EC_{50} 理论值，单位为毫克每升（mg/L）；

X_2——混剂的 EC_{50} 实测值，单位为毫克每升（mg/L）。

2 结果与分析

利用生长速率法测定氟啶胺与氰霜唑不同配比对马铃薯晚疫病菌的毒力。结果表明，97%氟啶胺原药与96.8%氰霜唑原药均表现出较强抑菌活性（表2）。

氟啶胺与氰霜唑的5个不同混配组合对马铃薯晚疫病菌均有较好的抑制效果，防效理想。其中以23%氟啶胺+7%氰霜唑的混配效果最明显，EC_{50} 为0.48μg/mL。另外，21%氟啶胺+9%氰霜唑、25%氟啶胺+5%氰霜唑、27%氟啶胺+3%氰霜唑、29%氟啶胺+1%氰霜唑这四个混配组合的 EC_{50} 分别为0.57、0.55、0.64、0.77μg/mL。

表2 氟啶胺与氰霜唑不同配比对马铃薯晚疫病菌的室内毒力

药剂名称	毒力回归方程	相关系数（R）	EC_{50}（μg/mL）
97%氟啶胺原药	$y=0.9753x+5.1506$	0.9923	0.70

（续表）

药剂名称	毒力回归方程	相关系数（R）	EC_{50}（μg/mL）
96.8%氰霜唑原药	$y=0.9996x+5.3425$	0.9978	0.45
21%氟啶胺+9%氰霜唑	$y=0.8080x+5.1977$	0.9979	0.57
23%氟啶胺+7%氰霜唑	$y=0.7766x+5.2490$	0.9795	0.48
25%氟啶胺+5%氰霜唑	$y=0.9847x+5.2586$	0.9873	0.55
27%氟啶胺+3%氰霜唑	$y=0.9325x+5.1806$	0.9922	0.64
29%氟啶胺+1%氰霜唑	$y=0.8777x+5.0992$	0.9883	0.77

根据增效系数（SR）的公式计算得到所有配比增效系数（SR）在0.5~2。其中25%氟啶胺+5%氰霜唑混配组合的SR为1.52，大于1.5，表现出显著的增效作用。另外，21%氟啶胺+9%氰霜唑、23%氟啶胺+7%氰霜唑、27%氟啶胺+3%氰霜唑混配组合的SR分别为1.06、1.30、1.04，均在0.5~1，即此3组不同混配组合对抑制马铃薯晚疫病菌菌丝生长有明显的相加作用（表3）。

表3　各个处理不同浓度梯度的抑制率、EC_{50}值及增效系数SR值

药剂名称	浓度（μg/mL）	菌落增长平均直径（mm）	相对抑制率（%）	EC_{50}（μg/mL）	SR
CK	0	37.06			
97%氟啶胺原药	0.5	19.97	46.12	0.70	—
	1	17.25	53.45		
	2	12.05	67.49		
	4	8.63	76.71		
	8	5.45	85.29		
96.8%氰霜唑原药	0.25	22.61	38.98	0.45	—
	0.5	17.40	53.06		
	1	13.61	63.28		
	2	9.84	73.45		
	4	6.34	82.89		
21%氟啶胺+9%氰霜唑	0.5	19.19	48.21	0.57	1.06
	1	15.39	58.47		
	2	12.43	66.47		
	4	9.39	74.67		
	8	6.38	82.80		

（续表）

药剂名称	浓度（μg/mL）	菌落增长平均直径（mm）	相对抑制率（%）	EC_{50}（μg/mL）	SR
23%氟啶胺+7%氰霜唑	0.5	17.54	52.67	0.48	1.30
	1	15.14	59.14		
	2	12.45	66.41		
	4	9.19	75.21		
	8	5.78	84.40		
25%氟啶胺+5%氰霜唑	0.5	18.21	50.87	0.55	1.52
	1	15.26	58.84		
	2	11.42	69.19		
	4	7.57	79.57		
	8	4.25	88.54		
27%氟啶胺+3%氰霜唑	0.5	19.33	47.85	0.64	1.04
	1	16.33	55.94		
	2	12.45	66.40		
	4	8.66	76.63		
	8	5.35	85.56		
29%氟啶胺+1%氰霜唑	0.5	20.42	44.89	0.77	0.89
	1	17.40	53.06		
	2	13.42	63.78		
	4	10.54	71.56		
	8	6.38	82.80		

3 结论与讨论

3.1 从室内毒力测定的结果来看，试验所设的7个处理对马铃薯晚疫病菌的菌丝生长均有一定的抑制作用。96.8%氰霜唑原药的抑菌效果是最好的。5个不同混配组合中又以23%氟啶胺+7%氰霜唑混配组合的抑菌效果最为显著，EC_{50}值达到0.48μg/mL。另外，21%氟啶胺+9%氰霜唑、23%氟啶胺+7%氰霜唑、27%氟啶胺+3%氰霜唑、29%氟啶胺+1%氰霜唑4组混配组合的EC_{50}值分别为0.57、0.55、0.64、0.77μg/mL，都表现出较好的抑菌效果。

由试验结果可知，25%氟啶胺+5%氰霜唑混配组合的EC_{50}值为0.55，增效系数（SR）最大，为1.52，大于1.5。23%氟啶胺+7%氰霜唑混配组合的EC_{50}值最小，为0.48μg/mL，但此配比下的增效系数（SR）为1.30，小于1.5。从用量上考虑，23%氟啶胺+7%氰霜唑混配组合要略微优于25%氟啶胺+5%氰霜唑，但从混配的增效作用上来看，25%氟啶胺+5%氰霜唑混配组合的增效作用显著，因此选用25%氟啶胺+5%氰霜唑的混配组合是科学合理的。

本项研究结果表明，氟啶胺与氰霜唑混配后表现出明显的增效作用，因此可以降低应用成本，减少化学农药的使用量，有助于延缓病原菌抗药性的产生，延长化学杀菌剂的商品寿命。此外，本实验中的数据均为室内毒力测定结果，在大田的实际防治效果还需通过试验予以验证。

参考文献

［1］ 张志铭，曹克强，张红，等. 中国马铃薯晚疫病流行和预测预报的研究进展［C］// 中国作物学会马铃薯专业委员会 2003 年学术年会. 2004.

［2］ 刘小林. 榆林市马铃薯晚疫病的发生规律及综合防治技术［J］. 现代农业科技，2011（4）：175.

［3］ 魏亚雯. 马铃薯晚疫病［J］. 农业科技与信息，2009（1）：28.

［4］ 李慧. 氟啶胺对马铃薯晚疫病的防治效果［J］. 农业与技术，2012，32（10）：110.

［5］ 刘长令. 世界农药大全 - 杀菌剂卷［M］. 北京：化学工业出版社，2006.

［6］ Nasa，Komyoji，Suzuki，et al. Imidazole compounds and biocidal compositions comprising the same［P］. EP0298196A1，1989 - 11 - 01

［7］ 程志明. 杀菌剂氰霜唑的开发［J］. 世界农药，2005（3）：1 - 4.

［8］ 王迪轩. 蔬菜常用杀菌剂——氰霜唑的使用与注意事项［J］. 农药市场信息，2015（6）：47.

［9］ 黄彰欣，黄瑞平，郑仲，等. 植物化学保护实验指导［M］. 北京：农业出版社，1993：52 - 61.

［10］ 刘学敏，李立军. 杀菌剂混配的增效作用［J］. 农药科学与管理，2002，23（5）：12 - 15.

8种杀菌剂对苦瓜炭疽病菌的毒力测定*

吴凤芝[1]**，曾向萍[2]，王会芳[2]，肖　敏[2]
（1. 海南省农业科学院热带果树研究所，海口　571100；
2. 海南省农业科学院植物保护研究所，海口　571100）

摘要：选择8种不同杀菌剂，采用菌丝生长抑制法对来自海南文昌的苦瓜炭疽病菌 *Colletotrichum gloeosporioides* 进行了的室内毒力测定。结果表明，供试药剂中50%咪鲜胺锰盐WP的抑菌效果最好，其 EC_{50} 和 EC_{90} 分别为0.139 6μg/mL和9.361 4μg/mL。10%苯醚甲环唑WG和25%吡唑醚菌酯EC也表现了良好的抑菌效果，其 EC_{50} 值分别为0.043 1μg/mL和1.828 9μg/mL，其 EC_{90} 值分别为116.824 5μg/mL和11.411 4μg/mL。

关键词：杀菌剂；苦瓜；胶孢炭疽菌；毒力

Detected Toxicity of Different Fungicides Against *Colletotrichum gloeosporioides* on Balsam Pear

Wu Fengzhi[1], Zeng Xiangping[2], Wang Huifang[2], Xiao Min[2]
(1. *Institute of Tropical Fruiter*, *Hainan Academy of Agricultural Sciences*, *Haikou* 571100, *China*; 2. *Institute of Plant Protection*, *Hainan Academy of Agricultural Sciences*, *Haikou* 571100, *China*)

Abstract: The detected toxicity experiments, using the method of mycelial growth inhibition, were conducted with 8 selected fungicides against anthracnose pathogen *Colletotrichum gloeosporioides* on balsam pear. The results showed that prochloraz - manganese chloride complex 50% WP had the greatest inhibiting efficiency among the tested fungicides with EC_{50} of 0.136 9μg/mL and EC_{90} of 9.361 4μg/mL. Difenoconazole 10% WG and pyraclostrobin 25% EC were revealed good inhibiting effects as well, with EC_{50} of 0.141μg/mL and 0.409 8μg/mL, EC_{90} of 954.80μg/mL and 207.776 9μg/mL, respectively.

Key words: fungicide; balsam pear; *Colletotrichum gloeosporioides*; toxicity

苦瓜俗称凉瓜，属葫芦科攀缘草本作物，因其是人们喜爱的一种蔬菜，已成为海南岛冬季北运蔬菜种植的主要种类。苦瓜炭疽病是苦瓜生产上为害最严重的一种真菌病害[1-2]，在整个生长季节里均可为害苦瓜的器官和组织，尤其以高温高湿季节蔓延速度快，常常造成茎蔓龟裂干枯或果实腐烂[3]，是制约苦瓜产量和质量的主要因子。病害严重发生情况下，杀菌剂应用是控制该病发生和蔓延的有效途径。为此，笔者选用了不同作用机理和不同作用方

* 基金项目：海南省省属科研院所技术开发研究专项（KYYS－2015－01），公益性行业（农业）科研专项（201303023）

** 第一作者：吴凤芝，女，高级农艺师，主要从事农药应用技术研究；E-mail：420550252@qq.com

式的杀菌剂对苦瓜炭疽病菌进行了室内毒力测定，旨在通过筛选和获得安全高效药剂种类，以期为病害防治和科学用药提供理论依据。

1 材料与方法

1.1 供试病原菌

从海南省文昌市潭牛镇采集疑似病害样本，按常规方法进行组织分离，并在 PDA 培养基上培养，获得纯培养菌株[4]。经回接实验，通过形态学观察，确认诱发病害的病原菌为胶孢炭疽菌 *Colletotrichum gloeosporioides*。纯化后的菌株于低温冰箱下保存备用。

1.2 供试药剂及使用浓度

选择来自不同农药生产企业的杀菌剂商品 8 种，按表 1 设计使用浓度。

表 1 供试药剂及使用浓度

药剂名称	生产企业	剂型	使用浓度（mg/L）
25% 吡唑醚菌酯	德国巴斯夫公司	EC	125、60、30、15、8.3
10% 苯醚甲环唑	瑞士先正达作物保护有限公司	WG	200、100、50、16.7、5
50% 咪鲜胺锰盐	德国拜耳作物科学公司	WP	100、50、10、5、2.5
75% 百菌清	北京华戎生物激素厂	WP	937.5、625、156.25、93.75、62.5
25% 嘧菌酯	瑞士先正达作物保护有限公司	SC	250、125、50、31.25、16.67
12.5% 腈菌唑	北京华戎生物激素厂	EC	62.5、31.25、12.5、4.17、2.08
50% 醚菌酯	德国巴斯夫公司	WG	200、100、55.56、33.33、14.28
70% 甲基托布津	江苏龙灯化学有限公司	WP	875、350、140、70、23.33

1.3 菌丝生长速率测定

将上述药剂分别配成所需浓度 50 倍，待培养基溶化后，分别用移液器吸取不同浓度的药液 1mL 注入 49mL 培养基中，摇匀后倒入 3 个直径为 9cm 的灭菌培养皿中，使其形成薄厚均匀的含药培养基平板，以 1mL 无菌水代替药液作对照。用直径为 5mm 的打孔器将活化好的供试菌种（PCA 培养基 28℃培养 5d）打取菌饼接入培养皿中央位置，置于 28℃恒温培养箱中。再培养 5d 后用十字交叉法测量菌落直径，求出相对抑制率[5]。用 DPS 软件进行统计分析，求出各药剂对红麻炭疽病菌的毒力回归方程、抑制中浓度（EC_{50}）及相关系数（R）。

菌丝生长抑制率（%）=（对照组菌落平均净生长量 × 处理组菌落平均净生长量）/ 对照组菌落平均净生长量 ×100

2 结果与分析

不同药剂对苦瓜炭疽病菌菌丝生长的抑制效果见表 2。结果表明，10% 苯醚甲环唑 WG 和 50% 咪鲜胺锰盐 WP 表现出良好的抑菌效果，其 EC_{50} 分别为 0.043 1mg/L 和 0.139 6mg/L；其次为 25% 吡唑醚菌酯 EC，其 EC_{50} 为 1.828 9mg/L。25% 嘧菌酯 SC 和 12.5% 腈菌唑 EC 的抑菌作用基本一致，其 EC_{50} 分别为 11.439 0mg/L 和 7.276 7mg/L，但是 25% 嘧菌酯 SC 表现不稳定，其 EC_{90} 达到 310 147.14mg/L。75% 百菌清 WP 和 70% 甲基托布津 WP 对苦瓜炭疽菌的抑菌效果最差，其 EC_{50} 分别为 106.543 1mg/L 和 4 159.480 7mg/L。

表2　8种杀菌剂对苦瓜炭疽病菌菌丝生长抑制作用

药剂名称	毒力回归方程	相关系数	F检验值	EC_{50} (mg/L)	EC_{90} (mg/L)
25%吡唑醚菌酯 EC	$y=4.5774+1.6117x$	0.949 5	27.456 6	1.828 9	11.411 4
10%苯醚甲环唑 WG	$y=5.5097+0.3733x$	0.954 6	30.785 2	0.043 1	116.824 5
50%咪鲜胺锰盐 WP	$y=5.6000+0.7016x$	0.985 4	100.248	0.139 6	9.361 4
75%百菌清 WP	$y=3.5864+0.6972x$	0.978 0	65.908 7	106.543 1	7 339.608 2
25%嘧菌酯 SC	$y=4.694+0.2891x$	0.951 6	28.780 1	11.439 0	310 147.140 0
12.5%腈菌唑 EC	$y=3.8411+1.3446x$	0.987 2	115.218	7.276 7	65.323 6
50%醚菌酯 DF	$y=4.3422+0.4399x$	0.952 9	29.608 5	31.288 3	25 619.359 0
70%甲基托布津 WP	$y=0.2423+1.3146x$	0.978 6	67.735 4	4 159.480 7	39 253.277 0

3　结论与讨论

化学防治是控制苦瓜炭疽病为害的有效技术手段。本实验选择有代表性的8种化学杀菌剂，利用生长速率法对苦瓜炭疽病菌（*Colletotrichum gloeosporioides*）的菌丝生长进行了测定，10%苯醚甲环唑WG、50%咪鲜胺锰盐WP和25%吡唑醚菌酯EC表现出理想的抑菌作用，其EC_{50}仅分别为0.043 1、0.139 6和1.829 8mg/L；具有良好抑菌效果的还有12.5%腈菌唑EC、25%嘧菌酯SC和50%醚菌酯DF。通过进一步的田间验证试验，这些杀菌剂可作为生产上推荐的农药品种。而75%百菌清WP和70%甲基托布津WP对苦瓜炭疽病菌的抑菌效果较差，这可能与其在海南菜田中大量长期使用而已经产生了抗药性有关，建议生产上不用于苦瓜炭疽病的防治。

分生孢子是苦瓜炭疽病传播和侵入的主要载体，分生孢子萌发与否，直接影响到苦瓜炭疽病的发生和再侵染。本实验仅仅是基于药剂对菌丝生长的室内毒力测定，表明了控制病害的一定理论基础，但是对指导大田生产有其局限性。因此，选择不同作用机理和作用方式的化学杀菌剂，进行其对苦瓜炭疽病菌的分生孢子萌发与侵入的抑制效果测定与评价极为必要。

参考文献

[1] 房德纯，刘秋，蒋玉文，等. 瓜类蔬菜病虫害防治［M］. 北京：农业出版社，2000.
[2] 翁祖信，冯兰香，李宝栋，等. 蔬菜病虫害诊断与防治［M］. 天津：天津科学技术出版社，1994.
[3] 王会芳，曾向萍，芮凯，等. 苦瓜炭疽病病原鉴定及生物学特性初步研究［J］. 中国农学通报，2012，28（7）：141－145.
[4] 方中达. 植病研究方法［M］. 北京：中国农业出版社，1998.
[5] 涂勇，姚昕，余前媛，等. 不同杀菌剂对青枣炭疽病菌的室内毒力测定［J］. 江苏农业科学，2009，(2)：136－137.

几种杀菌剂对瓜类土传病害的防治*

姚玉荣**，郝永娟***，霍建飞，刘春艳，王万立

（天津市植物保护研究所，天津　300381）

摘要：利用5种土壤消毒剂、11种化学杀菌剂和7种生物药剂测定对黄瓜枯萎病菌、甜瓜枯萎病菌以及甜瓜蔓枯病菌等瓜类土传病害病原菌的抑菌活性。试验结果表明土壤消毒剂 ClO_2 和甲醛具有较好的抑菌效果；化学杀菌剂中97%咪鲜胺对和98.05%戊唑醇对黄瓜枯萎病菌和甜瓜蔓枯病菌具有很好的抑菌效果，而85%代森锰锌和92%百菌清对黄瓜枯萎病菌和甜瓜蔓枯病菌几乎没有抑制作用；7种生物药剂中2亿/g木霉WDG和1 000亿枯草芽孢杆菌对黄瓜枯萎病菌和甜瓜蔓枯病菌抑制效果明显。

关键词：黄瓜枯萎病；甜瓜枯萎病；甜瓜蔓枯病；土壤消毒剂；化学杀菌剂；生物药剂

Abstract: Five kinds of soil disinfectants, 11 kinds of fungicides and 7 kinds of biological agents were used for inhibitory effect on soil-borne diseases such as *Fusarium oxysporum* (Schl.) f. sp. Cucumerinum Owen, *Fusarium oxysporum* f. sp. Melonis, *Didymella bryoniae*. The results showed that ClO_2 and formaldehyde had excellent control effect. Among 11 kinds of fungicides, 97% prochloraz and 98.05% tebuconazole showed good results on *Fusarium oxysporum* (Schl.) f. sp. Cucumerinum Owen and *D. bryoniae*, while 85% mancozeb and 92% chlorothalonil had almost no inhibitory effects on the two pathogens. In addition, 200 million/g *Trichoderma* sp. water dispersible granule and 100 billion *Bacillus subtilis* had showed good control efficiency to pathogens.

Key words: Cucumber fusarium wilt; Melon fusarium wilt; Melon tendril blight; Soil disinfectants; Fungicides; Biological agents

植物土传病害是指生活史中一部分或大部分存在于土壤中的病原物在条件适宜时萌发并侵染植物而导致的病害。土传病原菌在土壤中存活时间长、适应性强、繁殖快，危害严重，难以根除[1]。

葫芦科瓜类蔬菜在我国的蔬菜生产中占有重要的地位，随着人们生活水平的提高，以及农业产业化结构调整，瓜类蔬菜种植面积逐年增长，同一块土地连年种植同一品种导致土壤病原菌积累，成为严重制约其产业发展的重要因素。据报道瓜类蔬菜土传病害主要包括枯萎病、蔓枯病、猝倒病、立枯病、疫病等[2]。

本试验测定5种土壤消毒剂对黄瓜枯萎病、甜瓜枯萎病以及甜瓜蔓枯病等瓜类重要的土传病原菌的抑菌作用，并测定11种化学杀菌剂以及7种生物药剂对黄瓜枯萎病和甜瓜蔓枯病的抑菌活性，为今后生产上防治瓜类土传病害提供理论和技术支持。

---

* 基金项目：天津市科技支撑项目（13ZCZDNC00100）

** 第一作者：姚玉荣，女，助理研究员，专业方向：植物病理学；E-mail：yyr1012@126.com

*** 通讯作者：郝永娟，女，研究员，主要从事蔬菜病害防治研究及杀菌剂应用技术研究；E-mail：tjzbshyj@163.com

1 材料与方法

1.1 供试菌株、土壤消毒剂和杀菌剂

供试菌株：黄瓜枯萎病 *Fusarium oxysporum*（Schl.）f. sp. *cucumerinum* Owen，甜瓜枯萎病 *Fusarium oxysporum* f. sp. *melonis*，甜瓜蔓枯病 *Didymella bryoniae*，均保存在天津市植物保护研究所蔬菜病害实验室。

供试土壤消毒剂：ClO_2，漂白粉，石灰氮，95%酒精，甲醛。

供试化学杀菌剂：97%多菌灵，92%百菌清，95%福美双，85%代森锰锌，98%噁霉灵，95.1%甲托，97%咪鲜胺，98.05%戊唑醇，95%吡唑醚菌酯，50%嘧菌酯，96.8%苯醚甲环唑；供试生物药剂：2亿/g木霉水分散粒剂，3%多抗霉素可湿性粉剂，4%春雷霉素可湿性粉剂，10^6孢子/g寡雄腐霉，1 000亿枯草芽孢杆菌，中生菌素以及申嗪腐霉。

1.2 土壤消毒剂对瓜类土传病害的室内抑菌效果测定

将土壤消毒剂按表1中的稀释倍数依次进行稀释，然后按土壤消毒剂与培养基1∶9混合均匀后倒入直径9cm培养皿中，每皿10mL，配制成含系列浓度的培养基平板。

抑菌效果采用菌落生长测定法进行测定。将培养好的菌株用直径5mm的打孔器沿菌落边缘打取菌块，分别接种到含系列浓度土壤消毒剂的培养基平皿上，每皿测定2个菌株，重复4次。置于25℃培养48h量取菌落直径并记录数据。

1.3 化学杀菌剂以及生物药剂对瓜类土传病害的抑菌活性测定

含药培养基的配制：将咪鲜胺、戊唑醇、吡唑醚菌酯、嘧菌酯、苯醚甲环唑、百菌清、福美双、代森锰锌、噁霉灵、甲托原药用少量丙酮或二甲基甲酰胺溶原药用少量丙酮或者二甲基甲酰胺溶解（溶剂的最终含量<0.5%），然后用含0.05%吐温80的无菌水稀释成10 000μg/mL的母液，依次稀释成系列浓度；多菌灵原药则用1%盐酸溶液溶解，配制成10 000μg/mL的母液，再用无菌水稀释至系列浓度。按药液与培养基1∶9混合均匀后倒入直径9cm的培养皿中，每皿10mL，配制成系列浓度的含药培养基。

抑菌活性的测定同样采用菌落生长测定法进行测定，具体步骤参见1.2。

1.4 数据分析

按照下列公式计算菌丝生长抑制率：

$$\text{菌丝生长抑制率（\%）}=\frac{\text{（空白对照菌落直径增加值－处理菌落直径增加值）}}{\text{空白对照菌落直径增加值}}\times 100$$

上述所有试验数据用DPS数据软件进行处理，计算不同药剂的毒力回归方程、相关系数、致死中浓度EC_{50}值等。

2 结果与分析

2.1 土壤消毒剂对瓜类土传病害的室内抑菌结果

由表1可以看出，试验中所用的土壤消毒剂对黄瓜枯萎病菌、甜瓜枯萎病菌以及甜瓜蔓枯病菌均有一定的抑制作用。综合几种土壤消毒剂对黄瓜枯萎病、甜瓜枯萎病以及甜瓜蔓枯病的抑菌作用，效果较好的为ClO_2和甲醛。

表 1　土壤消毒剂对瓜类土传病害的室内抑菌结果

Table 1　Efficacy of soil disinfectants in suppressing melons soil borne diseases

试验药剂	稀释倍数	黄瓜枯萎病		甜瓜枯萎病		甜瓜蔓枯病	
		菌落半径	抑菌率（%）	菌落半径	抑菌率（%）	菌落半径	抑菌率（%）
ClO_2	250	15	50	9.25	73	9.75	66.1
	500	19	36.7	17.75	48.2	15.5	46.1
	1 000	29.75	0.83	19.25	43.8	18.25	36.5
漂白粉	250	18.5	38.3	16.25	52.6	10.25	64.3
	500	24	20	19.75	42.3	13	54.8
	1 000	26	13.3	24.5	28.5	15.75	45.2
石灰氮	250	11.25	62.5	24.75	27.7	13.25	53.9
	500	21	30	25.5	25.5	16	44.3
	1 000	25	16.7	26.75	21.9	20.25	29.6
95%酒精	500	27	10	22.75	33.6	13	54.8
	1 000	27	10	23.5	31.4	18	37.4
	2 000	29	3.3	22	35.8	20	30.4
甲醛	500	12.5	58.3	5	85.4	9	68.7
	1 000	23.5	21.7	13	62	12	58.3
	2 000	24.75	17.5	15.75	54	18	37.4
CK		30		34.25		28.75	

2.2　化学杀菌剂以及生物药剂对瓜类土传病害的室内抑菌结果

2.2.1　化学杀菌剂对黄瓜枯萎病和甜瓜蔓枯病的的室内抑菌结果

表 2 结果显示试验的 11 种化学杀菌剂中 97% 咪鲜胺对黄瓜枯萎病菌的抑菌效果最好，其 EC_{50} 值为仅 0.93mg/L；其次按照抑菌效果从大到小为 95% 吡唑醚菌酯、98.05% 戊唑醇和 97% 多菌灵；再次为 95% 福美双、98% 噁霉灵、50% 醚菌酯和 95.1% 甲托；最后 96.8% 苯醚甲环唑、85% 代森锰锌和 92% 百菌清的抑菌效果最差，对于黄瓜枯萎病菌的生长几乎无抑制作用。

试验药剂对甜瓜蔓枯病的抑菌效果测定结果显示 97% 咪鲜胺对甜瓜蔓枯病菌的抑制效果最好，EC_{50} 值为 0.72mg/L，97% 多菌灵仅次之；其次抑菌效果从大到小为 98.05% 戊唑醇、96.8% 苯醚甲环唑、95% 福美双、95.1% 甲托；再次之为 98% 噁霉灵和 95% 吡唑醚菌酯；最后 85% 代森锰锌、92% 百菌清和 50% 醚菌酯对于甜瓜蔓枯病菌的抑制效果最差，EC_{50} 值均在 100mg/L 以上，甚至达到 200mg/L 以上，证明这 3 种药剂对于甜瓜蔓枯病菌没有明显的抑制效果。

表2　11种化学杀菌剂对黄瓜枯萎病菌室内毒力测定

Table 2　Indoor toxicity regressions of 11 kinds of fungicides against *Fusarium oxysporum*（Schl.）f. sp. *Cucumerinum* Owen and *Didymella bryonia*

菌株	试验药剂	毒力回归方程	r	EC_{50}/（mg/L）	EC_{90}/（mg/L）
黄瓜枯萎病菌	97%咪鲜胺	$y=5.0228+0.8648x$	0.998 4	0.93	28.09
	98.05%戊唑醇	$y=4.4139+1.0327x$	0.988 9	3.69	64.35
	95%吡唑醚菌酯	$y=4.9381+0.2255x$	0.993 1	1.88	909 626.86
	50%醚菌酯	$y=4.1787+0.4289x$	0.906 5	82.22	80 004.27
	96.8%苯醚甲环唑	$y=3.8531+0.5411x$	0.985 0	131.66	30 745.6
	97%多菌灵	$y=4.2637+0.7490x$	0.987 2	9.61	494.24
	92%百菌清	$y=2.6816+0.9914x$	0.980 9	218.09	4 279.66
	95%福美双	$y=2.7065+1.3864x$	0.996 5	45.11	379.09
	85%代森锰锌	$y=3.4986+0.7047x$	0.931 2	135.06	8 895.02
	98%噁霉灵	$y=3.4344+0.8986x$	0.997 1	55.23	1 473.54
	95.1%甲托	$y=2.7815+1.1501x$	0.938	84.89	1 104.5
甜瓜蔓枯病菌	97%咪鲜胺	$y=5.2711+1.9232x$	0.933 8	0.72	3.35
	98.05%戊唑醇	$y=4.7768+0.8880x$	0.969 8	1.78	49.5
	95%吡唑醚菌酯	$y=4.4666+0.4038x$	0.971 3	20.93	31 220.98
	50%醚菌酯	$y=4.1827+0.3361x$	0.977 1	270.19	1 756 961.44
	96.8%苯醚甲环唑	$y=4.7087+1.0363x$	0.976 2	1.91	32.95
	97%多菌灵	$y=4.9989+2.5813x$	0.973 4	1.00	3.14
	92%百菌清	$y=2.6198+1.0867x$	1.000 0	155.02	2 343.14
	95%福美双	$y=4.7582+0.6461x$	0.898 6	2.37	227.91
	85%代森锰锌	$y=-4.2260+4.3573x$	0.996 9	131.02	257.91
	98%噁霉灵	$y=4.0031+0.8045x$	0.997 8	17.35	676.69
	95.1%甲托	$y=3.7555+1.6533x$	0.990 2	5.66	33.72

2.2.2　生物药剂对黄瓜枯萎病和甜瓜蔓枯病的室内抑菌结果

表3结果显示7种生物药剂对黄瓜枯萎病菌的抑菌活性依次为2亿/g木霉菌水分散粒剂、1 000亿枯草芽孢杆菌、3%多抗霉素可湿性粉剂、中生菌素、申嗪腐霉、4%春雷霉素WP、10^6孢子/克寡雄腐霉。2亿/g木霉菌水分散粒剂对黄瓜枯萎病的抑菌率达到100%。4%春雷霉素WP，10^6孢子/g寡雄腐霉的抑菌率为负，即对病原菌的生长起到一定的促进效果。

试验的7种生物药剂对于甜瓜蔓枯病菌的抑菌率从大到小依次为2亿/g木霉菌水分散粒剂，1 000亿枯草芽孢杆菌，中生菌素，3%多抗霉素可湿性粉剂，4%春雷霉素可湿性粉剂，10^6孢子/g寡雄腐霉。其中2亿/g木霉菌水分散粒剂的抑菌率达到100%，10^6孢子/g寡雄腐霉对甜瓜蔓枯病菌的抑制率为负，对甜瓜蔓枯病菌的生长有一定促进作用。

表 3　7 种生物药剂对甜瓜蔓枯病的抑菌活性测定

Table 3　Biological activity test of 7 kinds of biological agents against *Fusarium oxysporum* (Schl.) f. sp. *Cucumerinum* Owen and *Didymella bryoniae*

试验药剂	黄瓜枯萎病菌			甜瓜枯萎病菌		
	稀释倍数	菌落半径（mm）	抑菌率	稀释倍数	菌落半径（mm）	抑菌率
2 亿/g 木霉菌	3 000	0	100%	3 000	0	100%
3% 多抗霉素	3 000	13.25	38.73%	3 000	10.5	55.47%
4% 春雷霉素	3 000	22.67	-4.83%	3 000	10.5	55.47%
10^6孢子/g 寡雄腐霉	3 000	23.5	-8.67%	3 000	24	-1.78%
1 000 亿枯草芽孢杆菌	3 000	4	81.50%	3 000	6.75	71.37%
中生菌素	3 000	16.5	23.67%	3 000	10	57.59%
申嗪腐霉	3 000	18.25	15.61%	3 000	24.75	-4.96%

3　总结和讨论

通过室内菌丝生长速率法检测了 5 种土壤消毒剂、11 种化学杀菌剂以及 7 种生物药剂对瓜类土传病害的抑菌活性，结果显示土壤消毒剂 ClO_2 和甲醛抑菌作用较好。ClO_2 具有较强的氧化性，几乎对所有的病原微生物都有强烈的杀灭作用，由于其消毒的终产物为无机离子，在病原物体内不产生相应抗体而被世界卫生组织列为 A1 类消毒剂[3]，是一种具有良好应用前景的土壤消毒剂。

试验的 11 种化学杀菌剂中 97% 咪鲜胺对黄瓜枯萎病以及甜瓜蔓枯病的抑菌效果最好，EC_{50}值均小于 1mg/L；98.05% 戊唑醇对这两种病害也具有很好的防治效果；而 85% 代森锰锌和 92% 百菌清对两种病害几乎没有抑制作用。考虑到化学杀菌剂对环境和食品的安全性影响，一些生物源杀菌剂也逐步被应用到土传病害的防治中。试验中 2 亿/g 木霉水分散粒剂以及 1 000亿枯草芽孢杆菌对黄瓜枯萎病和甜瓜蔓枯病抑菌效果明显。本研究对土壤消毒剂、化学杀菌剂以及生物农药对瓜类土传病害的防治仅进行了室内抑菌活性测定，还有待于结合田间防治效果对这些药剂防治瓜类土传病害的效果进行综合评价。

目前，防治土传病害的的方法主要包括轮作、种植抗性品种、土壤消毒、施用化学杀菌剂主要包括咯菌腈、铜制剂、噁霉灵以及一些生物制剂等产品[4-6]。其中化学杀菌剂使用简便、效果显著，但是由于土传病害种类繁多，复合侵染普遍，显症时应用效果不佳[7]，此外化学杀菌剂的大量使用导致不仅病原菌产生抗性的现象越来越强，而且危害食品安全，破坏生态平衡。我国政府已高度重视农药使用所带来的系列问题，农业部 2015 年提出“到 2020 年化肥、农药使用量零增长”方案，引导农业“绿色化”发展。因此开发具有新作用位点或作用机制的高效低毒低残留农药以及生物农药等新型产品用于瓜类土传病害的防治，并与其他药剂组合实现综合治理。

参考文献

[1] 郝永娟，王万立，刘春艳，等. 设施蔬菜土传病害的综合调控及防治进展［J］. 天津农业科学，2006，12（1）：31－34.

[2] 梁建根，竺利红，施跃峰. 瓜类蔬菜土传病害的重要病原菌及其生态学研究进展［J］. 现代农业科技，2008，21：156－157.

[3] 王洪山. 二氧化氯消毒剂在温室蔬菜土壤消毒中的应用［J］. 中国蔬菜，2016（7）：92－93.

[4] 张晓波，茹李军，郑雪松，等. 杀菌剂 SYP－4288 对常见土传病害的防治效果［J］. 植物保护，2016，42（3）：255－260.

[5] 武泽民，于振茹，卢立华，等. 设施蔬菜土传病害的诊断与中和防控措施［J］. 安徽农业科学，2013，41（31）：12 320－12 323.

[6] 葛红莲，赵红六，郭坚华. 植物土传病害微生物农药的研究开发进展［J］. 安徽农业科学，2004，32（1）：269－277.

[7] 李清飞，赵承美，于国忠，等. 蔬菜土传病害生态防控技术研究［J］. 北方园艺，2011（14）：192－194.

7 种药剂对黄瓜炭疽病的防治效果比较*

肖　敏**，赵志祥，曾向萍，何　舒，严婉荣，王会芳，芮　凯，符美英
（海南省农业科学院植物保护研究所，海南省植物病虫害防控重点实验室，海口　571100）

摘要：通过不同类型成分化学药剂防治黄瓜炭疽病田间效试验，了解其在本地区的防治效果。结果表明，供试的 7 种药剂中，以 43% 戊唑醇悬浮剂 4 000倍液、50% 咪鲜胺锰盐可湿性粉剂 1 500倍液、10% 苯醚甲环唑水分散粒剂 1 000倍液和 40% 多・福・溴菌腈可湿性粉剂 1 500倍液处理防治黄瓜炭疽病的效果较好，第三次药后 7d，防效达 77.45% ~80.02%，相互之间的差异不显著。建议在黄瓜炭疽病发生初期施用，施药次数为 3 次，间隔 7 d。不同类型药剂轮换使用，既能更好保证防效，同时也能减缓病原菌抗药性产生。
关键词：黄瓜；炭疽病；防效比较

炭疽病属高温高湿条件下易发生的病害，广泛发生在蔬菜、果树和园林花卉上[1]。在海南，黄瓜的种植有一定的规模，由于区域性的特殊气候条件，极有利于炭疽病的发生，炭疽病已成为黄瓜生产过程中重要病害之一。据报道黄瓜炭疽病的病原菌为瓜类炭疽菌（*Colletotrichum orbiculare*），可引起黄瓜叶、蔓和果实受害，影响产量和品质[2]，产量损失 40% 以上，甚至绝收[3]。对黄瓜炭疽病防治的报道较多[4-7]，主要为化学防治，也有用生物药剂[8]和拮抗乳酸菌[9]防治的研究，而对于苯并咪唑类、甲氧基丙烯酸酯类、三唑类、咪唑类、有机硫类、二硫代氨基甲酸酯类不同类型药剂对黄瓜炭疽病的防治比较未有相关报道。因此笔者选择不同类型成分的化学药剂进行黄瓜炭疽病田间防效试验，以期为大田生产提供借鉴。

1　材料与方法

1.1　试验材料

供验黄瓜品种：博青 169。

供试药剂：70% 甲基硫菌灵可湿性粉剂（江苏龙灯化学有限公司），50% 咪鲜胺盐锰可湿性粉剂（德国拜耳作物科学公司），10% 苯醚甲环唑水分散粒剂（瑞士先正达作物保护有限公司），250g/L 嘧菌酯悬浮剂（河北威远生化农药有限公司），80% 代森锰锌可湿性粉剂（陶氏益农农业科技有限公司），40% 多・福・溴菌腈可湿性粉剂（中国农业科学院植物保护研究所廊坊农药中试厂），43% 戊唑醇悬浮剂（上海惠光化学有限公司）。

1.2　试验地点

试验设在海南省澄迈县永发镇排坡洋蔬菜地，面积约 1 000m^2，壤土，土壤有机质含量中等，排灌设施较好，管理水平为当地中上水平，田间管理一致，前茬作物为辣椒，2016

* 基金项目：海南省省属科研院所技术开发专项（KYYS - 2016 - 06），公益性行业（农业）科研专项（201303023），海南省省属科研院所技术开发专项（KYYS - 2015 - 01）
** 作者简介：肖敏，女，研究员，硕士，主要从事植物病害病理及防控技术研究；E-mail：xiaominnky@21cn.com

年4月1日移栽黄瓜。

1.3 试验设计

试验设70%甲基硫菌灵可湿性粉剂500倍液、50%咪鲜胺盐锰可湿性粉剂1 500倍液、10%苯醚甲环唑水分散粒剂1 000倍液、250g/L嘧菌酯悬浮剂1 500倍液、80%代森锰锌可湿性粉剂500倍液、40%多·福·溴菌腈可湿性粉剂1 500倍液、43%戊唑醇悬浮剂4 000倍液和清水对照共8个处理，每处理4次重复，共32个小区，小区面积约30m²，随机区组排列。试验共施药3次，分别于5月6日、13日和20日施用。第一次施药时炭疽病初发生，施药当天均无降雨天气。

1.4 药效调查和计算方法

每小区随机5点取样，每点调查3株，每株自上而下查10片叶。药前调查病情基数，第3次药前和第3次药后7 d各各调查病情，记录病叶数及病级，计算病指和防效，并采用DMRT法进行显著性测定。

分级标准按照标准[10]进行：0级，无病；1级：病斑面积占整个叶面积的5%以下；3级：病斑面积占整个叶面积的6%～10%；5级：病斑面积占整个叶面积的11%～25%；7级：病斑面积占整个叶面积的26%～50%；9级：病斑面积占整个叶面积的51%以上。

病情指数＝∑（各级病叶数×相对级数值）/（调查总叶数×9）×100

防治效果（%）＝［1－（空白对照区药前病情指数×处理区药后病情指数）/（空白对照区药的病情指数×处理区药前病情指数）］×100

2 结果与分析

试验结果见下表，第三次药前，各小区病指均有不同程度上升，其中清水处理病指达14.61。43%戊唑醇悬浮剂4 000倍液、50%咪鲜胺盐锰可湿性粉剂1 500倍液、10%苯醚甲环唑水分散粒剂1 000倍液和40%多·福·溴菌腈可湿性粉剂1 500倍液的防效分别为73.66%、71.48%、71.05%和70.29%，4个处理相互之间防效差异不显著，但都好于250g/L嘧菌酯悬浮剂1 500倍液（68.05%）、70%甲基硫菌灵可湿性粉剂500倍液（65.61%）和80%代森锰锌可湿性粉剂500倍液（64.60%）三个处理的防效，差异均达显著水平；而250g/L嘧菌酯悬浮剂1 500倍液、70%甲基硫菌灵可湿性粉剂500倍液和80%代森锰锌可湿性粉剂500倍液三个处理防效相当，差异不显著。

第三次药后7 d，各处理的防效从高至低依次为43%戊唑醇悬浮剂4 000倍液、50%咪鲜胺盐锰可湿性粉剂1 500倍液、10%苯醚甲环唑水分散粒剂1 000倍液、40%多·福·溴菌腈可湿性粉剂1 500倍液、250g/L嘧菌酯悬浮剂1 500倍液、70%甲基硫菌灵可湿性粉剂500倍液和80%代森锰锌可湿性粉剂500倍液。经方差分析，43%戊唑醇悬浮剂4 000倍液、50%咪鲜胺盐锰可湿性粉剂1 500倍液、10%苯醚甲环唑水分散粒剂1 000倍液和40%多·福·溴菌腈可湿性粉剂1 500倍液和250g/L嘧菌酯悬浮剂1 500倍液5个处理彼此间防效相当，差异不显著，但都优于70%甲基托布津可湿性粉剂500倍液和80%代森锰锌可湿性粉剂500倍液处理的防效，差异达极显著水平。而70%甲基硫菌灵可湿性粉剂500倍液和80%代森锰锌可湿性粉剂500倍液处理的防效相当，差异不显著。

表　7 种药剂防治黄瓜炭疽病试验结果

处理	药前病指	第三次药前		第三次药后 7 d	
		病指	防效（%）	病指	防效（%）
70% 甲基硫菌灵可湿性粉剂 500 倍液	2.56	4.69	65.61 de C	6.96	69.93 d C
50% 咪鲜胺锰盐可湿性粉剂 1 500倍液	2.81	4.28	71.48 ab AB	5.30	79.30 ab A
10% 苯醚甲环唑水分散粒剂 1 000 倍液	2.09	3.22	71.05 ab AB	4.04	78.72 ab A
250g/L 嘧菌酯悬浮剂 1 500倍液	1.96	3.37	68.05 cd BC	4.54	74.95 c B
80% 代森锰锌可湿性粉剂 500 倍液	3.17	5.98	64.60 e C	9.09	68.43 d C
40% 多・福・溴菌腈可湿性粉剂 1 500倍液	2.80	4.41	70.29 bc AB	5.72	77.45 b AB
43% 戊唑醇悬浮剂 4 000 倍液	2.59	3.65	73.66 a A	4.70	80.02 a A
CK（清水）	2.74	14.61		24.85	

注：小写字母 $P<0.05$，大写字母 $P<0.01$。

3　结论与讨论

整个试验过程中各药剂处理对黄瓜安全，并且均有一定的防治效果。在发病时期，可选用43% 戊唑醇悬浮剂、50% 咪鲜胺锰盐可湿性粉剂、10% 苯醚甲环唑水分散粒剂和 40% 多・福・溴菌腈可湿性粉剂喷雾防治，并建议轮换使用，达到减缓病原菌抗药性的产生，也能更保证田间的防治效果。结合生产成本，在预防阶段，也可使用 70% 甲基硫菌灵可湿性粉剂和 80% 代森锰锌可湿性粉剂。由于各地区的病因复杂，用药水平有差异，生产中可结合实际，在病害的防控上进行更科学的选择。

参考文献

[1] 姚红燕，赵健，谌江华. 可替代咪鲜胺类防治黄瓜炭疽病的高效药剂筛选［J］. 宁波农业科技，2012，4：8－10.

[2] 陆家云. 植物病害诊断［M］. 中国农业出版社，2004：185.

[3] 韩先旭. 第 58 节葫芦科蔬菜炭疽病［M］. 中国农作物病虫害（第三版中册），2015，3：189－191.

[4] 胡敏，张强. 20% 氟硅唑・咪鲜胺水乳剂防治黄瓜炭疽病［J］. 农药，2006，45（1）：58，64－65.

[5] 鲁建华，罗源华，刘建宇，等. 50% 咪鲜胺可湿性粉剂防治黄瓜炭疽病田间药效试验［J］. 湖南农业科学，2008（6）：74－75.

[6] 魏明山，彭玉祥，吴崇文，等. 不同药剂防治黄瓜炭疽病药效试验［J］. 上海蔬菜，2011（6）：63－64.

[7] 王丽颖，刘君丽，孙芹，等. 唑菌酯与苯醚甲环唑混剂对瓜类炭疽病的防治 [J]. 农药，2015，54（7）：530－532.

[8] 岳瑾，董杰，乔岩，等. 几种生物药剂对黄瓜炭疽病、细菌性角斑病的防治效果研究 [J]. 北京农业，2015，11：52－54.

[9] 许筱，施艳，高书锋，等. 拮抗乳酸菌的筛选及其对黄瓜炭疽病的防治效果 [J]. 河南农业科学，2012，41（5）：87－91.

[10] 农业部农药检定所生测室. 中华人民共各国国家标准：农药田间药效试验准则（二）[M]. 北京：中国标准出版社，2004：349－353.

29%吡唑萘菌胺·嘧菌酯悬浮剂对西瓜白粉病的田间防治效果

张艳华，刘　明，亓文德，王吉强，郭志刚，詹苏文
（先正达（中国）投资有限公司保定基地，保定　071000）

摘要：29%吡唑萘菌胺·嘧菌酯悬浮剂（商品名绿妃），由先正达公司生产，由SDHI类杀菌剂吡唑萘菌胺和甲氧基丙烯酸酯类嘧菌酯复配而成，对瓜类白粉病有特效，目前已登记在黄瓜白粉病上。本文探索了29%吡唑萘菌胺·嘧菌酯悬浮剂在西瓜白粉病上的防效。河北省保定试验结果：使用300mL/hm²、600mL/hm²、900mL/hm²防治西瓜白粉病，施药间隔7~10d，最后一次药后18天防效分别为88.7%，88.7%和85.6%，显著优于对照药剂30%醚菌酯·啶酰菌胺悬浮剂。辽宁省沈阳试验结果：使用29%吡唑萘菌胺·嘧菌酯悬浮剂450mL/hm²、675mL/hm²防治对西瓜白粉病，间隔7~10d用药，最后一次药后15d防效为81.4%、86.8%，并且可以延缓植株衰老，增产10%以上，提高商品果率，显著优于对照药剂30%醚菌酯·啶酰菌胺悬浮剂。山东省潍坊试验结果：使用29%吡唑萘菌胺·嘧菌酯悬浮剂450mL/hm²和675mL/hm²防治西瓜白粉病，间隔7~10d施药连续施药3次，最后一次药后26d对西瓜白粉的防效达到了95.21%和94.35%，与对照药剂50%氟吡菌酰胺·肟菌酯悬浮剂防效相当，显著优于当地常规对照药剂30%醚菌酯·15%大黄素甲醚·10%多抗霉素悬浮剂。

关键词：吡唑萘菌胺·嘧菌酯；西瓜；白粉病；防效

Efficacy of 29% Isopyrazam·Azoxystrobin SC Against Powdery Mildew of Watermelon in Field

Zhang Yanhua, Liu Ming, Qi Wende, Wang Jiqiang, Guo Zhigang, Zhan Suwen
(*Syngenta (China) Investment Company Limited*, *Baoding hub*, *Hebei*, 071000)

Abstract: Reflect Xtra, 29% isopyrazam·azoxystrobin SC, has been registered to control powdery mildew on cucumber in China. Isopyrazam is one of the SDHI (succinate dehydrogenase inhibitors) fungicides which has new chemical structure, developed by Syngenta. This study explored the efficacy of 29% isopyrazam·azoxystrobin SC against powdery mildew on watermelon. One trial in Baoding, Hebei province results showed that the efficacy of 29% isopyrazam·azoxystrobin SC at the rate of 300mL/hm², 600mL/hm² and 900mL/hm² were 88.7%, 88.7% and 85.6% respectively after 18days of the last application, which were better than 30% kresoxim-methyl·boscalid. The spraying interval was 7~10 days. One trial in Shengyang, Liaoning province results showed that the efficacy of 29% isopyrazam·azoxystrobin SC at the rate of 450mL/hm² and 675mL/hm² were 81.4% and 86.8% respectively after 15days of the last application, which were better than30% kresoxim-methyl·boscalid. Moreover, 29% isopyrazam-azoxystrobin SC can delay plant senescence, increase yield by 10%, and improve the rate of marketable fruits. Another trial in Weifang, Shandong province results showed that the efficacy of 29% isopyrazam·azoxystrobin SC at the rate of 450mL/hm² and 675mL/hm² were 95.21% and 94.35%, which were similar with the efficacy of 50% fluopyram·trifloxystrobin SC, but obviously better than local standard.

Key words: Isopyrazam·azoxystrobin; watermelon; powdery mildew; control

白粉病是西瓜的常见病害，全国各个西瓜产区均有分布。病原为苍耳叉丝单囊壳（*Sphaerotheca cucurbitae*）和菊科高氏白粉菌（*Erysipe cichoracearum*）。从全国看，该病在南方春西瓜和秋西瓜上均可发生，以秋西瓜发生较重。在北方，露地西瓜在夏季发病，温室及塑料大棚等保护地西瓜全年都可发病。此病在西瓜整个生育期都可染病，但以生长中后期发生重。发病严重时，叶片枯黄，植株干枯，产量和质量显著下降[1]。

此病主要为害叶片，其次是叶柄和茎，一般不为害果实。发病初期叶面或叶背产生白色近圆形星状小粉点，以叶面居多，当环境条件适宜时，粉斑迅速扩大，连接成片，成为边缘不明显的大片白粉区，上面布满白色粉末状霉（即病菌的菌丝体、分生孢子梗和分生孢子），严重时整叶布满白粉。叶柄和茎上的白粉较少。病害逐渐由老叶向新叶蔓延。发病后期，白色霉层因菌丝老熟变为灰色，病叶枯黄、卷缩，一般不脱落。当环境条件不利于病菌繁殖或寄主衰老时，病斑上出现成堆的黄褐色的小粒点，后变黑（即病菌的闭囊壳）[2-3]。

由于长期以来连续使用多菌灵、三唑酮、百菌清等药剂，在西瓜上形成了一定的抗药性，甚至无效[4-5]。因此，研究和开发新型杀菌剂防治西瓜白粉病具有重要的现实意义。吡唑萘菌胺是具有新型化学结构的吡唑菌胺类化合物，它是通过抑制琥珀酸脱氢酶（SDH），阻止细胞的有氧呼吸和电子传递，从而抑制病原菌的生长或杀死病原菌[6]。本试验使用29%吡唑萘菌胺·嘧菌酯悬浮剂于2013年和2015年在西瓜上进行了白粉病防治的药效试验，为该产品在将来的大面积应用和推广提供科学根据。

1 材料与方法

1.1 供试药剂

29%吡唑萘菌胺·嘧菌酯悬浮剂，瑞士先正达作物保护有限公司生产；30%醚菌酯·啶酰菌胺悬浮剂，德国巴斯夫股份有限公司产品；50%氟吡菌酰胺·肟菌酯悬浮剂，拜耳作物科学公司生产；30%醚菌酯·15%大黄素甲醚·10%多抗霉素悬浮剂，北京农利丰生物科技有限公司生产。

1.2 供试作物

京欣2号，万冠4号和早春红玉。

1.3 防治对象

西瓜白粉病。

1.4 试验点情况

试验点A：河北省保定市徐水县于坊村，春季露地西瓜，品种为京欣2号，种植密度850株/亩；2013年2月13日播种，3月19日移栽，早春覆小拱膜。土壤为沙壤土，灌溉方便，田间各小区作物长势均匀，肥水管理一致。

试验点B：辽宁省沈阳新民市柳河镇小朱屯村，拱棚栽培，品种为万冠4号，2013年7月12日育苗土壤为沙壤土，采用滴灌，人工防除杂草，虫害各试验小区统一用药，作物长势均匀，肥水管理一致。

试验点C：山东省潍坊市昌乐县宝城街道尧西村，拱棚种植，品种为早春红玉，种植密度550株/亩；2015年1月3日播种，2月15日移栽。施药时用的第二茬果，第六雌花。田间各小区作物长势均匀，肥水管理一致。

1.5 试验设计与安排

试验点A：试验共设5个处理，3次重复，试验小区按完全随机区组排列；小区面积为

16m^2，在西瓜白粉发病前，分3次预防施药，间隔7~10d，时间分别为2013年5月31日、6月8日和6月15日，用水量为675L/hm^2（表1）。

表1　试验A供试药剂处理设计

供试药剂	供试剂量	
	mL/hm^2	g a. i. /hm^2
29%吡唑萘菌胺·嘧菌酯悬浮剂	300	97.5
29%吡唑萘菌胺·嘧菌酯悬浮剂	600	195
29%吡唑萘菌胺·嘧菌酯悬浮剂	900	293
30%醚菌酯·啶酰菌胺悬浮剂	675	202.5
清水对照		

试验点B：试验共设4个处理，3次重复，试验小区按完全随机区组排列；小区面积为12m^2，第一次用药2013年8月20日，第二次用药于8月27日，第三次用药于9月3日。喷药均在天气晴朗，微风情况下用药。喷雾器械采用新加坡生产的“利农牌”DH400型可控压手动喷雾器，用水量750L/hm^2，全株均匀喷施（表2）。

表2　试验B供试药剂处理设计

供试药剂	供试剂量	
	mL/hm^2	g a. i. /hm^2
29%吡唑萘菌胺·嘧菌酯悬浮剂	450	146
29%吡唑萘菌胺·嘧菌酯悬浮剂	675	219
30%醚菌酯·啶酰菌胺悬浮剂	675	202.5
清水对照		

试验点C：此试验在试验点A后两年进行，根据之前的试验结果，对于试验药剂的用量进行了调正。试验共设7个处理，3次重复，试验小区按完全随机区组排列；小区面积为31.6 m^2，在西瓜白粉发病初期，分3次施药，间隔7d，时间分别为2015年5月8日、5月15日和5月22日，用水量为675L/hm^2（表3）。

表3　试验B供试药剂处理设计

供试药剂	供试剂量	
	mL/hm^2	g a. i. /hm^2
29%吡唑萘菌胺·嘧菌酯悬浮剂	450	146
29%吡唑萘菌胺·嘧菌酯悬浮剂	675	219
25%吡唑萘菌胺悬浮剂·苯醚甲环唑	450	113
25%吡唑萘菌胺悬浮剂·苯醚甲环唑	675	169
50%氟吡菌酰胺·肟菌酯悬浮剂	450	225

（续表）

供试药剂	供试剂量	
	mL/hm²	g a. i. /hm²
30%醚菌酯·15%大黄素甲醚·10%多抗霉素悬浮剂	675	371.25
清水对照		

1.6 试验调查

试验调查主要调查药效和安全性两个指标。药效调查为每次施药前和最后一次药后7、11、15（18）及26d，对整个小区的病害严重度进行目测打分，并计算防治效果。安全性调查在每次药后3、7d，与对照区西瓜比较，以百分率计，0%为无药害发生，100%为植株死亡。

2 结果与分析

2.1 安全性

在整个试验过程中，安全性调查是至关重要的一个环节，每次药后3、7d，对比处理区与空白对照区的生长状况，尤其关注长势、花、生长点、新叶及果实，三个试验各剂量对不同西瓜品种均未发现药害。说明29%吡唑萘菌胺·嘧菌酯悬浮剂对西瓜有非常好的安全性。

2.2 防治效果

2.2.1 试验点A防治效果

试验结果表明（表4），29%吡唑萘菌胺·嘧菌酯悬浮剂对西瓜白粉病有非常好的防效，并显著优于对照药剂30%醚菌酯·啶酰菌胺悬浮剂。29%吡唑萘菌胺·嘧菌酯悬浮剂300mL/hm²、600mL/hm²和900mL/hm²对西瓜白粉病连续3次施药后11d调查发现，防效均为100%；第三次药后18d调查发现，防效效果分别为88.74%、88.74%和85.55%，均优于对照农药30%醚菌酯·啶酰菌胺悬浮剂处理区的防效43.71%。由于29%吡唑萘菌胺·嘧菌酯悬浮剂三个剂量的防治效果没有显著差异，所以有必要降低其用量。

表4 吡唑萘菌胺混剂对西瓜白粉病的防效（河北2013年）

供试药剂	供试剂量		防治效果（%）			
	mL/hm²	g a. i. /hm²	8DAA1	7DAA2	11DAA3	18DAA3
29%吡唑萘菌胺·嘧菌酯悬浮剂	300	97.5	100	100	100	88.74
29%吡唑萘菌胺·嘧菌酯悬浮剂	600	195	100	100	100	88.74
29%吡唑萘菌胺·嘧菌酯悬浮剂	900	292.5	100	100	100	85.55
30%醚菌酯·啶酰菌胺悬浮剂	675	202.5	100	85.84	62.41	43.71
清水对照（严重度%）			0	13.3	23.3	53.3

注：8DAA1表示第一次药后8d，11DAA3表示第三次药后11d

2.2.2　试验点B防治效果

2.2.2.1　防治效果

试验结果表明（表5），29%吡唑萘菌胺·嘧菌酯悬浮剂对西瓜白粉病有很好的防效，并且明显优于对照30%醚菌酯·啶酰菌胺悬浮剂。最后一次施药后15d，在450mL/hm^2和675mL/hm^2剂量下，29%吡唑萘菌胺·嘧菌酯悬浮剂对于西瓜植株下部叶片的防效为54.7%和69.3%，对上部叶片（新叶）的防效为81.4%和86.8%，明显优于30%醚菌酯·啶酰菌胺悬浮剂在675mL/hm^2剂量下的防效（49.2%和55.2%）。可见，29%吡唑萘菌胺·嘧菌酯悬浮剂对西瓜白粉病具有保护和治疗作用，其中以保护作用为主，田间表现一定的治疗活性。

表5　29%吡唑萘菌胺·嘧菌酯悬浮剂对西瓜白粉病防效（辽宁2013年）

供试药剂	供试剂量		防治效果（%）							
	mL/hm^2	g a.i./hm^2	下部叶片				上部叶片			
			7DAA1	7DAA2	7DAA3	15DAA3	7DAA1	7DAA2	7DAA3	15DAA3
29%吡唑萘菌胺·嘧菌酯SC	450	146	74.1	71.3	64.5	54.7	88.4	81.8	85.7	81.4
29%吡唑萘菌胺·嘧菌酯SC	675	219	92.2	84.0	75.3	69.3	100	86.4	92.1	86.8
30%醚菌酯·啶酰菌胺SC	675	202.5	54.8	61.6	54.5	49.2	47.8	66.2	64.2	55.2
CK			33.2	66.4	77.8	91.6	13.8	40.6	60.4	70.6

2.2.2.2　增产效果

西瓜成熟后，将每个小区的所有瓜采摘进行测产。测量结果表明（表6），用药剂处理过的小区较未处理过的小区均有10%以上的增产效果。29%吡唑萘菌胺·嘧菌酯悬浮剂在450mL/hm^2和675mL/hm^2处理剂量下，较未处理的空白小区产量分别增加11.9%和23.6%，而对照药剂30%醚菌酯·啶酰菌胺悬浮剂的处理则增产16.9%。

表6　西瓜白粉试验测产结果（辽宁2013年）

供试药剂	供试剂量		总产量（kg/36m^2）	增产率（%）
	mL/hm^2	g a.i./hm^2		
29%吡唑萘菌胺·嘧菌酯SC	450	146	174.6	11.9
29%吡唑萘菌胺·嘧菌酯SC	675	219	192.8	23.6
30%醚菌酯·啶酰菌胺SC	675	202.5	182.3	16.9
CK			156	

试验发现，经29%吡唑萘菌胺·嘧菌酯悬浮剂处理的小区植株长势更旺盛，田间持绿性更好，一方面是该药剂能有效控制白粉病的发生，使得植株健康生长，相对空白对照产量

增加明显。另外，吡唑萘菌胺为琥珀酸脱氢酶抑制剂（SDHI），具有光调节作用，嘧菌酯可以调节作物内部生长环境，两者均有延缓衰老、提高抗逆能力，也是能明显增产的原因之一[6]。

2.2.3　试验点 C 防治效果

试验结果表明（表7），29% 吡唑萘菌胺·嘧菌酯悬浮剂 450mL/hm^2 和 675mL/hm^2 在最后一次药后 26d 对西瓜白粉的防效仍然能够达到 91.3% 和 95.21%；25% 吡唑萘菌胺·苯醚甲环唑悬浮剂 450mL/hm^2 和 675mL/hm^2 在最后一次药后 26d 对西瓜白粉的防效也达到了 95.21% 和 94.35%；另外对照药剂 50% 氟吡菌酰胺·肟菌酯悬浮剂对西瓜白粉病的防效与吡唑萘菌胺防效相当；但当地常规对照 30% 醚菌酯·15% 大黄素甲醚·10% 多抗霉素在最后一次药后 26d 防效只有 63.05%，明显低于吡唑萘菌胺混剂。

表 7　吡唑萘菌胺混剂对西瓜白粉病的防效（山东 2015 年）

供试药剂	供试剂量		防治效果（%）				
	mL/hm^2	g a. i. /hm^2	7DAA1	7DAA2	11DAA3	18DAA3	26DAA3
29% 吡唑萘菌胺·嘧菌酯悬浮剂	450	146	91.43	98.20	100.00	97.61	91.30
29% 吡唑萘菌胺·嘧菌酯悬浮剂	675	219	97.17	100.00	100.00	99.04	95.21
25% 吡唑萘菌胺·苯醚甲环唑悬浮剂	450	113	94.26	96.34	100.00	100.00	95.21
25% 吡唑萘菌胺·苯醚甲环唑悬浮剂	675	169	82.86	98.20	100.00	100.00	94.35
50% 氟吡菌酰胺·肟菌酯悬浮剂	450	225	80.03	96.34	100.00	99.04	95.21
30% 醚菌酯·15% 大黄素甲醚·10% 多抗霉素悬浮剂	675	371.25	85.69	83.63	82.30	89.04	63.05
清水对照（严重度%）			11.67	18.33	43.33	70	76.67

3　小结与讨论

田间试验结果表明，吡唑萘菌胺是防治西瓜白粉病的优秀药剂，在发病初期或发病前施药，持效期可达 15 ~ 26d，对西瓜安全。29% 吡唑萘菌胺·嘧菌酯防治西瓜白粉病，推荐在发病初期叶面喷施，剂量为 450 ~ 675mL/hm^2，喷液量 675mL/hm^2，施药间隔期建议 7 ~ 14d。由于 SDHI 类药剂作用位点单一，比较容易产生抗药性[7]，建议西瓜整个生育期喷施最多 3 次，最好与其他不同作用机理药物交替使用。具体使用方法，以取得西瓜上的登记后产品的标签为准。

参考文献

[1]　徐暄，孙其文，王永强. 几种新型药剂对西瓜白粉病的防治效果研究 [J]. 安徽农学通报，2015 (15)：83 - 84.

[2]　张存松，霍治邦，刘宏，等. 瓜类白粉病的发生与防治 [J]. 西北园艺·蔬菜专刊，2005，

4：33.

[3] 康晨辉，马兴华，黄冬梅，等.大棚西瓜白粉病综合防治技术［J］. 长江蔬菜，2013（1）：47－48.

[4] 梁萍，韦广天.大棚甜瓜、西瓜白粉病防治试验［J］. 广西农学报，2002（5）：6－7.

[5] 韩欢欢，马韬，谢冰.瓜类蔬菜白粉病抗性诱导及其抗病机制研究进展［J］. 中国农学通报，2012，28（25）：124－128.

[6] Harp T L，Godwin J R，Scalliet G，et al. Isopyrazam，a new generation cereal fungicide［J］. Aspects of Applied Biology［J］. 2011（106）：113－120.

[7] 王勇，段亚冰，王建新，等.琥珀酸脱氢酶类杀菌剂的研究进展［C］//中国植物病害化学防治研究（第九卷）. 北京：中国农业科学技术出版社，2014：20－25.

400g/L 氟吡菌酰胺·戊唑醇 SC 防治西瓜蔓枯病药效评价*

霍建飞**，郝永娟***，姚玉荣，刘春艳，王万立
（天津市植物保护研究所，天津 300381）

摘要：为明确 400g/L 氟吡菌酰胺·戊唑醇 SC 对西瓜蔓枯病的防治效果，对其进行 2 年田间药效试验，结果表明 400g/L 氟吡菌酰胺·戊唑醇 SC 对西瓜蔓枯病有较好的防治效果：2012 年试验其 100 g a. i. /hm^2、150g a. i. /hm^2、200 g a. i. /hm^2 处理的防效分别为 78. 39%、81. 58% 和 82. 36%，2013 年试验其 100 g a. i. /hm^2、150 g a. i. /hm^2、200 g a. i. /hm^2 处理的防效分别为 77. 43%、80. 57% 和 82. 74%。综合两年试验结果，400g/L 氟吡菌酰胺·戊唑醇防治西瓜蔓枯病效果较好，是防治西瓜蔓枯病的一种较好的复配杀菌剂。

关键词：400g/L 氟吡菌酰胺·戊唑醇 SC；西瓜蔓枯病；药效评价

The Efficacy Evaluation of 400g/L Fluopyram · Tebuconazole SC on Gummy stem blight of watermelon

Huo Jianfei, Hao Yongjuan, Yao Yurong, Liu Chunyan, Wang Wanli
(*Tianjin Plant Protection Institute*, *Tianjin* 300381)

Abstract: To identify the effect of fluopyram · tebuconazole suspension concentrate to control gummy stem blight of watermelon, the concentration of 400g/L were done in the field two years. The results showed that fluopyram · tebuconazole had a good effect to gummy stem blight of watermelon, the control effect of 100 g a. i. /hm^2、150 g a. i. /hm^2、200g a. i. /hm^2 was 78. 39%、81. 58% and 82. 36% respectively after the cucumber was treated in 2012, and the control effect of 100 g a. i. /hm^2、150 g a. i. /hm^2、200 g a. i. /hm^2 was 77. 43%、80. 57% and 82. 74% respectively after the cucumber was treated in 2013. Comprehensive two years of research results showed that fluopyram · tebuconazole had a good effect against gummy stem blight of watermelon. So it is a good compound fungicide against gummy stem blight of watermelon.

Key words: 400g/L fluopyram · tebuconazole SC; gummy stem blight of watermelon; efficacy evaluation

西瓜蔓枯病又称为黑腐病、褐斑病，是西瓜上一种重要的土传病害，在西瓜的各个生育阶段均可发生。其病原为瓜类黑腐球壳菌（*Didymella bryoniae*），可侵染多种葫芦科作物，如黄瓜，甜瓜、西葫芦等[1]。随着日光温室和塑料大棚等保护设施的应用，国内西瓜、甜瓜因重茬种植频繁，致使西瓜蔓枯病发生与危害逐年加重[2]。据统计，在西瓜上一般田块发病率为 15% ~25%，严重时高达 60% ~80%[3]。目前国内生产上防治西瓜蔓枯病的主要方法为化学防治，据报道，西瓜蔓枯病菌已对嘧菌酯[4]、甲基硫菌灵、苯菌灵[5]和多菌灵[6]产生抗药性。随着化学药剂的不断使用，西瓜蔓枯病的抗药性也会愈加严重。

氟吡菌酰胺（fluopyram）是一种新型吡啶基乙基苯甲酰胺类杀菌剂[7]，用于防治多种

---

* 基金项目：天津市科技支撑项目（13ZCZDNC00100）

** 作者简介：霍建飞，男，河北承德人，硕士，助理研究员，主要从事蔬菜病害研究；E-mail：hjf2203@163. com

*** 通讯作者：郝永娟，研究员，主要从事蔬菜病害防治研究及杀菌剂应用技术研究；E-mail：tjzbshyj@163. com

作物上的病害，通过阻碍呼吸链中琥珀酸脱氢酶的电子转移而抑制线粒体呼吸[8]，其中对白粉病菌和灰霉病菌引起的病害防效显著[9-10]，目前国内外尚未发现植物病原真菌对氟吡菌酰胺产生抗性[11]。戊唑醇属于三唑类杀菌剂，其主要作用机制是抑制细胞的麦角甾醇的生物合成，具有广谱、保护、治疗及内吸作用，由于其作用位点多，病原菌抗药性的产生较缓慢。氟吡菌酰胺与戊唑醇混配使之增加了作用位点和作用途径，一段时期内病原菌的简单变异不足以适应全部作用位点。本试验进行复配药剂氟吡菌酰胺·戊唑醇 SC 的田间药效试验，以筛选出防治西瓜蔓枯病的有效剂量，为氟吡菌酰胺·戊唑醇 SC 防治西瓜蔓枯病提供科学数据，并为其在田间大面积推广使用提供理论依据。

1 材料与方法

1.1 供试药剂

400g/L 氟吡菌酰胺·戊唑醇（fluopyram·tebuconazole）SC、500g/L 氟吡菌酰胺（fluopyram）SC 和 430g/L 戊唑醇（tebuconazole）SC 均为拜耳作物科学公司产品，10% 苯醚甲环唑（difenoconazole）WG 为瑞士先正达作物保护有限公司产品。

1.2 供试材料

试验作物为西瓜，两年试验品种均为秋丰 2 号，试验对象为由 *Mycosphaerella melonis* 侵染引起的西瓜蔓枯病。

1.3 试验设计

1.3.1 环境条件

2012 年试验在天津市武清区南蔡镇大张庄露地西瓜上进行，2013 年试验在天津市西青区杨柳青镇大柳滩村露地西瓜上进行，两块试验地土壤均为沙壤土，肥力较好，排灌条件较好，西瓜直播时间分别为 2012 年 5 月 11 日和 2013 年 5 月 15 日，密度分别为 11 250株/hm^2 和 10 500株/hm^2，覆盖地膜。两块试验地其他管理基本一致。试验开始时处于西瓜果实膨大期，试验期间未施其他药剂。

1.3.2 试验处理、施药方法及时间

试验共设 7 个处理，分别为：400g/L 氟吡菌酰胺·戊唑醇 SC 100g a. i. /hm^2，400g/L 氟吡菌酰胺·戊唑醇 SC 150 g a. i. /hm^2，400g/L 氟吡菌酰胺·戊唑醇 SC 200 g a. i. /hm^2，500g/L 氟吡菌酰胺 SC 75 g a. i. /hm^2，430g/L 戊唑醇 SC 75 g a. i. /hm^2，250g/L 嘧菌酯 SC 225 g a. i. /hm^2，清水对照等 7 个处理，小区面积 16m^3，4 次重复，各小区随机区组排列。

在西瓜蔓枯病发病初期开始施药，配制药液时，先用小水量溶解药剂，然后按 1 125L/hm^2 的药液量对植株茎蔓、叶柄和叶片正反两面均匀喷雾。空白对照喷等量清水。

施药时间：2012 年试验施药时间为 7 月 10 日、7 月 17 日；2013 年试验施药时间为 7 月 8 日、7 月 15 日。

1.3.3 试验调查和统计

第 1 次施药前（2012 年试验时间为 7 月 10 日、2013 年试验时间为 7 月 8 日）调查发病率和病情指数。此后调查于第 2 次药后 7d（分别为 2012 年 7 月 24 日和 2013 年 7 月 22 日）进行。每小区随机 4 点取样，每点调查 2 株，每株调查全部叶片。每片叶按病斑面积占叶表面积的百分率，并根据以下标准分级：

0 级：没有叶片发病；

1 级：植株发病叶片占总叶片数 10% 以下；

3 级：植株发病叶片占总叶片数的 10% ~25%；
5 级：植株发病叶片占总叶片数的 25% ~50%；
7 级：植株发病叶片占总叶片数的 50% ~75%；
9 级：植株发病叶片占总叶片数的 75% ~100%。

根据调查叶片的病级计算病情指数及防效。

$$病情指数 = \Sigma（发病叶数 \times 相应病级数）\times 100/（调查总叶数 \times 9）$$

$$防治效果（\%）= [1 - CK_0 \times pt_1/（CK_1 \times pt_0）] \times 100$$

CK_0：空白对照区药前病情指数，CK_1：空白对照区药后病情指数；
pt_0：药剂处理区药前病情指数，pt_1：药剂处理区药后病情指数。

1.4 数据分析

采用 Duncan's 新复极差法对试验数据进行统计分析和比较。

2 结果与分析

2.1 不同处理防效间比较

由表可知，2012 年用药 2 次后 7d 调查，400g/L 氟吡菌酰胺 · 戊唑醇 SC100g/hm^2、150g/hm^2 和 200g/hm^2 的防效分别为 78.39%、81.58% 和 82.36%。对照药剂 500g/L 氟吡菌酰胺 75g/hm^2 的防效为 81.76%，430g/L 戊唑醇 SC 75g/hm^2 的防效为 83.71%，250g/L 嘧菌酯 SC 225g/hm^2 的防效为 75.33%。

表 400g/L 氟吡菌酰胺 · 戊唑醇 SC 防治西瓜蔓枯病田间药效试验结果

处理（有效成分）	2012 年试验结果			2013 年试验结果		
	基数病指	药后病指	防效（5%）	基数病指	药后病指	防效（5%）
400g/L 氟吡菌酰胺 · 戊唑醇 SC 100g/hm^2	2.51	3.39	78.39bc	1.62	3.59	77.43ab
400g/L 氟吡菌酰胺 · 戊唑醇 SC 150g/hm^2	2.70	3.11	81.58ab	1.53	3.12	80.57ab
400g/L 氟吡菌酰胺 · 戊唑醇 SC 200g/hm^2	2.79	3.21	82.36ab	1.74	3.09	82.74a
500g/L 氟吡菌酰胺 SC 75g/hm^2	2.83	3.30	81.76ab	1.70	3.50	80.87ab
430g/L 戊唑醇 SC 75g/hm^2	2.53	2.56	83.71a	1.62	2.97	83.11a
250g/L 嘧菌酯 SC225g/hm^2	3.07	4.80	75.33c	1.55	4.08	75.55b
空白对照	2.26	14.43	—	1.67	18.14	—

注：表中同列不同小写字母表示在 5% 水平上差异显著（Duncan's 新复极差法）。

方差分析和多重比较结果表明，试验药剂 400g/L 氟吡菌酰胺 · 戊唑醇 SC 高、中剂量的防效均显著高于对照药剂 250g/L 嘧菌酯 SC 225g/hm^2 的防效；与对照药剂 500g/L 氟吡菌酰胺 75g/hm^2 和 430g/L 戊唑醇 SC 75g/hm^2 的防效无显著差异。试验药剂 400g/L 氟吡菌酰胺 · 肟菌酯三个剂量之间均无显著差异。

2013 年用药 2 次后 7d 调查，400g/L 氟吡菌酰胺 · 戊唑醇 100g/hm^2、150g/hm^2 和 200g/hm^2 的防效分别为 77.43%、80.57% 和 82.74%。对照药剂 500g/L 氟吡菌酰胺 75g/hm^2 的防效为 80.87%，430g/L 戊唑醇 SC 75g/hm^2 的防效为 83.11%，250g/L 嘧菌酯 SC 225g/hm^2 防效为 75.55%。

方差分析和多重比较结果表明，试验药剂 400g/L 氟吡菌酰胺·戊唑醇高剂量的防效显著高于对照药剂 250g/L 嘧菌酯 SC 225g/hm^2 的防效，与对照药剂 500g/L 氟吡菌酰胺 SC 75g/hm^2 和 430g/L 戊唑醇 SC75g/hm^2 的防效无显著差异。试验药剂中、低剂量的防效与三个对照药剂的防效均无显著差异。试验药剂 400g/L 氟吡菌酰胺·戊唑醇 SC 三个剂量的防效间无显著差异。

2.2 对西瓜及其他生物的影响

两年试验期间观察，试验药剂 400g/L 氟吡菌酰胺·戊唑醇 SC 各剂量处理对西瓜植株均无药害等不良影响，也未发现对其他非靶标生物有影响。

3 结论与讨论

目前在抗西瓜蔓枯病品种缺乏的情况下，选择适当杀菌剂并进行科学复配是防治西瓜蔓枯病的重要措施。为延缓抗性形成，通常采用多位点、具有不同作用机理杀菌剂进行混配。本试验开展了 400g/L 氟吡菌酰胺·戊唑醇 SC 对西瓜蔓枯病的田间药效试验，两年试验结果表明 400g/L 氟吡菌酰胺·戊唑醇 SC 对西瓜蔓枯病有较好的防治效果，在 2 次药后 7d 各剂量处理防效亦在 77.43% 以上，3 个剂量处理对黄瓜植株无药害等不良影响，两年试验结果基本一致。建议 400g/L 氟吡菌酰胺·戊唑醇 SC 防治西瓜蔓枯病，在发病初期使用，推广剂量（有效成分）为 100～200g/hm^2，生产应用时每个生长季节不宜超过 3 次，且应与其他类型药剂交替使用。

参考文献

[1] Li P F, Ren R S, Yao X F, et al. Identification and characterization of the causual agent of gummy stem blight from muskmelon and watermelon in east China [J]. Journal of Phytopathology, 2015, 163 (4): 314 – 319.

[2] 郑雪松，茹李军，张智能，等. 烯肟菌胺与苯醚甲环唑对西瓜蔓枯病菌的联合毒力及防效 [J]. 农学学报，2016，6（5）：28 – 32.

[3] 李雨，王少秋，谭蕊，等. 西瓜蔓枯病有效药剂筛选及药效评价 [J]. 农药，2016，55（6）：460 – 462.

[4] Stevenson K L, Langston D B, Seebold K W. Resistance to Azoxystrobin in the Gummy Stem Blight Pathogen Documented in Georgia [J]. Plant Health Progress, 2004 (204): 1 – 8.

[5] Keinath A P, Zitter T A. Resistance to Benomyl and Thiophanate – methyl in *Didymella bryoniae* from South Carolina and New York [J]. Plant Disease, 1998, 82 (5): 479 – 484.

[6] Liu S T, Yang Y H, Li Y, et al. Baseline Sensitivity and Resistance Monitoring of *Didymella bryoniae* to Carbendazim [C] //中国植物病理学会 2015 年学术年会. 北京：中国农业科学技术出版社，2015：52.

[7] 严智燕. 拜耳公司介绍新型杀菌剂 fluopyram [J]. 农药应用与研究，2009，13（5）：46.

[8] 胡进，吴进龙，陈铁春. 氟吡菌酰胺 500g/L 悬浮剂的气相色谱分析方法研究 [J]. 农药科学与管理，2008，29（11）：19 – 21.

[9] Klages U. New Fungicidal Active Lngredient Fluopyram Strengthens Peoduct Portfolio [EB/QL]. [2010 – 03 – 10].

[10] Fought L, Musson G H, Bloomberg J R, et al. Fluopyrama New Active Ingredient from Bayer Crop Science [J]. Phytopathology, 2009, 99: S36.

[11] 李良孔. 黄瓜白粉病菌对氟吡菌酰胺敏感基线的建立及其抗药性风险评估 [D]. 吉林：吉林大学，2011.

杀菌剂防治石斛炭疽病药物筛选初步试验

和理淮[1]，张巧玲[2]，潘毅铖[3]，斯那达吉[4]，何玉琼[5]
（1. 丽江市古城区农业技术推广中心，丽江 674100；2. 丽江市古城区植保站，丽江 674100；3. 丽江市古城区农村合作经济经营管理站，丽江 674100；4. 迪庆州德钦县奔子栏镇农技站，迪庆 674400；5. 丽江市古城区园艺站，丽江 674100）

摘要：在石斛炭疽病发病初期，选择农药福星乳油（40%氟硅唑 EC）、富力库水分散粒剂（43%戊唑醇 WG）、扑海因可湿性粉剂（50%异菌脲 WP）喷雾，隔 10d 防治 2 次；第二次喷药后 10d 调查。结果表明，用福星防治效果最好，达 71.28%，对石斛安全，可在生产中推广应用。
关键词：石斛炭疽病；化学防治

Abstract: At the early stage of Dendrobium anthracnose disease, 40% flusilazole EC, 43% tebuconazole WG, 50% iprodione WP were sprayed twice with the interval of 10 days, and efficacy of the fungicides were investigation 10 days after the second spraying. The results showed that efficacy (71.28%) of Fuxing was best and was safe to Dendrobium. Therefore, the fungicide could be applied to control the disease in future.
Key words: Dendrobium anthracnose; Chemical control

石斛叶斑类病害主要是石斛炭疽病，属真菌病害，夏秋季高温高湿时，管理粗放的棚室，或因种植过密，通风不良，植株瘦弱者易发生。石斛炭疽病最早报道在广西地区野生驯化的铁皮石斛和广东华南植物园的石斛兰上发生[1]，病原菌有性阶段为小丛 *Glomerella cingulata*（Stonem）Spaulding et Schlrenk，无性阶段为胶孢炭疽菌 *Colletotrichum gloeosporides* Penz。发病适温 22～28℃，相对湿度 95%以上。以分生孢子，一般通过伤口侵染。但分生孢子附在健康的兰叶上，即使侵入也不发病，而是潜伏在组织内，当寒害、药害、日灼、肥力不足等情况发生时，潜伏菌丝开始活动并引起病害。主要为害叶片，初时在叶面上出现若干浅黄色、黑褐色或深灰色的小区。内有许多黑色斑点，有时聚生成若干带，当黑色病斑发展时，周围组织变成黄色或灰绿色，下陷。严重时可导致整株死亡[2]。基地当前大范围使用的药剂为 50%异菌脲，仅有保护作用，防治效果不好，特别是病害高发时，需要试验推广应用更好的药剂。为了让种植户和企业少走弯路，减少化学农药的使用量、农药残留。课题组在前期室内毒力测定的基础上，筛选低毒、低残留的化学农药进行田间试验。

1 材料与方法

1.1 试验地概况

试验地选在丽江市古城区开南街道漾西村委会王家庄村，丽江金水中药材种植销售有限公司，基地面积 70 亩，有 13 个连栋大棚温室。海拔 2 400m，年平均降雨量 1 100mm，年平均气温 11℃。地势平坦，有节水喷灌设施和灌溉条件。为高墒基质苗床栽培，基质肥力高，有机质含量 60%以上，pH 值 5.5 左右，种植品种为铁皮石斛。组培苗移栽期为 2014 年 5

月，2015 年 11 月留 5～8cm 剪茎，翌年 4 月起萌发新枝。基地按石斛大棚标准化栽培技术措施进行田间管理。

1.2 试验设计与方法

1.2.1 试验设计

试验设 4 个处理，3 次重复，3 个空白对照。共 12 个小区，随机区组排列，周围设置保护行。小区面积 $16m^2$。茎叶喷雾，共喷药 2 次，即石斛炭疽病发病初期喷药，间隔 10d 第二次喷药（表 1）。

表 1 杀菌剂使用情况

处理	药剂名称	稀释浓度	处理次数	有效成分	生产厂商
A	福星 EC	8 000	2	40% 氟硅唑	天津久日化学工业有限公司
B	富力库 WG	3 000	2	43% 戊唑醇	拜耳作物科学（中国）有限公司
C	扑海因 WP	1 000	2	50% 异菌脲	富美实中国投资有限公司
清水（CK）					

1.2.2 调查方法

药前及第二次喷药后 10d 各查 1 次病情。每小区随机五点取样，每点定点调查 5 株，每株调查 10 片叶，记录总叶数及各级病叶数。记载发病率和发病程度情况。病情分级文献进行[3]，计算病情指数和相对防效，数据统计软件采用 SPSS。病情指数调查用五点取样法：在发病棚（田）选取五点，每点 5 株，每株 10 张叶片，分级记载发病情况（表 2）。

病情指数 = Σ（各级病叶数 × 各级代表值）/（调查总叶数 × 最高级代表值）×100

表 2 叶部病害病情指数分级标准

病级	发病程度	代表数值	病株
0	无病	0	
1	病斑面积占叶面积的 1% 以下	1	
3	病斑面积占叶面积的 2% ～5%	3	
5	病斑面积占叶面积的 6% ～10%	5	
7	病斑面积占叶面积的 11% ～20%	7	
9	病斑面积占叶面积 21% 以上至全叶枯死	9	

防治效果（%）=（对照病情指数 - 处理病情指数）/对照病情指数 ×100

1.2.3 定点调查

每片区定五点，定点、定时（10d）调查病害的发病率及病情指数，同时记录温度、湿度、光照。

1.3 试验实施

2016 年 5 月 1 日第一次喷药，石斛剪了一次，开春后重新发芽。新叶上的叶斑为初发生期，斑点散而小，未形成大病斑。按试验设计用药量，采用二次稀释法配药，使用农成牌 16 型电动喷雾器喷雾。

2 结果与分析

2.1 药害观察

喷药后 15d 观察，如未发生药害，即可证明供试药剂对石斛生产安全。

2.2 防治效果

从表 3 可见，40% 氟硅唑防治效果最好，达 71.28%，可在石斛栽培中推广使用，还有望兼治根腐病。43% 戊唑醇 WG 和目前大范围使用的 50% 异菌脲 WP 的差异不显著。在 $P_{0.05}$ 水平，40% 氟硅唑和目前大范围使用的 50% 异菌脲具有显著差异。

表 3 石斛炭疽病药效防治试验结果

处理	药前病情指数	药后病情指数	防治效果（%）	5% 显著性
40% 氟硅唑 EC	10.93	10.02	71.28	b
43% 戊唑醇 WG	10.13	16.29	53.29	ab
50% 异菌脲 WP	10.64	18.93	45.72	a
清水（CK）	9.46	34.88	0	

说明：用 Duncan 法统计，表中最后一列不同字母表示为 $P<0.05$ 差异显著水平

3 讨论

除了夏季高温高湿季节外，铁皮石斛受蛞蝓、红蜘蛛、蜗牛等害虫为害形成许多伤口，因此试验地技术人员需要及时打药防控炭疽病的传染[4]。40% 氟硅唑的最佳用药倍数、对疫病根腐的防治效果和是否有比它更好的生物农药等药剂等问题有待于进一步试验研究。种植户在合理栽培的基础上，应优先选用生物农药，必须用化学农药防治时优先选用低毒农药，以确保石斛高产优质[5]。

特别要强调的是，药物防治要与其他农艺措施、物理措施等组成综合治理方案。首先，做好铁皮石斛栽培“三透”（透风、透气、透水）管理，创造不利于病虫害滋生和暴发的环境；其次，加强观测做好病虫害预测预报工作，在病害发生初期、虫害幼龄期使用药物。这两个条件是所选药物发挥作用的重要前提[6]。

参考文献

[1] 李戈，李荣英，高微微，等. 药用植物石斛规模化种植中的病害问题及防治策略 [J]. 中国中药杂志，2013，38（4）：485－487.

[2] 曾宋君，刘东明. 石斛兰的主要病害其防治 [J]. 中药材，2003，26（7）：471－474.

[3] 农业部农药检定所生测室. 农药田间药效试验准则（一）[M]. 北京：中国标准出版社，2000，9：107－110.

[4] 姜泽海，黄志，王力前. 铁皮石斛规模化种植技术 [J]. 热带农业工程，2013，37（3）：9－12.

[5] 胡永亮，白学慧，李桂琳，等. 石斛锈病病原初步鉴定及其防治药剂筛选 [J]. 热带农业科学，2013，33（10）：53－55.

[6] 卢振辉，李明炎，王伟杰，等. 铁皮石斛主要病虫害及其非化学农药防治 [J]. 浙江农业科学，2016，57（1）：123－126.

苯醚甲环唑与甲基硫菌灵混配对梨黑星病菌的联合毒力及田间防效*

赵建江**，王文桥，马志强，韩秀英***

（河北省农林科学院植物保护研究所，河北省农业有害生物综合防治工程技术研究中心，农业部华北北部作物有害生物综合治理重点实验室，保定 071000）

摘要：为了明确苯醚甲环唑与甲基硫菌灵混配对梨黑星病菌的联合毒力，采用菌丝生长速率法测定了苯醚甲环唑、甲基硫菌灵及其不同配比对梨黑星病菌的毒力，以 Wadley 公式进行评价，并通过田间试验验证其对梨黑星病的防效。结果表明：苯醚甲环唑与甲基硫菌灵质量比为 1∶10、1∶15、1∶20、1∶25 和 1∶30 进行混配对菌丝生长均表现出很强的抑制作用，当二者以 1∶30 混配时，对梨黑星病菌毒力增效作用最明显，增效系为 3.19。在田间药效试验中，10% 苯醚甲环唑水分散粒剂与 70% 甲基硫菌灵可湿性粉剂，以质量比 1∶30 进行桶混，在 340～440mg/kg 的浓度下对梨黑星病的防治效果均在 80% 以上，与其单剂 10% 苯醚甲环唑水分散粒剂 17mg/kg 对梨黑星病的防治效果相当。苯醚甲环唑与甲基硫菌灵以 1∶30 进行桶混，可以在田间推广使用。

关键词：苯醚甲环唑；甲基硫菌灵；梨黑星病菌；增效作用；田间防效

Joint-toxicity and field efficacy of mixtures of difenoconazole and thiophanate-methyl against *Venturia nashicola* *

Zhao Jianjiang**, Wang Wenqiao, Ma Zhijiang, Han Xiuying***

(*Plant Protection Institute*, *Hebei Academy of Agricultural and Forestry Sciences*, *IPM Center of Hebei Province*, *Key Laboratory of Integrated Pest Management on Crops in Northern Region of North China*, *Ministry of Agriculture*, *Baoding* 071000, *China*)

Abstract: In order to clarify the synergistic interaction of difenoconazole and thiophanate-methyl against *Venturia nashicola*, the toxicity of difenoconazole, thiophanate-methyl and their mixtures at different ratios against *V. nashicola* was determined by mycelial growth rate test, the joint-toxicity of the two fungicides was assessed with Wadley formula, and the efficacy in controlling pear scab was assessed through the field trials. The results showed that the mixtures of 1∶10, 1∶15, 1∶20, 1∶25 and 1∶30 exhibited obvious inhibiting effect against the mycelial growth of *V. nashicola*. The synergistic effects of the mixture (1∶30) were most obvious, with synergistic ratio of 3.19. The results of field trials on controlling efficacy showed that the mixture of difenoconazole with thiophanate-methyl (1∶30, W/W) at the concentration of 340～440mg/kg exhibited a control efficacy of greater than 80%, which showed the same efficacy to difenoconazole 100 EC at the concentration of 17mg/kg. The mixture of difenoconazole with thiophanate-methyl (1∶30, W/W) can be widely used in the field.

---

* 基金项目：国家重点研发计划项目（果树）（2016YFD0200505）

** 第一作者：赵建江，男，助理研究员，主要从事杀菌剂抗药性及其应用技术研究；E-mail：chillgess@163.com

*** 通讯作者：韩秀英，女，研究员，主要从事杀菌剂抗药性及其应用技术研究；E-mail：xiuyinghan@163.com

Key words: Difenoconazole; Thiophanate-methyl; *Venturia nashicola*; Synergistic interaction; Field efficacy

梨黑星病（*Venturia nashicola* Tanaka et Yamamoto）是我国梨区普遍发生的重要病害。从落花到梨果采收期间均可发病，如防治不当，将会严重影响梨的品质和产量[1]。目前，通过遗传改良和培育抗病品种增强梨树对黑星病的抗性已取得一定进展[2-3]，但在生产上，仍以化学防治为主。

苯醚甲环唑是一种高效、广谱的麦角甾醇生物合成抑制剂，对梨黑星病具有良好的防治效果[4]，是防治该病害的主要药剂。河北省农科院植保所杀菌剂课题组在研究中发现，从河北辛集市采集的梨黑星病菌对苯醚甲环唑的敏感性呈下降趋势。农药的合理复配，不但可以提高防效，扩大杀菌谱，而且还可以延缓病菌抗药性的产生[5-6]。如氟硅唑与代森锰锌1∶25（质量比）混配对防治梨黑星病具有增效作用[7]。甲基硫菌灵是一种广谱内吸低毒杀菌剂，对多种病害具有预防和治疗作用，与多种杀菌剂混配具有增效作用[8]。为了明确苯醚甲环唑与甲基硫菌灵混配对梨黑星病菌是否具有毒力增效作用，笔者进行了苯醚甲环唑与甲基硫菌灵混合物对梨黑星病菌联合毒力的测定，并进行了该混合物对梨黑星病的田间药效评价。

1 材料与方法

1.1 供试培养基及菌株

马铃薯梨块培养基（PDA + L′）：马铃薯200g，鸭梨150g，葡萄糖20g，琼脂20g，蒸馏水1 000mL，pH值为6.0～6.2。

梨黑星病菌（*Venturia nashicola*）采自河北省辛集市垒头镇北小陈村，采用单孢分离法进行分离和纯化，分离纯化后将菌株转接到“PDA + L′”平板上[9]，于20℃保存备用。

1.2 供试药剂

95%苯醚甲环唑原药，江苏耕耘化工有限公司生产提供；95%甲基硫菌灵原药，山东华阳农药化工集团有限公司生产提供；40%氟硅唑乳油美国杜邦公司生产提供；70%甲基硫菌灵可湿性粉剂，山东省青岛瀚生生物科技股份有限公司生产提供；10%苯醚甲环唑水分散粒剂，瑞士先正达作物保护有限公司提供。

1.3 室内毒力测定

采用菌丝生长速率法[7]测定供试药剂对梨黑星病菌的抑制作用。分别用适量丙酮将苯醚甲环唑和甲基硫菌灵原药溶解，然后配制成质量浓度为1×10^4μg/mL的母液，并按苯醚甲环唑和甲基硫菌灵的不同质量比例（1∶10、1∶15、1∶20、1∶25和1∶30）混合，得到不同配比混合液，用无菌水进行系列稀释，按药液与培养基1∶9的体积比配制成最终含药量为100、50、10、5、1μg/mL的含药平板。单剂苯醚甲环唑和甲基硫菌灵含药平板的含药量分别为5、1、0.5、0.1、0.05和500、100、50、25、10、5μg/mL。同时设丙酮和清水对照。用直径5mm的打孔器在预培养40 d的菌落边缘打取菌饼，正面朝下接种到含药平板上，每处理重复3次，置于20℃培养箱中，黑暗培养40 d后，采用十字交叉法测量各处理的菌落直径。采用DPS7.05软件，求出苯醚甲环唑、甲基硫菌灵及其混合物的毒力回归方程，EC_{50}值和相关系数。试验重复2次。采用Wadley法[6]进行联合毒力评价。

EC_{50}（th）=（a + b）/［a/（EC（A）$_{50}$ + b/EC（B）$_{50}$］；增效系数（SR）= EC_{50}（th）/EC_{50}（ob），式中：A、B分别代表两种药剂在混剂中所占比例，ob为实际观察值，

th 为理论值。SR > 1.5 为增效作用；SR = 0.5～1.5 为相加作用；SR < 0.5 为拮抗作用。

1.4 田间药效试验

田间试验分别于 2012 年和 2013 年，在河北省辛集市垒头镇北小陈村同一梨园进行。梨树品种为鸭梨，树龄 20 年，株行距为 3m×6m，梨园土壤肥力中等，梨树长势及管理水平一致，黑星病历年发生。试验设 10% 苯醚甲环唑水分散粒剂与 70% 甲基硫菌灵可湿性粉剂的混合物（有效成分质量比 1：30）340、390 和 440mg/kg、10% 苯醚甲环唑水分散粒剂 17mg/kg、70% 甲基硫菌灵可湿性粉剂 700mg/kg、40% 氟硅唑乳油 50mg/kg 和空白对照共计 7 个处理。每处理 4 次重复，共 28 个小区，小区随机区组排列，每小区 2 株梨树。采用泰山－18 型机动高压喷雾器均匀喷施，至流失为止，株施药液 5 L 左右。梨树落花后、病害发生前用第一次药，以后根据病情发展和气候条件用药。2012 年用药时间为 5 月 2 日、5 月 18 日、6 月 5 日、7 月 2 日、7 月 19 日、8 月 10 日和 8 月 29 日，2013 年用药时间为 4 月 29 日、5 月 14 日、6 月 1 日、6 月 27 日、7 月 20 日、8 月 5 日和 8 月 25 日。

施药前的病情基数为零，最后一次用药后 14 d 调查病情。每小区调查 2 株，分东、南、西、北、中方向 5 点取样，每点取当年生枝条的 40 片叶，按病叶上病斑面积占整个叶片面积的百分率分级[10]，计算病叶率、病情指数及防治效果。数据采用 Fisher's LSD 法进行差异显著性分析。

$$病叶率（\%）=（病叶数/调查总叶片数）\times 100$$

$$病情指数 = [\sum（病叶数\times 相对级数）/（调查总叶片数\times 最高级数）]\times 100$$

2 结果与分析

2.1 苯醚甲环唑与甲基硫菌灵混配对梨黑星病菌菌丝生长的抑制作用

从表 1 可知，苯醚甲环唑与甲基硫菌灵按质量比 1：10、1：15、1：20、1：25、1：30 进行混配对梨黑星病菌的菌丝生长均表现出良好的抑制作用，当苯醚甲环唑与甲基硫菌灵以 1：15 和 1：30 进行混配时，对梨黑星病菌显示出毒力增效作用，增效系分别为 1.93 和 3.19。

Table 1 Inhibitory action of mixtures（difenoconazole + hiophanate-methyl）against mycelial growth of *V. nashicola*

Fungicide	Massage ratio	EC_{50}（μg/mL）		Synergy ratio		Biological response
		Range	Mean	Range	Mean	
Difenoconazole	—	0.131～0.134	0.133			
Hiophanate-methyl	—	257.294～309.964	283.629			
Difenoconazole + Hiophanate-methyl	1：10	1.248～2.243	1.745	0.65～1.18	0.91	additive
	1：15	0.928～1.361	1.144	1.56～2.29	1.93	synergistic
	1：20	2.482～2.566	2.524	1.09～1.12	1.10	additive
	1：25	3.313～3.580	3.447	0.96～1.04	1.00	additive
	1：30	1.232～1.341	1.286	3.06～3.33	3.19	synergistic

2.2 田间防治效果

从表2可知，10%苯醚甲环唑水分散粒剂与70%甲基硫菌灵可湿性粉剂，以有效成分质量比1：30进行桶混，对梨黑星病表现出良好的防治效果。该混合物在340～440mg/kg的使用浓度下，对梨黑星病的防治效果均在80%以上，与其单剂10%苯醚甲环唑水分散粒剂17mg/kg，对梨黑星病的防治效果（90.1%）没有显著差异；该混合物在390～440mg/kg的浓度下对梨黑星病的防治效果（89.8%～96.4%）与对照药剂40%氟硅唑乳油50mg/kg对梨黑星病的防治效果（94.6%）没有显著差异，但显著高于对照药剂70%甲基硫菌灵可湿性粉剂700mg/kg对梨黑星病的防治效果（72.3%）；该混合物在340mg/kg的浓度下对梨黑星病的防治效果（80.9%）显著低于对照药剂40%氟硅唑乳油50mg/kg对梨黑星病的防治效果，但与对照药剂70%甲基硫菌灵可湿性粉剂700mg/kg对梨黑星病的防治效果没有显著差异。2013年与2012年得到了相似的试验结果。

Table 2 Efficacy of different fungicides in controlling pear scab

Fungicide	Duluted time	2012			2013		
		Rate of diseased leaves (%)	Disease index	Controlling efficacy (%)	Rate of diseased leaves (%)	Disease index	Controlling efficacy (%)
Difenoconazole 100 WG + Thiophanate-methyl 700 WP (w/w 1：30)	440	2.7	0.6	96.4 a	4.1	1.4	92.5 a
	390	6.3	1.5	89.8 ab	7.9	2.6	86.6 ab
	340	11.3	3.0	80.9 bc	11.8	3.6	81.2 bc
Flusilazole 400 EC	50	3.8	0.8	94.6 a	5.2	1.8	90.7 a
Difenoconazole 100 WG	17	5.2	1.6	90.1 ab	5.3	2.0	89.7 ab
Thiophanate-methyl 700 WP	700	15.2	4.3	72.3 c	16.4	5.3	73.3 c
CK	—	36.0	15.8		34.7	19.2	

Note：Date in the same row followed by the same letter indicate no significant different（at the level of $P<0.05$）according to Fisher's LSD test.

3 结论与讨论

本研究结果表明，苯醚甲环唑与甲基硫菌灵以质量比1：30混配后对梨黑星病菌的菌丝生长表现出良好的毒力增效作用。该混合物340～440mg/kg对梨黑星病的田间防治效果与10%苯醚甲环唑水分散粒剂17mg/kg的防效相当。因此，苯醚甲环唑与甲基硫菌灵以质量比1：30进行桶混，可以在田间推广使用。

苯醚甲环唑属于麦角甾醇合成抑制剂，具有杀菌谱广、防效显著等优点而被广泛用于病原真菌病害的防治。但由于该类药剂的长期广泛使用，致使病原菌对该类药剂中多种药剂产生了抗性[11-12]。本研究结果表明，苯醚甲环唑与甲基硫菌灵以质量1：30桶混，在最高使用浓度440mg/kg时，对梨黑星病的防效高达96.4%，与其单剂10%苯醚甲环唑WG 17mg/kg的防效相当，但10%苯醚甲环唑WG在桶混使用中的浓度为14.19mg/kg，有效降低了其使用浓度。因此，苯醚甲环唑与甲基硫菌灵以1：30混配使用，可以有效的降低苯醚甲环唑的使用量，从而降低其对病原菌的药剂选择压，进而延缓梨黑星病菌对苯醚甲环唑抗性的产生。

苯醚甲环唑通过抑制麦角甾醇的合成，进而导致病原菌的细胞膜结构破坏和细胞死亡[13]。甲基硫菌灵是经植物体代谢为多菌灵后起作用，作用机制是通过阻碍病菌细胞有丝分裂而达到抑菌目的[14]。这两种作用机理截然不同的杀菌剂混配后增效作用明显，但是其增效机理还有待于进一步研究。

参考文献

[1] Sun A Q, Hu J C. Study on occurrence and control of pear scab [J]. Journal of Anhui Agricultural Sciences（安徽农业科学）, 2010, 38 (1): 76 - 77, 133.

[2] Wang J Z, LI X P. A new late chinese pear cultivar with long keeping quality [J]. Journal of Fruit Science（果树学报）, 2004, 21 (4): 391 - 392.

[3] Liu Z C, Liao M A, Bao D E. On resistance to black star disease of Jinhua pear variation strains [J]. Journal of Northwest Forestry University（西北林学院学报）, 2006, 21 (2): 108 - 109.

[4] Wu Y X, Li M N, Zhou Z S, et al. Controlling test of eight fungicides against pear scab [J]. China Fruits（中国果树）, 2008, (4): 37 - 40.

[5] Zhu W G, Hu W Q, Chen D H, et al. Co-toxicity of propiconazol mixed difenoconazole to *Rhizoctonia solani* [J]. Agrochemicals（农药）, 2008, 47 (5): 365 - 366.

[6] Fan Z Y, Wang W Q, Meng R J, et al. Joint-toxicity of mixtures of pyraclostrobin with difenoconazole against *Alternaria solani* and effect of their synergistic mixture on potato yield [J]. Chinese Journal of Pesticide Science（农药学学报）, 2011, 13 (6): 591 - 596.

[7] Zhao J J, Zhang X F, Wang W Q, et al. Joint-toxicity and field efficacy of mixtures of flusilazole and mancozeb against Venturia nashicola [J]. Plant Protection（植物保护）, 2014, 40 (3): 195 - 198.

[8] Geng Z Y, Zhao J L, Sun G B, et al. Studies on the synergistic effects of myclobutanil and thiophanate-methyl to three main pathogens of apple [J]. Chinese Agricultural Science Bulletin（中国农学通报）, 2010, 26 (18): 297 - 300.

[9] Niu X Y, Li L Q, Zhao H, et al. Isolation and medium selection of *Venturia prina* [J]. Journal of Northwest Sci-Tech University of Agriculture and Forestry（西北农林科技大学学报（自然科学版））, 2003, 31 (5): 80 - 82.

[10] Zhao X J, Zhou J B, Li X, et al. Field control test of fungicides to *Venturia pirina* in Shanxi [J]. Journal of Shanxi Agricultural Sciences（山西农业科学）, 2008, 36 (7): 71 - 73.

[11] Kunz S, Deising H, Mendgen K. Acquisition of resistance to sterol demethylation inhibitors by populations of *Venturia inaequalis* [J]. Phytopathology, 1997, 87 (12): 1 272 - 1 278.

[12] Kwon S M, Yeo M I, Choi S H, et al. Reduced sensitivities of the pear scab fungus (*Venturia nashicola*) collected in ulsan and naju to five ergosterol-biosynthesis-inhibiting fungicides [J]. Research in Plant Disease, 2010, 16 (1): 48 - 58.

[13] Ye T, Ma Z Q, Bi Q Y, et al. Research advances on the resistance of plant pathogenic fungi to SBIs fungicides [J]. Chinese Journal of Pesticide Science（农药学学报）, 2012, 14 (1): 1 - 16.

[14] Fan J Y, Fang Y L, Guo L Y. Sensitivity of *Monilinia fructicola* isolates to thiophanate-methyl and boscalid [J]. Acta Phytophylacica Sinica（植物保护学报）, 36 (3): 251 - 256.

基于天然抑菌活性倍半萜 Drimenol 的酯类化合物与生物活性探索

李挡挡，张莎莎，李圣坤
（南京农业大学植物保护学院农药系，南京 210095）

摘要：利用廉价易得的天然产物香紫苏醇完成了手性倍半萜类天然产物（-）-drimenal 和（-）-drimenol 的合成，采用菌丝生长速率法，测定了目标化合物在 50μg/mL 下对水稻纹枯病菌（*R. solani*）；油菜菌核病菌（*S. scleotiorum*）；小麦赤霉病菌（*F. graminearum*）；番茄灰霉病菌（*B. cirerea.*）；小麦全蚀病菌（*G. graminis*）；水稻恶苗病菌（*F. fujikuroi*）；马铃薯干腐病菌（*F. sulphureum*）；辣椒疫霉病菌（*P. capsici*）；黄瓜炭疽病菌（*C. lagenarium*）等 9 种常见病原菌的抑制活性。结果表明合成的手性倍半萜类天然产物对供试病原菌均表现出不错的抑制活性，且 drimenal 的活性要优于 drimenol。在手性倍半萜类天然产物 Drimenol 结构的基础上，设计合成了一系列酯类共计 14 个新的衍生物（4a ~ 4n），所有化合物结构均通过 ^{1}H-NMR、^{13}C-NMR 和 ESI-MS 确认，并且测定了酯类衍生物在 100μg/mL 下对水稻纹枯病菌（*R. solani*）；油菜菌核病菌（*S. scleotiorum*）；小麦赤霉病菌（*F. graminearum*）和番茄灰霉病菌（*B. cirerea.*）4 种常见病原菌的抑制活性。初步抑菌实验结果表明，大多数衍生物对上述四种菌都有一定的抑制活性，其中化合物 4l 对番茄灰霉病菌（*Botrytis cinerea*）的抑制率为 56.87%，明显优于其他衍生物。基于 drimane 骨架的倍半萜有待进一步优化，以探索其为新型杀菌剂的潜力。

关键词：倍半萜；杀菌剂；新农药；抑菌活性

Synthesis and Fungicidal Activity of the Esters of Natural Drimenol

Li Dangdang, Zhang Shasha, Li Shengkun*
(*Department of Pesticide, College of Plant Protection, Nanjing Agricultural University, Nanjing* 210095, *Jiangsu*)

Abstract: The chiral drimane sesquiterpenoids, drimenol and drimenal, were synthesized with the readily available natural product (-)-Sclareol as starting material. The fungicidal activities of the drimenol and drimenal against *R. solani*, *S. scleotiorum*, *F. graminearum*, *B. cirerea.*, *G. graminis*, *F. fujikuroi*, *F. sulphureum*, *P. capsici* and *C. lagenarium* were investigated through mycelium growth rate. The results demonstrated that the synthesized chiral drimanes showed good activities and drimenal possessed potential advantage. A series of novel ester substrates (4a ~ 4n) were designed and synthesized based on the lead compound (-)-drimenol. All the structures were characterized by ^{1}HNMR, ^{13}CNMR and ESI-MS. The fungicidal activities of the ester derivatives against *R. solani*, *S. scleotiorum*, *F. graminearum* and *B. cirerea* were investigated through mycelium growth rate. Preliminary antifungal activity in vitro showed that the synthesized chiral drimanes showed good activity and drimenal possessed potential advantage, the majority of the derivatives possessed low fungicidal activity against four phytopathogenic fungi above at 100μg/mL. Notably, compound 4l showed good fungicidal activity. It exhibited the inhibition rate of 56.87% to *Botrytis cinerea*, better obviously than others, which indicating that chiral drimane sesquiterpenoid may be optimized as a fungicidal lead compound, and be worthy of further study.

Key words: sesquiterpenoid; fungicide; novel pesticide; fungicidal activity

天然产物是新农药创制的重要资源[1]，以天然产物为先导化合物进行优化，在农药创制研究中仍将是一种有效的方法。倍半萜是分子中含 15 个碳原子的天然萜类化合物，其中，Drimane 骨架的倍半萜在高等植物，真菌，海洋微生物以及苔藓类生物中均有分布。近年来在医药领域发现 Drimane 类型的化合物具有抗癌活性，抑制胆固醇蛋白酶活性及抗细菌抗真菌活性；另外还可以作为名贵的香料等；在农药领域也有杀虫活性、拒食活性以及植物生长调节作用[2]。Drimane 倍半萜类化合物在农用杀菌活性方面的报道仍然比较少。2004 年 Becker 等从地钱类植物鞭苔 *Bazzania trilobata*（L.）*S. Gray* 中分离得到（－）－drimenal 及其类似物（－）－drimenol，生物活性筛选表明这两个倍半萜具有显著的抑菌活性，（－）－drimenal 对马铃薯晚疫病（*P. infestans*）的 IC_{50} 小于 0.03μg/mL[3]。为了提高这类天然产物的生物活性，以及解决其含量低、不稳定而难于进行进一步的生物活性评价这些问题，进行有效的化学合成及进一步的结构优化就显得尤为重要。

本文基于香紫苏醇完成了手性倍半萜类天然产物（－）－drimenal 和（－）－drimenol（结构见图 1）。对（－）－drimenol 的结构进行了进一步优化，设计合成了一系列新的酯类衍生物，并且对其抑菌活性进行了初步研究。

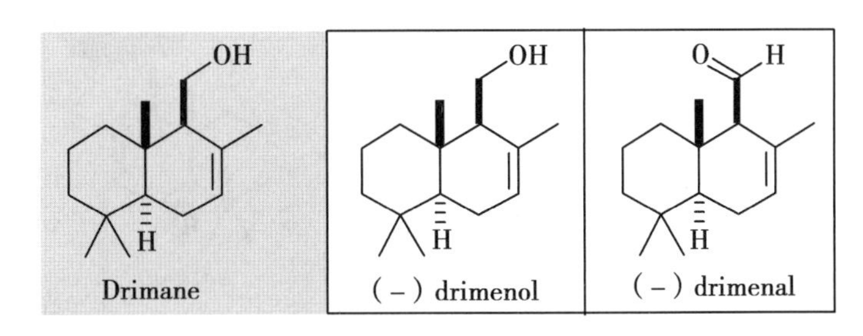

图 1 Drimane 类倍半萜（－）－drimenol 和（－）－drimenal 的结构

1 材料与方法

1.1 仪器与试剂

手提式两用紫外分析仪；X－4 数字显示显微熔点测定仪（未校正）；MP200A 型电子天平；IKA RV8V 旋转仪；SHZ－D（Ⅲ）循环水式真空泵；LCQ Advantage MAX 质谱仪；Bruker Avance 400/500 MHz 核磁共振仪（TMS 为内标）；KQ－250 型超声波清洗器；LS－B50L 型立式压力蒸汽灭菌器；AIRTECH 型超净工作台；SPX 型智能生化培养箱等；柱层析硅胶（200～300 目），薄层层析硅胶 GF_{254}，其他所用试剂均为市售分析纯试剂。对照药剂为多菌灵（Carbendazim）原药，纯度 98.5%（天津市均凯化工科技有限公司）。

1.2 化合物合成

1.2.1 化合物（－）－drimenol 的合成

参考文献[4]和[5-6]的方法合成中间体（＋）－Sclareolide 和甲基酮（化合物 1），Bayer－Villager 反应合成化合物 2，具体步骤如下。

将 50.0mL 的 H_2O_2（30% in H_2O，457 mmol）逐滴加入到乙酸酐（50.0 mL，529 mmol）的二氯甲烷（70 mL）溶液中，控制滴加速度，使体系温度不超过 5℃。滴加完毕后，在 5℃ 下磁力搅拌 1h。在不超过 8℃ 的条件下向体系中分批加入马来酸酐（30.0 g，306 mmol），加完后在此温度下磁力搅拌 1h，然后自然升至室温。向反应体系中逐滴加入甲基酮

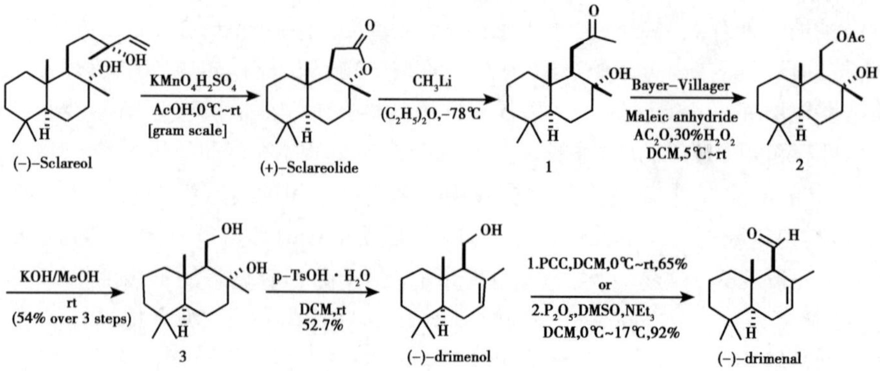

图 2　目标化合物 Drimenol 和 Drimenal 的合成

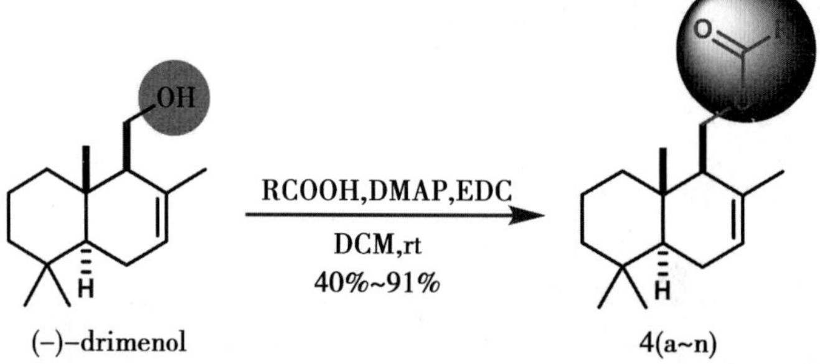

图 3　目标化合物 4a～4n 的合成路线

a：R = benzene；b：R = 2 – fluorophenyl；c：R = 2 – iodinephenyl；d：R = 3 – chlorphenyl；e：R = 4 – bromophenyl；f：R = 4 – nitrophenyl；g：R = 4 – methylphenyl；h：R = 4 – trifluoromethylphenyl；i：R = 3 – methoxyphenyl；j：R = 4 – fluorophenyl；k：R = 2 – methoxylpyridyl；l：R = trifluoromethyl；m：R = 2 – bromopropyl；n：R = butyl.

（化合物 1）的二氯甲烷溶液（10.6 g，39.9 mmol），室温下搅拌反应，TLC 跟踪监测。反应完毕后，加入 60mL 二氯甲烷稀释分液出有机相。二氯甲烷相依次用水（50 mL ×2），饱和碳酸氢钠溶液（50 mL × 2）和饱和氯化钠（40 mL ×2）洗涤，无水硫酸钠干燥，减压蒸除溶剂，硅胶柱层析（$V_{石油醚}/V_{乙酸乙酯}$ = 5 ：1）后获得乙酸酯中间体 2，白色固体，产率 78%。

手性倍半萜（－）－drimenol 的合成：

称取二醇中间即体化合物 3（1.20 g，5.0 mmol，1.0equiv）溶于 20mL 二氯甲烷中，冰浴条件下，向体系中分批加入对甲苯磺酸水合物（0.95 g，5.0 mmol，1.0equiv），磁力搅拌下自然恢复到室温。TLC 跟踪监测．反应完毕后，加入 20mL 二氯甲烷稀释反应体系。有机相分别用饱和碳酸氢钠溶液（20 mL × 3）、水（20 mL × 3）和饱和食盐水（20 mL）洗涤，无水硫酸钠干燥，减压蒸除溶剂，硅胶柱层析（$V_{石油醚}/V_{乙酸乙酯}$ = 4 ：1）后获得天然产物（－）－drimenol，白色固体，产率 58%。

1.2.2 化合物（-）-drimenal 的合成

称取倍半萜（-）-drimenol（0.444g，2.0mmol，1.0equiv）于干净干燥的 50mL 茄形瓶中，加入 15mL 无水二氯甲烷溶解，冰浴条件下向其中加入用 3A 分子筛干燥过的 DMSO（1mL，14.08mmol，7.0equiv），搅拌 5 分钟，自然恢复至 15~17℃，在此温度下继续搅拌 30min，重新恢复至 0℃，加入三乙胺（1.2mL，8.5mmol，4.25equiv），自然恢复至 15~17℃，TLC 跟踪监测至反应完全。将反应体系冷却至 0℃，逐滴加入水（5mL）稀释，饱和氯化铵（5mL）酸化，再用二氯甲烷萃取（15 mL ×3），合并的有机相用饱和氯化钠溶液洗涤（10 mL ×3），无水硫酸钠干燥，减压蒸除溶剂，快速硅胶柱层析（$V_{石油醚}/V_{乙酸乙酯}$ = 15：1）后，获得天然产物（-）-drimenal，无色油状物，产率 92%。

1.2.3 化合物（4a~4n）的合成（合成路线见图 3）

以 4a 为例，称取倍半萜（-）-drimenol（0.222g，1.0mmol，1.0equiv）于干净干燥的 50mL 茄形瓶中，加入 10mL 无水二氯甲烷溶解，依次加入苯甲酸（1.2 mmol，1.2equiv），DMAP（1.5 mmol，1.5equiv），EDC（0.2 mmol，0.2equiv）于室温下搅拌，TLC 跟踪监测至反应完全。加入 20mL 二氯甲烷稀释反应体系。有机相分别用饱和碳酸氢钠溶液（20 mL ×3）、水（20 mL ×3）和饱和食盐水（20 mL）洗涤，无水硫酸钠干燥，减压蒸除溶剂，硅胶柱层析（$V_{石油醚}/V_{乙酸乙酯}$ =20：1）后获得化合物 4a，白色固体，产率 55%。

1.3 抑菌活性测定

采用菌丝生长速率法[7]。供试病原菌：水稻纹枯病菌（*R. solani*）；油菜菌核病菌（*S. scleotiorum*）；小麦赤霉病菌（*F. graminearum*）；番茄灰霉病菌（*Botrytis cirerea*）；小麦全蚀病菌（*G. graminis*）；水稻恶苗病菌（*F. fujikuroi*）；马铃薯干腐病菌（*F. sulphureum*）；辣椒疫霉病菌（*P. capsici*）；黄瓜炭疽病菌（*C. lagenarium*）等均由南京农业大学植保学院提供。

先将待测化合物用 DMSO 配制成 20 000μg/mL 的母液（阳性对照药剂多菌灵用 0.1M HCl 配制成 20 000μg/mL 的母液），在无菌条件下用 PDA 培养基将母液稀释成 100μg/mL 的含药培养基，以不含药剂的处理为空白对照，各处理重复 3 次。具体数据见表 1 和表 2。

2 结果与讨论

2.1 化合物的合成

本研究利用廉价易得的天然产物香紫苏醇（-）-Sclareol 完成了手性倍半萜类天然产物（-）-drimenal 和（-）-drimenol 的合成，反应每一步均可以在克到几十克以上的水平制备，为进一步的生物活性评价打下基础。反应过程中利用甲基锂进行选择性的亲核开环和马来酸酐介导的 Bayer-Villager 重排反应是构建 Drimane 倍半萜骨架的关键。化合物 3 的区域选择性脱水也是很值得探索的一个环节。

合成手性倍半萜（-）-drimenal 和（-）-drimenol 及其相应衍生物的结构，通过 ^{1}H-NMR、^{13}C-NMR、DEPT135 和 ESI-MS 等现代波谱技术进行确认。部分结果如下：

手性倍半萜（-）-drimenol：产率 58%，白色固体，m.p.：70.7℃。^{1}H NMR（500MHz，$CDCl_3$）δ 5.54（d，J = 4.2 Hz，1H），3.85（dd，J = 11.3，3.1 Hz，1H），3.73（dd，J = 11.3，4.9 Hz，1H），1.98（m，2H），1.91~1.82（m，2H），1.78（s，3H），1.62~1.37（m，3H），1.33（br s，1H），1.17（td，J = 12.1，4.0 Hz，2H），1.07（td，J = 13.1，3.5 Hz，1H），0.89（s，3H），0.87（s，3H），0.86（s，3H）.^{13}C NMR

（126MHz，$CDCl_3$）δ 132.89，124.13，60.94，57.28，49.89，42.14，39.89，36.07，33.37，32.92，23.58，22.06，21.96，18.76，14.94.。质谱检测显示准分子离子峰：ESI－MS（m/z）223.21［M＋H］$^+$，245.34［M＋Na］$^+$。

手性倍半萜（－）－drimenal：产率65%，无色油状物，（含有约4%的双键异构体）。^{1}H NMR（400 MHz，$CDCl_3$）δ 9.69（d，J = 5.2 Hz，1H），5.70（m，1H），2.59（s，1H），2.08（m，1H），1.97（m，1H），1.66（dd，J = 13.0，2.3 Hz，1H），1.62（brs，3H），1.57～1.50（m，1H），1.50～1.41（m，3H），1.30（dd，J = 13.2，3.2 Hz，1H），1.24（dd，J = 7.7，3.2 Hz，1H），1.20（d，J = 3.3 Hz，1H），1.15（dd，J = 11.9，4.9 Hz，1H），1.07（s，3H），0.92（s，3H），0.88（s，3H）.^{13}C NMR（101 MHz，$CDCl_3$）δ 206.74，127.79，125.49，67.59，49.05，42.00，40.37，37.01，33.31，33.03，23.66，22.09，21.63，18.29，15.73 . LC－MS：$C_{15}H_{24}O$，质谱检测显示准分子离子峰：ESI－MS（m/z）：［M＋H］$^+$ 221.32，［M＋Na］$^+$ 243.30.

酯类衍生物（4a～4n）部分结构鉴定数据：

4a：产率55%，白色固体，m. p.：106.9～107.8（℃）。^{1}H NMR（400 MHz，$CDCl_3$）δ 8.02（d，J = 7.1 Hz，2H），7.55（t，J = 7.4 Hz，1H），7.44（t，J = 7.6 Hz，2H），5.55（s，1H），4.56（dd，J = 11.7，3.4 Hz，1H），4.32（dd，J = 11.7，6.0 Hz，1H），2.18（s，1H），2.03（d，J = 12.8 Hz，2H），1.97～1.86（m，1H），1.75（s，3H），1.61～1.55（m，1H），1.54～1.48（m，1H），1.45（dd，J = 11.2，8.4 Hz，1H），1.29～1.21（m，2H），1.21～1.11（m，2H），0.91（s，6H），0.88（s，3H）.

4d：产率60%，浅黄色油状。^{1}H NMR（400 MHz，$CDCl_3$）δ 7.97（d，J = 8.4 Hz，1H），7.89（t，J = 6.4 Hz，1H），7.52（t，J = 7.5 Hz，1H），7.37（dd，J = 15.1，7.3Hz，1H），5.56（s，1H），4.55（dd，J = 11.7，3.3 Hz，1H），4.33（dd，J = 11.7，6.1 Hz，1H），2.19（s，1H），2.03（t，J = 14.9 Hz，2H），1.97～1.84（m，1H），1.73（s，3H），1.57（dd，J = 9.6，6.5 Hz，2H），1.54～1.48（m，1H），1.48～1.40（m，1H），1.29～1.21（m，2H），1.20～1.10（m，2H），0.91（s，3H），0.90（s，3H），0.88（s，3H），0.83（d，J = 4.4 Hz，1H）.

4f：产率62%，白色固体，m. p.：98.8～100.4（℃）。^{1}H NMR（400 MHz，$CDCl_3$）δ 8.28（t，J = 8.5 Hz，2H），8.21～8.14（m，2H），5.58（s，1H），4.60（dd，J = 11.7，3.4 Hz，1H），4.39（dd，J = 11.7，5.8 Hz，1H），2.20（s，1H），2.03（t，J = 16.6 Hz，2H），1.97～1.85（m，1H），1.74（s，3H），1.59（s，2H），1.55～1.49（m，1H），1.46（dd，J = 11.3，8.9 Hz，1H），1.29～1.21（m，2H），1.21～1.11（m，2H），0.91（s，3H），0.91（s，3H），0.89（s，3H），0.84（d，J = 2.4 Hz，1H）.

4g：产率58%，白色固体，m. p.：142.0～142.4（℃）。^{1}H NMR（400 MHz，$CDCl_3$）δ 7.90（t，J = 7.8 Hz，2H），7.22（t，J = 7.2 Hz，2H），5.55（s，1H），4.53（dd，J = 11.7，3.3 Hz，1H），4.30（dd，J = 11.7，6.0 Hz，1H），2.40（s，3H），2.18（s，1H），2.02（d，J = 11.6 Hz，2H），1.97～1.84（m，1H），1.74（s，3H），1.59（s，3H），1.53～1.40（m，3H），1.28～1.21（m，2H），1.21～1.10（m，2H），0.91（s，3H），0.90（s，3H），0.88（s，3H），0.83（d，J = 2.6 Hz，1H）.

4h：产率61%，浅黄色油状。^{1}H NMR（400 MHz，$CDCl_3$）δ 8.12（t，J = 7.9 Hz，2H），7.70（t，J = 8.1 Hz，2H），5.57（s，1H），4.58（dd，J = 11.7，3.4 Hz，1H），

4.37 (dd, J = 11.7, 5.9 Hz, 1H), 2.19 (s, 1H), 2.03 (t, J = 14.2 Hz, 2H), 1.91 (dd, J = 21.9, 8.1 Hz, 1H), 1.74 (s, 3H), 1.58 (s, 3H), 1.51 (ddd, J = 10.1, 5.6, 2.9 Hz, 1H), 1.48 ~ 1.40 (m, 2H), 1.30 ~ 1.21 (m, 2H), 1.20 ~ 1.11 (m, 2H), 0.91 (s, 6H), 0.89 (s, 3H), 0.84 (d, J = 2.0 Hz, 1H).

2.2 抑菌活性

本论文通过菌丝生长速率法测定了合成的倍半萜（－）－drimenal 和（－）－drimenol 在 50μg/mL 下对水稻纹枯病菌（*R. solani*）；油菜菌核病菌（*S. scleotiorum*）；小麦赤霉病菌（*F. graminearum*）；番茄灰霉病菌（*B. cirerea.*）；小麦全蚀病菌（*G. graminis*）；水稻恶苗病菌（*F. fujikuroi*）；马铃薯干腐病菌（*F. sulphureum*）；辣椒疫霉病菌（*P. capsici*）；黄瓜炭疽病菌（*C. lagenarium*）等 9 种常见病原菌的抑制活性。结果如表 1 所示，（－）－drimenal 和（－）－drimenol 对供试菌种均表现出一定的抑制活性，且（－）－drimenal 的活性优于（－）－drimenol。两者对番茄灰霉病菌、水稻纹枯病菌和油菜菌核病菌的抑制活性要优于其他病害。

选择上述活性相对较好的 4 种菌：水稻纹枯病菌（*R. solani*）；油菜菌核病菌（*S. scleotiorum*）；小麦赤霉病菌（*F. graminearum*）和番茄灰霉病菌（*B. cirerea.*）对结构进行优化的衍生物 4a ~ 4n 测定了离体抑菌活性，结果如表 2 所示，不幸的是整体活性下降，均明显低于先导化合物（－）－drimenol 和阳性对照药剂多菌灵。除化合物 4l 对番茄灰霉病菌的抑制率超过 50%，其他化合物均表现出较差的抑菌活性。可以看出，酯键官能团的引入未能明显提高先导化合物对上述病原菌的抑制率，但是仍然能够看出不同取代基的引入会对化合物的抑菌活性产生明显的差异。综上可以得出，Drimane 类倍半萜可以作为一个天然先导结构，进一步的结构优化和构效关系研究以及对映异构体毒力差异性研究是值得进一步深入探讨的。

表 1　（－）－drimenal 和（－）－drimenol 在 50μg/mL 时对 9 种病原菌的离体抑制率

Table 1　Inhibitory percentage of (－) －drimenal and (－) －drimenol against nine plant pathogens in vitro (at 50μg/mL)

化合物	R. S.	S. S.	F. G.	B. C.	G. G.	F. F.	F. S.	P. C.	C. L.
（－）－drimenol	43.98 ±1.34	40.12 ±0.76	30.37 ±1.37	55.55 ±0.49	17.19 ±2.29	41.66 ±0.70	21.4 ±1.14	42.03 ±3.12	15.57 ±1.79
（－）－drimenal	62.26 ±0.10	61.37 ±2.05	35.52 ±2.82	65.08 ±0.18	24.47 ±2.52	34.88 ±1.38	23.49 ±1.02	45.57 ±2.58	20.79 ±1.27

Note：All values are means of three replicates. R. S.：水稻纹枯病菌（*R. solani*）；S. S. 油菜菌核病菌（*S. scleotiorum*）；F. G.：小麦赤霉病菌（*F. graminearum*）；B. C.：番茄灰霉病菌（*B. cirerea.*）；G. G.：小麦全蚀病菌（*G. graminis*）；F. F.：水稻恶苗病菌（*F. fujikuroi*）；F. S.：马铃薯干腐病菌（*F. sulphureum*）；P. C.：辣椒疫霉病菌（*P. capsici*）；C. L.：黄瓜炭疽病菌（*C. lagenarium*）

表2 目标化合物在100μg/mL时对四种病原菌的离体抑制率

Table 2 Inhibitory percentage of the target compounds against four plant pathogens in vitro (at 100μg/mL)

编号	R. S.	S. S.	F. G.	B. C.
4a	33.8 ±0.7	30.19 ±0.47	10.48 ±4.25	—
4b	25.48 ±7.74	19.56 ±5.12	10.86 ±1.09	16.02 ±4.12
4c	22.9 ±3.73	25.38 ±4.99	—	—
4d	24.27 ±0.41	42.38 ±4.74	2.03 ±0.85	5.3 ±2.77
4e	17.82 ±8.87	29.18 ±1.95	9.76 ±0.88	0.02 ±2.83
4f	33.41 ±12.21	26.85 ±3.94	7.18 ±0.51	—
4g	7.82 ±13.73	19.58 ±2.62	9.03 ±2.12	10.64 ±1.13
4h	28.58 ±3.88	13.65 ±0.95	2.38 ±3.66	—
4i	16.56 ±0.76	24.93 ±4.93	9.02 ±0.32	8.01 ±0.33
4j	17.63 ±1.36	15.64 ±0.96	11.59 ±4.25	—
4k	30.37 ±0.8	25.32 ±5.6	13.08 ±0.89	0.00
4l	23.26 ±0.83	22.81 ±4.73	9.75 ±2.99	56.87 ±3.19
4m	33.34 ±4.37	27.55 ±1.04	12.71 ±1.51	1.39 ±4.03
4n	36.59 ±0.8	43.44 ±4.3	14.18 ±0.89	7.44 ±0.98
(－) －drimenol	55.2 ±0.95	53.37 ±4.21	49.71 ±0.07	43.90 ±0.66
Carbendazim	100.00	100.00	100.00	75.27 ±1.73

Note：All values are means of three replicates. R. S.：水稻纹枯病菌（*R. solani*）；S. S：. 油菜菌核病菌（*S. scleotiorum*）；F. G.：小麦赤霉病菌（*F. graminearum*）；B. C.：番茄灰霉病菌（*B. cirerea.*）；“—”代表基本无抑菌活性

参考文献

[1] 吴文君. 从天然产物到新农药创制－原理［M］. 北京：化学工业出版社，2006.

[2] Jansen B J M, Groot A D. Occurrence, biological activity and synthesis of drimane sesquiterpenoids [J]. Nat. Prod. Rep., 2004, 21: 449－477.

[3] Scher J M, Speakman J B, ZAPP J. Bioactivity guided isolation of antifungal compounds from the liverwort *Bazzania trilobata* (L.) S. F. Gray [J]. Phytochemistry, 2004, 65: 2 583－2 588.

[4] Barrero A F, Alvarez Anzaneda E J. Degradation of the Side Chain of (－) －Sclareol: A Very Short Synthesis of norAmbreinolide and Ambrox [J]. Synthetic Communications, 2011, 34 (19): 3 631－3 643.

[5] Kuchova K I, Chumakov Y M, Simonov Y A, et al. A Short Efficient Synthesis of 11－monoacetate of Drimane－8alpha, 11－diol from Norambreinolide [J]. Synthesis, 1997, 9: 1 045－1 048.

[6] Sudhakarrao V, Charles T K. Improved Synthesis of 11－Acetoxy－8α－Drimanol [J]. Organic Preparations and Procedures Int., 2008, 40 (2), 201－213.

[7] 慕立义，吴文君，王开运. 植物化学保护研究方法［M］. 北京：中国农业出版社，1991：37－82.

19.5%咯菌腈·精甲霜灵·噻菌灵悬浮种衣剂对玉米出苗及产量影响田间试验

刘 聃*，王吉强**

（先正达（中国）投资有限公司哈尔滨基地，哈尔滨 150000）

摘要：19.5%咯菌腈·精甲霜灵·噻菌灵悬浮种衣剂是先正达公司即将上市的具有3种活性成分的种衣剂。试验结果表明，在病害压力非常大的情况下，19.5%咯菌腈·精甲霜灵·噻菌灵悬浮种衣剂100mL/100kg种子、200mL/100kg种子、300mL/100kg种子处理的出苗率分别为83.7%、81.2%和84.9%，显著高于空白对照的出苗率47.4%，并能有效降低小苗数，做到保苗齐苗，增产达到20%以上。

关键词：咯菌腈·精甲霜灵·噻菌灵；镰刀菌；粉籽；玉米；出苗率；增产

Safety of 19.5% fludionil – metalaxyl – M – thiabendazole SC Seed Coating on the Control of Seedling and Yield of Corn in Field

Liu Dan，Wang Jiqiang

（*Syngenta*（*China*）*Investment Company Limited*，*Hairbin hub*，*Hairbin Heilongjiang* 150000）

Abstract：19.5% fludionil – metalaxyl – M – thiabendazole SC is a seedcare with three active ingredient，which will be launched soon. Under the heavy disease pressure，the results showed that the seedling rate of 19.5% fludionil – metalaxyl – M – thiabendazole SC Seed Coating at the rate of 100mL/100kg seed，200mL/100kg seed，300mL/100kg seed were 83.7%、81.2% and 84.9% respectively，which were better than 47.4% untreated check. Moreover，19.5% fludionil – metalaxyl – M – thiabendazole SC seed coating can decrease the number of small plants and keep seedling emergence even，increase yield by 20%.

Key words：fludionil – metalaxyl – M – thiabendazole；*Fusarium*；corn；seedling rate；increase yield

黑龙江玉米播种期为每年4月末至5月上旬[1]，在这段时间内温度较低，如果遇到长时期干旱或雨水频繁天气，很容易造成粉籽，致使种子出苗参差不齐，出苗率差[2-3]。在这种情况下，土壤中的种子极易受到大量积累的病原菌如镰刀菌和丝核菌等侵入[4]，严重影响玉米产量。

种子包衣技术是防治玉米病虫害最有效的技术之一[5]，优秀的种衣剂可以提高种子出苗率，做到保苗齐苗，最终达到增产作用，因此，研究和开发新型多元种衣剂具有重要意义。

咯菌腈是一种非内吸性苯基吡咯类杀菌剂，对子囊菌、担子菌、半知菌等病原菌有良好的防效[6]。精甲霜灵为酰苯胺类杀菌剂，具有优良的保护、治疗、铲除活性，对卵菌中的腐霉属、疫霉属和许多霜霉病菌有特效。噻菌灵为苯并咪唑类杀菌剂，为安全广谱内吸性杀

* 第一作者：刘聃，男，黑龙江哈尔滨人，主要从事种衣剂和除草剂的田间药效试验工作

** 通讯作者：王吉强，男，硕士，主要从事农药田间试验及试验员管理工作；E-mail：jiqiang.wang@syngenta.com

菌剂，对大多数子囊菌、半知菌、担子菌引起的植物病害有特效。以上三种活性成分作用机理不同，但都可用于种子处理。本试验使用19.5%咯菌腈·精甲霜灵·噻菌灵对玉米（先玉335）进行了包衣后的出苗和产量测试，为该产品在将来的大面积应用和推广提供科学根据。

1 材料与方法

1.1 供试药剂

19.5%咯菌腈·精甲霜灵·噻菌灵悬浮种衣剂，瑞士先正达作物保护有限公司生产；3.5%咯菌腈·精甲霜灵悬浮种衣剂，瑞士先正达作物保护有限公司生产；35%精甲霜灵种子处理乳剂，瑞士先正达作物保护有限公司生产；12.5%咯菌腈·精甲霜灵·嘧菌酯悬浮种衣剂，瑞士先正达作物保护有限公司生产。

1.2 供试作物

春玉米（先玉335）。

1.3 防治对象

玉米苗期病害，镰刀菌，低温，粉籽。

1.4 试验地点以及田块情况

试验设在黑龙江省道里区新农镇，土壤为黑钙土，播种均匀一致，杂草采用化学统一防除，虫害各试验小区统一用药，作物长势均匀，肥水管理一致。

1.5 试验设计和安排

供试作物于2015年4月21日播种，每个小区固定播种量，播种深度保持一致。试验设有7个处理，每个处理设4次重复，试验小区按完全随机区组排列，包衣方法为先正达SCI试验室包衣机统一包衣，试验小区面积为26.8m^2。

1.6 试验调查

药效调查为播种并齐苗后调查玉米出苗率和小苗率，小苗即为小于正常植株1/3或小于正常植株2片叶。产量调查为收获小区中间两行玉米，晾干后对玉米棒进行脱粒、称重和测水，最后换算成玉米14%含水量的产量。

2 结果与分析

2.1 出苗率

齐苗后对每个小区进行出苗数调查，调查结果显示，空白对照处理出苗率只有47.4%，而19.5%咯菌腈·精甲霜灵·噻菌灵悬浮种衣剂100mL/100kg、200mL/100kg、300mL/100kg三个剂量处理出苗率分别为83.7%、81.2%和84.9%，说明19.5%咯菌腈·精甲霜灵·噻菌灵悬浮种衣剂可以显著提高玉米种子出苗率。3.5%咯菌腈·精甲双灵处理与12.5%咯菌腈·精甲霜灵·嘧菌酯悬浮种衣剂处理出苗率分别为78.4%和83.5%。35%精甲霜灵单剂处理出苗率只有54%，出苗率高于空白对照，但显著低于其他处理（表1）。

2.2 小苗率

齐苗后对每个小区进行了小苗数调查，试验结果表明，19.5%咯菌腈·精甲霜灵·噻菌灵悬浮种衣剂100mL/100kg、200mL/100kg、300mL/100kg都可以降低小苗率，三个剂量处理的小苗率分别为2.9%、2.3%和3.7%，都显著低于空白对照处理小苗率12.9%。3.5%咯菌腈·精甲霜灵悬浮种衣剂和12.5%咯菌腈·精甲霜灵·嘧菌酯悬浮种衣剂处理的小苗率分别为3.8%和3.1%，小苗率也显著低于空白对照处理。35%精甲霜灵种子处理乳剂小苗率达到9%（表1）。

表1　19.5%咯菌腈·精甲霜灵·噻菌灵悬浮种衣剂处理出苗率和小苗率

供试药剂	供试剂量	出苗率（%）		平均小苗数	小苗率（%）	显著性
	mL/100kg	平均值	显著性			
空白对照		47.4	C	10.8	12.9	A
19.5%咯菌腈·精甲霜灵·噻菌灵悬浮种衣剂	100	83.7	Ab	4.3	2.9	Cd
19.5%咯菌腈·精甲霜灵·噻菌灵悬浮种衣剂	200	81.2	Ab	3.3	2.3	D
19.5%咯菌腈·精甲霜灵·噻菌灵悬浮种衣剂	300	84.9	Ab	5.5	3.7	Bcd
3.5%咯菌腈·精甲霜灵悬浮种衣剂	150	78.4	Ab	5.3	3.8	Bcd
35%精甲霜灵种子处理乳剂	11.4	54.0	Bc	8.5	9.0	Abc
12.5%咯菌腈·精甲霜灵·嘧菌酯悬浮种衣剂	200	83.5	Ab	4.5	3.1	Cd

注：小苗即为小于正常植株1/3或小于正常植株2片叶

2.3　产量

产量调查为收获小区中间两行玉米，晾晒半个月后对玉米棒进行脱粒、称重和测水，换算成14%含水量的理论产量。测产结果表明，19.5%咯菌腈·精甲霜灵·噻菌灵悬浮种衣剂100mL/100kg、200mL/100kg、300mL/100kg处理的干重产量分别11 531kg/hm²、11 642kg/hm²和11 674kg/hm²，其增产率分别为21.4%、22.5%和22.9%。3.5%咯菌腈·精甲霜灵与12.5%咯菌腈·精甲霜灵·嘧菌酯悬浮种衣剂处理干重产量（14%含水量）分别为10 810kg/hm²和12 278kg/hm²，增产率分别为13.8%和29.2%，12.5%精甲霜灵处理干重产量为9 214kg/hm²，其产量低于空白对照处理3.0%（表2）。

表2　玉米测产结果

供试药剂	供试剂量（mL/100kg）	湿重产量（kg/hm²）	产量（kg/hm²）	增产率（%）	显著性
空白对照		12 887	9 500		C
19.5%咯菌腈·精甲霜灵·噻菌灵悬浮种衣剂	100	13 046	11 531	21.1	Ab
19.5%咯菌腈·精甲霜灵·噻菌灵悬浮种衣剂	200	15 690	11 642	22.1	Ab
19.5%咯菌腈·精甲霜灵·噻菌灵悬浮种衣剂	300	14 443	11 674	23.2	Ab
3.5%咯菌腈·精甲霜灵悬浮种衣剂	150	14 683	10 810	13.7	B

（续表）

供试药剂	供试剂量（mL/100kg）	湿重产量（kg/hm²）	产量（kg/hm²）	增产率（%）	显著性
35%精甲霜灵种子处理乳剂	11.4	13 260	9 214	-3.2	C
12.5%咯菌腈·精甲霜灵·嘧菌酯悬浮种衣剂	200	13 461	12 278	29.5	A

3 讨论

田间试验结果表明，在病害压力非常大的情况下，19.5%咯菌腈·精甲霜灵·噻菌灵悬浮种衣剂与12.5%咯菌腈·精甲霜灵·嘧菌酯悬浮种衣剂在推荐剂量下可以显著提高玉米出苗率，并能降低大小苗数，做到保苗齐苗，最终达到增产作用。

本试验播期较早，播种后遇低温干旱，故出苗晚，有粉籽现象。试验中的35%精甲霜灵处理出苗率差，小苗数量多，而含有噻菌灵和嘧菌酯成分的处理出苗率都达到80%以上，小苗数量少。含咯菌腈成分的处理出苗率和小苗率也有良好表现，所以初步分析该试验缺苗的主要原因是由土壤中的高等真菌镰刀菌或丝核菌等引起的，同时也说明在咯菌腈·精甲霜灵配方基础上增加噻菌灵或嘧菌酯都能提供更好和更稳定的保苗和增产效果。

生产上建议使用19.5%咯菌腈·精甲霜灵·噻菌灵悬浮种衣剂剂量为200mL/100kg，既能提高玉米种子出苗率并降低小苗数，同时又能降低土壤中病原菌的侵染机会，提高玉米产量。

参考文献

[1] 姜峰. 玉米发生粉籽和烂芽的原因及应对措施［J］. 农民致富之友，2016（11）：159.

[2] 弓晓峰，杨永强. 玉米播种后出现粉籽和烂芽的主要原因与防治措施［J］. 种子科技，2011（10）：36.

[3] 李欢. 造成玉米播种后粉籽和烂芽的原因及预防措施［J］. 农民致富之友，2015（16）：76.

[4] 晋齐鸣，骈跃斌，宋淑云，等. 玉米苗期病害诊断与防治技术研究［J］. 吉林农业大学学报，2004，26（4）：355－359.

[5] 马建仓，李文明，杨鹏，等. 种衣剂对玉米种子出苗率的影响及对苗枯病和顶腐病的防治效果［J］. 甘肃农业大学学报，2010（5）：51－55.

[6] 杨玉柱，焦必宁. 新型杀菌剂咯菌腈研究进展［J］. 现代农药，2007（6）：5.

抗倒酯在水稻上的施用方法初探

文君慧*，罗　沙，蔡健英，梁远成，程　永**
（先正达（中国）投资有限公司，增城　511358）

摘要： 倒伏是影响水稻生产上常见的减产因素之一，抗倒酯是先正达公司开发生产的植物生长调节剂，能明显增强植株的抗倒伏能力。本文通过不同时期施药，不同品种及不同剂量下对水稻生长的影响的研究表明：抗倒酯对水稻有明显的矮化作用，在水稻的产量、株高、穗长、剑叶长等方面的调节存在明显的剂量反应。施药时期试验结果表明在 PI（圆锥花序起始时期）时施用比在 ME（剑叶的领飞与倒二叶齐平时期）时喷施矮化效果更明显，且对水稻的更友好。通过在不同水稻品种上进行过测试，结果表明抗倒酯对 5 个水稻品种均表现出较好的矮化作用。通过剂量反应试验表明 30g a. i. /hm^2 的抗倒酯在水稻 PI 时使用，对水稻的友好，且水稻的株高得到有效的控制，抗倒伏的能力增强。因此，我们推荐在水稻 PI 时期，叶面喷施，剂量为 30g a. i. /hm^2。

关键词： 抗倒酯；水稻；倒伏

Preliminary Study on Application Method of Trinexpac－ethyl on Rice

Wen Junhui，Luo Sha，Cai Jianying，Liang Yuancheng，Cheng Yong
（*Syngenta*（*China*）*Investment Co. Ltd.*，*Zengcheng Guangdong* 511358）

Abstract： Trinexpac－ethyl is a plant growth regulator which was produced by Syngenta. Test results showed that Trinexpac－ethyl had an obvious stunt effect on rice，but it had some other effects on yield，plant height，spike length，sword leaf length etc. These kind effects would be higher as the higher use rate. Using Trinexpac－ethyl at different development stages of rice，also can produce different effects，test result indicated that using it at the stage of PI（panicle initiation period）is safer than using it at the stage of ME（sword leaves fly and fall two leaves level period）．The dwarfing effect was stable from 5 test varieties. 30 g a. i. /hm^2 of Trinexpac－ethyl had less effect on the yield，and the height of rice could be controlled.

Key words： Trinexapac－ethyl；Rice；Lodging

抗倒酯（trinexapac－ethyl）属于环已烷二酮类植物生长调节剂，为赤霉素生长合成抑制剂，可有效降低向日葵、春小麦、山地水稻等作物的植株高度，并且在牧草上施用不受干旱气候的影响，还可与除草剂、杀菌剂和其他植物生长调节剂复配使用，有报道指出抗倒酯和调环酸钙复配可有效降低双子叶作物的高度，减少苹果黑星病的危害[1-2]。目前，抗倒酯在我国登记的制剂品种只有瑞士先正达作物保护有限公司的 250g/L 乳油和 11.3g/L 可溶液

* 作者简介：文君慧，男，广东增城人，主要从事杀菌剂田间药效试验工作
** 通讯作者：程永，男，研究方向为产品开发；E-mail：Yong. Cheng@ Syngenta. com

剂[3]。抗倒酯对鱼、鸟、蜜蜂、家蚕均为低毒。抗倒酯可被植物茎、叶迅速吸收，根部吸收很少。可在多种作物上使用[4]。水稻倒伏是水稻生长过程中在风雨，地形，土壤环境，耕作措施等外在因子与植物自身抗倒伏性等内在因子相互作用下，茎秆从自然直立状态发生歪斜甚至全株匍倒在地上的现象。水稻的倒伏是造成水稻产量，品质下降的主要因素之一，20 世纪 50 年代以来，对矮化基因的挖掘和育种的利用，在很大程度上避免了栽培高秆水稻品种带来的减产风险，基本解决了水稻倒伏问题，但是限制了水稻高产潜力的进一步发展[5]。倒伏使得水稻光合产物的形成，运输和储存受阻，结实率明显降低，限制产量潜力的发挥，同时收获损失加重，导致产量下降，遇到梅雨天气引起霉变和穗发芽，影响稻米品质，因此矮杆品种被用于水稻抗倒伏品种的选育，随着产量水平的提高，倒伏对水稻的产量影响越来越大，特别直播和抛秧等轻型水稻栽培技术的推广和超高产水稻面积扩大，倒伏问题日趋严重[6]。抗倒品种也限制了水稻高产潜力的进一步发展，通过化学农药降低倒伏对产量的影响研究较少，化学农药匮乏，而抗倒酯是通过降低植株高度，增强茎秆厚度，从而提高植株对倒伏的抵抗作用的药剂，但是在国内很少看到有抗倒酯在水稻上的应用研究，因此本文从施药时期/剂量及品种等方面对抗倒酯对水稻植株高度的调节活性进行研究，并测定其对产量的影响，以求指导水稻科学生产。

1 材料和方法

1.1 试验药剂

抗倒酯 25%乳油，先正达（中国）投资有限公司提供。

1.2 水稻品种

华航 31，华航丝苗，穗珍香，粤晶丝苗，黄壳占。

1.3 研究对象

水稻抗倒伏。

1.4 试验田情况

试验地点为广东省增城市新塘镇宁西华南农业大学教学科研基地，土壤为砂壤土，灌溉方便；田间各试验小区用相同的药剂及时防治杂草以及病害虫；作物长势均匀，肥水管理一致。

1.5 试验设计

不同剂量活性及不同生育期施用对比试验：试验设 10 个处理，分别为 PI 时期施药的 5 个处理 CK、30g a. i. /hm^2，60g a. i. /hm^2，120g a. i. /hm^2，180g a. i. /hm^2；ME 时期施药的 5 个处理 CK、30g a. i. /hm^2，60g a. i. /hm^2，120g a. i. /hm^2，180g a. i. /hm^2，喷液量为 450L/hm^2（表 1），试验采用随机区组设计，3 次重复，小区面积为 40m^2。

品种试验：试验设置 5 个品种，分别是：华航丝苗，穗珍香，华航 31，黄壳占，粤晶丝苗（表 2）；喷液量为 450L/hm^2。试验采用随机区组设计，3 次重复，小区面积为 40m^2。

表 1　不同剂量和不同水稻生育期对比试验

施药时生育期	药剂名称	剂型	剂量 g a. i. /hm^2
PI	抗倒酯 25% EC	EC	0
			30
			60
			120
			180
ME	抗倒酯 25% EC	EC	0
			30
			60
			120
			180

表 2　不同品种对比试验

项目	华航丝苗	穗珍香	华航 31	黄壳占	粤香丝苗
剂量（g a. i. /hm^2）	30	30	30	30	30

1.6　试验调查

所有试验均在水稻的 HE（80% 花穗可见）、AR（乳熟中期）时期分别目测调查各处理水稻的生物量（作物生物量 = ［（灌丛结构分数 + 叶色分数 + 作物密度分数）/15］×100）、长势。在 HE 时期调查各小区的有效分蘖数，在 AR 时期调查各小区的株高、穗长、剑叶长，12 株水稻分蘖每节的节长；在 HA（收获期）时期调查每个小区 3 个 5m^2 的样方，测量每个样方的谷粒鲜重、水分含量、水稻的兜数；还有每个小区 3 点，每点 12 兜水稻的有效分蘖数。

2　结果与分析

2.1　安全性

试验表明，抗倒酯对水稻的株高、生物量、剑叶长、穗长等具有较高的活性，且存在明显的剂量反应，说明抗倒酯的剂量越高对水稻的调节作用越强。在多次试验过程中，在抗倒酯 30～180g a. i. /hm^2 的条件下均为发现调节作用，同时抗倒酯剂量越高，水稻生物量、株高、穗长、产量等反应越强（表 3 和表 4），存在明显的剂量反应。且根据各生理指标数据发现在 30g a. i. /hm^2 的剂量下，水稻已有明显矮化，且对剑叶长穗长等影响较小，因此拟进一步测定 30g a. i. /hm^2 调节植株抗倒伏能力及最佳施用时期。

表3　不同剂量对比（PI 时期施药）

剂型名称	剂量（g a. i./hm²）	株高（cm）	穗长（cm）	剑叶长（cm）	HE 长势	AR 长势	HE 生物量	AR 生物量
抗倒酯 25% EC	0	113 a	29 a	33.6 a	100.0	100.0	81.3	83.3
	30	108 a	28 a	31.7 ab	88.3	93.3	77.8	80.0
	60	106 b	28 a	29.8 b	76.7	85.0	73.6	78.9
	120	95.8 c	26 b	24.9 c	66.7	66.7	70.7	74.4
	180	90.3 d	25 c	24.9 c	60.0	60.0	69.8	74.4

注：上表数据是单个试验，3 次重复的平均值

表4　不同剂量对比（ME 时期施药）

剂型名称	剂量（g a. i./hm²）	株高（cm）	穗长（cm）	剑叶长（cm）	HE 长势	AR 长势	HE 生物量	AR 生物量
抗倒酯 25% EC	0	114 a	29 a	33.0 a	100.0	100.0	81.6	81.1
	30	108 a	28 a	33.3 a	91.7	95.0	78.4	80.0
	60	103 b	27 a	28.7 b	80.0	85.0	75.1	76.7
	120	103 b	28 a	35.0 a	73.3	85.0	75.3	75.6
	180	94 c	26 b	33.4 a	73.3	78.3	75.6	72.2

注：表中数据是单个试验，3 次重复的平均值

2.2　抗倒伏比较试验

通过单一剂量不同品种的抗倒伏研究表明：抗倒酯对水稻植株具有明显的矮化作用（表5），从而明显提高了植株的抗倒伏能力。由表5可知，试验1、3、5施用抗倒酯后，明显降低了倒伏，从而降低了倒伏带来的产量损失，保证水稻的产量。试验2、4，由于没有遇到台风、大暴雨等恶劣天气，没有发生倒伏，但施用抗倒酯处理和对照处理产量差异不显著。

表5　抗倒伏比较

水稻品种	试验编号	药剂名称	剂量（g a. i./hm²）	施药时期	倒伏率（%）	产量（kg/hm²）	株高（cm）
华航丝苗	1	抗倒酯 25% EC	0	PI	60	3 220 a	118.2 a
	1	抗倒酯 25% EC	30	PI	5	5 213 b	112.3 a
穗珍香	2	抗倒酯 25% EC	0	PI	0	6 452 a	113.5 a
	2	抗倒酯 25% EC	30	PI	0	6 123 a	101.5 b
华航 31	3	抗倒酯 25% EC	0	PI	12	4 264 a	114.3 a
	3	抗倒酯 25% EC	30	PI	5	5 866 b	108.2 b

（续表）

水稻品种	试验编号	药剂名称	剂量（g a. i. / hm^2）	施药时期	倒伏率（%）	产量（kg/hm^2）	株高（cm）
黄壳占	4	抗倒酯 25% EC	0	PI	0	5 920 a	111. 4 a
	4	抗倒酯 25% EC	30	PI	0	6 000 a	108. 3 a
粤香丝苗	5	抗倒酯 25% EC	0	PI	30	4 133 a	116. 7 a
	5	抗倒酯 25% EC	30	PI	5	5 604 b	108. 5 b

注：表中是 5 个试验点，2 年的数据的平均值

2. 3　不同水稻生育期施药对比试验

由表 6 可知，抗倒酯在相同剂量、相同水稻品种，在 PI 时期施药在抗倒、产量等指标上比在 ME 时期施药效果好（表 6）。

表 6　不同生育期对比试验

剂型名称	施药时期	剂量（g a. i. / hm^2）	产量（kg/hm^2）	株高（cm）	穗长（cm）	剑叶长（cm）	HE 长势	AR 长势	HE 生物量	AR 生物量
抗倒酯 25% EC	PI	30	6 375 abc	107. 9 b	28. 1 a	31. 7 b	88. 3	93. 3	77. 8	80. 0
抗倒酯 25% EC	ME	30	6 186 abcd	108. 3 b	27. 8 a	33. 3 a	91. 7	95. 0	78. 4	80. 0
抗倒酯 25% EC	PI	60	6 063bcd	106. 3bc	27. 9 a	29. 8bc	76. 7	85. 0	73. 6	78. 9
抗倒酯 25% EC	ME	60	5 729 de	103. 4 cd	26. 8 b	28. 7 c	80. 0	85. 0	75. 1	76. 7
抗倒酯 25% EC	PI	120	5 976cde	95. 8 e	26. 4 b	24. 9 d	66. 7	66. 7	70. 7	74. 4
抗倒酯 25% EC	ME	120	5 548ef	102. 5 d	27. 7 a	35. 0 a	73. 3	85. 0	75. 3	75. 6
抗倒酯 25% EC	PI	180	5 190 f	90. 3 f	24. 8 c	24. 9 d	60. 0	60. 0	69. 8	74. 4
抗倒酯 25% EC	ME	180	5 147 f	94. 2 e	26. 4 b	33. 4 a	73. 3	78. 3	75. 6	72. 2

注：表中数据是单个试验，3 次重复的平均值

2. 4　不同品种对比试验

试验结果表明，抗倒酯在水稻的矮化还是相对稳定的，剂量 30g a. i. /hm^2 在水稻 PI 时期施药，5 个试验品种中，有 4 个品种株高都比空白对照矮 5cm 以上（粤晶丝苗矮了 2cm）。而对产量的影响，5 个品种都在能接受的范围内，其中有两个品种（华航丝苗、黄壳占）还出现增产的情况（表 7）。

表 7　不同品种对比试验

水稻品种	剂型名称	剂量（g a. i. /hm^2）	产量（kg/hm^2）	株高（cm）
华航丝苗	抗倒酯 25% EC	0	4 770 a	118. 0 a
	抗倒酯 25% EC	30	4 932 a	112. 2 a

（续表）

水稻品种	剂型名称	剂量（g a. i. /hm²）	产量（kg/hm²）	株高（cm）
穗珍香	抗倒酯 25% EC	0	6 452 a	111.0 a
	抗倒酯 25% EC	30	6 123 a	99.9 b
华航 31	抗倒酯 25% EC	0	6 263a	114.0 a
	抗倒酯 25% EC	30	5 967 a	107.9 b
黄壳占	抗倒酯 25% EC	0	5 920 a	112.2 a
	抗倒酯 25% EC	30	6 000 a	107.3 a
粤晶丝苗	抗倒酯 25% EC	0	6 113 a	114.4 a
	抗倒酯 25% EC	30	5 904 a	112.9 a

注：表中数据是单个试验，3 次重复的平均值

3 小结与讨论

试验表明，抗倒酯对不同品种水稻均有明显的矮化作用，对水稻株高、穗长、剑叶长等方面具有一定的调节作用，且这种调节具有明显的剂量反应。在水稻的不同生育期使用抗倒酯，会有明显不一样的效果，试验结果表明在 PI（圆锥花序起始时期）时使用，比在 ME（剑叶的领飞与倒二叶齐平时期）时使用对水稻抗倒效果好，且对水稻更友好。随着我国水稻种植的机械化的推广，水稻精耕细作的情况会越来越少，大面积、大规模的种植模式将会增加。这样的情况下，水稻倒伏对水稻产量的影响将会更加严重，水稻倒伏造成的经济损失（水稻因倒伏造成的稻谷霉变，品质降低，采收问题，人力投入）也会更加严重。因此，我们推荐使用抗倒酯 25% EC 来降低水稻倒伏的风险。在水稻上推荐剂量为：30g a. i. /hm²，施药方法为：叶面喷施，喷液量为：450L/hm²，喷施时期为：水稻 PI 时期（圆锥花序起始时期）。为节省人工成本，可与其他药剂桶混喷施，但是桶混其他药剂喷施，其对药效的表现以及对水稻其他指标的调节，有待以后进一步探索。

参考文献

[1] 杨鹏，阳鹏. 25% 抗倒酯乳油的制备研究［J］. 广东化工，2013，40（2）：18 - 19.

[2] 朱长松，黄斌，刘骏结，等. 抗倒酯的合成研究［J］. 2011，19（8）：11 - 13.

[3] 杨鹏，陈龙然，刘雪粉. 25% 抗倒酯水分散粒剂的研制［J］. 农药，2012，51（12）：872 - 877.

[4] 佚名. 抗倒酯［J］. 农药科学与管理，2008，29（12）：53.

[5] 顾铭洪. 水稻高产育种中一些问题的讨论［J］. 作物学报，2010，36（9）：1 431 - 1 439.

[6] 邓文，青先国，马国辉，等. 水稻抗倒伏研究进展［J］. 杂交水稻，2006，2（6）：6 - 10.

稻清对水稻稻瘟病的防治效果及其施乐健植物健康功能简介

周美军[1]，陶龙兴[2]，唐　设[3]，金丽华[1]，陆悦健[1]
（1. 巴斯夫（中国）有限公司，上海　200137；2. 中国水稻科学研究所，杭州　310006；3. 南京农业大学，南京　210095）

摘要：稻清© 100g/L 的吡唑醚菌酯微胶囊悬浮剂是德国巴斯夫于 2016 年在中国上市的水稻稻瘟病专利杀菌剂。稻清独特的创新剂型具有良好的环境相容性，能使有效成分随着环境的变化而逐步释放，为水稻稻瘟病提供稳定高效的防效，同时具有施乐健植物健康作用，促进水稻健壮，收获更多健康稻米。

Abstract: Seltima is a specific F500 formulation for rice farmers to control rice diseases, in particular rice blast (*Pyricularia orgzae*). The formulation behaves differently when reaching the leaf surface or the paddy water and creates the necessary safety for aquatic organisms in the paddy water.

1　稻清产品介绍

稻清© 100g/L 的吡唑醚菌酯微胶囊悬浮剂是德国巴斯夫于 2016 年在中国上市的水稻稻瘟病专利杀菌剂。稻清独特的创新剂型具有良好的环境相容性，能使有效成分随着环境的变化而逐步释放，为水稻稻瘟病提供稳定高效的防效，同时具有施乐健植物健康作用，促进水稻健壮，收获更多健康稻米。

2　稻瘟病介绍

稻瘟病又称稻热病，是水稻上危害最严重的病害之一，以日照少，雾露持续时间长的山区和气候温和的沿江、沿海地区为重，产量损失达数亿千克，同时严重影响稻米品质。病原菌为半知菌亚门灰梨孢，学名 *Pyricularia orgzae*（无性阶段）。病菌以分生孢子或菌丝体在病谷和病稻草上越冬。种子上的病菌在室温或薄膜育秧的条件下容易诱发苗瘟，露天堆放的稻草为第二年发病的主要侵染源。病菌发病的最适温度为 25～28℃，高湿有利于分生孢子形成、飞散和萌发；高湿度持续达一昼夜以上，有利于病害的发生与流行。

3　稻清防治水稻穗颈瘟表现

目前，市场上防治水稻稻瘟病的药剂主要有三环唑、稻瘟灵、咪鲜胺。近年来也出现了三唑类＋甲氧基丙烯酸酯类的复配制剂。从 2011 年开始，巴斯夫公司开始在中国进行稻清防治水稻稻瘟病的试验，取得了优异的防效表现。2015 年，稻清在上海、安徽、广东、广西、黑龙江和江苏的植保机构进行了水稻穗颈瘟的防治试验，在水稻破口初期第一次施药，齐穗期第二次施药，稻清 60mL/亩的平均防效在 80% 以上（末次药后 14 天调查），显著好于其他常用药剂（图 1）。

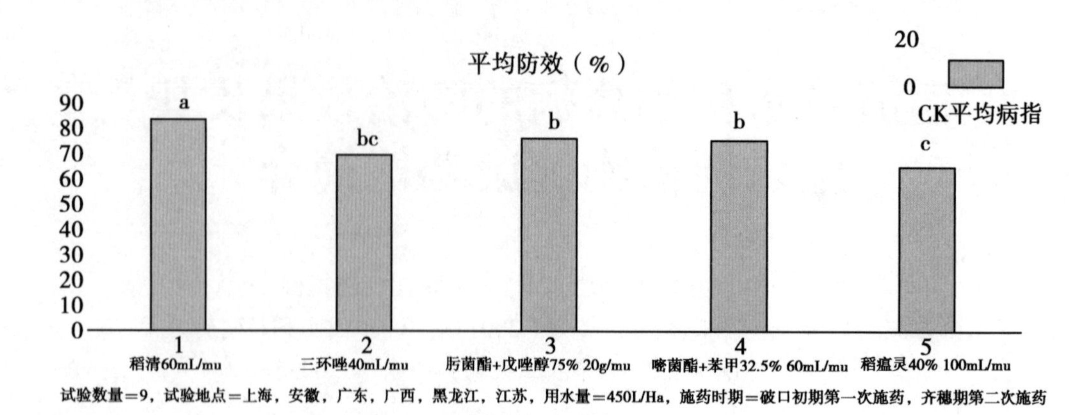

图1　稻清防治水稻穗颈瘟防治效果

4　施乐健功能

稻清是巴斯夫施乐健产品家族在水稻上的最新产品，该产品除了具有优异的病害防治效果之外，在施乐健植物健康方面也有非常明显的表现。

4.1　提高氮肥利用效率

氮肥是含有作物营养元素氮的化肥。元素氮对作物生长起着非常重要的作用，它是植物体内氨基酸的组成部分、是构成蛋白质的成分，也是植物进行光合作用起决定作用的叶绿素的组成部分。氮还能帮助作物分殖，施用氮肥不仅能提高农产品的产量，还能提高农产品的质量。中国水稻研究所及印度尼西亚顶级科学家项目研究表明，在水稻生长过程中，在分蘖末期施用一次稻清（可视情况 10～12d 后第二次施药）能够有效地提高水稻对氮肥的利用效率，在中、高氮条件下水稻有显著的增产作用（图2）。

4.2　增加水稻植株叶绿素含量，增强光合作用

叶绿素是一类与光合作用有关的最重要的色素。光合作用是通过合成一些有机化合物将光能转变为化学能的过程。叶绿素实际上存在于所有能营造光合作用的生物体，包括绿色植物、原核的蓝绿藻（蓝菌）和真核的藻类。叶绿素从光中吸收能量，然后能量被用来将二氧化碳转变为碳水化合物。试验表明，在分蘖末期一次施用稻清 50～66.6mL/亩，能显著增加叶片的叶绿素含量，增强水稻植株的光合作用（图3）。

4.3　应对高温胁迫

近年来，气候变化对水稻生产的影响不断加大，高温胁迫尤为突出。1970 年以来，水稻热害问题已有较多报道，研究认为高温导致水稻不结实的关键期在花期前后，光合作用则是对高温最敏感的过程之一。高温一方面影响水稻开花受精过程，导致空粒率增加；另一方面，高温伤害水稻的灌浆过程，导致秕粒率增加，千粒重下降。为此，巴斯夫与中国水稻研究所开展了稻清抵抗水稻高温胁迫的研究。研究表明，稻清能有效降低水稻穗部温度从而减轻高温对灌浆的不利影响。

试验中，先将开花到灌初期的水稻连续 10d 置于 41～45℃的温度下，再分别施用稻清 50、56.6 和 66.6mL/亩。采用红外热像仪（Therma CAMTMS65，灵敏度 0.05℃）于第 3 天及第 6 天测定水稻穗温。结果如下：

（1）3d 后测得稻清各浓度处理的穗温没有显著差异。

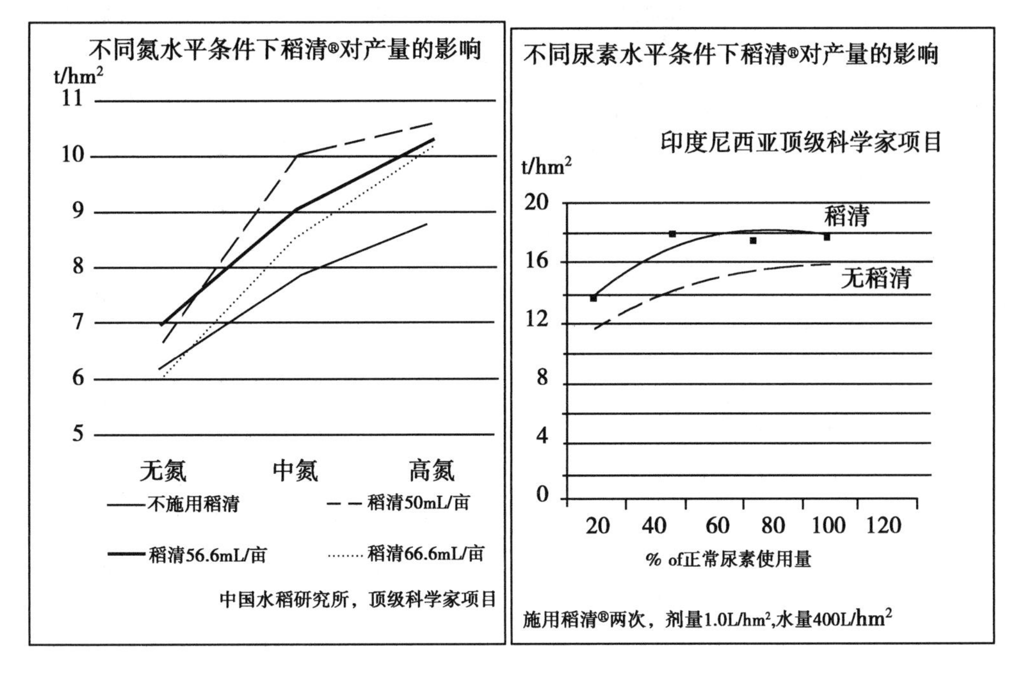

图2　不同氮水平和尿素水平下稻清对产量的影响

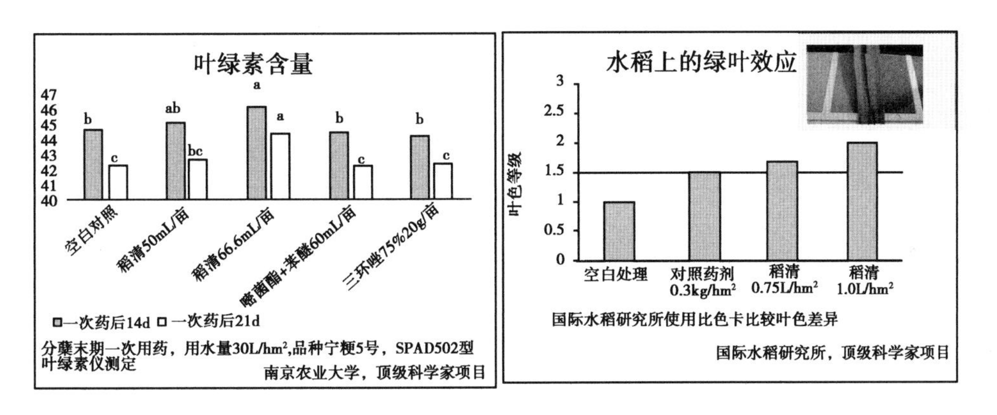

图3　施用稻清后水稻叶绿素变化和效果

（2）6d后测得稻清各浓度处理的穗温较未喷稻清的穗温降低1.6～3.4℃（图4）。说明稻清能有效降低水稻穗部温度从而减轻高温伤害。

5　结语

稻瘟病一直是水稻三大病害之一，对水稻生产造成的危害极大。稻清是巴斯夫公司最新推出的含有吡唑醚菌酯的专利创新制剂，成功解决了吡唑醚菌酯原药对水田生物毒性过高的问题，对水稻稻瘟病有着优异的防效，其施乐健作用即使在复杂的环境条件下也能帮助种植者提高水稻的产量与品质。

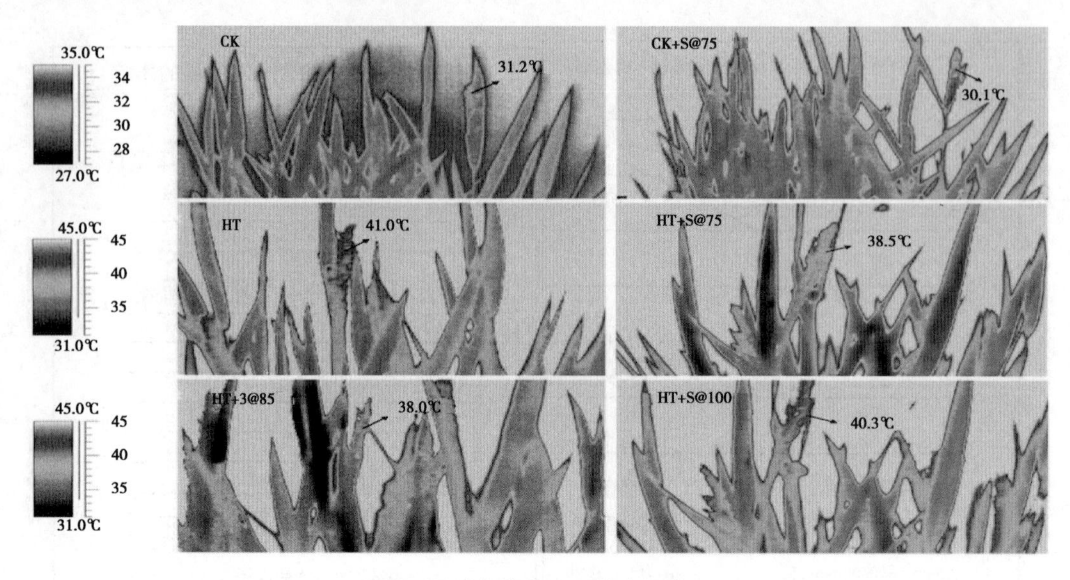

图 4　稻清各浓度处理的穗温情况

参考文献（略）

健攻药效试验综述

金丽华[1]，冯希杰[1]，王绍敏[2]，王培松[3]，范　坤[4]，时春喜[5]
（1. 巴斯夫（中国）有限公司，上海　200137；2. 山东农业工程学院，
济南　250100；3. 山东省烟台市农业科学研究院，烟台　265500；
4. 山东省果树研究所，泰安　271000；5. 西北农林科技大学，杨凌　712100）

摘要：健攻是巴斯夫开发的新一代 SDHI 类杀菌剂，由氟唑菌酰胺和苯醚甲环唑混配而成，为 125g/L 悬浮剂，兼有保护和治疗活性，并且非常适于抗性管理和病害综合治理。健攻应用作物种类多，杀菌谱广，与苯醚甲环唑单剂相比，两种成分相互协同的增效作用使其杀菌谱更广、药效更突出，是出色的抗性管理工具。复配制剂对由子囊菌、担子菌或半知菌引起的多种病害如白粉病、锈病、早疫病、蔓枯病、斑点落叶病、黑星病均有很好的防效，在生产上适用于各种蔬菜、果树的病害防治。健攻具有保护和治疗活性，使用适期长，持效期长，更适用于白粉病反复侵染的葫芦科作物和各种病害混合发生的果树。健攻渗透性和向顶传导性好，对新叶保护能力强。而且得益于其超高的表面活性，耐雨水冲刷能力好。

Abstract：Fluxapyroxad is one of the newest generation SDHI fungicides which is developed by BASF. Sercadis Plus is the mixture of Fluxapyroxad and Difenoconazole，with the formulation 125g/L SC. Sercadis Plus provides excellent control against various fungi diseases with protectant and curative properties. And it is a very good resistance management tool.

Sercadis Plus provides wide rangeof diseases control on many crops. Compared with Difenoconazole solo，the mixture shows synergy effect and wider disease range. Sercadis Plus can control many diseases such as powdery mildew，rust，early blight，gummy stem blight，apple *Alternaria* leaf spot and scab and so on，which are caused by ascomycetes，basidiomycetes and adelomycete. It can be used in different stage of crops and shows long lasting effect. It is fit for high intensity powdery mildew segments like Cucurbits and pome fruits. Sercadis Plus provides good penetrability and acropetal translocation，so can provide good protection on new leaves. Sercadis Plus performs better rain fastness due to its higher surfactant.

1　健攻防治番茄灰叶斑病

1.1　试验设计

1.1.1　药剂处理与小区排列

供试药剂：健攻 125g/L 悬浮剂，剂量分别为 53. 3 、66. 7 、80mL/亩。

对照药剂：① 世高 10% 水分散粒剂 66. 7g/亩；② 万兴 20. 67% 乳油 21 mL/亩；③加瑞农 47% 可湿性粉剂 100 mL/亩；④ 拿敌稳 75% 可湿性粉剂 15g/亩；⑤ 宝丽安（多抗霉素）10% 可湿性粉剂 120g/亩。每处理重复 3 次，共 27 个小区，小区面积 27. 6 m^2，区组随机排列。

试验地点：山东省济阳县。

1.1.2 供试作物及品种

供试作物为番茄，品种为抗热粉王。

1.1.3 试验地基本情况

土质为壤土，中性，肥力较好，土壤有机质含量1.2%，管理水平中等，排灌条件良好。7月10日移栽，大小行种植，株距为35cm，小行行距为50cm，大行行距为70cm。施药时番茄为结果盛期，长势基本一致，棚内相对湿度70%～90%，棚温16～36℃。

1.1.4 施药时间与调查方法

于2012年9月15日，番茄灰叶斑病发病初期开始施药，连续施药2次，间隔14d。配制药液时先用少量水稀释药剂，喷施药液量以叶片充分着药而不滴药液为宜。每亩用药液量60L。调查方法为每个小区固定2点取样，每点调查5株，每株从上到下调查10片复叶，以复叶病斑面积占整个叶面积百分率分级。

1.2 试验结果与分析

试验结果表明（表1），健攻125g/L悬浮剂对番茄灰叶斑病有良好的防治效果，在试验剂量内对番茄安全无药害，试验期间观察，健攻66.7 mL/亩和80 mL/亩处理番茄长势旺，叶色浓绿，表明药剂有刺激番茄生长作用。健攻125g/L悬浮剂的持效期长达21d。健攻125g/L悬浮剂推荐使用浓度为有效成分100～150g/hm²，在番茄灰叶斑病发病初期使用，连续用药2次，间隔14d。

表1 健攻防治番茄灰叶斑病防治效果

编号	处理	药前病情指数	2次药后7d		2次药后14d		2次药后21d	
			防效（%）	差异显著性	防效（%）	差异显著性	防效（%）	差异显著性
1	空白对照（病指）	8.9	19.6		24.0		27.0	
2	健攻53.3 mL/亩	9.3	50	c	61	cd	64	bc
3	健攻66.7 mL/亩	8.2	53	b	63	bc	67	b
4	健攻80 mL/亩	9.5	57	a	67	a	71	a
5	世高66.7 g/亩	8.4	43	e	47	g	49	f
6	万兴21 mL/亩	9.4	40	f	40	i	41	i
7	加瑞农100 mL/亩	8.8	49	cd	57	e	58	d
8	拿敌稳15 g/亩	9.0	43	e	45	gh	46	gh
9	宝丽安120 g/亩	8.7	47	d	53	f	54	e

2 健攻防治番茄叶霉病

2.1 试验设计

2.1.1 药剂处理与试验地点

供试药剂：健攻125g/L悬浮剂，剂量分别为40、53.3、66.7、80 mL/亩。

对照药剂：①世高10%水分散粒剂66.7 g/亩；②万兴20.67%乳油21 mL/亩；③加瑞农47%可湿性粉剂100 mL/亩；④拿敌稳75%可湿性粉剂15 g/亩；每处理重复3次，

共27个小区，区组随机排列。

试验地点：上海市奉贤区。

2.1.2 供试作物及品种

供试作物为番茄，品种为欧迪斯。

2.1.3 施药时间与调查方法

于2013年3月22日，4月4日连续用药两次，用药时番茄处于结果初期，叶霉病开始发生。每亩用药液量60L。调查方法为每小区固定2点取样，每点调查5株，每株从上到下调查10片复叶，以复叶病斑面积占整个叶面积百分率分级。

2.2 试验结果与分析

试验结果表明（表2），健攻125 g/L悬浮剂对番茄叶霉病有良好的防治效果，在试验剂量内对番茄安全无药害。健攻125g/L持效期长。健攻推荐使用浓度为有效成分100～150g/hm^2，在番茄叶霉病发病初期使用，连续用药2～3次，间隔7～10d。

表2 健攻防治番茄叶霉病防治效果

编号	处理	药前病情指数	1次药后13d		2次药后8d		2次药后13d	
			防效（%）	差异显著性	防效（%）	差异显著性	防效（%）	差异显著性
1	空白对照（病指）	0.4	4.6		20.2		28.9	
2	健攻40 mL/亩	0.9	87	a	92	a	90	a
3	健攻53.3 mL/亩	0.9	85	ab	89	b	90	a
4	健攻66.7 mL/亩	0.8	71	cde	83	c	90	a
5	健攻80 mL/亩	1.0	82	b	89	b	92	a
6	世高66.7 g/亩	1.0	78	c	87	bc	70	b
7	万兴21 mL/亩	1.0	66	e	82	c	68	b
8	加瑞农100 mL/亩	1.2	67	e	81	c	53	c
9	拿敌稳15 g/亩	0.9	67	e	63	d	49	d

3 健攻防治番茄早疫病

3.1 试验设计

3.1.1 药剂处理与试验地点

供试药剂：健攻125g/L悬浮剂，剂量分别为40、53.3、66.7mL/亩。

对照药剂：①世高10% 水分散粒剂83.3 g/亩；②阿米西达250g/L悬浮剂24mL/亩；每处理重复4次，共24个小区，小区面积为15m^2，区组随机排列。

试验地点：山东烟台。

3.1.2 供试作物及品种

供试作物为番茄，品种为毛粉802。

3.1.3　施药时间与调查方法

试验期间番茄生育期为开花期至果实膨大期。番茄开花期，早疫病发生前首次用药，以后间隔7～10d用药1次，连续用药3次。具体时间：2013年8月1日，8月10日，8月19日。每亩用药液量60L。调查方法为每小区固定2点取样，每点调查5株，每株从上到下调查10片复叶，以复叶病斑面积占整个叶面积百分率分级。

3.2　试验结果与分析

试验结果表明（表3），健攻能有效防治番茄早疫病，在试验剂量内对番茄安全无药害。健攻推荐使用浓度为有效成分75～125g/hm^2，在番茄早疫病发病初期使用，连续用药2～3次，间隔7～10d。

表3　健攻防治番茄早疫病防治效果

编号	处理	2次药后9d		末次药后9d	
		防效（%）	差异显著性	防效（%）	差异显著性
1	空白对照（病指）	9.8		22.1	
2	健攻40 mL/亩	77	d	83	d
3	健攻53.3 mL/亩	80	bcd	85	cd
4	健攻66.7 mL/亩	87	a	90	a
5	世高83.3 g/亩	80	cd	85	cd
6	阿米西达24mL/亩	82	abc	86	bc

4　健攻防治黄瓜白粉病

4.1　试验设计

4.1.1　药剂处理与试验地点

供试药剂：健攻125g/L悬浮剂，剂量分别为53.3、66.7mL/亩。

对照药剂：① 翠泽300g/L悬浮剂45mL/亩；② 乙嘧酚250g/L悬浮剂60mL/亩；③ 拿敌稳75%水分散粒剂10g/亩；④氟菌唑30%可湿性粉剂18g/亩；⑤ 福星400g/L乳油7.5mL/亩 。每处理重复3次，共24个小区，小区面积8.25m^2，区组随机排列。

试验地点：上海市奉贤区。

4.1.2　供试作物及品种

供试作物为黄瓜，品种为毛龙绿之春。

4.1.3　施药时间与调查方法

试验期间黄瓜生育期为花果期。用药时间：2013年5月17日，5月23日，6月1日。每亩用药液量45L。调查方法为每小区取10～12株，调查整株罹病度（所有粉状物面积占整株所有叶片的面积,%）。

4.2 试验结果与分析

试验结果表明（表4），健攻能有效防治黄瓜白粉病；健攻 推荐使用浓度为有效成分100～125g/hm^2 处理无白粉病侵染的新生叶片，叶片健康叶色油绿，表明其具有良好的向顶输导性和新生组织保护能力。

表4　健攻防治黄瓜白粉病防治效果

编号	处理	2次药后7d		3次药后9d		3次药后13d		3次药后20d	
		防效（%）	差异显著性	防效（%）	差异显著性	防效（%）	差异显著性	防效（%）	差异显著性
1	空白对照（罹病度）	4		12		16		33	
2	健攻 53.3 mL/亩	76	a	97	a	99	a	99	a
3	健攻 66.7 mL/亩	68	bc	97	a	98	a	98	a
4	翠泽 45mL/亩	61	c	89	bc	88	b	85	c
5	乙嘧酚 60mL/亩	63	c	87	c	75	c	70	d
6	拿敌稳 10g/亩	63	c	92	ab	91	b	90	b
7	氟菌唑 18g/亩	45	d	75	d	59	d	52	e
8	福星 7.5mL/亩	74	ab	96	a	98	a	99	a

5 健攻防治苹果斑点落叶病

5.1 试验设计

5.1.1 药剂处理与试验地点

供试药剂：健攻 125g/L 悬浮剂，稀释倍数分别2 500倍，2 000倍，1 600倍，1 350倍。

对照药剂：① 世高 10% 水分散粒剂 2 000倍液；②好力克 43% 水分散粒剂 500 倍液；③安泰生 70% 可湿性粉剂 600 倍液；每小区两株成年果树，重复 3 次，区组随机排列。

试验地点：济宁市曲阜市。

5.1.2 供试作物及品种

试树品种为新红星，树龄 25 年，株距×行距＝3m×5m，亩栽树 44 株，肥水管理水平一般，亩产苹果约为 2 500kg，斑点落叶病历年发生较重。

5.1.3 施药时间与调查方法

于春梢初发病时开始施药，5 月 10 日、5 月 24 日、6 月 7 日、6 月 21 日进行喷药，共施药 4 次。每小区两株均调查，每株分东、西、南、北、中五个方向各固定 2 个春梢，定期调查其全部叶片，记录总叶数、各级病叶数。

5.2 试验结果与分析

试验结果表明（表5），健攻125g/L悬浮剂用于防治苹果斑点落叶病，对果树安全，生产上若每10～15d喷药一次，推荐健攻用药浓度为1 350～1 600倍。

表5　健攻防治苹果斑点落叶病防治效果

编号	处理	4次药后7d		4次药后14d		4次药后21d	
		防效（%）	差异显著性	防效（%）	差异显著性	防效（%）	差异显著性
1	空白对照（病指）	7.9	—	6.4	—	7.1	—
2	健攻2 500倍液	82	d	83	d	82	d
3	健攻2 000倍液	83	bcd	87	bcd	85	bcd
4	健攻1 600倍液	87	a	89	a	87	a
5	健攻1 350倍液	90	cd	92	cd	90	cd
6	世高2 000倍液	82	abc	85	abc	78	abc
7	好力克5 000倍液	84	bcd	86	bcd	83	d
8	安泰生600倍液	86	a	89	a	85	bcd

6　健攻防治苹果轮纹病

6.1　试验设计

6.1.1　药剂处理与试验地点

供试药剂：健攻125g/L悬浮剂，稀释倍数分别2 500倍，2 000倍，1 600倍，1 350倍。

对照药剂：①世高10%水分散粒剂2 000倍液；②安泰生70%可湿性粉剂600倍液；每小区两株成年果树，重复4次，区组随机排列。

试验地点：陕西省咸阳市。

6.1.2　供试作物及品种

供试品种为秦冠，树龄为4年生，株行距2m×3m，土壤类型为土娄土，pH值中性，施有机肥2 000kg左右，浇水2次，管理水平一般。

6.1.3　施药时间与调查方法

试验于分别于2012年5月26日、6月11日、7月5日、7月25日、8月9日喷药，共喷药5次。每公顷喷液量约为2 000L。苹果采收期：每处理调查发病均匀而且具有代表性的4株苹果树上的全部果实及落地果（每小区1株），调查果实总数不低于1 000个，记录总果数、病果数。

6.2　试验结果与分析

试验表明（表6），健攻125g/L悬浮剂对苹果轮纹病有较好的防治效果，其1 350～2 000倍液的药剂在采收期的防治效果达83%～93%，贮藏期30d的防治效果达88%～100%，而且对苹果树生长发育安全，无不良影响。建议对其进行大面积推广，推荐浓度稀释倍数1 350～2 000倍液，于苹果谢花后7～10d开始喷药防治，间隔14～21d喷药一次，连续喷施5次左右为宜，同时应与其他不同作用机理的杀菌剂品种交替使用，以提高防治效果。

表6　健攻防治苹果轮纹病防治效果

处理	采收期			调查总果数	病果数（个）	病果率（%）	防效（%）
	平均病果率（%）	平均防效（%）	差异显著性5%				
清水对照	18.6	—	—	200	17	8.5	—
健攻2 500倍液	3.9	79	c	200	3	1.5	82
健攻2 000倍液	3.2	83	c	200	2	1	88
健攻1 600倍液	2.0	89	b	200	1	0.5	94
健攻1 350倍液	1.4	93	ab	200	0	0	100
世高2 000倍液	3.1	83	c	200	1	0.5	94
安泰生600倍液	3.5	81	c	200	2	1	88

7　结论

从试验结果看，健攻对防治番茄灰叶斑病、叶霉病和早疫病、黄瓜白粉病和苹果斑点落叶病和轮纹病均表现出比较理想的防治效果，优于或相当于常规药剂处理的效果。推荐于病害发生初期，病害压力较轻时开始用药防治。防治番茄灰叶斑病、叶霉病采用100～150g/hm^2的剂量，连续用药2～3次，间隔期7～10d；防治番茄早疫病采用75～125g/hm^2的剂量，连续用药2～3次，间隔期7～10d；防治黄瓜白粉病采用100～125g/hm^2的剂量，连续用药3次，间隔期7～10d；防治苹果斑点落叶病采用1 350～1 600倍液，连续用药3～4次，间隔期10～15d；防治苹果轮纹病采用1 350～2 000倍液，于苹果谢花后7～10d开始喷药防治，间隔14～21d左右喷药一次，连续用药5次左右为宜。建议在使用健攻防治病害时，同时与其他不同作用机理的杀菌剂品种交替使用，延缓抗药性的产生，提高防效。

参考文献（略）

氟唑环菌胺·咯菌腈·噻虫嗪种子包衣处理对冬小麦纹枯病防效探索

柴延生[1]*，马向峰[2]，赵其森[1]，武传志[3]，杨洪李[3]，李　峰**
（1. 山东省德州市齐河县良种棉加工厂，德州　251100；2. 山东省德州市齐河县农业局，德州　251100；3. 先正达（中国）投资有限公司，上海　200120）

摘要：在山东省齐河县先正达试验基地，本文研究了氟唑环菌胺·咯菌腈·噻虫嗪种子包衣处理对冬小麦纹枯病的防效和对冬小麦的安全性。结果表明：氟唑环菌胺·咯菌腈·噻虫嗪在93.8g a. i./100kg 种子剂量下对冬小麦纹枯病防效可达80%以上，持效期可达179天以上。同时，氟唑环菌胺混剂对冬小麦的出苗时间没有影响，提高了冬小麦出苗率。氟唑环菌胺混剂对冬小麦株高、根长、茎粗、分蘖数、茎叶鲜重和根重等生理指标都没有显著影响。氟唑环菌按混剂对冬小麦安全。本试验未进行测产。

关键词：氟唑环菌胺·咯菌腈·噻虫嗪；冬小麦；纹枯病；安全性

Efficacy of Seed Treatment of Sedaxane · Fludioxonil · Thiamethoxam Against Sheath Blight and Crop Tolerance on Winter Wheat

Chai Yansheng[1], Ma Xiangfeng[2], Zhao Qisen[1], Wu Chuanzhi[3], Yang Hongli[3], Li Feng
(1. *Qihe Improved Cotton Seed Multiplication Farm*, 2. *Qihe Agricultural Bureau*, *Qihe Shandong* 25110. 3. *Syngenta* (*China*) *Investment Co. Ltd. Shanghai* 200120)

Abstract: The trial of efficacy of seed treatment of sedaxane · fludioxonil · thiamethoxam against sheath blight and crop tolerance on winter wheat was conducted in Qihe Shandong province. The result showed that sedaxane · fludioxonil · thiamethoxam at 93.8 g a. i./100 kg seeds can provide good efficacy against *Rhizoctonia cerealis*. The efficacy at 179DAS (179 days after sowing) was more than 80% which was based on disease severity. The persistence of sedaxane · fludioxonil · thiamethoxam at 93.8 g a. i./100 kg seeds was more than 179 days. Mixtures of sedaxane at all test rates had no significant impact on germination date, germination rate, shoot height, root height, stem diameter, tillers, shoot weight and root weight, they were safe to winter wheat. No yield data in this trial.

Key words: Sedaxane · fludioxonil · thiamethoxam; Wintet wheat; Sheath blight; Crop Tolerance

小麦纹枯病是小麦上普遍发生，危害严重的一种土传性真菌病害。近年来，由于全球气候变暖、化肥用量增多、秸秆还田以及小麦播期提前等因素的影响，小麦纹枯病发生逐年严重，对小麦的产量也产生较大影响[1-2]。在我国小麦纹枯病主要发生在江淮流域和黄河中下游冬麦区，麦田病株率在10%～20%，严重的超过60%，引起的产量损失在5%～10%，严重的超过20%[3]。

* 第一作者：柴延生，男，农艺师，常年从事棉花、小麦品种选育及农化产品田间试验工作；E-mail：chaiyansheng@126. com
** 通讯作者：李峰，男，主要从事种衣剂产品开发工作；E-mail：feng. li@ syngenta. com

小麦纹枯病的防治方法主要有叶面喷雾和种衣剂种子包衣两种方法，主要药剂有三唑酮、丙环唑、苯醚甲环唑、戊唑醇、噻氟酰胺、井冈霉素、福美双和咯菌腈等[4-7]。其中种衣剂种子包衣法因为便捷、省时省力和防效优秀而应用越来越广，主要药剂有福美双、戊唑醇和咯菌腈等。

氟唑环菌胺（Sedaxane）是先正达公司开发的一种新型种子处理杀菌剂，属于 SDHI 类杀菌剂，化学名称为 2’–［（1RS，2RS）–1，1’–联环丙烯–2–基］–3–（二氟），1–甲基吡唑–4–羧酸苯胺，分为顺式结构和反式结构（图 1 和图 2），英文通用名称为：Sedaxane。分子式为 $C_{18}H_{19}F_2N_3O$。对玉米、麦类、水稻等作物上的丝黑穗病有良好的防治效果，用作种衣剂时，对多种种传、土传病害有较好的防治效果，还可促进作物根系的生长[8]。

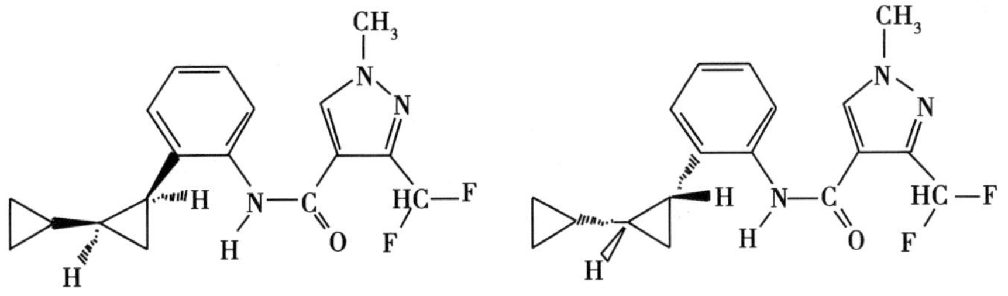

图 1　氟唑环菌胺顺式结构

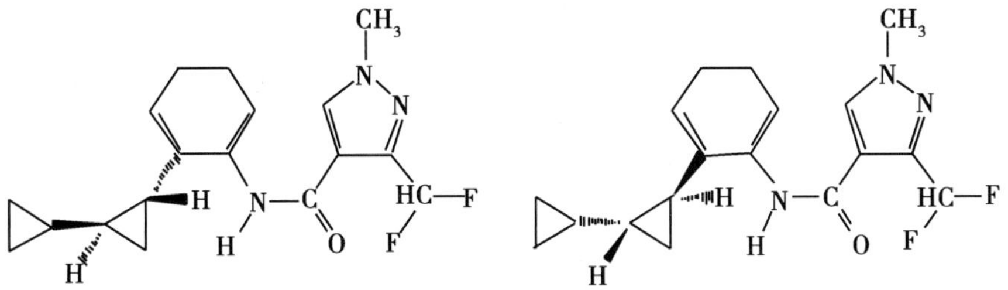

图 2　氟唑环菌胺反式结构

咯菌腈（fludioxonil）是一种非内吸性苯基吡咯类杀菌剂，化学结构式如图 3。1984 年由瑞士 Ciba–Geigy 公司（现先正达公司）研发成功，能专一性的抑制霉菌生长，对子囊菌、半知菌、担子菌等病原菌，有良好防效，广泛应用于小麦种子包衣处理防治小麦纹枯病[8-11]。但随着咯菌腈的大量应用，小麦纹枯对其具有一定的抗性风险。王成凤[12]等研究表明，咯菌腈对小麦纹枯具有低到中等抗性风险。

为提高提高种衣剂对小麦纹枯病的防效，延缓小麦纹枯病抗性的产生，本文研究了氟唑环菌胺·咯菌腈·噻虫嗪混剂种子包衣处理对冬小麦纹枯病的防治效果和对冬小麦的安全性。

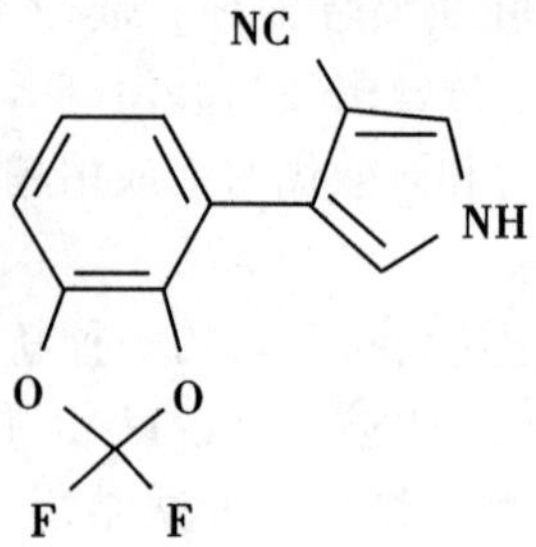

NC NH O O F F

图3　咯菌腈的结构式

1　材料与方法

1.1　供试药剂

312.5 g a.i./L 氟唑环菌胺·咯菌腈·噻虫嗪种子处理悬浮剂（312.5FS），瑞士先正达作物保护有限公司生产；350 g a.i./L 噻虫嗪种子处理悬浮剂（锐胜，350FS），瑞士先正达作物保护有限公司生产；60 g a.i./L 戊唑醇种子处理悬浮剂（立克秀，60FS），拜耳作物科学公司生产；600 g a.i./L 吡虫啉种子处理悬浮剂（高巧，600FS），拜耳作物科学公司生产。

1.2　供试作物

冬小麦，品种为良星77。

1.3　防治对象

小麦纹枯病（*Rhizoctonia cerealis*）。

1.4　试验设计和安排

本试验的药剂设计如表1所示，共开展了1个试验。试验地点为山东省齐河县焦庙镇先正达试验基地。试验共设6个处理，3次重复，试验小区按完全随机区组排列；小区面积为22m^2，播种量150kg/hm^2。人工播种，播种时间2014年10月10日。

表1　处理列表

处理	制剂用量（mL/100kg 种子）	有效成分用量（g a.i./100kg 种子）
CK		
氟唑环菌胺·咯菌腈·噻虫嗪312.5FS	100	31.3
氟唑环菌胺·咯菌腈·噻虫嗪312.5FS	200	62.5
氟唑环菌胺·咯菌腈·噻虫嗪312.5FS	300	93.8
戊唑醇60FS + 吡虫啉600FS	40 + 600	2.4 + 360
噻虫嗪350FS	150	52.5

1.5　调查方法

安全性调查，调查小麦出苗时间和出苗率，评估药剂对小麦的安全性；每个小区内小麦出苗率达到50%时视为出苗时间。

播种后 30d，调查小麦株高、根长、茎粗、分蘖数、茎叶鲜重和根鲜重，评估药剂对小麦生理指标的影响。每个小区调查 5 点，每点调查 10 株，取平均值。

防效调查，分别在冬前、拔节期和乳熟期调查纹枯病发病率和严重度。计算防效。调查方法为：每小区五点取样，每点随机取样 20 株，共 100 株，计算整个小区 100 株的发病率和平均严重度。严重度 0～100%。

0：不发病；

10%：叶鞘面积的 10% 发病，茎秆不发病；

20%：叶鞘面积的 50% 发病，茎秆不发病；

30%：茎秆刚刚发病，不足环茎的 5%；

40%：茎秆病斑环茎不足 30%；

50%：茎秆病斑环茎不足 50%；

70%：茎秆病斑环茎超过 50%，但不倒伏；

90%：枯死，倒伏，枯白穗。

1.6 数据统计

所有数据采用 Duncan's 法进行方差分析和差异显著性检验。所有表格中数据为 3 次重复的平均数，同列数据后不同小写字母表示差异显著（$P<0.05$）。

2 结果与分析

2.1 安全性

2.1.1 氟唑环菌胺·咯菌腈·噻虫嗪对冬小麦出苗时间、出苗率的影响

如表 2 所示，氟唑环菌胺混剂对冬小麦出苗时间没有显著影响，各处理都在 10d 左右。同时，从表中可以看出，氟唑环菌胺混剂能显著增加小麦的出苗率，但与锐胜处理相比没有显著差异。因此，在出苗安全性上，氟唑环菌胺混剂对冬小麦是安全的。

表 2 氟唑环菌胺混剂对冬小麦出苗时间和出苗率的影响

处理	制剂用量（mL/100kg 种子）	有效成分用量（g a. i. /100kg 种子）	出苗时间（d）	出苗率（%）
CK			10a	79b
氟唑环菌胺·咯菌腈·噻虫嗪 312.5FS	100	31.3	10.5a	91.3a
氟唑环菌胺·咯菌腈·噻虫嗪 312.5FS	200	62.5	10a	87.5a
氟唑环菌胺·咯菌腈·噻虫嗪 312.5FS	300	93.8	10.3a	87.3a
戊唑醇 60FS＋吡虫啉 600FS	40＋600	2.4＋360	10.5a	85.8ab
噻虫嗪 350FS	150	52.5	10.3a	84.5ab

2.1.2 氟唑环菌胺·咯菌腈·噻虫嗪对冬小麦株高和根长的影响

如表 3 所示，播种后 30d，氟唑环菌胺混剂和锐胜单剂对冬小麦的根长有促进作用，但没有达到显著水平。氟唑环菌胺混剂对冬小麦株高和根长未产生显著影响。

表 3　氟唑环菌胺混剂对冬小麦株高和根长的影响

处理	制剂用量（mL/100kg 种子）	有效成分用量（g a. i. /100kg 种子）	株高（cm）	根长（cm）
CK			18. 8a	10. 5a
氟唑环菌胺 · 咯菌腈 · 噻虫嗪 312. 5FS	100	31. 3	19. 6a	11. 5a
氟唑环菌胺 · 咯菌腈 · 噻虫嗪 312. 5FS	200	62. 5	18. 1a	11. 0a
氟唑环菌胺 · 咯菌腈 · 噻虫嗪 312. 5FS	300	93. 8	19. 2a	11. 2a
戊唑醇 60FS + 吡虫啉 600FS	40 + 600	2. 4 + 360	17. 3a	11. 0a
噻虫嗪 350FS	150	52. 5	17. 6a	11. 6a

2. 1. 3　氟唑环菌胺 · 咯菌腈 · 噻虫嗪对冬小麦茎粗和分蘖数的影响

从表 4 中可以看出，播种后 30d，氟唑环菌胺混剂对冬小麦的根长和株高也未产生显著影响。

表 4　氟唑环菌胺混剂对冬小麦茎粗和分蘖数的影响

处理	制剂用量（mL/100kg 种子）	有效成分用量（g a. i. /100kg 种子）	茎粗（mm）	分蘖数（个）
CK			2. 5 a	1. 3 a
氟唑环菌胺 · 咯菌腈 · 噻虫嗪 312. 5FS	100	31. 3	2. 6 a	1. 6 a
氟唑环菌胺 · 咯菌腈 · 噻虫嗪 312. 5FS	200	62. 5	2. 4 a	1. 2 a
氟唑环菌胺 · 咯菌腈 · 噻虫嗪 312. 5FS	300	93. 8	2. 6 a	1. 4 a
戊唑醇 60FS + 吡虫啉 600FS	40 + 600	2. 4 + 360	2. 4 a	1. 2 a
噻虫嗪 350FS	150	52. 5	2. 4 a	1. 3 a

2. 1. 4　氟唑环菌胺 · 咯菌腈 · 噻虫嗪对冬小麦茎叶鲜重和根重的影响

如表 5 所示，播种后 30d，氟唑环菌胺混剂的各个剂量对冬小麦的根长和株高未观察到显著的影响，只有低剂量氟唑环菌胺 · 咯菌腈 · 噻虫嗪 31. 3g a. i. /100kg 种子有一定的促进作用，但也未达到显著水平。

表 5　氟唑环菌胺混剂对冬小麦茎叶鲜重和根重的影响

处理	制剂用量（mL/100kg 种子）	有效成分用量（g a. i. /100kg 种子）	茎叶鲜重（g）	根重（g）
CK			6. 7 a	0. 4 a
氟唑环菌胺 · 咯菌腈 · 噻虫嗪 312. 5FS	100	31. 3	8. 1 a	0. 5 a
氟唑环菌胺 · 咯菌腈 · 噻虫嗪 312. 5FS	200	62. 5	6. 4 a	0. 4 a
氟唑环菌胺 · 咯菌腈 · 噻虫嗪 312. 5FS	300	93. 8	7. 3 a	0. 4 a
戊唑醇 60FS + 吡虫啉 600FS	40 + 600	2. 4 + 360	6. 4 a	0. 3 a
噻虫嗪 350FS	150	52. 5	6. 3 a	0. 4 a

2.2 防治效果

从表6可以看出，在各处理中，氟唑环菌胺混剂最高剂量93.8 g a.i./100kg种子对小麦纹枯病防效最好，在179DAS（179 days after sowing，播种后179d）还能取得75%以上的防效。远优于立克秀混剂处理，但氟唑环菌胺混剂中剂量和低剂量的防效较差。立克秀混剂处理仅在前期有一定防效。

表7中按照发病严重度计算的防效也得到了同样的结果，氟唑环菌胺混剂最高剂量93.8 g a.i./100kg种子在179DAS防效达到80%以上。

表6　氟唑环菌胺混剂对冬小麦纹枯病发病率防效

处理	制剂用量（mL/100kg种子）	有效成分用量（g a.i./100kg种子）	防效（%）		
			31DAS	179DAS	223DAS
CK			0a	0a	0a
氟唑环菌胺·咯菌腈·噻虫嗪312.5FS	100	31.3	54.2a	48.2a	0.8a
氟唑环菌胺·咯菌腈·噻虫嗪312.5FS	200	62.5	66.7a	49.4a	15.2a
氟唑环菌胺·咯菌腈·噻虫嗪312.5FS	300	93.8	75a	75.3a	11.2a
立克秀+吡虫啉600FS	40+600	2.4+360	33.3a	0a	0a
噻虫嗪350FS	150	52.5	20.8a	54.1a	20a

表7　氟唑环菌胺混剂对冬小麦纹枯病严重度防效

处理	制剂用量（mL/100kg种子）	有效成分用量（g a.i./100kg种子）	防效（%）	
			179DAS	223DAS
CK			0a	0a
氟唑环菌胺·咯菌腈·噻虫嗪312.5FS	100	31.3	56.4a	1a
氟唑环菌胺·咯菌腈·噻虫嗪312.5FS	200	62.5	55.5a	11.2a
氟唑环菌胺·咯菌腈·噻虫嗪312.5FS	300	93.8	81a	17.4a
戊唑醇60FS+吡虫啉600FS	40+600	2.4+360	0a	0a
噻虫嗪350FS	150	52.5	58.3a	16a

3 讨论

在华北地区冬小麦种植区，纹枯病发生普遍，是冬小麦田的主要病害；由于连年种植，纹枯病发年发生都较严重。

本文研究了氟唑环菌胺和麦田包衣杀菌剂咯菌腈、主流杀虫剂噻虫嗪预混包衣处理，对冬小麦的安全性和纹枯病的防效；结果表明，氟唑环菌胺混剂对冬小麦安全性良好，对冬小麦出苗时间没有影响，对出苗率有一定的促进作用。氟唑环菌胺混剂对冬小麦的株高、根长、茎粗、分蘖数、茎叶鲜重和根重等各项生理指标有一定促进作用，但都没有显著的影响。

本文同时研究了氟唑环菌胺混剂对冬小麦纹枯病的防效，观察了在冬小麦苗期（30DAS），返青拔节期（179DAS）和灌浆期（223DAS）对纹枯病的防治效果。结果表明，氟唑环菌胺混剂的最高剂量93.8 g a.i./100kg种子对小麦纹枯病具有较好的防效，防效达80%左右；同时氟唑环菌胺混剂对小麦纹枯病具有较长的持效期，其最高剂量播种后179d防效还能达到80%以上，优于戊唑醇和吡虫啉的混剂。同时，我们观察到，噻虫嗪单剂处理对小麦纹枯病也有一定防效，这可能跟锐胜具有一定的壮苗效果，提高了作物自身抵抗力有关系。

尽管有研究报道，6%戊唑醇悬浮种衣剂3～4 g a.i./100kg种子对小麦纹枯病具有很好的防效[13]，但本研究中戊唑醇2.4 g a.i./100kg种子未能观察到很好的防效。这可能与在本地区戊唑醇使用频繁，纹枯病产生抗药性有关系。

因此，在防治冬小麦纹枯病时，氟唑环菌胺·咯菌腈·噻虫嗪93.8 g a.i./100kg种子以上剂量有较好的防治效果。

参考文献

［1］ 张会云，陈荣振，冯国华，等. 中国小麦纹枯病的研究现状与展望［J］. 麦类作物学报，2007，6：1 150－1 153.

［2］ 陈健华，张炽昌，徐东方，等. 小麦纹枯病的研究进展［J］. 现代农业科技，2011，1：169－170.

［3］ 檀根甲，季伯衡. 小麦纹枯病的研究进展（综述）［J］. 安徽农业大学学报，1998，1：72－77.

［4］ 史建荣，王裕中，陈怀谷，等. 戊唑醇种子处理防治小麦纹枯病［J］. 植物保护学报，2000，3：231－237.

［5］ 孙炳剑，雷小天，袁虹霞，等. 小麦纹枯病化学防治药剂的筛选［J］. 麦类作物学报，2007，5：914－918.

［6］ 齐永霞，陈莉，丁克坚. 小麦纹枯病田间药剂防治技术研究［J］. 麦类作物学报，2015，4：577－583.

［7］ 任学祥，叶正和，丁克坚，等. 噻呋酰胺种衣剂防治小麦纹枯病效果及安全性研究［J］. 麦类作物学报，2015，11：1 588－1 591.

［8］ 佚名. 氟唑环菌胺［J］. 中国农药，2015，1：60.

［9］ 杨玉柱，焦必宁. 新型杀菌剂咯菌腈研究进展［J］. 现代农药，2007，5：35－39.

［10］ Schirra M，D'Aquino S，Palma A，et al. Residue level，persistence，and storage performance of citrus fruit treated with fludioxonil［J］. Journal of Agricultural & Food Chemistry，2005，53（17）：6 718－6 724.

［11］ 王中信，马晓燕，蔡宏芹. 2.5%适乐时种衣剂拌种对小麦纹枯病的防治效果［J］. 安徽农业科学，2001，5：626.

［12］ 程水明，宋家永，夏国军，等. 复方适乐时拌种防治小麦纹枯病和全蚀病的试验研究［J］. 麦类作物学报，2002，1：76－79.

［13］ 王成凤. 小麦纹枯病菌（*Rhizoctonia cerealis*）对噻呋酰胺和咯菌腈的抗性风险评估［D］. 南京：南京农业大学，2014.

313g/L 咯菌腈·氟唑环菌胺·噻虫嗪种衣剂包衣对小麦散黑穗病的田间防治效果

郭志刚*，张艳华，李 峰**

（先正达（中国）投资有限公司 保定基地，保定 071000）

摘要：313g/L 咯菌腈·氟唑环菌胺·噻虫嗪 是先正达公司即将上市的具有全新化学结构并且对麦类黑穗病有特效的新型三元复配种衣剂。试验结果表明，使用313g/L 咯菌腈·氟唑环菌胺·噻虫嗪 50mL/100kg seeds、100mL/100kg seeds、200mL/100kg seeds 包衣小麦防治小麦散黑穗病，防效分别为98%，99.8%和100%，优于对照药剂600g/L 吡虫啉·60g/L 戊唑醇衣剂。并且使用313g/L 咯菌腈·氟唑环菌胺·噻虫嗪，可以使植株更健壮，增产10%左右。

关键词：氟唑环菌胺；冬小麦；散黑穗病；防效

Efficacy of 313g/L Fludioxonix – Sedaxane – Thiamethoxam FS Against Loose Smut of Winter Wheat in Field

Guo Zhigang，Zhang Yanhua，Li Feng

（*Syngenta*（*China*）*Investment Company Limited*，*Baoding hub*，*Hebei*，071000）

Abstract：313g/L fludioxonix – sedaxane – thiamethoxam FS is a seedcare with new chemical structure against loose smut，which will be launched soon. The results showed that the efficacy of 313g/L fludioxonix – sedaxane – thiamethoxam FS at the rate of 50mL/100kg seeds，100mL/100kg seeds and 200mL/100kg seeds were 98%，99.8% and 100% respectively，which were better than the mixture of 60g/L tebuconazole and 600g/L imidachloprid. Moreover，313g/L fludioxonix – sedaxane – thiamethoxam FS can delay plant senescence，increase yield at 10%.

Key words：Fludioxonix – sedaxane – thiamethoxam；Winter wheat；Loose smut；control

小麦散黑穗又称菌，属担子菌亚门真菌[1]。异名厚垣孢子球形，褐色，一边颜色稍浅，表面布满细刺，20~25℃最宜发病，萌发先有菌丝，但不产生担孢子。侵害小麦，引起散黑穗病，该菌有寄主专化性，小麦上的病菌不能侵染大麦，但大麦上的病菌能侵染小麦。散黑穗病为花器侵染病害，一年只能侵染一次。带菌种子是病害传播的唯一途径[2-3]。

小麦散黑穗主要为害穗部，病株抽穗较健穗早，最初病穗外面包一层灰色薄膜，成熟后破裂，散出黑粉（即病菌的厚垣孢子），黑粉吹散后，只残留裸露的穗轴[4]。小麦上主茎、分蘖都出现病穗，但有报道称一些抗病品种上有的分蘖不发病。

氟唑环菌胺这类杀菌剂通过作用于病原菌线粒体呼吸电子传递链上的蛋白复合体Ⅱ（即琥珀酸脱氢酶）影响病原菌线粒体呼吸电子传递系统，阻碍其能量的代谢，抑制病原菌

* 第一作者：郭志刚，男，河北承德人，主要从事杀虫剂和种衣剂的田间药效试验工作

** 通讯作者：李峰，男，硕士，主要从事种衣剂产品开发工作；E-mail：feng. li@ syngenta. com

的生长、导致其死亡[5]，该物质低毒，用作种衣剂时，对多种种传、土传病害有较好的防治效果，还可促进作物根系的生长[6]。本试验使用313g/L咯菌腈·氟唑环菌胺·噻虫嗪种衣剂包衣用于防治小麦散黑穗的药效试验，为该产品在将来的大面积应用和推广提供科学根据。

1　材料与方法

1.1　供试药剂

313g/L咯菌腈·氟唑环菌胺·噻虫嗪种衣剂，瑞士先正达作物保护有限公司生产；60g/L立克秀，德国拜耳作物科学有限公司产品。

1.2　供试作物

冬小麦（冀麦823）。

1.3　防治对象

小麦散黑穗病菌（*Ustilago nuda*（*USTINH*，*USTIHO*，*USTINT*））。

1.4　试验地点以及田块情况

小麦散黑穗病试验设在河北省保定市東鹿园村进行，土壤为砂壤土，灌溉方便，杂草采用封闭药剂处理，虫害各试验小区统一用药，作物长势均匀，肥水管理一致。

1.5　试验设计和安排

供试作物于2014年10月6日播种。试验设有7个处理，每处理设4次重复，试验小区按完全随机区组排列，施药方法为种子包衣。试验小区面积为32m^2，播种前 一天进行包衣，阴凉处晾干后进行人工播种。

1.6　试验调查

散黑穗发病率调查：共调查1次，在乳熟期至成熟期进行。五点取样，每点调查1.5m^{2}内总株数和病株数，计算病株率。安全性调查在出苗后3、7d进行。产量调查为收获时调查每小区的全部产量，计算其增产率。

药效计算方法：

病株率（%）＝病株数/调查总株数×100

防治效果（%）＝（空白对照区药后病株率－处理区病株率）/空白对照区病株率×100

2　结果与分析

2.1　安全性

在整个试验过程中，安全性调查是至关重要的一个环节，每次药后3、7d，对比处理区与空白对照区的生长状况，尤其关注长势、分蘖，各剂量均未发现药害。说明313g/L咯菌腈·氟唑环菌胺·噻虫嗪种衣剂对小麦具有非常好的安全性。

2.2　防治效果

试验结果表明，313g/L咯菌腈·氟唑环菌胺·噻虫嗪种衣剂对小麦散黑穗病有非常好的防效，并优于对照药剂60g/L立克秀种衣剂。313g/L咯菌腈·氟唑环菌胺·噻虫嗪种衣剂50mL/100kg seeds、100mL/100kg seeds、200mL/100kg seeds对小麦散黑穗病包衣后调查发现：防效分别为98%、99.8%和100%，优于对照药剂60g/L戊唑醇处理区的防效97.5%（表1）。

表 1　313g/L 咯菌腈·氟唑环菌胺·噻虫嗪种衣剂对小麦散黑穗病防效

供试药剂	供试剂量（g a. i. /100kg seed）					防效			
	噻虫嗪	氟唑环菌胺	咯菌腈	吡虫啉	戊唑醇	调查株数（个）	发病株数（个）	发病率（%）	防效（%）
空白对照·噻虫嗪（发病严重度）	52.5	—	—	—	—	1 000	121.6	12.16	—
313g/L 咯菌腈·氟唑环菌胺·噻虫嗪	13.13	1.25	1.25	—	—	1 000	2.4	0.24	98
313g/L 咯菌腈·氟唑环菌胺·噻虫嗪	26.25	2.5	2.5	—	—	1 000	0.2	0.02	99.8
313g/L 咯菌腈·氟唑环菌胺·噻虫嗪	52.5	5	5	—	—	1 000	0	0	100
吡虫啉·戊唑醇	—	—	—	67.2	2.4	1 000	3	0.3	97.5

注：1. 防效调查只在小麦乳熟期进行病穗率调查，并计算其防效；2. 各处理均使用种衣剂包衣防治小麦蚜虫，如噻虫嗪与高巧

2.3　增产效果

在整个试验过程中，收获时按小区收获，将每个小区的小麦全部收获，并分别称重，然后得出所测的总产量，计算出相对于空白对照的增产率。测产结果表明，313g/L 咯菌腈·氟唑环菌胺·噻虫嗪种衣剂 50mL/100kg seeds、100mL/100kg seeds、200mL/100kg seeds 种子包衣对小麦的增产率分别为 7.0%、9.8% 和 15.7%，而对照药剂吡虫啉·戊唑醇种衣剂处理区的增产率为 6.6%，与最低剂量的 313g/L 咯菌腈·氟唑环菌胺·噻虫嗪种衣剂的增产率相当（表 2）。

表 2　小麦散黑穗试验测产结果

供试药剂	供试剂量（g a. i. /100kg seed）							
	噻虫嗪	氟唑环菌胺	咯菌腈	吡虫啉	戊唑醇	小区产量（kg）	亩产量（kg）	增产率（%）
空白对照·噻虫嗪（发病严重度）	52.5	—	—	—	—	20.48	427	—
313g/L 咯菌腈·氟唑环菌胺·噻虫嗪	13.13	1.25	1.25	—	—	21.91	456.7	7.0
313g/L 咯菌腈·氟唑环菌胺·噻虫嗪	26.25	2.5	2.5	—	—	22.49	468.8	9.8
313g/L 咯菌腈·氟唑环菌胺·噻虫嗪	52.5	5	5	—	—	23.7	494	15.7
吡虫啉·戊唑醇	—	—	—	67.2	2.4	21.8	454.4	6.6

注：测产为小区全部采收，保证数据准确有效

在试验过程中，发现 313g/L 咯菌腈·氟唑环菌胺·噻虫嗪种衣剂处理过的小麦叶片更加浓绿，植株长势旺盛，衰老慢，增产的原因有以下两点。

（1）氟唑环菌胺作为新型种子处理剂，能在各种新环境下增强根系吸收，利用根部水分和营养吸收能力，提高产量，防治来自土壤、空气中真菌和种子感染性真菌引起的病害，

对其植物根系起到良好的保护能力，并提供早期叶面病害的防治作用。

（2）SDHI 类杀菌剂具有光调节作用，可以增强植物变绿和适应能力[6]，植株长势旺盛，延缓衰老，故而产量增加。

3 讨论

田间试验的结果表明，在空白对照病害压力非常大的情况下，施用 313g/L 咯菌腈·氟唑环菌胺·噻虫嗪种衣剂包衣防效能够达到 98% 以上，而且能够延缓植株衰老，保持叶片浓绿，提高产量。因此，313g/L 咯菌腈·氟唑环菌胺·噻虫嗪种衣剂是防治小麦散黑穗病非常安全，且防效十分优秀的理想药剂。

生产上建议使用 313g/L 咯菌腈·氟唑环菌胺·噻虫嗪种衣剂的剂量为 50 ~ 100mL/100kg 在小麦播种前进行包衣处理，在药剂拌种时应做好安全防护工作，避免发生人畜中毒事故。

参考文献

[1] 王锁牢，刘建，郝彦俊，等. 烯唑醇对小麦散黑穗病防治效果及对小麦出苗和生长的影响 [J]. 新疆农业科学，2001，38（5）：260 - 261.

[2] 肖红，曹春梅，李子钦，等. 药剂防治小麦散黑穗病 [J]. 内蒙古农业科技，2000（S1）：129 - 130.

[3] 郑君民，李秀芹. 立克秀拌种防治小麦散黑穗病试验 [J]. 中国农技推广，2002（6）：37.

[4] 张贵，侯生英，王爱玲，等. 30g/L 苯醚甲环唑悬浮种衣剂防治小麦散黑穗病效果 [J]. 江苏农业科学，2012（5）：95，98.

[5] Harp T L，Godwin J R，Scalliet G，et al. Isopyrazam，a new generation cereal fungicide. Aspects of Applied Biology [J]. 2011（106）：113 - 120

[6] FRAC. List of fungal species with resistance reports towards SDHI fungicides and mutations in the succinate dehydrogenase gene [EB/OL]. [2014 - 09 - 03].

欧帕防治小麦锈病、小麦赤霉病、玉米大斑病及花生褐斑病效果研究报告

周美军，金丽华，陆悦健

（巴斯夫（中国）有限公司，上海　200137）

摘要：欧帕是巴斯夫公司在2015年推出的旱田作物杀菌剂，含有133g/L的吡唑醚菌酯和50g/L的氟环唑。多年的田间试验证明其对小麦锈病、小麦赤霉病、玉米大斑病和花生褐斑病具有优异的防治效果，还具有独特的施乐健植物健康功能，能改善植物生理机能，提高作物抗逆性，有效提升产量。

Abstract: Opera is a fungicide launched in China market by BASF in 2015, with active ingredient pyraclostrobin 133g/L and epoxiconazole 50g/L. Opera showed excellent efficacy for controlling of wheat rust and head scab, corn northern leaf blight and peanut brown leaf spot, and also showed good Agcelence effect which can help crop grow better for higher yield.

1　欧帕产品及病害介绍

欧帕是巴斯夫公司在2015年推出的旱田作物杀菌剂，含133g/L的吡唑醚菌酯和50g/L的氟环唑，剂型是安全环保的悬乳剂。氟环唑能通过木质部传导到植株顶端，具备优秀的内吸性能，能为整株植物提供保护，还能杀死已进入植株内的病原真菌；吡唑醚菌酯随时间向附近扩散，渗透进入植株，为整片叶片提供有效保护。

小麦锈病又叫黄疸，有条锈、叶锈、秆锈3种，是我国小麦上发生最广，为害最重的一类病害。条锈：主要在西北、华北、淮北冬麦区和西南冬麦区和西北春麦区发生为害。叶锈：主要在长江中下游麦区和四川、贵州发生多，近年华北、东北麦区也有上升趋势。秆锈：主要在东北、内蒙古春麦区和华东沿海冬麦区为害。

小麦赤霉病是麦类作物上的一种流行病害，尤以小麦受害最重。主要发生在穗期，造成穗腐；在扬花期发生，致小穗枯死，形成干瘪粒；后期在小穗基部出现粉红色胶状霉层，高湿条件下，粉红色霉层处产生蓝黑色小颗粒，即子囊壳；也可在苗期引起苗枯、基腐等症状；病麦粒中含有脱氧雪腐镰刀菌烯醇、玉蜀黍赤霉烯酮等多种毒素，人、畜误食后可发生中毒，怀孕母畜中毒后可导致流产。

玉米大斑病又称条斑病、煤纹病、枯叶病、叶斑病等。主要为害玉米的叶片、叶鞘和苞叶。叶片染病先出现水渍状青灰色斑点，然后沿叶脉向两端扩展，形成边缘暗褐色、中央淡褐色或青灰色的大斑。后期病斑常纵裂。严重时病斑融合，叶片变黄枯死。潮湿时病斑上有大量灰黑色霉层。下部叶片先发病。在单基因的抗病品种上表现为褪绿病斑，病斑较小，与叶脉平行，色泽黄绿或淡褐色，周围暗褐色。有些表现为坏死斑。

花生褐斑病主要为害花生叶片，初为褪绿小点，后扩展成近圆形或不规则形小斑，病斑较黑斑病大而色浅，叶正面呈暗褐或茶褐色，背面呈褐或黄褐色，病斑周围有亮黄色晕圈。

湿度大进病斑上可见灰褐色粉状霉层，即病菌分生孢子梗和分生孢子。叶柄和茎秆病斑长椭圆形，暗褐色。

2 欧帕防治小麦锈病试验

2.1 试验设计（表1）

表1 欧帕防治小麦锈病试验设计

编号	处理	喷药时间	药剂用量（mL/亩）	用水量（L/亩）	试验面积（亩）
1	空白对照				0.2
2	欧帕 183g/L	拔节末期孕穗初期（挑旗初期）	50	30	0.5
3	欧帕 183g/L	孕穗末期（麦芒显现）	50	30	0.5
4	欧帕 183g/L	扬花初期	50	30	0.5
5	欧帕 183g/L	扬花初期	50	30	0.5
		扬花盛期	50	30	
6	欧帕 183g/L	扬花盛期	50	30	0.5
7	47%多·酮	扬花初期 扬花盛期	100	30	0.5

2.2 试验时间及方法

试验地选择在小麦主产区真州镇农歌村小农场，其地块平整，土壤质地、肥力等均匀、田间长势一致，周围无树木、建筑等遮挡物。

施药方法：按30L/亩对水叶面喷雾处理。

施药时间：小麦拔节末期孕穗初期、孕穗末期、扬花初期、扬花盛期、各处理按试验设计时间要求打药。

2.3 试验结果及分析

示范结果表明，在这几个不同的生育时期用欧帕183g/L预防锈病，效果均不错，防效都在80%以上，但同时也表明用得早效果更好；在适期内用药，用两次比用一次的效果提高不了多少，因此用一次药即可；在扬花盛期每亩用欧帕50mL比大面积用47%多·酮100g要好不少，防效高出24个百分点（图1）。

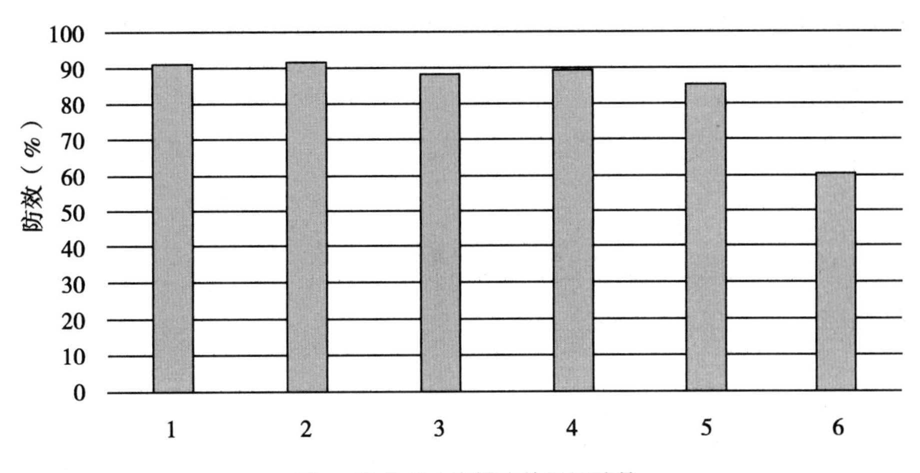

图 1　欧帕对小麦锈病的田间防效

注：1. 欧帕 50mL/亩，拔节末期施药一次；2. 欧帕 50mL/亩，孕穗末期一次用药；3. 欧帕 50mL/亩扬花初期一次用药；4. 欧帕 50mL/亩，扬花初期和扬花盛期两次用药；5. 欧帕 50mL/亩，扬花盛期一次用药；6. 多酮 47% WP 100mL/亩，扬花初期和扬花盛期两次用药

3　欧帕防治小麦赤霉病试验

3.1　试验设计（表 2）

表 2　欧帕防治小麦赤霉病试验设计

编号	处理	喷药时间	药剂用量（mL/亩）	用水量（L/亩）	试验面积（亩）
1	空白对照				0.5
2	欧帕	拔节末期孕穗初期（挑旗初期）82	50	30	0.5
3	欧帕	孕穗末期（麦芒显现）	50	30	0.5
4	欧帕	扬花初期	50	30	0.5
5	欧帕 L	扬花初期	50	30	0.5
		扬花盛期	50	30	
6	欧帕	扬花盛期	50	30	0.5
7	戊唑·多菌灵 40% SC	扬花初期	75	30	0.5
		扬花盛期	75	30	

3.2　试验时间及方法

本试验设在安徽省定远县严桥乡官东村大盛村民组，进行本试验的官东村地处定远县南，地势平坦，土壤肥力中等，肥力均匀，土质为水稻土，耕层深约 15cm，小麦于 2014 年 11 月 5 日播种，品种为淮麦 28，其他栽培条件与当地农业实践一致。处理 2 于小麦挑旗初期 1 次施药；处理 3 于小麦孕穗末期 1 次施药；处理 4 于小麦扬花初期 1 次施药；处理 5 于小麦扬花初期、扬花盛期 2 次施药；处理 6 于小麦扬花盛期 1 次施药；处理 7 于小麦扬花初

期、扬花盛期2次施药。使用人工背负式喷达3WBD 16L型电动喷雾器，喷雾器的工作压力为0.15～0.4MPa，喷片孔径为0.7mm，每亩用30kg药液均匀喷雾。

3.3 试验结果及分析

在7个处理中，以扬花初期施用欧帕1次和扬花初期、扬花盛期施用欧帕2次的防效最好，分别为92.21%、95.71%，达到优秀水平；农户自防处理2次用药防效达到了82.34%；在孕穗末期、扬花盛期用欧帕防治1次的防效分别为78.55%、75.35%；小麦挑旗初期施用欧帕1次的防效最差，为65.08%（图2）。

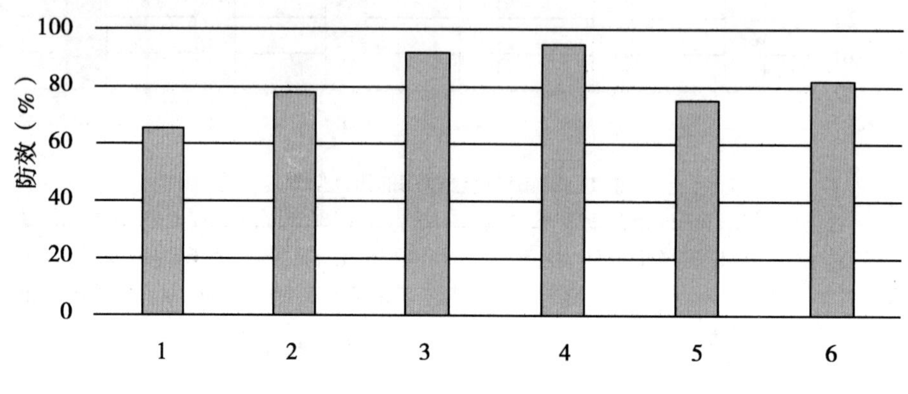

图2 欧帕防治小麦赤霉病的药效试验

注：1. 欧帕50mL/亩，拔节末期施药一次；2. 欧帕50mL/亩，孕穗末期一次用药；3. 欧帕50mL/亩扬花初期一次用药；4. 欧帕50mL/亩，扬花初期和扬花盛期两次用药；5. 欧帕50mL/亩，扬花盛期一次用药；6. 戊多40% SC 75mL/亩，扬花初期和扬花盛期两次用药。

4 欧帕防治玉米大斑病试验

4.1 试验设计（表3）

表3 欧帕防治玉米大斑病试验设计

处理编号	药剂	施药剂量（制剂g/亩）	有效成分量（g/hm^2）
1	欧帕	40	103.2
2	欧帕	50	129.0
3	欧帕	60	154.8
4	25%吡唑醚菌酯乳油	26.7	100
5	12.5%氟环唑悬浮剂	120	135
6	25%丙环唑乳油	45	168.75
7	45%代森铵水剂	100	675.0
8	空白对照		

4.2 试验时间及方法

试验安排在林甸县天弘种业试验园区的玉米田中进行，该园区内历年在种植玉米期间均有大斑病发生。所有试验小区的栽培条件一致，符合当地科学的农业实践。土壤类型：碳酸盐黑钙土，亩施农肥2 000kg，二胺40kg，试验地面积1 000m²，每个小区面积为30m²，采用垄作种植。垄宽0.65m。在玉米8～10叶期第一次施药，11～12叶期第二次施药，13～14叶期第三次施药，用水量450L/hm²。

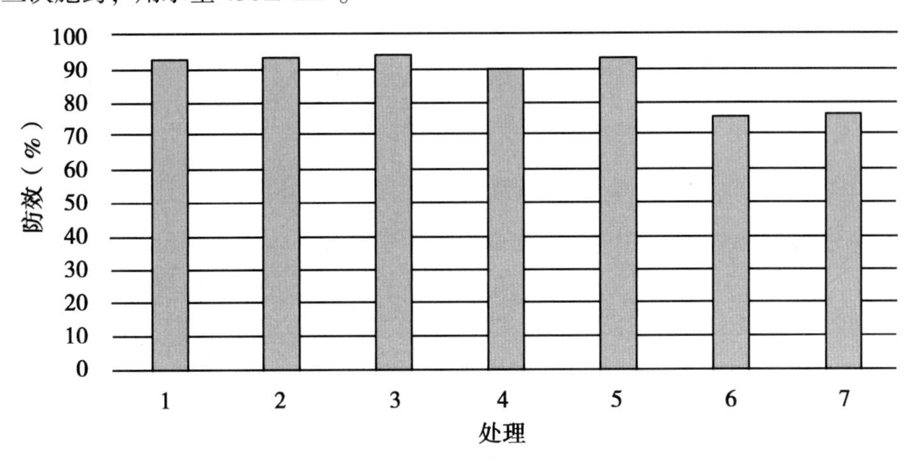

图3　欧帕防治玉米大斑病试验

注：1. 欧帕40mL/亩，3次用药；2. 欧帕50mL/亩，3次用药；3. 欧帕60mL/亩，3次用药；4. 25%吡唑醚菌酯乳油26.7mL/亩，3次用药；5. 12.5%氟环唑悬浮剂120mL/亩，3次用药；6. 25%丙环唑乳油45mL/亩，3次用药；7. 45%代森铵水剂100mL/亩，3次用药

4.3　试验结果与分析

欧帕在40～60g/亩剂量范围内，对玉米生长安全、无要害、保产效果明显。对大斑病具有较好的防治效果，连续三次用药后15d防效为93.0%～94.1%，可以在生产中推广使用（图3）。

5　欧帕防治花生褐斑病试验

5.1　试验设计（表4）

表4　欧帕防治花生褐斑病试验设计

处理编号	药剂	施药剂量（g/亩）	有效成分（g/hm²）
1	欧帕	40	103.2
2	欧帕	50	129
3	欧帕	60	154.8
4	25%吡唑醚菌酯乳油	26.7	100
5	12.5%氟环唑悬浮剂	120	135
6	40%百菌清悬浮剂	250	1 500
7	30%苯甲+丙环唑乳油	30	135
8	50%戊唑醇水分散粒剂	11.25	84.4
9	空白对照		

5.2 试验时间及方法

花生于2013年6月8日播种，试验地属壤土，肥力中等，排灌方便，土壤pH值在7.1左右，试验地常规管理。共施药3次，分别在8月30日，9月6日，9月17日进行。小区随机区组排列，重复4次。

5.3 试验结果及分析

末次施药后10d的调查结果表明，欧帕有效成分用量154.8g/hm^{2}的防治效果最好，为81.9%，与其他各处理差异极显著。欧帕有效成分用量129g/hm^{2}处理防效为76.3%，欧帕有效成分用量103.2g/hm^{2}处理的防效为73.3%，对照药剂12.5%氟环唑悬乳剂有效成分用量135g/hm^{2}处理的防效为75.6%，对照药剂30%苯甲+丙环唑乳油有效成分用量135g/hm^{2}处理的防效为75.2%，对照药剂25%吡唑醚菌酯乳油有效成分用量100g/hm^{2}处理的防效为75.1%，5个处理间差异不显著。对照药剂40%百菌清可湿性粉剂有效成分用量1 500 g/hm^{2}处理的防效为72.2%，对照药剂50%戊唑醇水分散粒剂有效成分用量84.4g/hm^{2}处理的防效为70.3%，各处理与清水对照呈极显著差异。通过田间试验表明，欧帕对花生褐斑病有明显的防治效果，使用欧帕应在花生褐斑病发生初期用药（图4）。

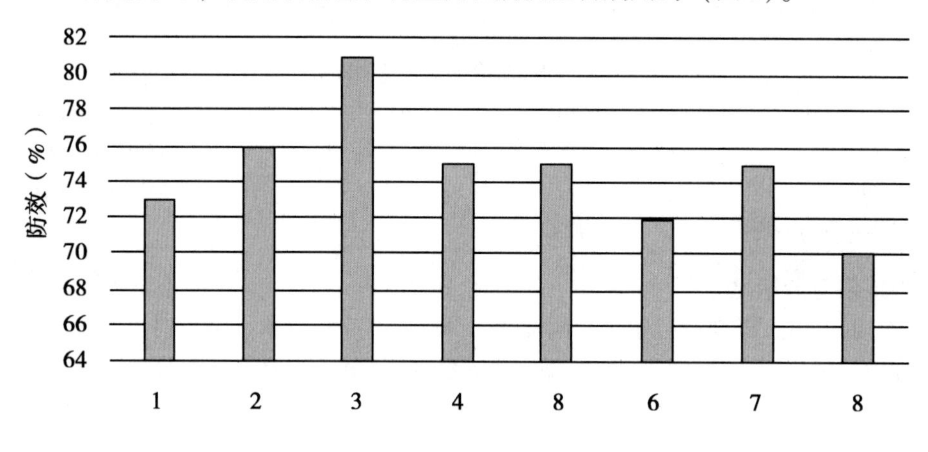

图4 欧帕防治花生褐斑病试验

注：1. 欧帕40mL/亩，3次用药；2. 欧帕50mL/亩，3次用药；3. 欧帕60mL/亩，3次用药；4. 25%吡唑醚菌酯乳油26.7mL/亩，3次用药；5. 12.5%氟环唑悬浮剂120mL/亩，3次用药；6. 40%百菌清悬浮剂250mL/亩，3次用药；7. 30%苯甲+丙环唑乳油30mL/亩，3次用药；8. 50%戊唑醇水分散粒剂11.25g/亩。

6 结语

欧帕是巴斯夫公司推出的含有133g/L的吡唑醚菌酯和50g/L氟环唑的复配悬浮剂，对小麦锈病、小麦赤霉病、玉米大斑病和花生褐斑病具有优异的防治效果；同时，欧帕还含有施乐健植物健康功能，能改善作物生理机能，增强作物的抗逆性，有效提升产量。

英腾42%悬浮剂（SC）防治白粉病药效试验结果

苏瑞霭，冯希杰
（巴斯夫（中国）有限公司，上海 200137）

摘要：英腾42%苯菌酮悬浮剂是巴斯夫公司开发的，在中国第一个登记的二苯菌酮类杀菌剂。其具有优异的预防和保护作用，主要用于防治瓜果类白粉病，通过作用于病菌细胞骨架——肌动蛋白，造成细胞不稳定，干扰其形成及分解，具有抑制病菌的菌丝生长，孢子形成的作用。通过田间对草莓白粉病、甜瓜白粉病和苦瓜白粉病的试验结果表现，英腾1 500倍液处理3次药后7d对草莓白粉病的预防效果优秀，达到97.4%，与对照药剂硝苯菌酯36%EC 1 500倍液和乙嘧酚25%SC750倍液的防效相当；英腾1 875倍液和1 500倍液对甜瓜上部叶片白粉病的预防效果优秀，三次药后7d防效分别达到91.9%和95.7%，明显优于对照药剂处理的效果；英腾2 500倍液、1 875倍液和1 500倍液对苦瓜白粉病的预防效果优秀，持效期长，3次药后11d防效分别达到94.1%、95.4%和97.8%，明显优于对照药剂处理。推荐在白粉病发生前或发生极早期（发病率0～1%）时开始用药，连续用药3次，间隔期7d。

Abstract: Metrafenone is developed by BASF, whose brand name is Vivando, it is the first benzophenone fungicide registered in China. Metrafenone shows excellent preventive activity against powdery mildew on cucurbits and bean vegetables. Identified by field trials, Vivando 42% SC 1 500 × showed excellent control efficacy against strawberry powdery mildew, which reached 97.4% 7 days after the 3rd application, no significant difference with Mepthldinocap 36% EC 1 500 × and ethirimol 25% SC 750 ×. Vivando 1 875 × and 1 500 × showed excellent control efficacy against powdery mildew on upside leaves of sweet melon 7 days after the 3rd application, which reached 91.9% and 95.7%, superior to local standards Mepthldinocap 36% EC 1 500 × and ethirimol 750 ×. For bitter gourd powdery mildew, Vivando 2 500 ×, 1 875 × and 1 500 × showed very good performance, which reached 94.1%, 95.4% and 97.8%.

1 英腾42%SC防治草莓白粉病试验

1.1 试验设计

1.1.1 药剂处理与试验地情况

供试药剂：英腾42%SC，设3个处理浓度，分别为2 500倍、1 875倍和1 500倍；对照药剂：①硝苯菌酯36%EC 1 500倍；②乙嘧酚25%SC 750倍，另设空白对照。每处理4次重复，共20小区，小区面积2.5m^2。

试验地点：上海奉贤。

试验品种：红霞，生长期果实膨大期至成熟期。

1.1.2 施药和调查方法

于2015年3月13日至4月9日施药，共4次，间隔期7d，用水量675L/hm^2。分别于第3次药后7d，第4次药后7d，14d和19d调查白粉病发病情况。采用全小区调查总果数和病果数，计算各处理的病果率和相对防效。药前各处理果实白粉病发病率1%～2%，叶片未有白粉病发生。

1.2 试验结果与分析（表1）

表1 英腾42%SC防治草莓白粉病试验防治结果

编号	处理	三次药后7d		四次药后7d		四次药后14d		四次药后19d	
		防效（%）	差异显著性	防效（%）	差异显著性	防效（%）	差异显著性	防效（%）	差异显著性
1	空白对照（病果率）	11		22.8		34.1		45.9	
2	英腾42%SC 2 500倍液	80.5	a	78.4	a	73.6	a	68.6	a
3	英腾42%SC 1 875倍液	88.3	a	82.1	a	80.9	a	76.7	a
4	英腾42%SC 1 500倍液	97.4	a	75.9	a	78.2	a	80.9	a
5	硝苯菌酯36% EC 1 500倍液	94.8	a	89	a	87	a	78.5	a
6	乙嘧酚25% SC 750倍液	98.7	a	78.4	a	76.5	a	68.1	a

试验结果表明，3次药后7d，英腾1 500倍液对草莓白粉病具有优秀的预防效果，优于2 500倍液和1 875倍液处理。英腾1 500倍液处理稍优于硝苯菌酯36% EC 1 500倍液，但稍差于乙嘧酚25% SC 750倍液处理，各处理间的差异性不显著。

4次药后19d，英腾1 500倍液对草莓白粉病防效达到80.9%，优于2 500倍液和1 875倍液处理，也优于对照药剂硝苯菌酯36% EC 1 500倍液和乙嘧酚25% SC 750倍液处理，各处理间的差异性不显著。

2 英腾42%SC防治甜瓜白粉病试验结果

2.1 试验设计

2.1.1 药剂处理与试验地情况

供试药剂：英腾42% SC，设3个处理浓度，分别为2 500倍、1 875倍和1 500倍。

对照药剂：①硝苯菌酯36% EC 1 500倍；②乙嘧酚25% SC 750倍，另设空白对照。每处理4次重复，共20小区，小区面积7m^2。

试验地点：上海奉贤。

试验品种：金香玉，生长期为花期。

2.1.2 施药和调查方法

于2015年5月27日至6月10日施药，共3次，间隔期7d，用水量900 L/hm^2。分别于第一次药后7d，第二次药后7d，第三次药后7d和14d调查白粉病发病情况。第一次药后7d调查全部叶片，其他时间分别区分上部叶片和下部叶片调查。药前总体发病病指在0～0.5%。

2.2 试验结果与分析

2.2.1 对甜瓜上部叶片的防治效果（表2）

表2 英腾42%SC防治甜瓜白粉病上部叶片

编号	处理	一次药后7d		二次药后7d		三次药后7d		四次药后14d	
		防效（%）	差异显著性	防效（%）	差异显著性	防效（%）	差异显著性	防效（%）	差异显著性
1	空白对照（病指）	6.7		28.3		35		62	
2	英腾42%SC 2 500倍液	75	b	91.8	b	88.6	b	80.1	b
3	英腾42%SC 1 875倍液	80.1	bc	92.3	bc	91.9	b	85.4	b
4	英腾42%SC 1 500倍液	85	bcd	97.1	bc	95.7	b	89.8	b
5	硝苯菌酯36% EC 1 500倍液	75	b	70.6	a	74.3	a	54.1	a
6	乙嘧酚25% SC 750倍液	60	a	75.3	a	75.2	a	78.4	b

试验结果表明，英腾42%SC 1 500～2 500倍液对甜瓜上部叶片白粉病的预防效果明显，三次药后7d的防效达到88.6%～95.7%，三次药后14d达到80.1%～89.8%。均明显优于对照药剂硝苯菌酯36%EC 1 500倍液和乙嘧酚25%SC 750倍液处理。

3.2.2 对甜瓜下部叶片的防治效果（表3）

表3 英腾42%SC防治甜瓜白粉病下部叶片防治效果

编号	处理	二次药后7d		三次药后7d	
		防效（%）	差异显著性	防效（%）	差异显著性
1	空白对照（病指）	15.7		40	
2	英腾42%SC 2 500倍液	27.7	a	16.7	a
3	英腾42%SC 1 875倍液	74.5	b	41.7	ab
4	英腾42%SC 1 500倍液	66.0	b	62.5	b
5	硝苯菌酯36%EC 1 500倍液	53.2	c	50	ab
6	乙嘧酚25%SC 750倍液	55.3	c	50	ab

试所有药剂处理对甜瓜下部叶片白粉病的防治效果均差于上部叶片的防治效果；英腾1 500倍液处理的防效在第三次药后7d明显优于英腾2 500倍液和1 875倍液处理，也优于对照药剂硝苯菌酯36%EC 1 500倍液和乙嘧酚25%SC750倍液处理。

3 英腾42%SC防治苦瓜白粉病试验结果

3.1 试验设计

3.1.1 药剂处理与试验地情况

供试药剂：英腾42%SC，设3个处理浓度，分别为2 500倍液、1 875倍液和1 500倍液。

对照药剂：①硝苯菌酯 36% EC 1 500倍液；②乙嘧酚 25% SC 750 倍液，另设空白对照。每处理 4 次重复，共 20 小区，小区面积 3.5m^2。

试验地点：上海奉贤。

试验品种：粤华，生长期侧枝形成期至花期。

3.1.2 *施药和调查方法*

于白粉病发病前开始用药，全程共用药 3 次，分别是 2015 年 9 月 28 日、10 月 8 日和 10 月 23 日，用水量 900L/hm^2。分别于第二次药后 11d，第三次药后 4d，11d 和 18d 调查白粉病发病情况，调查方法采用全株调查。

3.2 试验结果与分析

试验结果表明（表 4），英腾 2 500倍液、1 875倍液和 1 500倍液处理对苦瓜白粉病表现出优异的预防效果，3 次药后 4～11d，均明显优于对照药剂硝苯菌酯 36% EC 1 500倍液处理；优于乙嘧酚 25% SC 750 倍液处理，但差异性不显著。

表 4 英腾 42%SC 防治苦瓜白粉病防治效果

编号	处理	三次药后 4d		三次药后 11d		三次药后 18d	
		防效（%）	差异显著性	防效（%）	差异显著性	防效（%）	差异显著性
1	空白对照（病指）	20.9		41.1		55.6	
2	英腾 42% SC 2 500 倍液	88.6	b	94.1	b	82.5	bc
3	英腾 42% SC 1 875 倍液	94.6	b	95.4	b	89.1	c
4	英腾 42% SC 1 500倍液	94.7	b	97.8	b	90.9	c
5	硝苯菌酯 36% EC 1 500倍液	31.0	a	61.6	a	64	a
6	乙嘧酚 25% SC 750 倍液	84.1	b	84.6	b	67.6	ab

英腾 1 875 倍液、1 500倍液处理，3 次药后 18d 的防治效果均优于其他处理，与对照药剂硝苯菌酯 36% EC 1 500倍液和乙嘧酚 25% SC 750 倍液处理比较，达到极显著；与英腾 2 500倍液处理比较达到显著性。

4 结论

从试验结果看，英腾对草莓白粉病、甜瓜白粉病和苦瓜白粉病均表现出理想的预防和保护效果，持效期长达 7～14d。英腾 1 500倍液对草莓果实白粉病的预防效果，与常规处理药剂硝苯菌酯 36% EC 1 500倍液和乙嘧酚 25% SC750 倍液的效果相当；英腾 2 500倍液、1 875 倍液和 1 500倍液对甜瓜白粉病和苦瓜白粉病的预防效果明显优于对照药剂硝苯菌酯 36% EC 1 500倍液和乙嘧酚 25% SC750 倍液。建议在白粉病发生前或发生极早期（发病率 0～1%）时开始用药，连续用药 3 次，间隔期 7d，防治草莓白粉病推荐 1 500倍液，防治甜瓜白粉病和苦瓜白粉病推荐 1 500～1 875倍液。

参考文献（略）

2－酰氧基乙磺酰胺衍生物合成与杀菌活性研究

王闽龙*，芮　朋，纪明山，祁之秋，李兴海**
（沈阳农业大学植物保护学院，沈阳　110866）

摘要：为了获得具有高效杀菌活性的先导化合物，探索合成了结构新颖的2－酰氧基乙磺酰胺衍生物。以*N*－（2－三氟甲基－4－氯苯基）－2－（3，5－二氟苯基）－2－羟基乙磺酰胺为原料，经过与酰氯在TMEDA和分子筛的催化下反应，制备*N*－（2－三氟甲基－4－氯苯基）－2－（3，5－二氟苯基）－2－酰氧基乙磺酰胺化合物（V）。再以采自辽宁省多个地区灰霉病病菌为供试靶标，采用菌丝生长速率法和活体盆栽试验对化合物进行活性筛选。合成了16个*N*－（2－三氟甲基－4－氯苯基）－2－（3，5－二氟苯基）－2－酰氧基乙磺酰胺衍生物，结构经^{1}H NMR，IR及元素分析确认。生测表明：化合物V－10对灰霉病病菌DL－11和HLD－15的EC_{50}为0.03μg/mL和3.23μg/mL，表现出极优异的抑菌活性。[结论]：该类化合物具有很好的杀菌效果，有进一步开发的价值。

关键词：苯乙酮；乙磺酰胺衍生物；灰霉病病菌；杀菌活性。

Synthesis and Fungicidal Activity of Novel 2－Acyloxy Ethyl Sulfonamide Derivatives

Wang Minlong，Rui Peng，Ji Mingshan，Qi Zhiqiu，Li Xinghai*
（*College of Plant Protection*，*Shenyang Agricultural University*，*Shenyang* 110866，*China*）

Abstract：To explore the novel compounds with high fungicidal activity，a series of novel 2－acyloxyethylsulfonamides were synthesized. *N*－（2－trifluoromethyl－4－chlorophenyl）－2－（3，5－difluorophenyl）－2－acyloxyethylsulfonamides were synthesized by the reaction of *N*－（2－trifluoromethyl－4－chlorophenyl）－2－（3，5－difluorophenyl）－2－hydroxylethylsulfonamides with acyl chloride in dichloromethane under the catalysis of TMEDA and molecular sieve. The fungicidal activities against Botrytis cinerea Pers. strains collected from different areas in Liaoning，China *in vitro* were evaluated by adopting mycelium growth rate method *in vitro* and houseplants method *in vivo*. [result] 16 2－acyloxyethylsulfonamides，whose structures were confirmed by IR，^{1}H NMR and element analysis，had been synthesized. The results of the bioassay showed that the compounds V－10 possessed excellent fungicidal activity against *Botrytis cinerea* DL－11 and HLD－15，the EC50 values of which were 0.03 and 3.23μg/mL，respectively. All the compounds show high fungicidal activities and can be used as possible new lead compounds for further developing novel fungicides against *B. cinerea*.

Key words：acetophenone；ethylsulfonamides；*Botrytis cinerea* Pers；fungicidal activities

磺酰胺类化合物具有广泛的生物活性，并在临床医学药物和农用化学品领域都扮演着重要角色[1-3]。由此，含磺酰基化合物的研发吸引了越来越多的关注，并且其研发进程也日益

* 作者简介：王闽龙，男，硕士，主要从事农药合成与杀菌活性研究；E-mail：wangminlong@163.com
** 通讯作者：李兴海，男，博士，副教授，主要从事农药合成与杀菌活性研究；E-mail：xinghai30@163.com

加快。在杀菌剂领域，磺酰胺类化合物也受到人们的关注[4-5]。随着研究的深入，市场上也已出现了商品化的磺酰胺类杀菌剂品种。且在近年来，一些含磺酰基的具有高效杀菌活性的新颖结构也相继被报道。含 1，2，3－三唑苯磺酰胺类化合物（A）在 50 mg/L 时对小麦锈病（*Puccinia recondite* f. sp. *tritici*）的抑制率达到了 80%[6]。含有吡喃环苯磺酰胺类化合物（B）在 50mg/L 浓度对芦笋茎枯病菌（*Phomopsis asparagi*（Sacc.）Bubak）和番茄灰霉病菌（*Botrytis cinerea*）的抑制率达 80% 以上[7]。含有 1，6－己内酰胺苯磺酰胺类化合物（C）在 50mg/L 浓度对番茄叶霉病菌（*Cladosporium fulvum*）的抑制率为 90%，与百菌清相当[8]。含噻唑烷酮苯磺酰胺类化合物（D）对棉花枯萎病菌（*Fusarium oxysporum* Schl. f. sp. *vasinfectum*(Atk.) Snyd & Hans）具有一定的杀菌活性[9]。香豆素磺酰胺类化合物（E）有良好的杀细菌活性[10]。一些杂环基磺酰胺类化合物也具有良好的杀菌活性，如 1，3，4－噻二唑磺酰胺（F）在 500 mg/L 浓度对烟草花叶病毒（TMV）的抑制率为 42.5 %[11]。

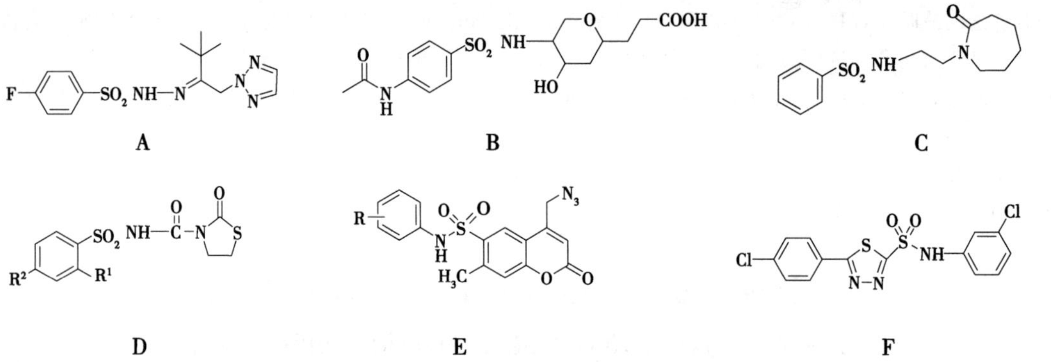

图 1　具有杀菌活性的磺酰胺类化合物

Figure 1　Structure of some sulfonyl－containing compounds（A－H）with antifungal activity

在我们先前的研究工作中，以 2－氧代环己烷基磺酰胺为先导化合物，系统研究了将羰基还原成羟基，而后进一步将羟基酯化，所得的这一系列化合物的生物活性与构效关系[12-14]。在本研究中，借鉴前期的研究工作[15-16]，为了寻找到更高杀菌活性的化合物，以 2－取代苯基－2－氧代乙磺酰胺为先导，将羰基还原而后酯化，合成一系列结构新颖的 2－酰氧基乙磺酰胺化合物（V），这类化合物都具有非常高的杀菌活性和宽广的杀菌谱。目标化合物的合成路线见图 2。

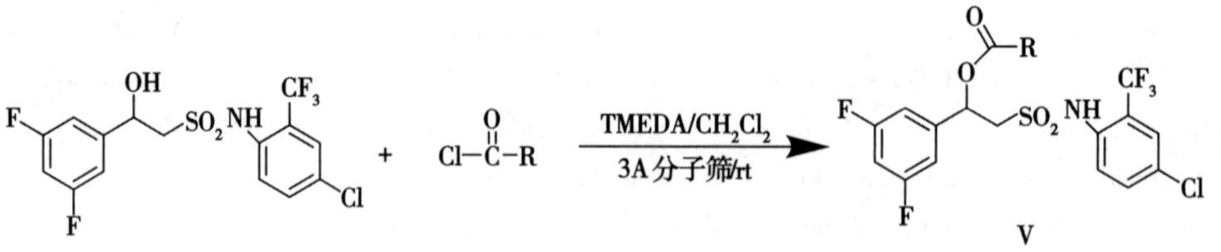

图 2　化合物的合成路线

Figure 2　Synthetic route of compounds V

1　材料与方法

1.1　仪器与药剂

X－5 型熔点测定仪（温度计未校正）；Bruker300－MHz 型核磁共振仪（TMS 为内标，溶剂为 CDC13）；岛津 IRAffinity－1 傅立叶红外变换光谱仪（KB 压片法）。R－210 旋转蒸发仪（Buchi Rotavapor）；德国 ELEMENTAR 公司 Vario ELIII 型元素分析仪。

1.2　供试菌株

供试植物病原真菌为黄瓜灰霉病菌（*Botrytis cinerea* Pers）：DG－1（采自辽宁东港）、CY－09（采自辽宁朝阳）、CY－14（采自辽宁朝阳）、DL－11（采自辽宁大连）、HLD－15（采自辽宁葫芦岛）、FS－09（采自辽宁抚顺）、FS－10（采自辽宁抚顺）均为沈阳农业大学植物保护学院农药学科教研室采集、保存的菌种，经活化后使用。黄瓜植株是由中国农业科学院出售的黄瓜种培育的。

1.3　合成试验

1.3.1　2－酰氧基乙磺酰胺衍生物（V）的合成

在氮气保护下，室温将干燥的 20 mL CH_2Cl_2，0.001 mol *N*－（2－三氟甲基－4－氯苯基）－2－（3，5－二氟苯基）－2－羟基乙磺酰胺、0.1 mL *N*，*N*，*N*'，*N*'－四甲基乙二胺（TMEDA）、1 g 分子筛先后加入 50 mL 三口瓶中，搅拌下加入酰氯，搅拌反应 4h，抽滤，滤液转入分液漏斗，快速用冰水（15 mL×2）洗涤后，用无水硫酸钠干燥 30min，再抽滤，用旋转蒸发仪减压浓缩得粗产物。

1.3.2　2－酰氧基乙磺酰胺化合物（V）的纯化与结构鉴定

采用硅胶柱层析与重结晶对化合物进行纯化，用^{1}H NMR、IR 和元素分析进行结构鉴定。

1.4　杀菌活性筛选方法

1.4.1　菌丝生长速率法

参照农药室内生物测定试验准则，采用平皿法（抑制菌丝生长实验），用不同浓度的含药培养基测定目标化合物 V 对病原真菌的杀菌活性，病原菌为采自辽宁省不同地区的黄瓜灰霉病菌（*Botrytis cinerea* Pers），筛选浓度为 50 mg/L。以腐霉利、百菌清、B－1（N－（2－三氟甲基－4－氯苯基）－2－氧代－2－苯基乙磺酰胺）为对照药剂，设置丙酮溶剂为空白对照，每个处理重复三次。在（23±2）℃培养 3～4d，待空白对照中的菌落充分生长后，以十字交叉法测量各处理的菌落直径，取其平均值，用以下公式计算抑制率，利用 Excel 软件计算 EC_{50}：

$$抑制率（\%）=\frac{空白菌落直径-含药菌落直径}{空白菌落直径-菌饼直径}\times 100$$

1.4.2　黄瓜活体叶片法

将根据《农药室内生物测定试验准则　杀菌剂》NY/T 1156.9—2008：采用活体盆栽茎叶喷雾法即黄瓜活体叶片法，首先将化合物 V 分别配制成 2% 的乳油溶液，方法如下：先准确称取 20 mg 的化合物，与 15 mg 吐温 20 混匀后用 0.1 mL 的 DMSO 将之溶解，然后将其与 7.5 mg 农乳 500 与 30 mg 农乳 600 溶解到 0.4 mL 二甲苯中，再用二甲苯补足到 1mL 即为 2% 的乳油溶液。

（1）取 2 % 乳油 100μL 用水 4 mL 稀释成 500 mg/L，喷雾。

（2）空白：为不加药的0%乳油溶液。在黄瓜子叶长至2叶期左右要进行药剂喷雾处理，待植株表面的药液自然风干并充分吸收后接种灰霉病菌菌饼，再移至低温高湿常规培养，定期观察空白对照组发病情况，待空白对照组发病后，记录黄瓜发病状态，病斑直径大小。

利用十字交叉法计算植株发病情况。利用Excel软件统计方法计算，比较化合物V的防效。

2 结果与分析

2.1 化合物的合成与结构鉴定

按照上述的方法合成得到2－（3，5－二氟苯基）－2－酰氧基乙磺酰胺类化合物（V）共16个，且所得化合物产率较高，收率大多数均在65%以上。其结构经过^{1}H NMR分析、IR分析及元素分析确认，所得波谱数据与其分子结构吻合，具体分析结果见表1和表2。磺酰胺中苯环上的H位移均在7.50ppm左右，处于磺酰基和羰基之间的CH_2上H位移在4.70mg/kg左右，为单重峰，NH上H位移在7.25左右。红外光谱中C＝O的伸缩振动吸收峰出现在1 680～1 730cm^{-1}，NH的伸缩振动吸收峰出现在3 250～3 350cm^{-1}。

表1 2－酰氧基乙磺酰胺（V）的物化数据及元素分析结果

Table 1 Physical and elemental data of 2－oxo ethyl sulfonamides（V）

序号 Compd.	R	收率（%） yield（%）	熔点（℃） m. p.（℃）	元素分析 elemental analysis C（理论值） C（calcd）	H（理论值） H（calcd）	N（理论值） N（calcd）
V－1	CH_3	65.5	96～98	44.42（44.60）	3.01（2.86）	3.25（3.06）
V－2	C_6H_5	51.8	93～95	50.67（50.83）	3.12（2.91）	2.48（2.69）
V－3	2－$CH_3C_6H_5$	56.3	69～71	51.92（51.74）	3.10（3.21）	2.81（2.62）
V－4	3－$CH_3C_6H_5$	81.1	114～116	51.59（51.74）	3.43（3.21）	2.75（2.62）
V－5	4－$CH_3C_6H_5$	81.8	137～139	51.88（51.74）	3.10（3.21）	2.57（2.62）
V－6	4－$OCH_3C_6H_5$	60.5	118～120	50.49（50.24）	2.94（3.12）	2.72（2.55）
V－7	2，4－$(CH_3)_2C_6H_5$	71.8	120～122	52.74（52.61）	3.41（3.50）	2.68（2.56）
V－8	$ClCH_2CH_2$	62.7	124～126	42.56（42.70）	2.91（2.79）	2.58（2.77）
V－9	Cl_2CH	51.4	150～152	38.86（38.77）	2.33（2.36）	2.48（2.66）
V－10	CCl_3	80.9	168～170	36.51（36.39）	2.02（1.80）	2.31（2.50）
V－11	2－FC_6H_5	75.6	118～120	49.32（49.13）	2.51（2.62）	2.84（2.60）
V－12	3－FC_6H_5	32.9	97～99	48.97（49.13）	2.54（2.62）	2.77（2.60）
V－13	2－ClC_6H_5	49.6	99～101	47.43（47.67）	2.38（2.55）	2.69（2.53）
V－14	3－ClC_6H_5	69.5	114～116	47.45（47.67）	2.78（2.55）	2.39（2.53）
V－15	2－$CF_3C_6H_5$	79.8	81～83	46.78（46.99）	2.64（2.40）	2.14（2.38）
V－16	3－$CF_3C_6H_5$	43.9	119～121	47.11（46.99）	2.26（2.40）	2.51（2.38）

表2 2－酰氧基乙磺酰胺衍生物的核磁与红外数据

Table 2 ^{1}H NMR and IR data of 2－acyloxy ethyl sulfonamides

序号 Compd.	^{1}H NMR（$CDCl_3$，400MHz，TMS，ppm），δ	IR，ν/ cm^{-1}
V－1	2.11（s，3H，CH_3），3.38－3.68（m，2H，CH_2），6.27－6.29（q，1H，CH），7.55－7.58（d，1H，NH），6.78－7.75（m，6H，$C_6H_3+C_6H_3$）	3350（NH），1761（C=O）
V－2	3.53－3.88（m，2H，CH_2），6.77－6.95（q，1H，CH），6.77－6.95（q，3H，CH_3），7.55－7.76（d，1H，NH），7.44－7.4（7t，2H，CH2），7.50－8.01（m，6H，$C_6H_3+C_6H_3$）	3308（NH），1748（C=O）
V－3	2.54（s，3H，CH_3），3.51－3.82（m，2H，CH_2），6.44－6.45（q，1H，CH），6.78－8.05（m，11H，$C_6H_3+C_6H_3+NH+C_6H_4$）	3325（NH），1748（C=O）
V－4	2.41（s，3H，CH_3），3.53－3.89（m，2H，CH_2），6.45－6.48（q，1H，CH），6.77－6.79（m，11H，$C_6H_3+C_6H_3+NH+C_6H_4$）	3254（NH），1748（C=O）
V－5	2.43（s，3H，CH_3），3.52－3.87（q，2H，CH_2），6.45－6.47（q，1H，CH），6.77－7.88（m，11H，$C_6H_3+C_6H_3+NH+C_6H_3$）	3308（NH），1748（C=O）
V－6	3.52－3.55（q，1H，CH），3.83－3.87（q，1H，CH），3.88（s，3H，CH_3），6.43－6.45（q，1H，CH），6.75－7.95（m，11H，$C_6H_3+C_6H_3+NH+C_6H_4$）	3308（NH），1717（C=O）
V－7	2.37（s，3H，CH_3），2.51（s，3H，CH_3），3.51－3.85（m，2H，CH_2），6.42－6.44（q，1H，CH），6.77－7.83（m，11H，$C_6H_3+C_6H_3+NH+C_6H_3$）	3308（NH），1746（C=O）
V－8	3.30－3.40（d，2H，CH_2），3.77－5.82（m，6H，$CH_2+CH_2+CH_2$），6.58－7.85（m，7H，$C_6H_3+C_6H_3+NH$）	3503（NH），1692（C=O）
V－9	3.93－4.62（m，2H，CH_2），6.53（d，1H，CH），5.99－6.02（d，1H，CH），6.35－7.88（m，7H，$C_6H_3+C_6H_3+NH$）	3328（NH），1748（C=O）
V－10	4.05－4.71（m，2H，CH_2），6.32－6.34（q，1H，CH），6.88－7.79（m，7H，$C_6H_3+C_6H_3+NH$）	3480（NH），1748（C=O）
V－11	3.52－3.88（m，2H，CH_2），6.49－6.51（q，1H，CH），6.78－7.93（m，11H，$C_6H_3+C_6H_3+NH+C_6H_4$）	3308（NH），1744（C=O）
V－12	3.51－3.97（m，2H，CH_2），6.45－6.47（q，1H，CH），6.77－7.81（m，11H，$C_6H_3+C_6H_3$+NH+C6H4）	3291（NH），1748（C=O）
V－13	3.52－3.87（m，2H，CH_2），6.45－6.48（q，1H，CH），6.87－7.90（m，11H，$C_6H_3+C_6H_3+NH+C_6H_4$）	3327（NH），1742（C=O）
V－14	3.52－3.88（m，2H，CH_2），6.45－6.48（q，1H，CH），6.78－7.95（m，11H，$C_6H_3+C_6H_3+NH+C_6H_4$）	3273（NH），1748（C=O）
V－15	3.51－3.84（m，2H，CH_2），6.45－6.48（q，1H，CH），6.75－7.90（m，11H，$C_6H_3+C_6H_3+NH+C_6H_4$）	3096（NH），1748（C=O）
V－16	3.52－4.15（m，2H，CH_2），6.49－6.51（q，1H，CH），6.79－8.27（m，11H，$C_6H_3+C_6H_3+NH+C_6H_4$）	3273（NH），1748（C=O）

2.2 化合物V对灰霉病病菌的杀菌活性

测定了2－酰氧基乙磺酰胺化合物（V）对CY－09（采自辽宁朝阳）、CY－14（采自辽宁朝阳）、DG－1（采自辽宁东港）、DL－11（采自辽宁大连）、FS－09（采自辽宁抚顺）、FS－10（采自辽宁抚顺）HLD－15（采自辽宁葫芦岛）7种灰霉病病菌（*Botrytis cine-*

rea）的抑制活性，采用采自不同地区的灰霉病菌菌株对同一药剂进行筛选，由于菌株抗病性的不同，将使药剂对于番茄灰霉病的活性测定结果更加可靠，而后进一步测定了所合成化合物对灰霉病菌菌株 DL－11 在黄瓜活体上的防治效果，试验结果见表3。

在利用菌丝生长速率法测试所合成化合物对多种灰霉病菌株的离体防效时，在50μg/mL的浓度下，新化合物 V 对多种灰霉病菌表现出了非常高的杀菌活性，分别有4个、2个、0个、2个、2个、3个、3个化合物对7种灰霉菌株的活性高于对照药剂腐霉利，最高抑制率分别是91.8%、95.3%、80.1%、90.6%、89.2%、89.2%、90.4%。初步构效关系表明，在酰氧基上所引入的基团越小，所合成化合物对灰霉病菌的抑制活性越好；当酰氧基上所引入基团为芳香取代基时，当苯环上无取代时，化合物所表现出的活性最高；含氟基团的引入同样对化合物的活性起到一定的提升效果。选取初筛效果较好的化合物进行进一步的精确毒力测定，所得受测试的化合物的 EC_{50}值见表4。受测试的五个化合物的 EC_{50}值均低于对照药剂腐霉利和嘧霉胺，特别是对于灰霉病菌株 DL－11，化合物表现出极优异的抑制活性，四个化合物的 EC_{50} 值均小于 1μg/mL，尤其是化合物 V－10 的 EC_{50} 值更是达到了0.03μg/mL。

在黄瓜活体盆栽试验中500μg/mL的浓度下，化合物 IV、V 对黄瓜灰霉病的抑制率最高达到了82.3%，好于对照药剂嘧霉胺。活体盆栽试验结果表明：不同目标化合物对于不同真菌活性存在较大差异，如在灰霉病菌的离体试验中对灰霉病有较高的杀菌活性，但在活体盆栽试验中结果并不理想。这可能是由于在活体盆栽试验中，药剂并未完全吸附在黄瓜的叶片上，而是有部分流失，导致测定结果不及离体试验。或者是在做活体试验时所用的剂型并不适合该类药剂。总体而言，2－取代苯基－2－羟基和2－酰氧基乙磺酰胺类化合物（IV、V）的杀菌活性尤为显著，有进一步研究的价值。

表3　2－酰氧基乙磺酰胺化合物（V）对灰霉病菌的杀菌活性

Table 3　Fungicidal activity of 2－acyloxyethylsulfonamides（V）against *Botrytis cinerea*

序号 Compd.	离体防效 Inhibition rate（%）							活体防效 Control efficiency（%）
	CY－09	CY－14	DG－1	DL－11	FS－09	FS－10	HLD－15	DL－11
V－1	89.7	78.5	57.7	86.3	85.9	72.8	86.6	45.5
V－2	61.1	79.6	25.3	40.2	45.2	10.0	48.1	29.8
V－3	59.3	55.2	25.0	45.5	53.2	26.3	56.0	2.9
V－4	56.7	9.9	40.9	50.1	53.2	28.8	44.6	20.3
V－5	54.4	53.9	44.9	23.2	35.8	22.6	7.9	—
V－6	28.1	13.5	30.2	30.0	49.9	23.5	25.7	48.2
V－7	38.9	47.3	34.1	36.1	51.0	14.6	22.2	20.3
V－8	80.2	86.4	68.5	74.3	56.0	46.4	70.3	45.5
V－9	86.3	56.5	64.5	84.0	70.1	77.1	80.8	—
V－10	91.8	95.3	80.1	90.6	89.2	89.2	90.4	48.2

（续表）

序号 Compd.	离体防效 Inhibition rate（%）							活体防效 Control efficiency（%）
	CY－09	CY－14	DG－1	DL－11	FS－09	FS－10	HLD－15	DL－11
V－11	41.2	24.5	7.4	33.6	51.0	20.8	30.9	—
V－12	51.0	64.9	17.7	55.0	43.8	6.1	47.5	17.4
V－13	53.1	27.7	13.1	28.5	40.8	11.9	35.9	82.3
V－14	22.7	61.7	17.1	22.6	28.8	18.5	16.0	69.4
V－15	50.5	35.0	31.9	34.6	49.6	11.9	52.8	2.9
V－16	49.7	42.3	11.1	46.3	44.9	22.8	40.2	14.5
腐霉利	75.3	81.1	97.7	84.0	76.2	63.8	75.2	58.8

表4　2－酰氧基乙磺酰胺化合物（V）对灰霉病菌的 EC_{50} 值测定

Table 4　The EC_{50} values of 2－acyloxyethylsulfonamides（V）against *Botrytis cinerea in vitro*

序号 Compd.	DL－11		HLD－15	
	EC_{50}（μg/mL）	95% CL（μg/mL）	EC_{50}（μg/mL）	95% CL（μg/mL）
V－1	0.68	0.25～1.84	5.18	3.51～7.65
V－8	0.55	0.03～11.38	4.93	3.26～7.46
V－9	0.76	0.43～1.36	3.46	2.34～5.14
V－10	0.03	0.00～1.20	3.23	2.42～4.33
V－15	4.23	3.10～5.78	2.47	1.85～3.31
腐霉利	4.84	3.62～6.46	7.84	6.48～9.48
嘧霉胺	12.93	7.97～20.97	12.38	9.41～31.79

3　结论

磺酰胺类化合物在农用化学品领域的应用大多集中在除草剂领域，目前已商品化的磺酰胺类杀菌剂品种仅有数种，因此，磺酰胺类杀菌剂的开发和研究还有很大空间。根据已有的研究，磺酰胺类化合物对多数病原真菌都有着良好的杀菌活性，但对磺酰胺类化合物的杀菌机理和构效关系研究还不够深入。本研究以前期研究中筛选出的具有高效抑菌活性的 *N*－（2－三氟甲基－4－氯苯基）－2－（3，5－二氟苯基）－2－羟基乙磺酰胺为先导，对其结构进行衍生，合成了16个 *N*－（2－三氟甲基－4－氯苯基）－2－（3，5－二氟苯基）－2－酰氧基乙磺酰胺类化合物，其对灰霉病表现出了较高的抑制活性。

通过化合物的初步构效关系与杀菌活性分析发现，在酰氧基上所引入的基团越小，所合成化合物对灰霉病菌的抑制活性越好；当酰氧基上所引入基团为芳香取代基时，当苯环上无取代时，化合物所表现出的活性最高。当目标化合物的取代基中含有氟原子或者含有氟的基团时也利于提高化合物杀菌活性，因为氟原子有电子效应和渗透效应，因此，将其引入到化合物的结构中可能使其化合物生物活性增强。总体而言，所合成的化合物 V 表现出较好的生物活性，有进一步研究的价值。

参考文献

[1] Alsughayer A, Elassar A Z A, Mustafa S, et al. Synthesis, structure analysis and antibacterial activity of new potent sulfonamide derivatives [J]. JBiomater Nanobiotechnol, 2011, 2 (02): 143.

[2] 王小玲，王宪龙，耿蓉霞，等. 含磺酰胺结构的抗微生物药物研究进展 [J]. 中国新药杂志, 2010 (22): 2 050 -2 059.

[3] Hen N, Bialer M, Wlodarczyk B, et al. Syntheses and evaluation of anticonvulsant profile and teratogenicity of novel amide derivatives of branched aliphatic carboxylic acids with 4 - aminobenzensulfonamide [J]. J Med Chem, 2010, 53 (10): 4 177 -4 186.

[4] Li X H, Pan Q, Cui Z N, et al. Synthesis and fungicidal activity of N - (2, 4, 5 - trichlorophenyl) - 2 - oxo - and 2 - hydroxycycloalkylsulfonamides [J]. Lett Drug Des Discov, 2013, 10 (4): 353 -359.

[5] 杨红业，闫晓静，袁会珠，等. 含氟 2 - 氧代环己基磺酰胺的合成及杀菌活性 [J]. 农药学学报, 2010, 12 (4): 449 -452.

[6] 曾东强，张耀谋，陈君. *N* - 取代苯磺酰基 - α - 三唑基片呐酮腙的合成及杀菌活性 [J]. 广西农业生物科学, 2004, 23 (4): 313 -315.

[7] Zhong Z M, Chen R, Xing R G, et al. Synthesis and antifungal properties of sulfanilamide derivatives of chitosan [J]. Carbohydr Res, 2007, 342 (16): 2 390 -2 395.

[8] 梁晓梅，方学琴，董燕红，等. *N* - (芳基磺酰氨基乙基) -1, 6 - 己内酰胺的合成及杀菌活性 [J]. 农药学学报, 2008, 10 (2): 156 -160.

[9] 陈庆悟，沈德隆. 2 - 噻唑烷酮磺酰脲衍生物的合成及生物活性 [J]. 浙江工业大学学报, 2008, 36 (5): 562 -564.

[10] Mahantesha Basanagouda K, Shivashankar Manohar V Kulkarni, et al. Synthesis and antimicrobial studies on novel sulfonamides containing 4 - azidomethyl coumarin [J]. European Journal of Medinal Chemistry, 2010, 45: 1 151 -1 157.

[11] Zhou C, Xu W M, Liu K M. Synthesis and Antiviral Activity of 5 - (4 - Chlorophenyl) -1, 3, 4 - Thiadiazole Sulfonamides [J]. Molecules 2010, 15: 9 046 -9 056.

[12] Li X H, Wu D C, Qi Z Q, et al. Synthesis, fungicidal activity, and structure - activity relationship of 2 - oxo and 2 - hydroxycycloalkylsulfonamides [J]. JAgric Food Chem, 2010, 58 (21): 11 384 -11 389.

[13] Li X H, Cui Z N, Chen X Y, Wu D C, Qi Z Q, Ji M S. Synthesis of 2 - Acyloxycyclohexylsulfonamides and Evaluation on Their Fungicidal Activity [J]. International Journal of Molecular Sciences, 2013, 14 (11): 22 544 -22 557.

[14] Li X H, Yang X L, Liang X M, et al. Synthesis and biological activities of 2 - oxocycloalkylsulfonamides [J]. Bioorg Med Chem, 2008, 16 (8): 4 538 -4 544.

[15] 李兴海，祁之秋，钟昌继，等. 苯甲酰基甲磺酰胺的合成与杀菌活性 [J]. 农药学学报, 2008, 10 (2): 136 -140.

[16] 李兴海，芮朋，潘强，等. 2 - 氧代 -2 - 苯基乙磺酰胺化合物组合合成与杀菌活性研究 [J]. 农药学学报, 2016, 18 (1): 28 -36.

几种生产上常用水稻纹枯病防治药剂药效的比较研究*

陈香华[1**]，李　茹[1]，熊战之[1]，付佑胜[1]，王宏宝[1]，
周长勇[1]，段亚冰[2]，周明国[2]，赵桂东[1***]
（1. 江苏徐淮地区淮阴农业科学研究所，淮安　223001；2. 南京农业大学植物保护学院，南京　210095）

摘要：以农业生产中长期单一使用的井冈霉素作为对照药剂，通过不同药剂处理进行水稻纹枯病田间防效比较试验，结果表明，施药后7～30d，24%噻呋酰胺悬浮剂300mL/hm²处理的防治效果和理论产量均显著高于其他5个处理。22d后对水稻纹枯病的防效达85.2%；收获时测产的增产率达12.6%，表明24%噻呋酰胺SC具有长效控制水稻纹枯病的作用，防治效果和增产作用显著，并对作物安全，具有较好的推广和应用价值。

关键词：水稻纹枯病；防治效果；增产作用；噻呋酰胺

由立枯丝核菌（*Rhizoctonia solani*）引起的水稻纹枯病（Rice sheath blight）又称云纹病，俗名花足秆，在全国各地均有发生。水稻纹枯病的发生不但影响水稻的产量而且还影响到水稻的品质，已严重影响到我国的水稻生产实践，已成为水稻三大病害之首。近年来，水稻纹枯病在我省随着水稻高产栽培技术的推广应用，该病害呈逐年加重趋势，但目前生产上防治水稻纹枯病的药剂还是以井冈霉素药剂为主，由于长年大量和单一使用，井冈霉素药剂防治水稻纹枯病的效果有所下降。水稻实际生产中井冈霉素的亩用量已明显增加，用药次数也在增多，浓度不断加大，认为可能会加速抗药性的产生，因此监测其抗性发展情况尤为重要。为了寻求防治水稻纹枯病的高效低毒、安全环保的新型药剂，来减少或延缓抗药性的产生，笔者于2015年对目前生产上防治水稻纹枯病的常用药剂进行了田间防效筛选试验，旨在为大面积推广和应用提供科学理论依据。

1　材料与方法

1.1　供试药剂

5%井冈霉素AS（钱江生物化学股份有限公司）；12.5%井·腊芽AS（浙江桐庐汇丰生物化工有限公司）；5%已唑醇EC（连云港立本农化有限公司）；43%戊唑醇（好力克）SC（拜耳作物科学有限公司）；30%苯醚甲环唑·丙环唑（爱苗）EC（先正达中国投资有限公司）；24%噻呋酰胺（满穗）SC（美国陶氏益农公司提供原药上海泰禾分装）。

1.2　供试作物

水稻，品种为淮稻5号。

---

* 基金项目：国家公益性行业（农业）科研专项（201303023）；江苏省农业科技自主创新资金CX（16）1001

** 作者简介：陈香华，男，江苏高淳人，副研究员，主要从事植物病害化学防治方面的研究

*** 通讯作者：赵桂东，男，研究员，主要从事植物病理学相关研究；E-mail：cxh13327972368@163.com

1.3 防治对象

水稻纹枯病（Rice sheath blight）。

1.4 试验地情况

本试验田设在淮安市金湖县金南镇－农户承包田，水稻为机插秧，前茬为小麦，该地块水稻纹枯病常年发生较重，种植品种为淮稻5号，2015年6月10日移栽。试验田地势平坦，肥力中等，质地中壤，pH值6.2，有机质含量2.3%。试验田水稻长势、栽培密度均匀一致。

1.5 试验设计

试验共设7个处理：Ⅰ：24%噻呋酰胺 SC 300mL/hm^2；Ⅱ：5%井冈霉素 AS 4L/hm^2；Ⅲ：12.5%井·腊芽 AS 3L/hm^2；Ⅳ：5%已唑醇 EC 1.2L/hm^2；Ⅴ：43%戊唑醇 SC 270mL/hm^2；Ⅵ：30%苯醚甲环唑·丙环 EC 300mL/hm^2；Ⅶ：空白对照（CK）。各处理小区面积30m^2，重复4次，随机区组排列。

1.6 试验方法

于水稻分蘖末期，2015年7月18日水稻纹枯病发病初期，各处理使用新加坡利农 Jacto16L型喷雾器按每公顷对水750kg均匀喷雾。调查方法采取每小区对角线5点取样法，每处理取5点，每点取100株，施药前调查病情基数，施药后7d、14d、22d和30d分别调查总株数、病株数、严重度；计算病株率、病情指数和防效；在水稻收获前1d，每个试验处理区选5点，每点拔取1丛，共5丛，测定各处理区的穗粒数、实粒数、结实率、千粒重，计算理论产量和增产率。并用邓肯氏新复极差法（DMRT）进行统计分析。

水稻纹枯病严重度分级方法：0级：全株健康无病；1级：在第4～6叶叶片和叶鞘上有病斑，即在倒4叶及以下叶鞘、叶片上发病；3级：在第3～4叶叶鞘和叶片上有病斑，即第3叶片及其以下各叶片、叶鞘发病；5级：病斑高度为剑叶叶鞘露出部分的1/2长度以内，整个剑叶叶鞘受害；7级：剑叶及其以下各叶片和叶鞘都发病，或剑叶叶片功能严重受损，物质转运受阻，叶片变红；9级：全株发病，植株提早枯死。

$$病株率（\%）=病株数/调查总株数\times 100$$

$$病指=\frac{\sum（各级病株数\times 相对级数值）}{调查总株数\times 9}\times 100$$

$$防效（\%）=（1-\frac{对照区药前病指\times 施药区药后病指}{对照区药后病指\times 施药区药前病指}）\times 100$$

2 结果与分析

2.1 防治效果

由下表可以看出，前6个处理在施药后7～30d期间，其防治效果均呈先增加后降低的趋势，其中处理Ⅰ、Ⅳ、Ⅵ、的最大防效出现在施药后第22d左右，其他3个处理Ⅱ、Ⅲ、Ⅴ的最大防效均出现在施药后第14d左右。

不同处理之间，处理Ⅰ的防治效果在施药后7～30d均显著高于其他7个处理，但其他7个处理之间在施药后不同时间表现为不同的差异顺序。施药7d，处理Ⅰ、Ⅱ之间差异不显著，但均显著高于处理Ⅲ、Ⅳ、Ⅴ、Ⅲ、Ⅳ的防效；施药后14d，处理Ⅲ、Ⅳ之间差异不显著，但均显著高于处理ⅤⅢ、Ⅳ的防效；施药后22d，处理Ⅰ显著高于处理Ⅱ、Ⅲ、Ⅳ、Ⅴ、Ⅵ的防效；施药后30d，处理Ⅰ的药效显著高于处理Ⅱ、Ⅲ、Ⅳ、Ⅴ、Ⅵ的防效。

2.2 增产效果

由下表可以看出，与清水对照相比，各处理的理论产量均显著增加，其中处理Ⅰ的测产产量和增产率均显著高于其他 5 个处理；处理Ⅰ、Ⅵ之间的理论产量差异不显著，但均显著高于处理Ⅱ、Ⅲ、Ⅳ、Ⅴ（处理Ⅳ、Ⅴ之间的理论产量差异不显著）。

表　不同药剂对水稻纹枯病的防治效果与增产效果的比较

Table　The control and yield – increasing effect of different agents on Rice sheath blight

处理	施药前病指	施药后 7d		施药后 14d		施药后 22d		施药后 30d		理论产量（t/hm²）	增产率（%）
		病指	防效（%）	病指	防效（%）	病指	防效（%）	病指	防效（%）		
Ⅰ	0.57	0.66	51.24a	2.87	76.63a	3.02	85.20a	7.27	75.76a	8.65a	12.60
Ⅱ	0.61	0.69	34.77c	4.67	52.21c	9.26	49.07d	16.38	34，45d	7.12c	6.78
Ⅲ	0.58	0.69	30.78c	4.78	51.76c	8.78	48.55d	16.57	32.01d	6.95c	3.75
Ⅳ	0.60	0.71	36.35b	4.68	60.07b	6.72	60.25c	12.67	52.03c	7.63b	8.23
Ⅴ	0.54	0.65	34.67b	4.55	62.53b	6.65	61.44c	11.89	53.88c	7.78b	9.79
Ⅵ	0.52	0.69	43.25a	3.13	68.27a	5.65	73.77b	10.54	68.22b	8.34a	11.67
Ⅶ	0.55	1.32	—	9.15	—	14.00	—	21.55	—	6.58d	—

注：竖向小写字母不同，表示差异达 0.05 显著水平

3　结论与讨论

水稻纹枯病是水稻生产上一种严重的土传病害，该病害在我省常年较重发生，由于目前水稻实际生产中还缺乏高抗品种，化学（药剂）防治仍然是一种主要的防治手段，是水稻生产实践中最直接、最有效的防治措施。本试验各药剂处理区均未发现药害症状（如矮化、褪绿、畸形等），未对水稻产生药害，表明各药剂在推荐使用剂量范围内无药害，对水稻安全。24%噻呋酰胺（满穗）SC 属苯酰胺类杀菌剂，具有高度选择性，防治水稻纹枯病的效果最好，持效期最长，对水稻保产、增产效果明显，在防治纹枯病方面具有较好的应用前景。建议水稻大田施药浓度为用 300mL/hm²，对水 600～750kg/hm²，采用二次稀释法，均匀喷雾。

近年来水稻纹枯病在我省呈现发病早、蔓延速度快、为害时间长、防治难等特征，轻则造成倒伏减产，重则冒穿失收。水稻纹枯病病情有横向和垂直发展两个高峰。7 月中下旬，随着气温的回升，病菌在稻株间传播扩散，病株率迅速增加，病情以横向发展为主，病情指数能上升至病初的 15～20 倍，形成第一发病高峰；8 月中下旬是水稻纹枯病的第二个发病高峰，随着植株基部节间的伸长，病菌逐渐侵染茎秆，病情主要为垂直发展，病情指数也上升较快，水稻易出现倒伏或冒穿现象。针对本地区水稻纹枯病发生和发展特点，笔者建议在水稻纹枯病大发生前，开展两次防治，防治时间于水稻分蘖末期的 7 月 15 日前后和水稻破口抽穗期的 8 月 15 日左右为佳。以“预防为主”的策略控制纹枯病的危害。

由于常年单一使用同类药剂，水稻纹枯病菌极易对其产生抗药性，为了有效控制纹枯病的危害，防治药剂可选择防治效果高，增产效果好的满穗、爱苗及井冈霉素等药剂，并注意不同类药剂的交替使用，以抑制水稻纹枯病抗性的产生。

参考文献

[1] Hide C C L, Bithell S L, Cromey M G. Disease progress of sharp eyespot in wheat fields [J]. New Zealand Plant Protection, 2006, 59: 371.

[2] 生越明. Rhizoctonia solani Kühn における菌丝融合群の诸性质 [J]. 日植病报, 1972, 38: 123-129.

[3] 徐雍皋，徐敬友. 农业植物病理学 [M]. 南京：江苏科技出版社，1996：86-90.

[4] 范怀忠，王焕如. 植物病理学 [M]. 北京：中国农业出版社，1999：88-93.

[5] 莫惠栋，农业试验统计：第2版 [M]. 上海：上海科学技术出版社，1992：256-260.

[6] 丁晓丽，易红娟. 江苏沿江地区水稻纹枯病发生特点及其综防技术 [J]. 中国植保导刊，2012，11（3）：14-16.

[7] 袁庆，张建军. 不同药剂处理对水稻纹枯病防治效果的研究 [J]. 金陵科技学院学报，2010，6（2）：43-45.

[8] 孟庆忠，刘志恒，王鹤影，等. 水稻纹枯病研究进展 [J]. 沈阳农业大学学报，2001-10，32（5）：376-381.

[9] 夏慧. 稻麦纹枯病菌对井冈霉素的敏感性及其影响因素的研究 [D]. 扬州：扬州大学，2004.

[10] 常望霓. 水稻纹枯病防治药剂的筛选及田间防效 [D]. 长沙：湖南农业大学，2011.

[11] 夏炜. 不同类型杀菌剂对水稻纹枯病菌的毒力及田间药效研究 [D]. 南京：南京农业大学，2013.

[12] 邓金花，顾俊荣，周新伟，等. 水稻纹枯病室内药剂筛选及田间防治效果 [J]. 江苏农业科学，2010（6）：196-197.

[13] 农业部农药检定所. 农药田间药效试验准则 GB/T 17980.20—2000 [S]. 北京：中国标准出版社，2000.

新杀菌剂烯肟菌酯在防治马铃薯晚疫病上的应用

刘君丽*，司乃国，王　斌
（沈阳中化农药化工研发有限公司，新农药创制与开发国家重点实验室，沈阳　110021）

摘要： 马铃薯在栽培作物中占有重要地位，在国家启动马铃薯主粮化战略后，对马铃薯病虫草害的防治再次成为人们关注热点，尤其是马铃薯晚疫病的防治。马铃薯晚疫病是一种流行速度快、危害范围广，能够造成马铃薯严重损失、甚至绝产的毁灭性病害，马铃薯的主栽品种大多不抗病，化学药剂仍然是马铃薯晚疫病最有效的防治措施。本文针对烯肟菌酯防治马铃薯晚疫病进行了广泛的试验研究，烯肟菌酯对马铃薯晚疫病有较好的防治效果，对马铃薯早疫病也有一定的兼治作用；烯肟菌酯与霜脲氰组成的混剂，以及烯肟菌酯与氟啶胺桶混使用，对马铃薯晚疫病的防治效果在75%以上，并能显著提高产量和品质，增产率超过10%。首次施药时期是影响马铃薯晚疫病防治效果的关键因素，在未发病或中心病株出现前施药会增强对病害的防控能力。

关键词： 烯肟菌酯；马铃薯晚疫病；防治效果；施药时期

The Application of Enostrobilurin Against Potato Blight Disease

Liu Junli, Si Naiguo, Wang Bin
(*Shenyang Sinochem Agrochemicals R&D Co. Ltd.*, *State Key Laboratory of the Discovery and Development of Novel Pesticide*; *Shenyang* 110021, *China*)

Abstract: Potato occupies an important place in cultivated crops. The control of the diseases, insect pests and weeds of cotton becomes hot topics again, especially potato blight disease, after the government initiate the strategy that potato turn into a staple food. Potato blight disease is a disease that its epidemic characteristics are fast, wide spread and serious infestation. It can cause death of plant and dramatical yield loss. In the market, the main cultivars of potato can not resist the disease effectively. Chemical agents are still the most effective measure controlling against potato blight disease. The study confirm the efficacy of enostrobilurin against potato blight disease by extensive test. The results showed that enostrobilurin achieve the effective control against potato blight disease and a certain degree of control against potato eariy blight. The mixed formulation contained enostrobilurin, cymoxanil, and the mixture of the preparations of enostrobilurin and fluazinam had control effect reaching more than 75% against potato blight disease, and increased yield and quality significantly, which yield improvement reached more than 10%. The spray period is a critical factor on controlling potato blight disease. It will reach better control effect that use pesticide before plants are infected.

Key words: Enostrobilurin; Potato blight disease; Control effect; Spray period

* 作者简介：刘君丽，教授级高工，主要从事新农药创制开发及农药应用技术研究；E-mail：liujunli@ sinochem. com

1 概述

1.1 马铃薯的种植状况

马铃薯在栽培作物中占有重要地位，在全世界150多个国家种植，总面积在2.8亿亩上下波动。中国是种植马铃薯面积最大的国家，2010年全国种植面积7 807.5万亩，产量8 153.5万吨，种植面积和产量均呈上升趋势。

2016年国家启动马铃薯主粮化战略，农业部23号正式发布《关于推进马铃薯产业开发的指导意见》，推进马铃薯主粮化战略，提出力争到2020年，马铃薯种植面积扩大到1亿亩以上，适宜主食加工的品种种植比例达到30%，主食消费占马铃薯总消费量的30%。种植面积较大的省份为内蒙古（1021.65万亩）、贵州（968.7万亩）、甘肃（968.25万亩）、四川（862.02万亩）、云南（739.65万亩）及重庆、陕西、黑龙江、宁夏等省。

1.2 马铃薯病害的发生及防治

在马铃薯的栽培中，由病原菌引起的植物病害是影响产量的关键因素，在马铃薯上经常发生的病害有晚疫病、早疫病、黑痣病、疮痂病、软腐病、黑斑病，以及根结线虫病，晚疫病是马铃薯上最重要的病害，也是关注的重点。对马铃薯病害的防治，化学药剂仍然是最有效和方便可行的方法。全球几大农药公司都将马铃薯作为农药研发的主要目标，近些年不断有新品种问世，如杜邦的增威赢绿（氟噻唑吡乙酮），每公顷仅用22.5～30g。马铃薯整体解决方案也是各公司关注的重点，先正达、拜耳、杜邦都以本公司的药剂为核心，把产品和技术柔和在一起，形成整体解决方案在马铃薯主要产区推广。

目前对马铃薯病害的防治，专业化大规模种植企业或种植大户，以及马铃薯加工企业普遍采取组合用药，以预防病害为主，在马铃薯一个生育期内按时间喷药，通常8～10次，这部分种植地块几乎不发生马铃薯晚疫病。马铃薯零散种植户，在偏远地区多数人在马铃薯上不采取防治措施，在河边、内蒙古、黑龙江、云南等种植水平和产值较高地区，散户通常喷3～4次药剂，主要是防治晚疫病。

1.3 新杀菌剂烯肟菌酯的开发及应用

烯肟菌酯隶属于甲氧基丙烯酸酯类杀菌剂，化学名称：3－甲氧基－2－［2－（（（（1－甲基－3－（4－氯苯基）－2－丙烯基叉）氨基）氧）－甲基）苯基］丙烯酸甲酯[7]。

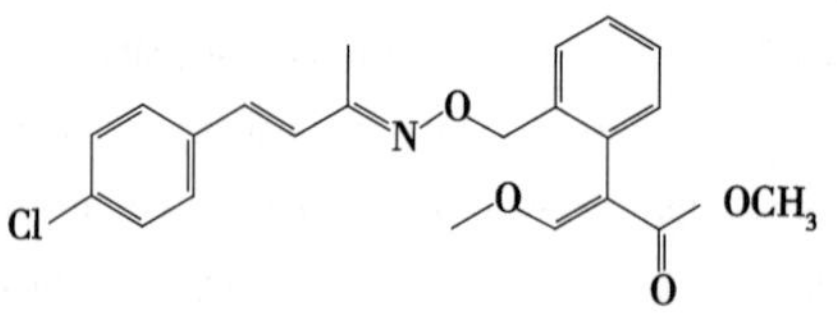

Cl　N　O　O　OCH_3　O

enstroburin

烯肟菌酯是沈阳化工研究院发现，国内第一个以天然抗生素Strbilurin为先导化合物的新型杀菌剂品种。其作用机理是通过与细胞色素bc_1复合体结合抑制线粒体的电子传递，进而破坏病菌能量合成而起到杀菌作用。目前已获得了中国（98113756－3）、美国（US1060139）、日本（JP11315057）和欧洲（EP936213）发明专利，2002年完成原药和25%乳油的农药临时登记。开发的制剂有：25%烯肟菌酯乳油、25%烯肟·霜脲可湿性粉剂、28%多菌灵·烯肟菌酯可湿性粉剂、18%氟环唑·烯肟菌酯悬浮剂，制剂在黄瓜霜霉病、葡萄霜霉病、小麦赤霉病、苹果斑点落叶病上获得农药临时登记。

烯肟菌酯具有杀菌谱广、活性高、毒性低，与环境相容性好等特点。对由鞭毛菌、结合菌、子囊菌、担子菌及半知菌引起的病害均有很好的防治作用，与苯基酰胺杀菌剂无交互抗性。田间试验结果显示：使用剂量 100 ~ 200g a. i. /hm^2 能有效地控制黄瓜霜霉病、葡萄霜霉病的危害，60 ~ 100g a. i. /hm^2 对小麦白粉病白粉病有良好的防效，由烯肟菌酯与多菌灵、氟环唑组成的混剂对小麦赤霉病、苹果斑点落叶病等有较好的防效，是具有广阔应用前景的杀菌剂新品种。

鉴于烯肟菌酯的广谱性和对卵菌病害的高活性，以及对烯肟菌酯对作物产品和品质的保健作用。笔者在 2013—2016 年，进行了烯肟菌酯混剂及烯肟菌酯桶混防治马铃薯病害的相关试验，现总结如下。

2 试验材料及方法

2.1 试验材料

2.1.1 试验药剂

97% 烯肟菌酯原药，沈阳化工研究院；98% 霜脲氰原药，利民化工股份有限公司；25% 烯肟菌酯乳油，沈阳科创化学品有限公司；25% 烯肟 · 霜脲可湿粉，沈阳科创化学品有限公司；50% 氟啶胺悬浮剂，江苏扬农股份有限公司；20% 氟吗啉可湿粉，沈阳科创化学品有限公司；72% 霜脲 · 锰锌可湿粉（克露），上海杜邦公司农化有限公司；10% 氰霜唑悬浮剂，日本石原产业株式会社；80% 代森锰锌可湿粉，利民化工股份有限公司；687. 5g/L 氟吡菌胺 · 霜霉威悬浮剂，拜耳作物科学有限公司。

2.1.2 防治对象

马铃薯晚疫病［*Phytophthora infestans*（Mont）De Bary］。

2.2 试验方法

2.2.1 烯肟菌酯与霜脲氰混剂筛选

烯肟菌酯与霜脲氰的复配比例为（4 : 1、2 : 1、1 : 1、1 : 2、1 : 4），97% 烯肟菌酯原药的处理浓度为 4、2、1、0. 5、0. 25mg/L，98% 霜脲氰原药和烯肟菌酯、霜脲氰混配的处理浓度为 2、1、0. 5、0. 25、0. 125mg/L，另设无菌水空白对照，共 36 个处理，每个处理 3 次重复。

将熔好的培养基冷却至 60 ~ 70℃，按所设浓度加入定量药剂，制成含有不同药量的含毒培养基，待其充分冷却后接种直径 0. 5cm 的供试病原菌菌片，另设空白对照，放置培养箱中培养（22 ± 1）℃。

依据菌落生长直径，计算各药剂处理的抑菌率，由求出毒力回归方程及 EC_{50}，由 Sun - Johnson 毒力指数计算法计算五个复配剂的共毒系数。

2.2.2 25% 烯肟 · 霜脲可湿粉田间试验

试验设在辽宁省阜新蒙古族自治县大固本镇进行。该试验地为多年马铃薯种植地区，马铃薯晚疫病、早疫病和疮痂病比较普遍。试验小区的栽培条件（土壤类型、水肥管理、移栽期、种植密度、生育期）均匀一致，并按当地农事操作方法管理。

供试药剂 25% 烯肟菌酯 · 霜脲氰可湿性粉剂的处理剂量为 100、150、200 g a. i. /hm^2，对照药剂 72% 霜脲 · 锰锌可湿粉的处理剂量 800 g a. i. /hm^2 的效果相当，25% 烯肟菌酯乳油的处理剂量为 200 g a. i. /hm^2，另设空白对照。在田间出现中心病株初期喷药，连续喷药 3

次，间隔7d，第3次喷药后10d调查防效，分级标准和计算方法如下，试验结果采用Duncan's新复极差法，进行差异显著性分析。

分级标准：

0级：全株无病；

1级：病斑面积点整个叶面积的5%以下；

3级：病斑面积点整个叶面积的6%～10%；

5级：病斑面积点整个叶面积的11%～20%；

7级：病斑面积点整个叶面积的21%～50%；

9级：病斑面积点整个叶面积的50%以上。

药效计算方法：

$$\text{病情指数}=\frac{\Sigma\ (\text{各级病叶数}\times\text{相对级数值})}{\text{调查总数}\times 9}\times 100$$

$$\text{防治效果（\%）}=\frac{\text{对照区病情指数}-\text{处理区病情指数}}{\text{对照区病情指数}}\times 100$$

2.2.3　烯肟菌酯与氟啶胺桶混田间试验

试验设在黑龙江绥化和辽宁省阜新进行。试验分为烯肟菌酯与氟啶胺桶混与其他马铃薯晚疫病防治药剂对比试验，以及烯肟菌酯与氟啶胺不同桶混比例对马铃薯晚疫病对比试验。

烯肟菌酯与氟啶胺桶混与其他马铃薯晚疫病防治药剂对比试验处理为：烯肟菌酯＋氟啶胺（50＋150）g a. i. /hm^2、对照药剂50%氟啶胺悬浮剂200 g a. i. /hm^2、20%氟吗啉可湿粉200 g a. i. /hm^2、10%氰霜唑悬浮剂100 g a. i. /hm^2、80%代森锰锌可湿粉1 000 g a. i. /hm^2、687. 5g/L氟吡菌胺·霜霉威悬浮剂/600 g a. i. /hm^2。在田间出现中心病株后喷药，喷药2次，间隔8d。调查方法、分级标准及统计分析同2. 2. 2。

烯肟菌酯与氟啶胺不同桶混比例为：烯肟菌酯＋氟啶胺（50＋150）g a. i. /hm^2、烯肟菌酯＋氟啶胺（100＋100）g a. i. /hm^2、烯肟菌酯＋氟啶胺（150＋50）g a. i. /hm^2，设10%氰霜唑悬浮剂100 g a. i. /hm^2、25%烯肟菌酯乳油200 g a. i. /hm^2、687. 5g/L氟吡菌胺·霜霉威悬浮剂/600 g a. i. /hm^2药剂对照，以及空白对照。在田间没有发病，气象进入多雨季节时喷药，共喷药3次，间隔7d。调查方法、分级标准及统计分析同2. 2. 2。在收获时，每小区单采单收，测定马铃薯产量。

3　试验结果

3.1　烯肟菌酯与霜脲氰混配筛选结果

混配筛选结果显示：烯肟菌酯、霜脲氰（4∶1、2∶1、1∶1、1∶2、1∶4）五个供试配比的EC_{50}值分别为0. 74、0. 55、0. 48、0. 46、0. 45mg/L，共毒系数分别为117. 49、136. 01、132. 60、120. 41、111. 55。烯肟菌酯、霜脲氰（2∶1、1∶1、1∶2、）三个供试配比共毒系数在120以上，表现为增效作用。详细结果见表1。

霜脲氰对菌丝生长有很好的抑制作用，加之烯肟菌酯对孢子萌发的抑制活性，二者复配可以做到很好的优势互补。

表 1　烯肟菌酯、霜脲氰对马铃薯晚疫病的联合毒力测定结果

处理	配比	毒力方程	相关系数	EC_{50} (mg/L)	共毒系数
烯肟菌酯：霜脲氰	4 : 1	$y=5.117+0.910x$	0.954	0.74	117.49
	2 : 1	$y=5.334+1.285x$	0.966	0.55	136.01
	1 : 1	$y=5.554+1.722x$	0.979	0.48	132.60
	1 : 2	$y=5.707+2.069x$	0.988	0.46	120.41
	1 : 4	$y=5.821+2.362x$	0.994	0.45	111.55
烯肟菌酯单剂	—	$y=4.950+0.981x$	0.981	1.15	—
霜脲氰单剂	—	$y=6.113+3.092x$	0.999	0.44	—

3.2　25%烯肟·霜脲可湿性粉剂田间试验结果

25%烯肟菌酯·霜脲氰可湿性粉剂防治马铃薯晚疫病田间小区试验结果显示：25%烯肟菌酯·霜脲氰可湿性粉剂对马铃薯晚疫病有较好防治效果，150～200 g a. i. /hm^2用量的防效与72%霜脲·锰锌可湿粉800 g a. i. /hm^2的效果相当，结果详见表2。

表 2　25%烯肟菌酯、霜脲氰可湿性粉剂防治马铃薯晚疫病试验结果

供试处理	剂量 (g a. i. /hm^2)	防治效果 (%)					显著性测定	
		I	II	III	IV	X	0.05	0.01
25%烯肟·霜脲氰可湿粉	200	93.28	91.28	92.77	88.63	91.53	a	A
	150	89.68	85.55	91.78	89.02	89.05	a	A
	100	73.08	76.61	80.31	77.58	76.97	b	B
72%霜脲·锰锌可湿粉	800	89.44	91.97	91.66	87.62	90.17	a	A
25%烯肟菌酯乳油	200	74.76	76.16	80.53	78.47	77.54	b	B
空白对照	(病指)	20.170	21.12	16.610	19.56	19.36		

3.3　烯肟菌酯与氟啶胺桶混药剂对比试验结果

在黑龙江绥化进行的烯肟菌酯与氟啶胺桶混与其他马铃薯晚疫病防治药剂对比试验结果显示，烯肟菌酯与氟啶胺桶混对马铃薯晚疫病表现出较好的防治效果，在200 g a. i. /hm^2的相同剂量下，其防效优于氟啶胺单剂和氟吗啉单剂的效果。结果见表3。

表 3　烯肟菌酯与氟啶胺桶混对防治马铃薯晚疫病对比试验结果（绥化）

供试处理	剂量 (g a. i. /hm^2)	防治效果 (%)					显著性测定	
		I	II	III	IV	X	0.05	0.01
氟啶胺+烯肟菌酯	150+50	63.93	63.76	63.36	63.34	63.60	b	B
氟吗啉	200	44.47	43.51	41.88	43.97	43.46	d	E
氟啶胺	200	53.80	53.72	53.77	56.47	54.44	c	C
银法利	600	52.13	49.95	52.54	55.49	52.53	d	C D
代森锰锌	1000	71.68	71.60	72.85	71.83	71.99	a	A
氰霜唑	100	52.20	51.54	49.44	52.27	51.36	d	CD

3.4 烯肟菌酯与氟啶胺不同桶混比例试验结果

在辽宁阜新进行的防治试验结果显示：使用烯肟菌酯与氟啶胺桶混，在马铃薯现蕾期开始喷药，间隔10d，连续喷药3次。烯肟菌酯+氟啶胺（50+150）g a. i. /hm^2、烯肟菌酯+氟啶胺（100+100）g a. i. /hm^2、烯肟菌酯+氟啶胺（150+50）g a. i. /hm^2桶混对马铃薯有较好的防治效果，防效在75%以上。其中，烯肟菌酯+氟啶胺（50+150）g a. i. /hm^2的防效较高，烯肟菌酯与氟啶胺桶混的优选比例为1：3。结果见表4。

表4 烯肟菌酯与氟啶胺桶混在马铃薯上的试验结果（阜新）

供试处理	剂量（g a. i. /hm^2）	防治效果（%）					显著性测定	
		Ⅰ	Ⅱ	Ⅲ	Ⅳ	X	0.05	0.01
烯肟菌酯+氟啶胺	100+100	78.26	81.89	79.91	82.01	80.52	a	A
烯肟菌酯+氟啶胺	50+150	83.33	80.09	82.45	83.69	82.39	a	A
烯肟菌酯+氟啶胺	150+50	74.35	76.61	75.52	74.00	75.12	b	B
烯肟菌酯	200	70.46	73.89	72.23	69.85	71.61	c	B
氟啶胺	200	72.89	74.33	71.56	74.16	73.24	bc	B
银法利	600	78.87	83.22	80.98	83.67	81.69	a	A

使用烯肟菌酯与氟啶胺桶混，在马铃薯现蕾期开始喷药，能明显提高马铃薯产量，其产量较空白对照增加10%以上。其中，烯肟菌酯+氟啶胺（50+150）g a. i. /hm^2的增产效果明显，结果见表5。

表5 烯肟菌酯与氟啶胺桶混对马铃薯产量的影响（阜新）

供试处理	剂量（g a. i. /hm^2）	产量影响（kg/24m^2）				增产率（%）	显著性测定	
		Ⅰ	Ⅱ	Ⅲ	Ⅳ		0.05	0.01
烯肟菌酯+氟啶胺	100+100	81.35	80.5	78.25	80.55	11.07	ab	AB
烯肟菌酯+氟啶胺	50+150	82.55	79.7	83.45	83.35	14.19	a	A
烯肟菌酯+氟啶胺	150+50	78.5	80.25	77.95	81.45	10.10	b	AB
烯肟菌酯	200	79.65	77.55	79.05	77.15	8.44	b	BC
氟啶胺	200	76.25	75.35	74.15	76.45	4.57	c	C
银法利	600	81.65	79.75	80.65	78.15	10.80	ab	AB
空白对照	清水	70.15	72.5	71.95	74.45		d	D

4 结果与讨论

（1）25%烯肟菌酯·霜脲氰可湿粉以及烯肟菌酯与氟啶胺桶混对马铃薯晚疫病具有较好的防治效果，在200g a. i. /hm^2的剂量下，能够控制马铃薯晚疫病的为害，并能够显著提高马铃薯产量。试验中还发现烯肟菌酯对马铃薯早疫病、黑斑病也有兼治作用。使用化学农药防治马铃薯晚疫病，首次施药时间对病害的控制和产量影响非常重要，施药的最佳时间是田间出现中心病株或出现病株之前。

（2）马铃薯晚疫病为害严重，病害发生和使用药剂防治的面积大，资料显示2015年在西南、西北、华北和东北主要产区，马铃薯晚疫病发生面积3 500万亩。人们对防治马铃薯晚疫病药剂的关注度高，登记和使用的品种多，产品竞争激烈。因此，防治马铃薯晚疫病产品的合理利用非常重要，在使用时要做到既能有效控制发生，又要降低用药成本。

（3）马铃薯虽然在全国均有种植，但种植和用药水平差异很大，如河北、内蒙古、黑龙江、云南在马铃薯上的投入明显高于其他地区，在河北、内蒙古的一些马铃薯种植企业或种植大户，每个生长季节在马铃薯上用药高达8～10次，而在青海、广西、辽宁在马铃薯上用药则较少，甚至在马铃薯上不使用农药。因此，要加强新杀菌剂的示范推广，基层一线的示范、推广很重要。只有让农民看到有“多好”，教会他们“怎么用”，才能实现防治马铃薯晚疫病的目的。

（4）建立马铃薯作物整体解决方案，是促进马铃薯产业健康发展的有效措施。作物整体解决方案备受推崇，拜耳、先正达等国外公司都先后推出“马铃薯整体解决方案”。其目的是更好控制马铃薯病虫草害的发生，为马铃薯产业提供从种到收的全程解决方案，进而通过产量、品质和收益。在建立马铃薯整体解决方案时，一定要以解决关键问题为突破口，除马铃薯晚疫病之外，马铃薯疮痂病、黑痣病、环腐病、线虫病都是毁灭性的病害，应着重考虑；再则，减少中间环节，降低整体解决方案成本，建立马铃薯大客户营销体系，更有利于整体解决方案的实施。

参考文献

[1] 朱杰华. 浅议马铃薯病害的发生几起防治药剂优选［J］. 农药市场信息，2015，536（17）：33－35.

[2] 罗彦涛，孟润杰，赵建江，等. 马铃薯晚疫病菌对不同药剂敏感性及相应药剂田间药效验证［J］. 农药，2016，56（2）：134－137.

[3] 刘波微，李洪浩，彭化贤，等. 防治马铃薯晚疫病新农药筛选及经济效益评价［J］. 西南农业学报，2013，26（2）：595－599.

[4] 司乃国，刘君丽，李志念，等. 创制杀菌剂烯肟菌酯生物活性及应用研究（Ⅰ）——黄瓜霜霉病［J］. 农药，2003，42（10）：36－38.

[5] 司乃国，刘君丽，张宗俭，等. 创制杀菌剂烯肟菌酯生物活性及应用研究（Ⅱ）——小麦白粉病［J］. 农药，2003，42（11）：39－40.

[6] 刘君丽，司乃国，陈亮，等. 创制杀菌剂烯肟菌酯生物活性及应用研究（Ⅲ）——葡萄霜霉病［J］. 农药，2003，42（12）：32－33.

杧果可可球二孢对多菌灵抗性及抗性机制初步探讨*

赵　磊[1**]，杨　叶[1***]，王　萌[1]，贺　瑞[1]，陈绵才[2]
（1. 海南大学环境与植物保护学院，海口　570228；2. 海南省农业科学院植物保护研究所，海南省植物病虫害防控重点实验室，海口　571100）

摘要：2016 年从海南杧果园采集分离可可球二孢菌株，室内进行致病力测定，采用区分剂量法对杧果可可球二孢多菌灵抗药性进行检测，并通过对不同抗性菌株的渗透压力及电渗率测定比较抗敏菌株的生理生化特性。结果表明：海南杧果果实采后蒂腐病的发病率为 26.2%，强致病力菌株比例达 69.56%，抗性菌株率为 74.42%。不同抗性水平菌株的渗透压没有差异，但是抗性菌株的相对电导率高于敏感菌株，显示细胞膜透性增大，胞内电解质渗出较多。

关键词：杧果蒂腐病；可可球二孢；多菌灵；抗药性；电导率

Primary Studies on the Resistance Evaluation and Resistance Mechanism of *Botryodiplodia theobromae* from Mango to Carbendazim

Zhao Lei[1], Yang Ye[1], Wang Meng[1], He Rui[1], Chen Miancai[2]
(1. *College of Environment and Plant Protection*, *Hainan University*, *Haikou*, *Hainan*, 570228; 2. *Institute of Plant Protection of Hainan Academy of Agricultural Sciences*, *Hainan Key Laboratory for Control of Plant Diseases and Insect Pests*, *Haikou*, *Hainan*, 571100)

Abstract: Strains of *Botryodiplodia theobromae* were collected from mango orchard in Hainan in 2016. Those strains were tested in laboratory to observe the pathogenicity to mango fruits and to detected the resistance to carbendazim using distinguishing dosage method. The physiological and biochemical characters were compared between resistant strains and sensitive strains of *B. theobromae* by osmotic pressure and electroosmosis rate of different resistant strains. The results showed that the stem – end rot of mango fruits incidence was 26.2%. The tested stains had different pathogenicity, the high virulent strains accounted for 69.6%, the resistance frequencie to carbendazim was 70.9%. The osmotic pressure was not different between resistant and sensitive strains of B. theobromae. The relative conductivity of resistant strains mycelia was higher than that of the sensitive strains, the cell membrane permeability increased, which led to cell electrolyte leakage.

Key words: mango stem – end rot; *Botryodiplodia theobromae*; carbendazim; fungicide resistance; conductivity

杧果蒂腐病是世界杧果产地仅次于炭疽病的重要采后病害。该病使果实在贮运期间严重腐烂[1-2]。可可球二孢菌（*Botryodiplodia theobromae*）是引起海南杧果蒂腐病的主要病原菌。

---

* 基金项目：国家自然科学基金（31560521），海南省属科研院所技术开发研究专项（No. KYYS – 2015 – 01）

** 作者简介：赵磊，男，在读研究生，林业专业；E-mail：1127951179@ qq. com
*** 通讯作者：杨叶，女，教授，主要从事植物保护及农药学相关研究工作；E-mail：yyyzi@ tom. com

该病菌还可以为害除了杧果以外的多种作物、果树及林木，造成田间及贮藏期病害[3]。目前，杧果生产上及采后防治病害大量应用苯并咪唑类杀菌剂（BMZs）和甾醇脱甲基抑制剂（DMIs）等防治杧果病害。但是由于该病原菌具有潜伏侵染的特性，防治时大量使用杀菌剂，给杧果蒂腐病病原菌造成了很大的选择压力[4]，2009 年胡美姣等[5]在海南首次报道对多菌灵产生抗性可可球二孢菌株，并指出杧果蒂腐病菌对多菌灵抗性以高抗为主；而 2015 年王萌等[6]则发现海南杧果蒂腐病菌可可球二孢对多种杀菌剂产生抗药性。研究数据显示，海南不同杧果主产区杧果可可球二孢对供试 2 种苯并咪唑类杀菌剂（BMZs）均出现十分严重抗性，并且抗药性在海南比较普遍。

海南杧果可可球二孢对多菌灵抗性水平极高且非常普遍，摸清海南杧果可可球二孢对多菌灵抗性水平变化及其抗性机理对多菌灵抗药性研究具有重要的意义和代表性。笔者分离纯化并检测了海南主要杧果生产区杧果可可球二孢对多菌灵抗药性，测定了菌株致病力及不同抗性菌株细胞膜通透性和渗透压敏感性等生理特性，以期为该病菌的抗性监测及抗性机制的研究奠定基础。

1　材料与方法

1.1　试验材料

分离菌株用杧果：于 2016 年 3—5 月在海南主要杧果产区乐东、黄流、三亚等 16 个样地采回，洗净晾干后室温保存。

供试菌株：采用组织分离法分离病原菌，并根据杧果产地与品种进行编号（表 1），并按科赫法则确定为致病菌，对分离的 86 个菌株杧果可可球二胞进行致病性及抗药性初步检测。另外，抗敏菌株生理生化特性实验所用部分菌株来自于 2014 年本实验室分离保存杧果可可球二孢菌株。

药剂：96.9 % 多菌灵原药（海南正业中农高科股份有限公司）

表 1　杧果蒂腐病可可球二孢采集信息

Table 1　The information of *B. theobromae* collected from mango

采集地点	菌株数	分离杧果品种	编号
昌江黎族自治县	40	台农、金煌、象牙	CJTN、CJJH、CJXY
乐东黎族自治县	15	台农、金煌	LDJH、JSTN、JSJH、HLTN、HLJH
三亚市	37	金煌、象牙、红金农	YCJH、YCXY、YCHJ
陵水黎族自治县	27	台农、金煌、红金农	YZTN、YZJH、YZHJ
东方市	5	台农	DFTN
儋州市	15	台农	DZTN

1.2　杧果蒂腐病可可球二孢的致病性测定

采用室内离体菌块接种法：将可可球二孢菌株于 28℃ 培养 4d 后，在菌落边缘打取菌饼（$\phi = 5$ mm），接种部位先刺伤，再接种菌饼。杧果材料为台农，每个果实接种 2～3 点，每处理接种 3 个果实，以接种无菌琼脂块为对照，室温下保湿培养。接种第 2 天开始检查杧果的发病情况，记录发病率并测量病斑直径大小，计算病情指数（DI）。

$$DI = \frac{\sum (s_i n_i)}{9N} \times 100$$

式中：DI——病情指数；

S——发病级别；

n——相应的发病级别数目；

i——病情分级的各个级别；

N——调查的总数目。

病情分级标准：参照杧果炭疽病害严重度分级标准[7]，以病斑直径进行分级，详细分级标准如下：0 级，X = 0；1 级，0 < X ≤ 0.5 cm；3 级，0.5 cm < X ≤ 1.0 cm；5 级，1.0cm < X ≤ 2.0 cm；7 级，2.0 cm < X ≤ 4.0 cm；9 级，X > 4.0 cm。X 为接种点病斑直径。

致病力划分：以相同时间内病果病情指数进行划分，强致病力：病情指数≥70；中致病力 40 ~ 69；弱致病力 < 40。

1.3 菌株抗药水平检测

采用用采用区分剂量法测定，多菌灵浓度分别为 1、10、100mg/L，以不含药培养基为对照，接种后放入 28℃ 恒温培养箱内培养 36h 后用十字交叉法测量菌落直径。参照 Kim 和乔广行的标准[8-9]，在含 1 mg/L（质量浓度）药剂平板上能正常生长，在 10mg/L不能正常生长（抑制率小于 50%）的菌株为敏感菌（S），在 10 mg/L 药剂平板上能生长，在 100 mg/L不能正常生长（抑制率小于 50%）的为中抗菌株（R），在 100 mg/L 药剂平板上正常生长的菌株为高抗菌株（HR）。

1.4 抗敏菌株生理生化特性

1.4.1 抗敏菌株对渗透压敏感性测定

将在 PDA 平板上长势良好的各菌株打成 5mm 菌饼接入到含有 10g/L、20g/L、40g/L、80g/L、100g/L、150g/L、300g/L 浓度葡萄糖的含糖培养基以及含 1.25g /L、2.5g /L、5g/L、10g /L、20 g /L、40g /L NaCl 的含 NaCl 培养基上，28℃下培养 36h 后，以十字交叉法，测量菌落直径。试验进行 3 次重复。根据葡萄糖浓度与菌落直径、NaCl 浓度与菌落直径绘制菌丝生长曲线。

1.4.2 抗敏菌株细胞膜通透性测定

将上述供试菌株接于 PDA 平板上，28℃ 培养 36h 后，以打孔器打取菌落边缘的菌丝，制成直径 5mm 菌碟，分别接入大量高温消毒后的 PDA 液体培养基上，28℃ 培养 2 ~ 3d，过滤菌丝，用磷酸缓冲液冲洗，再用重蒸水冲洗 2 次，真空抽滤近干后称取菌丝鲜重 2g，分别加入盛有由重蒸水稀释的不同梯度的多菌灵盐酸盐药液的试管中，多菌灵盐酸盐溶液浓度分别为 0.5、1、5、10、50μg/mL。25℃ 的恒温水浴中、以 120r/min 振荡 0、5、10、20、40、80、120、180、240、300、360min 后，用电导仪测定电导率。以菌丝加重蒸水为对照，最后煮沸（死处理），测定电导率。据公式计算每次测定的相对渗率，然后根据渗漏情况比较细胞膜的透性[10]。每个处理重复 3 次。

$$\text{相对渗率}(\%) = C_t - C_0 / C\text{死处理} \times 100$$

式中：C_t——某一时刻的电导率值；

C_0——最初时的电导率值；

C 死处理——死处理后电导率值。

2 结果与分析

2.1 杧果蒂腐病可可球二孢的分离纯化及致病性测定

2.1.1 杧果蒂腐病可可球二孢种群分布监测

本次采样分别于 2016 年 3 月、5 月从海南乐东黎族自治县、昌江黎族自治县、三亚市等地共 16 个样地采回 522 个杧果，感染可可球二孢发病病果为 137 个，发病率为 26.2 %；分离得到共 259 个菌株，经镜检鉴定 147 株为可可球二孢，检出率为 43.2%。

2.1.2　致病力测定

对 92 株分离出来的菌株进行致病力测定（表 2），测得强致病力菌株 64 株，中致病力菌株 15 株，弱致病力菌株 13 株。强致病力比例达到 69.56%。

表 2　海南杧果可可球二孢的致病力

Table 2　The pathogenicity of *Botryodiplodia theobromae* from mango fruits in Hainan

菌株	病情指数	致病力	菌株	病情指数	致病力	菌株	病情指数	致病力
YZTN70102	100.00	强	CJTN30101	100.00	强	YCTN70103	72.22	强
YZTN100504	100.00	强	CJTN20405	100.00	强	HLTN10104	72.22	强
YZTN100503	100.00	强	CJTN20403	100.00	强	YCXY70601	69.44	中
YZTN100303	100.00	强	CJTN20303	100.00	强	YCTN70102	69.44	中
YZTN100301	100.00	强	CJTN20302	100.00	强	YCXY70602	66.67	中
YZTH100101	100.00	强	CJJH20803	100.00	强	HLTN51001	66.67	中
YZHJ90402	100.00	强	CJJH20801	100.00	强	JSJH10101	66.66	中
YZHJ90401	100.00	强	CJJH20705	100.00	强	YCXY70703	61.11	中
YZHJ90103	100.00	强	CJJH20704	100.00	强	HLTN10301	61.11	中
YZHJ90102	100.00	强	CJJH20504	100.00	强	YCXY70706	54.17	中
YZHJ90101	100.00	强	CJJH20201	100.00	强	YCXY70704	54.17	中
YZHJ80203	100.00	强	YZJH100201	97.22	强	HLTN50805	54.17	中
YCXY70301	100.00	强	HLTN51002	97.22	强	YZJH100102	52.78	中
YCHJ80301	100.00	强	YCTN70101	96.00	强	YCJH70401	50.00	中
YCHJ80202	100.00	强	CJTN20301	96.00	强	TZJH100103	50.00	中
YCHJ80103	100.00	强	CJJH20502	96.00	强	DZTN110104	47.22	中
YCHJ80101	100.00	强	CJJH20103	96.00	强	DFTN40801	41.67	中
YCHJ70101	100.00	强	HLTN10401	94.44	强	CJTN10101	39.00	弱
WBHJ100101	100.00	强	YZTN100601	92.00	强	YZTN110304	19.44	弱
TCHJ80302	100.00	强	YCJH70402	91.67	强	YZHJ90305	19.44	弱
HLTN10402	100.00	强	YCHJ80302	91.67	强	TZHJ90305	19.00	弱
HLJH50501	100.00	强	CJJH20107	89.00	强	CJXY20102	19.00	弱
GJXY30601	100.00	强	HLTN50801	88.89	强	CJJH20505	16.67	弱
DFTN41002	100.00	强	HLTN50901	86.11	强	YZJH100103	12.50	弱
DFTN40802	100.00	强	CJJH20802	85.00	强	YZHJ90301	12.50	弱
CJXY30401	100.00	强	YCXY70603	83.33	强	WBHJ100103	9.72	弱
CJXY30103	100.00	强	CJTN20501	78.00	强	YZHJ90302	0.00	弱
CJXY30101	100.00	强	CJTN10501	78.00	强	YCHJ80401	0.00	弱
CJTN30205	100.00	强	LDJH	77.78	强	DZTN110602	0.00	弱
CJTN30204	100.00	强	JSTN10102	77.78	强	DZTN110301	0.00	弱
CJTN30203	100.00	强	JSTN10101	77.78	强			

2.2　菌株抗药水平初步检测

在分离得到的 86 个菌株进行的抗药性测定中 55 个菌株在 100mg/L 的浓度下正常生长，初步判断为高抗性菌株；25 个菌株在 1mg/L 浓度下被抑制，初步判断为敏感菌株；其余 6 个菌株为低抗或中抗菌株（表 3）。抗性菌株比率为 70.09%，而高抗性菌株比率达 64.00%。

表 3　2016 年海南芒果可可球二孢抗药性

Table 3　Resistance of isolates of *Botryodiplodia theobromae* to triadimefon from in Hainan in 2016

菌株	菌落直径（cm）				抗性	菌株	菌落直径（cm）				抗性	菌株	菌落直径（cm）				抗性
	CK	1mg/L	10mg/L	100mg/L			CK	1mg/L	10mg/L	100mg/L			CK	1mg/L	10mg/L	100mg/L	
YCXY70301	9.00	8.57	9.00	9.00	HR	YZHJ90301	8.77	8.68	8.52	7.97	HR	CJXY20102	5.47	1.93	0.78	0	S
YZHJ90403	9.00	8.12	9.00	9.00	HR	CJXY30601	8.67	8.57	8.41	7.97	HR	CJTN20403	7.83	0	0.62	0	S
YCTN70101	8.90	0	9.00	9.00	HR	CJTN30101	8.6	8.48	8.7	7.88	HR	YCHJ70101	8.87	7.75	0	0	S
CJJH20505	9.00	7.93	8.83	9.00	HR	YZJH100101	8.43	7.28	8.42	7.68	HR	HLJH50501	4.03	1.78	0	0	S
YCHJ80103	9.00	8.78	8.77	9.00	HR	CJJH20504	7.13	7.5	8.38	7.65	HR	CJTN20501	4.98	1.48	0	0	S
YZHJ90401	9.00	8.28	8.45	9.00	HR	YCXY70601	8.83	8.33	7.3	7.63	HR	CJTN20302	7.47	1.23	0	0	S
YZHJ90103	8.57	7.85	8.3	9.00	HR	CJSY30401	7.25	8.63	8.27	7.58	HR	CJJH20201	8.43	1.2	0	0	S
YCHJ80101	8.50	8.4	8.23	9.00	HR	HLTN50805	8.67	8.3	6.95	7.57	HR	CJJH20802	8.18	0.7	0	0	S
WBHJ100103	9.00	8.83	8.48	8.82	HR	CJJH20705	8.57	8.7	8.65	7.43	HR	CJTN20303	9	0	0	0	S
YCHJ80102	8.65	7.03	9	8.8	HR	HLTN50901	9	8.5	7.5	7.43	HR	CJJH20106	9	0	0	0	S
YZHJ90102	8.82	8.62	8.45	8.57	HR	YZTN100303	9	8.38	8.6	7.23	HR	DZTN110104	9	0	0	0	S
WBHJ100101	8.6	8.7	8.47	8.55	HR	YCJH70401	9	7.78	8.18	7.2	HR	CJTN20301	8.82	0	0	0	S
YZTN100301	8.7	8.68	8.68	8.52	HR	YCHJ80401	6.75	6.13	5.87	6.02	HR	CJJH20107	8.77	0	0	0	S
ぬYCHJ80202	8.6	8.42	8.27	8.52	HR	YZHJ90101	8.57	8.35	8.15	5.57	HR	CJTN20403	8.75	0	0	0	S
YCHJ80203	8.65	8.53	8.37	8.43	HR	YCHJ80301	8.82	8.3	8.35	5.53	HR	CJJH20801	8.7	0	0	0	S
YCXY70602	8.7	8.73	8.18	8.43	HR	CJTN10501	5.82	4.67	5.53	4.18	HR	YZTN100504	8.52	0	0	0	S
ぬCJXY30101	9	8.65	8.67	8.42	HR	YZHJ100102	8.4	8.1	8.33	2.65	R	DFTN41002	8.4	0	0	0	S
ぬCJJH20703	9	8.07	7.92	8.4	HR	YZHJ100103	8.53	9	8.33	2.37	R	CJTN20405	8.38	0	0	0	S
ぬCJJH20505	9	8.2	8.48	8.28	HR	CJTN30203	8.5	8.3	8.15	2.32	R	YZTN10601	7.7	0	0	0	S
ぬYZHJ90402	9	8.22	8.22	8.27	HR	YZJH100201	8.05	7.2	7.18	1.62	R	DZTN110301	7.33	0	0	0	S
YZHJ90302	9	8.73	8.43	8.18	HR	YCXY70703	4.8	0.95	0.95	1.2	R	DFTN40801	6.92	0	0	0	S
YCHJ80302	8.88	8.03	7.97	8.17	HR	YCXY70704	2.8	0.85	0.9	1.18	R	CJTN10101	6.8	0	0	0	S
ぬCJXY30103	9	8.65	8.58	8.15	HR	YCXY70702	4.3	1.02	0.63	1.07		DZTN110602	6.4	0	0	0	S
ぬYCTN70102	9	8.73	8.73	8.13	HR	CJTN30204	8.83	8.1	8.35	0		DFTN40802	6	0	0	0	S
YZHJ90305	9	8.47	8.1	8.05	HR	CJTN30205	9	8.38	8.18	0		YZTN100503	5.65	0	0	0	S

2.3 抗敏菌株生理生化特性试验

2.3.1 抗敏菌株对渗透压敏感性测定

不同抗性的菌株在渗透压的影响下菌丝生长速率差别不大。葡萄糖浓度在10~80g/L时对菌株的影响不大，当超过150g/L的时候会受到明显的抑制作用。NaCl浓度则在超过10g/L后所有菌株生长均受到抑制。所以抗敏菌株对渗透压的敏感性并无差异（图1、图2）。

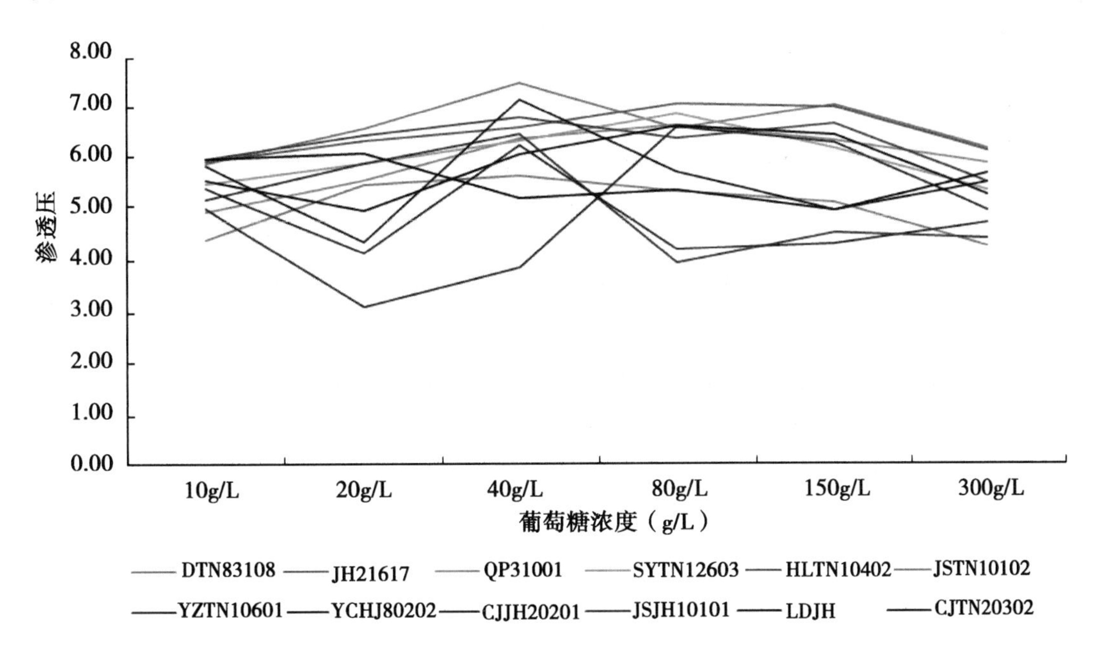

图1　不同抗性的可可球二孢对葡萄糖渗透压测定

Fig. 1　The osmotic pressure measure of different resistance strains to amylaceum

2.3.2 抗敏菌株电导率测定

结果如图3、图4所示，高抗菌株JH21615的电导率明显高于敏感菌株QP31001；不同多菌灵浓度处理下两个菌株均随着时间增加电导率升高，但敏感菌株的上升速率要比抗性菌株快，抗性菌株变化非常缓慢；在做过致死处理后，细胞死亡，电渗率迅速升高。

3　结论与分析

海南杧果蒂腐病可可球二孢潜伏期长、发病快，致病力高，给海南杧果生产造成了很大的危害。从实验结果上看，可可球二孢在海南各个杧果产区均有分布，且发病率高、致病力强、对多菌灵产生严重抗性。2014年本课题组从发病杧果上共分离获得132株致病菌，其中124株为杧果可可球二孢，分离率达93.9%，强致病力菌株占28.23%[11]。2016年，我们共分离得到259个菌株，经镜检鉴定147株为可可球二孢，检出率为43.2%，而强致病力比例高达69.56%。通过两年测定数据的对比，2016年的杧果可可球二孢的检出率明显降低，但致病力均很高。田间调研时发现，由于2016年初海南出现严重寒害，而进入5月后的持续高温，异常的天气导致今年杧果的病害及蓟马等危害加重，果农为了确保杧果产量，大量喷施各种杀菌剂、杀虫剂及植物生物生长调节剂，对于杧果蒂腐病的发生有一定的控制，可能导致杧果可可球二孢检出率的下降。因为今年杧果生产上大剂量频繁的施药，杧果

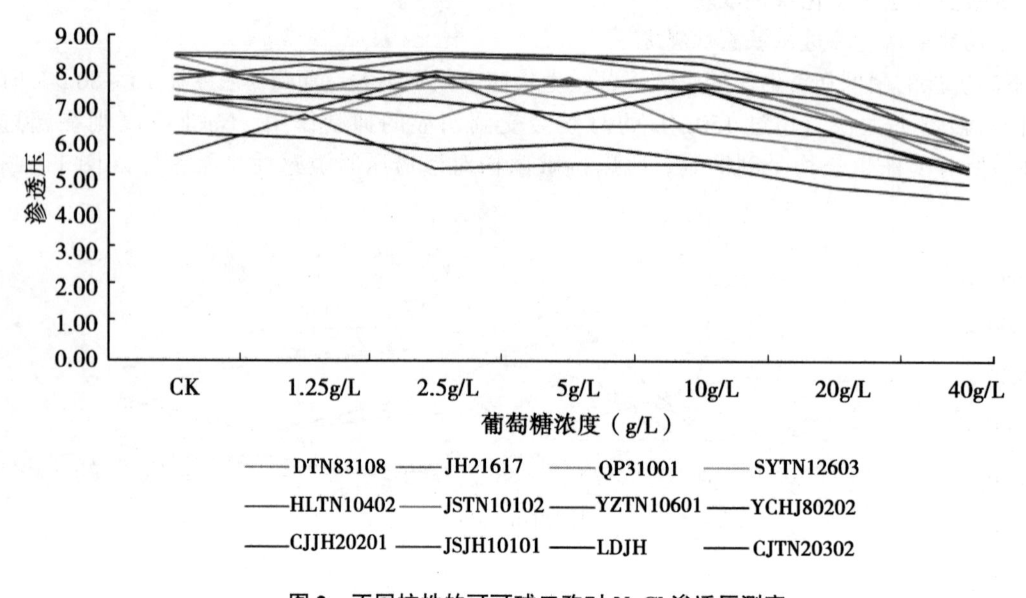

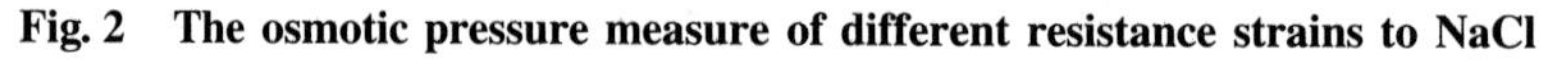

图 2 不同抗性的可可球二孢对 NaCl 渗透压测定

Fig. 2 The osmotic pressure measure of different resistance strains to NaCl

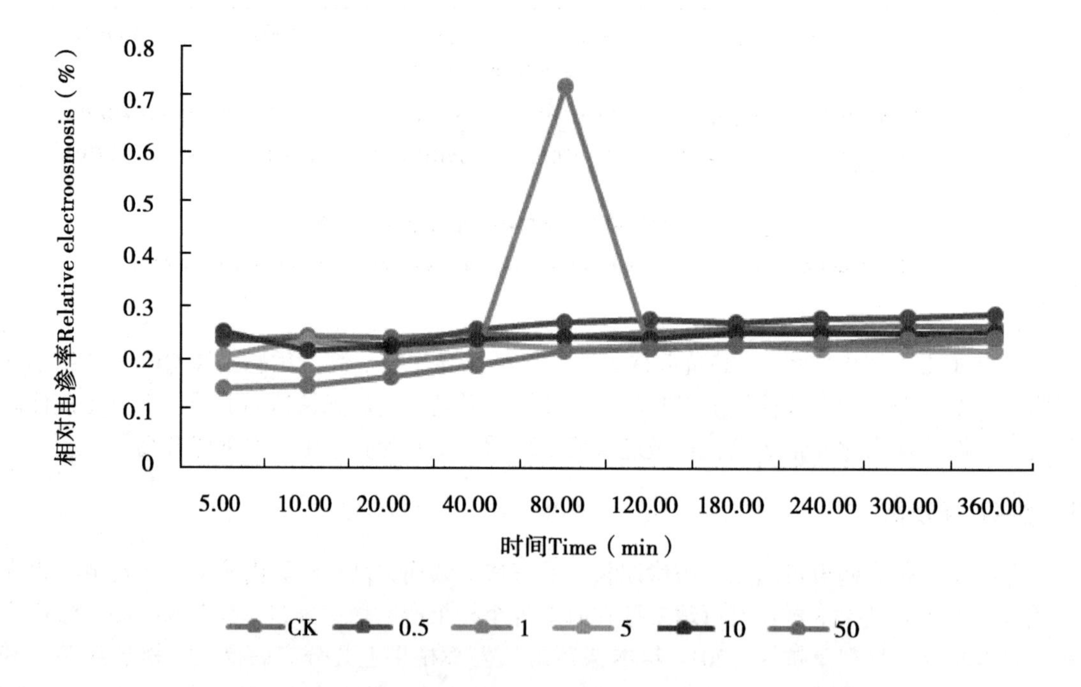

图 3 抗性菌株 JH21615 在不同浓度多菌灵作用下相对电渗率变化

Fig. 3 The relative electroosmosis rate of insensitive strains JH21615 in effect of different carbendazim concentration

果实的药害也非常严重，出现大量畸形果、转色不佳、不能正常后熟等现象。

而 2014 年对多菌灵的抗性测定表明，该菌对多菌灵、甲基硫菌灵等杀菌剂产生严重的抗药性，对多菌灵的抗性频率达 90.1%[6]，本次测定对多菌灵的抗性频率达到 70.9%。说

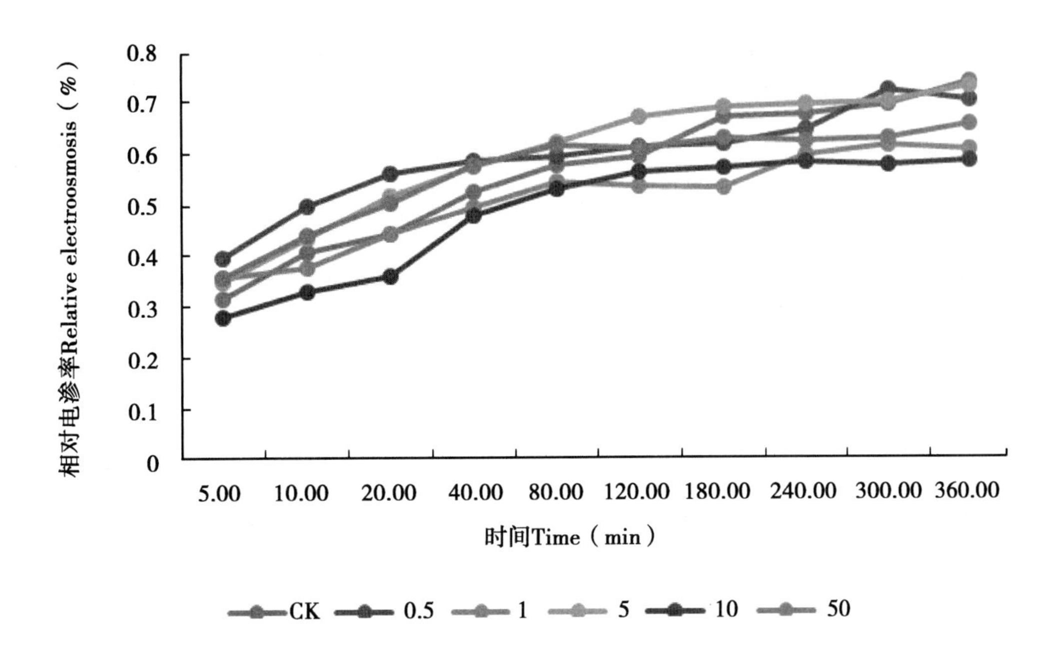

图 4　抗性菌株 QP31001 在不同浓度多菌灵作用下相对电渗率变化

Fig. 4　The relative electroosmosis rate of sensitive strains QP31001 in effect of different carbendazim concentration

明海南杧果可可球二孢对多菌灵的抗药性普遍且非常严重，这与海南杧果产区长期频繁施用上述杀菌剂有密切关系。大量研究表明，病原菌对药剂产生抗药性主要存在 3 种抗性机理，分别是脱毒作用增强、毒性代谢产物转换作用的减弱、靶标位点敏感性降低[12-13]。前两种抗性机理很有可能伴随着菌株体内生理生化特性的转变。本次研究表明，杧果蒂腐病可可球二孢的抗、敏菌株渗透压敏感性上没有表现出明显差异，却在细胞膜通透性上表现出了不同。抗性菌株 JH21617 的相对电导率明显高于敏感菌株 QP31001，而煮沸致死后电导率的同时上升则表明了在两个菌株在细胞活着的时候细胞膜通透性才会有差异，表明细胞膜通透性的差异很有可能与抗性产生的机制有关。

参考文献

[1]　邓泽明，胡美姣，白菊仙. 海南省杧果蒂腐病病原菌的初步研究Ⅲ［J］. 中国南方果树，2002，31（4）：39－41.

[2]　李敏，胡美姣，岳建军，等. 杧果可可球二孢蒂腐病菌生物学培养特性［J］. 热带作物学报，2009，30（11）：1 660－1 664.

[3]　胡美姣，李敏，高兆银，刘秀娟. 热带亚热带水果采后病害及防治［M］. 北京：中国农业出版社，2010：94.

[4]　胡美姣，师超，安勇，等. 杧果蒂腐病菌对多菌灵的抗药性测定及其杀菌剂筛选［J］. 果树学报，2009，26（5）：671－677.

[5]　胡美姣，高兆银，李敏，等. 杧果果实潜伏侵染真菌种类研究［J］. 果树学报，2012，29（1）：105－110.

[6]　王萌，陈小莉，杨叶，等. 海南杧果蒂腐病对 8 种杀菌剂的抗药性测定［J］. 农药，2015，54

(5)：384－386.

[7] 陈业渊，贺军虎. 热带、南亚热带果树种质资源数据质量控制规范［M］. 北京：中国农业出版社，2006：102－103.

[8] Byungsup K，Eunwoo P，Yun C K. Population Dynamics of Sensitive and Resistance Phenotypes of *Botrytis cinereato* Benzimidazole，Dicarboximide and N-phenylcarbamate Fungicides in Korea［J］. Journal of Pesticide Science，2000，25（4）：385－386.

[9] Ma H X，Zhou M G，et al. Detection of resistance to dimethachlon and carbendazim in Sclerotinia sclerotiorum in Jangsu Province［J］. Journal of plant pathology，2008，90（2，supplement）：143.

[10] 石志琦，周明国，叶钟音. 核盘菌对菌核净的抗药性机制初探［J］. 农药学学报，2000，20（2）：47－51.

[11] 陈小莉，杨叶，王萌，等. 杧果蒂腐可可球二孢菌致病力测定及杧果主要品种抗病性评价. 果树学报，2015，32（3）481－486.

[12] 周明国，叶钟音，刘经芬. 杀菌剂抗性研究进展［J］. 南京农业大学学报，1994，17（3）：33－41.

[13] 于永学，王英姿. 灰霉病菌抗药性发生概况及机理研究进展［J］. 现代农业技. 2009，37（11）：117－118.

烯肟菌胺、苯醚甲环唑与噻虫嗪混配生物活性研究

王军锋*，单中刚，李志念，丑靖宇，司乃国
（沈阳中化农药化工研发有限公司，新农药创制与开发国家重点实验室，沈阳 110021）

摘要：烯肟菌胺是沈阳中化农药化工研发有限公司创制的一个甲氧基丙烯酸酯类杀菌剂，苯醚甲环唑为三唑类杀菌剂，二者对多种病害具有良好的防治效果。室内联合毒力测定试验发现烯肟菌胺与苯醚甲环唑混配对禾谷丝核菌具有增效作用，且加入噻虫嗪混配不影响烯肟菌胺与苯醚甲环唑对禾谷丝核菌的抑菌作用。室内安全性测定结果表明45%烯·苯·噻FSC种子包衣及处理对小麦生长有一定的影响，低剂量下不明显。室内盆栽活性测定结果表明45%烯·苯·噻FSC种子包衣对小麦纹枯病具有良好的防治效果，具有良好的应用前景。

关键词：烯肟菌胺；苯醚甲环唑；禾谷丝核菌；生物活性

Biological Activity Study of the Mixture of Fenaminstrobin、Difenoconazole and Thiamethoxam

Wang Junfeng，Shan Zhonggang，Li Zhinian，Niu Jingyu，Si Naiguo
(*Shenyang Sinochem Agrochemicals R&D Co. Ltd. State Key Laboratory of the Discovery and Development of Novel Pesticide of Shenyang*，*Liaoning Shenyang* 110021)

Abstract：Fenaminstrobin is one of Methoxyl acrylic ester fungicide developed by Shenyang Sinochem Agrochemicals R&D Co. Ltd，difenoconazole belongs to Fungicide – triazole derivative，and they both have been commercially applied against many fungal pathogens. The joint activity measurement test showed the mixture of fenaminstrobin and difenoconazole had synergistic effect on *Rhizoctonia cerealis*，and adding thiamethoxam did not affect inhibiting effect on *Rhizoctonia cerealis*. The pot tests showed that wheat growth whose seeds coated by Mixture of Fenaminstrobin、Difenoconazole and Thiamethoxam was influenced adversely，but the influence did by low dose was distinct. The results displays Fenaminsrobin – Difenoconazole – thiamethoxam 45% FSC had good control effect on wheat sharp eyespot by seed coating，Fenaminsrobin – Difenoconazole – thiamethoxam 45% FSC exhibited a promising application.

Key words：Fenaminstrobin；Difenoconazole；Wheat sharp eyespot；Biological activity

烯肟菌胺是沈阳中化农药化工研发有限公司（原沈阳化工研究院农药研究所）创制的一个甲氧基丙烯酸酯类杀菌剂，苯醚甲环唑为三唑类杀菌剂，二者对多种病害具有良好的防治效果[1-6]。农药创制是一项高技术、高投资且高风险的系统工程，开发出高效化合物的难度越来越大，针对我国农药工业现状，进行科学合理的混剂开发，能够促进创新杀菌剂获得更好经济效益及社会效益[7]。近年来由气候变暖，化肥特别是氮肥施用量的提高，小麦播

* 作者简介：王军锋，男，高级工程师，主要从事创制杀菌剂生物活性筛选及植物病害化学防治技术研究；E-mail：wangjunfeng@ sinochem. com

期提前及播种量加大等原因，小麦纹枯病的发生呈现逐年加重趋势[8]，田间试验发现药剂种子包衣防治小麦纹枯病效果较好[9-11]。本文以通过室内离体试验及活体盆栽试验测定烯肟菌胺、苯醚甲环唑和噻虫嗪混配对禾谷丝核菌的联合抑菌作用、对小麦的安全性及纹枯病的盆栽防效。

1 材料与方法

1.1 材料

1.1.1 试验药剂

98%烯肟菌胺原药，96%苯醚甲环唑原药，96.2%噻虫嗪原药均由沈阳化工研究院有限公司提供，用于三者混配室内联合毒力测定。45%烯·苯·噻 FSC（烯肟菌胺：苯醚甲环唑 =1：3，两者总含量为2.4%）1.8%苯醚甲环唑 FSC，0.9%烯肟菌酯 FSC；所有药剂均由沈阳科创化学品有限公司提供。

1.1.2 试剂

丙酮 吐温 -80。

1.1.3 作物品种及靶标病害

小麦纹枯病（*Rhizotonia cerealis*），室内毒力测定及盆栽防效试验用；

小麦（*Triticum aestivum*），品种为辽春10号，小麦活体盆栽试验用。

1.2 方法

1.2.1 烯肟菌胺与苯醚甲环唑混配室内联合毒力测定

试验参考进行农业行业标准[12] NYT 1156.6—2006 农药室内生物测定试验准则进行。烯肟菌胺、苯醚甲环唑、烯肟菌胺与苯醚甲环唑混剂（配比为5：1、3：1、1：1、1：3、1：5）的分别设6个处理浓度，另设清水空白对照；将熔好的PDA培养基冷却至60~70℃，按所设浓度加入定量药剂，制成含有不同浓度药液的含毒培养基，待其充分冷却后，接种直径5mm的供试病原菌菌片，置于培养箱中培养（23±1）℃，2d后进行调查。调查时，分别测量每处理的供试病原菌菌落直径，根据下式计算抑菌率，求出毒力回归方程及 EC_{50} 值，由 Sun－Johnson 毒力指数计算方法，计算五个配比混剂的共毒系数（CTC）。CTC < 80 为拮抗作用；80≤CTC≤120 为相加作用；CTC > 120 为增效作用。

$$抑菌率（\%）=\frac{对照菌落直径-处理菌落直径}{对照菌落直径}\times 100$$

Sun－Johnson 毒力指数计算法

$$ATI=\frac{S}{M}\times 100 \quad (1)$$

式（1）中：ATI——混剂实测的毒力指数；

S——标准药剂的 EC_{50} 理论值；

M——混剂的 EC_{50} 理论值。

$$TTI=TI_A\times P_A+TI_B\times P_B \quad (2)$$

式（2）中：TTI——混剂的理论毒力指数；

TI_A——A药剂毒力指数；

P_A——A药剂在混剂中的百分含量，单位为百分率（%）；

TI_B——B药剂毒力指数；

P_B——B 药剂在混剂中的百分含量，单位为百分率（%）。

$$CTC = \frac{ATI}{TTI} \times 100 \quad (3)$$

式（3）中：CTC——共毒系数；

ATI——混剂实测毒力指数；

TTI——混剂理论毒力指数。

1.2.2 烯·苯与噻虫嗪混配对禾谷丝核菌室内联合毒力测定

烯肟菌胺与苯醚甲环唑（比例 1∶3 混配）作为一个整体以烯·苯表示，根据联合作用测定结果，综合市场开发需要选定烯肟菌胺与苯醚甲环唑一定比例组合再与噻虫嗪混配（配比为 9∶1、7∶3、5∶5、7∶3、1∶9）进行室内联合毒力测定，试验方法同 1.2.1。

1.2.3 药剂对小麦安全性测定

试验参照农业行业标准农药对作物安全性评价准则[13]进行。

1.2.3.1 种子包衣

45%烯·苯·噻 FSC（烯肟菌胺∶苯醚甲环唑 =1∶3，两者总含量为 2.4%）试验处理以烯肟菌胺与苯醚甲环唑总有效成分计算，所有药剂均设五个剂量，分别为 360、180、90、75、60 mg/kg；根据试验需要称取适量药剂加入小麦种子中进行包衣，晾干备用。

1.2.3.2 发芽率测定

将不同种衣剂包衣处理的小麦种子及空白对照（未进行包衣处理）直接播种在培养钵中，每钵播种 12 粒种子，每处理 4 钵，每钵为一重复，不同处理的重复分为四个区组，在温室中常规管理，分别于 7d 后调查出苗数，以式（4）计算发芽率，并进行邓肯氏差异显著性统计分析。

$$发芽率(\%) = \frac{出苗数}{播种数} \times 100 \quad (4)$$

1.2.3.3 对小麦植株生长影响测定

采用温室盆栽法。按 1.2.3.2 方法进行播种，7d 后分别测定各处理小麦株高、根长及鲜重，并进行邓肯氏差异显著性统计分析。

1.2.4 对小麦纹枯病防效试验方法

采用温室盆栽法。将土壤与小麦纹枯病菌以 10∶1 的比例混合均匀，按 1.2.3.2 方法进行播种，7d 后调查各处理发芽率，每钵播种 15 粒种子，据式（4）、式（5）和式（6）计算发病率及防治效果，并进行邓肯氏差异显著性统计分析。

$$发病率(\%) = 100 - 发芽率 \quad (5)$$

$$防效(\%) = \frac{对照发病率 - 处理发病率}{对照发病率} \times 100 \quad (6)$$

2 结果与分析

2.1 烯肟菌胺与苯醚甲环唑混配对禾谷丝核菌室内联合毒力测定结果

烯肟菌胺单剂、苯醚甲环唑单剂及五个供试混剂（配比为 5∶1、3∶1、1∶1、1∶3、1∶5）对禾谷丝核菌的 EC_{50} 分别为 1.745 1、0.092 5、0.108 8、0.063 1、0.028 6、0.024 2、0.044 1mg/L；烯肟菌胺与苯醚甲环唑五个供试混剂（配比为 5∶1、3∶1、1∶1、1∶3、1∶5）的共毒系数分别为 403.09、505.60、614.51、501.42、249.06，均远大于 120，试验结果详见表 1。

表 1　烯肟菌胺、苯醚甲环唑对禾谷丝核菌的联合毒力测定结果

药剂	配比	毒力方程	EC_{50}（mg/L）	相关系数	共毒系数
烯肟菌胺：苯醚甲环唑	5：1	$y=5.5538+0.5749x$	0.1088	0.9825	403.09
	3：1	$y=5.6564+0.5472x$	0.0631	0.9707	505.60
	1：1	$y=5.7724+0.5003x$	0.0286	0.9627	614.51
	1：3	$y=5.8703+0.5383x$	0.0242	0.9460	501.42
	1：5	$y=5.9416+0.6950x$	0.0441	0.9860	249.06
烯肟菌胺		$y=4.8090+0.7900x$	1.7451	0.9474	—
苯醚甲环唑		$y=5.5330+0.5156x$	0.0925	0.9605	—
空白对照			（菌落直径：50.5 mm）		

2.2　烯·苯与噻虫嗪混配对禾谷丝核菌室内联合毒力测定结果

烯肟菌胺与苯醚甲环唑（比例 1：3 混配）作为一个整体以烯·苯表示，分别以配比 9：1、7：3、5：5、7：3、1：9 进行室内联合毒力测定，五个混配组合的共毒系数分别为 96.01、96.47、96.57、85.33、88.70，均处于 80～120，试验结果详见表 2。

表 2　烯·苯与噻虫嗪混配对禾谷丝核菌的联合毒力测定结果

药剂	配比	毒力方程	EC_{50}（mg/L）	相关系数	共毒系数
烯·苯：噻虫嗪	9：1	$y=5.8294+0.5361x$	0.0280	0.9792	96.01
	7：3	$y=5.9051+0.6259x$	0.0358	0.9909	96.47
	5：5	$y=5.8078+0.6473x$	0.0500	0.9899	96.57
	3：7	$y=5.7036+0.6863x$	0.0940	0.9886	85.33
	1：9	$y=5.4333+0.7555x$	0.2670	0.9963	88.70
烯·苯		$y=5.8703+0.5383x$	0.0242	0.9460	
噻虫嗪			（大于 10mg/L，共毒系数以 10mg/L 计算）		

2.3　对小麦安全性测定结果

2.3.1　对发芽率影响

45% 烯·苯·噻 FSC 各处理的发芽率均在 89.58% 以上，统计分析结果表明，在 5% 及 1% 水平上各处理没有显著差异，试验结果详见表 3。

2.3.2　对小麦株高的影响

试验结果表明 45% 烯·苯·噻 FSC 高剂量的处量对小麦株高有一定的影响，邓肯氏差异显著性统计分析结果表明 45% 烯·苯·噻 FSC360mg/kg 剂量处理在 5% 水平上相对于空白对照有显著差异，其他各处理与空白对照没有显著差异，试验结果详见表 4。

表3　不同烯·苯·噻 FSC 对小麦种子包衣的发芽率测定结果

药剂	剂量（mg/kg）	发芽率（%）					显著性	
		I	II	III	IV	平均值	5%	1%
45%烯·苯·噻 FSC	360	91.67	100	83.33	83.33	89.58	a	A
	180	100	91.67	100	91.67	95.83	a	A
	90	100	100	91.67	83.33	93.75	a	A
	75	100	91.67	100	91.67	95.83	a	A
	60	100	91.67	100	83.33	93.75	a	A
1.8%苯醚甲环唑 FSC	360	100	100	100.0	91.67	97.92	a	A
	180	100	100	91.67	91.67	95.83	a	A
	90	100	100	91.67	83.33	93.75	a	A
	75	100	100	91.67	100	97.92	a	A
	60	100	83.33	91.67	100	93.75	a	A
0.9%烯肟菌酯 FSC	360	91.67	100	100	91.67	95.83	a	A
	180	91.67	100	91.67	100	95.83	a	A
	90	91.67	100	91.67	100	95.83	a	A
	75	100	91.67	100	100	97.92	a	A
	60	91.67	91.67	100	91.67	93.75	a	A
CK		100	100	91.67	100	97.92	a	A

2.3.3　对小麦根长的影响

试验结果表明45%烯·苯·噻 FSC 不同剂量处理对小麦的根生长有一定的影响，剂量越大影响越明显；邓肯氏差异显著性统计分析结果表明45%烯·苯·噻 FSC360mg/kg 剂量处理在5%水平上与空白对照有显著差异，但在1%水平上相对空白对照差异不显著，其他各处理与空白对照没有显著差异，试验结果详见表5。

2.3.4　对小麦鲜重的影响

试验结果表明45%烯·苯·噻 FSC 高剂量处理对小麦植株生长量有一定的影响，剂量越大影响越明显；邓肯氏差异显著性统计分析结果表明5%显著水平上45%烯·苯·噻 FSC180mg/kg 以上剂量的处量与空白对照相比有显著差异，但在1%水平上与空白对照差异不显著，其他各处理与空白对照在5%与1%水平上没有显著差异，试验结果详见表6。

表4　不同烯·苯·噻FSC对小麦株高的影响

药剂	剂量(mg/kg)	株高（cm）					显著水平	
		Ⅰ	Ⅱ	Ⅲ	Ⅳ	平均值	5%	1%
45%烯·苯·噻FSC	360	13.8	12.6	14.2	13.8	13.6	cde	BCD
	180	15.0	16.2	14.8	13.8	15.0	abc	ABC
	90	15.2	16.6	14.7	14.9	15.4	ab	ABC
	75	15.5	14.8	15.7	16.3	15.6	a	AB
	60	15.5	15.3	15.6	16.5	15.7	a	A
1.8%苯醚甲环唑FSC	360	13.7	15.0	14.7	14.3	14.4	abcd	ABCD
	180	14.4	15.3	14.4	14.0	14.5	abcd	ABCD
	90	15.2	15.8	15.8	14.7	15.4	ab	ABC
	75	16.6	16.6	14.4	14.2	15.5	ab	AB
	60	16.5	15.6	14.0	14.1	15.0	abc	ABC
0.9%烯肟菌酯FSC	360	13.3	11.7	12.8	13.9	12.9	e	D
	180	14.5	15.4	13.4	14.9	14.6	abcd	ABCD
	90	15.5	13.7	15.0	13.9	14.5	abcd	ABCD
	75	14.5	13.7	15.0	14.5	14.4	abcd	ABCD
	60	16.3	13.6	16.6	14.1	15.1	ab	ABC
CK		14.7	16.2	14.9	14.9	15.2	ab	ABC

表5　不同烯·苯·噻FSC对小麦根长的影响

药剂	剂量(mg/kg)	根长（cm）					显著水平	
		Ⅰ	Ⅱ	Ⅲ	Ⅳ	平均值	5%	1%
45%烯·苯·噻FSC	360	11.6	10.2	11.1	12.0	11.2	c	A
	180	14.6	10.8	10.9	11.5	11.9	abc	A
	90	13.2	11.4	11.0	14.0	12.4	abc	A
	75	11.6	13.3	11.7	13.2	12.5	abc	A
	60	12.1	13.3	14.2	12.7	13.1	ab	A
1.8%苯醚甲环唑FSC	360	11.0	11.9	11.4	10.5	11.2	c	A
	180	13.4	12.0	11.2	11.4	12.0	abc	A
	90	13.4	14.0	10.2	10.9	12.1	abc	A
	75	12.5	14.0	13.5	13.0	13.3	ab	A
	60	13.0	14.2	12.8	13.2	13.3	ab	A

（续表）

药剂	剂量 (mg/kg)	根长（cm）					显著水平	
		Ⅰ	Ⅱ	Ⅲ	Ⅳ	平均值	5%	1%
0.9%烯肟菌酯 FSC	360	12.1	11.4	9.8	11.8	11.3	c	A
	180	12.1	11.7	11.3	12.7	12.0	abc	A
	90	15.9	13.7	10.4	12.7	13.2	ab	A
	75	13.0	12.7	13.7	13.6	13.2	ab	A
	60	14.9	13.4	12.5	12.8	13.4	a	A
CK		12.8	13.1	12.8	14.1	13.2	ab	A

表6　不同烯·苯·噻 FSC 对小麦鲜重的影响

药剂	剂量 (mg/kg)	鲜重（g）					显著水平	
		Ⅰ	Ⅱ	Ⅲ	Ⅳ	平均值	5%	1%
45%烯·苯·噻 FSC	360	0.13	0.12	0.10	0.11	0.11	fg	BC
	180	0.10	0.12	0.14	0.10	0.12	efg	BC
	90	0.14	0.16	0.14	0.15	0.15	abc	AB
	75	0.12	0.15	0.14	0.14	0.14	abcde	AB
	60	0.15	0.13	0.15	0.17	0.15	a	A
1.8%苯醚甲环唑 FSC	360	0.12	0.11	0.14	0.12	0.12	cdefg	ABC
	180	0.17	0.13	0.17	0.13	0.15	ab	A
	90	0.15	0.14	0.14	0.15	0.14	abcd	AB
	75	0.15	0.13	0.16	0.18	0.15	a	A
	60	0.16	0.15	0.15	0.14	0.15	a	A
0.9%烯肟菌酯 FSC	360	0.11	0.10	0.10	0.10	0.10	g	C
	180	0.13	0.12	0.12	0.13	0.12	bcdefg	ABC
	90	0.13	0.15	0.11	0.14	0.13	abcdef	ABC
	75	0.13	0.14	0.12	0.15	0.13	abcdef	ABC
	60	0.14	0.12	0.14	0.11	0.13	abcdefg	ABC
CK		0.14	0.13	0.17	0.13	0.14	abcd	AB

2.4　对小麦纹枯病的防治效果

试验结果表明，45%烯·苯·噻 FSC 对小麦进行种子包衣对小麦纹枯病具有良好的防治效果，平均防效可达到50%以上；低剂量 60 mg/kg 的处理相对于其他处理防效较低，在5%显著水平上与180 mg/kg 高剂量处理有显著差别，但在1%显著水平上与180 mg/kg 高剂量处理没有显著差别，试验结果详见表7。

表7　不同烯·苯·噻FSC对小麦纹枯病防治效果

药剂	剂量 (mg/kg)	防效（%）					显著水平	
		Ⅰ	Ⅱ	Ⅲ	Ⅳ	平均值	5%	1%
45%烯·苯·噻FSC	360	66.67	63.64	57.14	55.56	60.75	abcdef	AB
	180	75.00	90.91	71.43	88.89	81.56	abc	A
	90	75.00	81.82	42.86	88.89	72.14	abcde	AB
	75	66.67	72.73	57.14	55.56	63.02	abcdef	AB
	60	66.67	54.55	42.86	44.44	52.13	def	AB
1.8%苯醚甲环唑FSC	360	91.67	63.64	100	77.78	83.27	a	A
	180	66.67	63.64	85.71	100	79.00	a	A
	90	66.67	90.91	57.14	77.78	73.12	abcd	AB
	75	91.67	72.73	28.57	44.44	59.35	abcdef	AB
	60	75.00	63.64	28.57	55.56	55.69	bcdef	AB
0.9%烯肟菌酯FSC	360	75.00	63.64	42.86	33.33	53.71	cdef	AB
	180	66.67	72.73	57.14	33.33	57.47	bcdef	AB
	90	66.67	54.55	42.86	44.44	52.13	def	AB
	75	66.67	54.55	14.29	33.33	42.21	ef	B
	60	66.67	54.55	28.57	22.22	43.00	f	B

3　小结

烯肟菌胺与苯醚甲环唑五个供试混剂（配比为5：1、3：1、1：1、1：3、1：5）的共毒系数均大于120，具有明显的增效作用；烯肟菌胺与苯醚甲环唑（比例1：3混配）作为一个整体，分别以配比9：1、7：3、5：5、7：3、1：9进行室内联合毒力测定，五个混配组合的共毒系数均处于80至120之间，说明噻虫嗪对烯肟菌胺与苯醚甲环唑混配没有拮抗作用。45%烯·苯·噻FSC种子包衣不同剂量对小麦种子发芽率有一定影响，随着剂量提高，发芽率降低；在360mg/kg以下时，发芽率均在89.58%以上，统计分析结果表明，在5%及1%水平上各处理没有显著差异。45%烯·苯·噻FSC种子包衣高剂量3对小麦的株高有一定的影响，低剂量影响不明显；对根长与鲜重的影响表现为剂量越大，影响越大，表现为根长与鲜重值越小。45%烯·苯·噻FSC进行小麦种子包衣对小麦纹枯病具有良好的防治效果，平均防效可达到50%以上，180 mg/kg高剂量的处理防效较其他剂量为好。本试验结果采用小麦辽春10号进行试验，对其他品种的效应还需进一步研究。

参考文献

［1］ 司乃国，刘君丽，杨春河，等.新型广谱杀菌剂-烯肟菌胺（SYP-1620）［C］//中国植物病害化学防治研究（第四卷）.北京：中国农业科学技术出版社，2004：31-42.

［2］ Zhang X K，Wu D X，Duan Y B，et al. Biological characteristics and resistance analysis of the novel fungicide SYP - 1620 against botrytis cinerea [J]. Pesticide Biochemistry and Physiology, 2014, 114 (167): 72 - 78.

[3] 王丽，周增强，侯珲. 三唑类杀菌剂对苹果主要病原菌的毒力及田间防效 [J]. 河南农业科学, 2016 (07): 82 - 86.

[4] 张国珍，张雪松，梁月，等. 10% 世高水分散粒剂防治西洋参黑斑病试验 [J]. 植物保护, 2005, 31 (3): 86 - 88.

[5] 叶长飞，温莉娜，林永红，等. 7 种药剂防治梨轮纹病、黑星病和黑斑病的效果 [J] 浙江农业科学, 2016, 57 (2): 248 - 250.

[6] 卢颖，杨念福，繁生. 世高防治西瓜炭疽病药效试验 [J]. 农药, 2001, 40 (4): 35 - 35.

[7] 刘君丽. 中国创新杀菌剂研发进展 [C] //中国植物病害化学防治研究（第五卷）. 北京：中国农业科学技术出版社, 2006: 8 - 14.

[8] 张会云，陈荣振，冯国华，等. 中国小麦纹枯病的研究现状与展望 [J]. 麦类作物学报, 2007, 27 (6): 1 150 - 1 153.

[9] 任学祥，叶正和，丁克坚，等. 噻呋酰胺种衣剂防治小麦纹枯病效果及安全性研究 [J]. 麦类作物学报, 2015, 35 (11): 1 588 - 1 591.

[10] 谢凤珍，王振河，李淑恒. 不同杀菌剂防治优质小麦纹枯病的研究 [J] 河南科技学院学报（自然科学版）2007, 35 (1): 25 - 26.

[11] 张友明，莫婷，史晓利，等. 6% 立克秀悬浮种衣剂防治小麦纹枯病试验研究 [J]. 2012 (10): 160, 166.

[12] NY/T 1156. 6—2006. 农药室内生物测定试验准则　杀菌剂　第 6 部分：防治小麦白粉病试验混配的联合作用测定 [S].

[13] NYT 1965. 1—2010 农药对作物安全性评价准则　第 1 部分：杀菌剂和杀虫剂对作物安全性评价室内试验方法 [S].

杀菌剂交替施用对马铃薯晚疫病的防治效果及经济效益评价*

台莲梅[1**]，靳学慧[1***]，张亚玲[1]，李海燕[1]，张宗敏[2]

（1. 黑龙江八一农垦大学，大庆 163319；2. 克山农场，克山 161600）

摘要：选用不同类型的药剂交替施用，研究其对马铃薯晚疫病的防治效果，结果表明，不同药剂交替施用处理对晚疫病都有明显的防治效果，防效在75%以上。其中，60%氟吗·锰锌与58%甲霜灵·锰锌交替用药，防效高，为95.3%，增产率和获得纯经济效益最高，分别为41%、357.1元/667 m^2，25%嘧菌酯与60%氟吗·锰锌、58%甲霜灵·锰锌交替用药对马铃薯晚疫病防效为81.4%，但增产和获得纯经济效益也较高。

关键词：马铃薯；晚疫病；杀菌剂；防治效果；经济效益

我国马铃薯生产是世界上的第一大国[1]，但马铃薯病害的发生限制了马铃薯产业的发展，特别是马铃薯晚疫病，发生普遍，危害严重，是马铃薯第一大病害。黑龙江省是全国重要的马铃薯种薯和商品薯生产基地之一[2]。2005年黑龙江省由于气象条件适宜马铃薯晚疫病的大发生，造成马铃薯大幅度减产，产量损失达30%以上，严重地块减产50%左右，严重地影响了马铃薯种植业发展。目前，对晚疫病的防治主要以抗病品种及化学药剂防治为主[3]。生产中广泛使用的多种杀菌剂也出现了不同程度的晚疫病菌株的抗药性[4]。为了更好地防治晚疫病同时避免病菌抗药性的产生，通过选取不同类型的药剂交替施用，研究其对马铃薯晚疫病的防治效果及经济效益，为马铃薯生产中晚疫病的防治提供参考。

1 材料与方法

1.1 供试品种

东农303，保苗6万株/hm^2。

1.2 供试药剂

25%嘧菌酯悬浮剂（先正达公司）；72%霜脲·锰锌可湿性粉剂（美国杜邦公司）；60%氟吗·锰锌可湿性粉剂（沈阳化工研究院）；50%三乙膦酸铝可溶性粉剂（江苏利民农化有限公司）；58%甲霜灵·锰锌可湿性粉剂（河北双吉农化有限公司）。

1.3 试验设计

①嘧菌酯480mL/ hm^2第一次喷施、氟吗·锰锌1 500g/ hm^2第二次喷施、甲霜灵·锰锌2 250g/hm^2第三次喷施；②霜脲·锰锌2 000g/ hm^2和甲霜灵·锰锌2 250g/hm^2交替施用；③霜脲·锰锌2 000g/hm^2和三乙膦酸铝2 000g/hm^2交替施用；④氟吗·锰锌1 800g/hm^2和甲霜灵·锰锌2 250g/hm^2交替施用；⑤氟吗·锰锌1 800g/hm^2和三乙膦酸铝2 250g/hm^2交

* 基金项目：黑龙江省农垦总局项目（HNKXIV－06－03B）

** 作者简介：台莲梅，女，主要从事植物真菌病害和病害综合防治研究工作；E-mail：tailianmei@ sina. com

*** 通讯作者：靳学慧，男，主要从事植物真菌病害和病害综合防治研究工作；E-mail：jxh2686@ 163. com

替施用；⑥清水空白对照6个处理，4次重复，随机区组排列。每小区为5行区、行长5m、行距0.8 m，小区面积20m²。

1.4 试验方法

于发病初期喷药，每隔7~10d喷药1次，共喷3次。山东卫士WS-16型手动喷雾器，扇形喷头，喷液量200kg/hm²。每次喷药后到下次喷药前进行调查，定点定株每区对角线5点取样，每点调查2株，记录病株发病情况，计算出发病率、病情指数和防治效果。于马铃薯收获时，每区取4m²，测定各处理的产量，并计算每667m²的总产量、增产率和经济效益等。

2 结果与分析

2.1 不同药剂交替处理防治效果

不同处理病害调查结果见表1。第2次喷药后霜脲·猛锌与甲霜灵·锰锌、氟吗·锰锌与甲霜灵·锰锌、嘧菌酯与氟吗·锰锌、甲霜灵·锰锌交替用药对马铃薯晚疫病防效高，均在95%以上，3个处理防效差异不显著，但与其他杀菌剂交替施用处理差异显著；霜脲·猛锌与乙膦铝、氟吗·锰锌与乙膦铝交替施用防效低些，但防效也高于85%。

第3次喷药后病害调查，霜脲·猛锌与甲霜灵·锰锌、氟吗·锰锌与甲霜灵·锰锌交替用药，防治效果仍在95%以上，与其他处理差异显著。霜脲·猛锌与乙膦铝、氟吗·锰锌与乙膦铝交替施用、嘧菌酯与氟吗·锰锌、甲霜灵·锰锌交替用施用处理，防效有所降低，但防治效果均在75%以上，3个处理之间差异不显著。用药处理叶枯率明显低于清水对照。

表1 杀菌剂交替施用对马铃薯晚疫病防治效果

交替用药处理	第2次喷药后调查		第3次喷药后调查		叶枯率（%）
	病情指数	防效（%）	病情指数	防效（%）	
霜脲·猛锌/甲霜灵·锰锌	0.34	98.4a	1.57	96.2a	15.1
氟吗·锰锌/甲霜灵·锰锌	0.59	96.0a	1.98	95.3a	17.3
嘧菌酯/氟吗·锰锌/甲霜灵·锰锌	0.74	94.9a	7.44	81.4b	12.3
霜脲·猛锌/三乙膦酸铝	1.51	89.8bc	8.13	79.9b	23.0
氟吗·锰锌/三乙膦酸铝	2.31	86.5c	10.5	75.1b	20.7
CK	17.29		41.2		34.0

注：表中小写字母为 $P=0.05$ 水平的显著性，下表同

2.2 不同药剂交替处理的产量

通过测产，结果见表2。不同杀菌剂交替施用处理区大中薯率高于对照区，产量也高于对照区（1 868.4kg/667m²）。氟吗·锰锌与甲霜灵·锰锌、嘧菌酯与氟吗·锰锌、甲霜灵·锰锌交替施用处理产量最高，分别为2 635.1kg/667m²、2 618.5kg/667m²，与其他处理差异显著，且与对照比增产率分别为41.0%和40.1%。霜脲·猛锌与乙膦铝、氟吗·锰锌与乙膦铝、霜脲·猛锌与甲霜灵·锰锌交替施用处理产量均高于对照，但差异不显著。

表 2 杀菌剂交替施用防治马铃薯晚疫病的产量

交替用药处理	结薯量（个）				大中薯率（%）	产量（kg/667m²）	增产（%）
	大	中	小	总计			
氟吗·锰锌/甲霜灵·锰锌	12.5	9.5	23.8	45.8	48.0	2 635.1a	41.0
嘧菌酯/氟吗·锰锌/甲霜灵·锰锌	15.5	8.0	22.3	45.8	51.3	2 618.5a	40.1
霜脲·猛锌/三乙膦酸铝	14.5	8.3	27.8	50.6	45.1	2 363.5ab	26.5
氟吗·锰锌/三乙膦酸铝	14.5	6.3	20.3	41.1	50.6	2 345.1ab	25.5
霜脲·猛锌/甲霜灵·锰锌	9.3	9.0	27.0	45.3	40.4	2 068.4b	10.7
CK（清水）	10.8	5.8	30.5	47.1	35.2	1 868.4b	

2.3 经济效益分析

对不同药剂交替施用防治马铃薯晚疫病进行经济效益分析，见表 3。在 5 个处理中，不同药剂交替施用均能使马铃薯生产增加效益。其中，氟吗·锰锌与甲霜灵·锰锌、嘧菌酯与氟吗·锰锌、甲霜灵·锰锌交替施用处理 667m² 纯收入分别比空白对照区多收入 357.1 元和 343 元，霜脲·猛锌与乙膦铝、氟吗·锰锌与乙膦铝交替施用防治的马铃薯每 667m² 纯收入分别增加 229.8 元、189.8 元。

表 3 杀菌剂交替施用防治晚疫病的经济效益分析

交替用药处理	产量/（kg/667 m²）	收入/（元/667 m²）	增加收入（元/667 m²）	用药成本（元/667 m²）	纯增加收入（元/667 m²）
氟吗·锰锌/甲霜灵·锰锌	2 635.1	1 475.6	429.4	72.3	357.1
嘧菌酯/氟吗·锰锌/甲霜灵·锰锌	2 618.5	1 466.4	420.1	77.1	343.0
霜脲·猛锌/三乙膦酸铝	2 363.5	1 323.6	277.3	47.5	229.8
氟吗·锰锌/三乙膦酸铝	2 345.1	1 313.3	266.9	77.1	189.8
霜脲·猛锌/甲霜灵·锰锌	2 068.4	1 158.3	112.0	44.0	68.0
CK（清水）	1 868.4	1 046.3			

注：农药成本按照当时零售价格，药剂按施用 3 次的总费用计算；马铃薯价格按照当时销售的平均价格计算

3 结论

几种杀菌剂交替施用对晚疫病均有较好的防治作用，防效在 75% 以上，较对照增产明显，增产率在 10% ~41.0%。根据防效、产量和经济效益分析，氟吗·锰锌与甲霜灵·锰锌药剂交替施用、嘧菌酯与氟吗·锰锌、甲霜灵·锰锌交替施用处理最好，可在生产上应用推广。

参考文献

[1] 屈冬玉，陈伊里. 马铃薯产业与中国式主食 [M]. 哈尔滨：哈尔滨地图出版社，2016：7－14.

[2] 李成军. 黑龙江省马铃薯产业的现状及发展思路 [J]. 作物杂志，2007 (4)：13－16.

[3] 金光辉，吕文河，白雅梅，等. 黑龙江省马铃薯晚疫病菌生理小种的鉴定 [J]. 东北农业大学学报，2009，40 (10)：13－17.

[4] 娄树宝. 马铃薯晚疫病菌抗药性研究现状 [J]. 黑龙江农业科学，2010 (7)：165－168.

N－（2，4，5－三氯苯基）－2－氧代环己烷基磺酰胺（SYAUP108）防治番茄灰霉病的内吸输导性研究*

祁之秋**，孙青彬，李兴海，张　杨，纪明山
（沈阳农业大学植物保护学院，沈阳　110866）

摘要：采用离体番茄叶片法，测定新型环烷基磺酰胺类化合物 N－（2，4，5－三氯苯基）－2－氧代环己烷基磺酰胺（简称 SYAUP108）防治番茄灰霉病的作用方式及内吸输导性能。结果表明当 SYAUP108 浓度为 80μg/mL 时，对番茄灰霉病的保护作用效果可达 70.30%，SYAUP108 具有一定的内吸活性，可通过叶柄吸收，向上输导至叶片，治疗效果达 62.69 %，药剂处理 7d 时，防效仍达 60% 以上。SYAUP108 表现出较强的保护、治疗和内吸活性。

关键词：灰霉病；环烷基磺酰胺类化合物；保护作用；治疗作用

Studies on theSystemic Translocation of N－（2，4，5－Trichlorophenyl－2－Oxocyclohexylsulfonamide Against Tomato Gray Mold

Qi Zhiqiu，Sun Qingbin，Li Xinghai，Zhang Yang，Ji Mingshan
（*College of Plant Protection*，*Shenyang Agricultural University*，*Shenyang* 110866，*China*）

Abstract：The action mode and systemic translocation of N－（2，4，5－trichlorophenyl）－2－oxocyclohexylsulfonamide（Short for SYAUP 108）were measured by excised leaf method in lab. When the concentration of SYAUP108 was at 8.00μg/mL，the protective efficacy and curative efficacy against tomato grey mould were 70.30% and 62.69%，respective. The control efficacy was more than 60% at 7d after treatment at 80μg/mL. SYAUP108 was absorpted by petioles and conducted upward to the blade. SYAUP108 shows stronger protective，curative and some systemic activity.

Key words：Tomato greymould；Cyclohexylsulfonamide；Protective efficacy；Curative efficacy；Systemic activity

番茄灰霉病为番茄设施栽培中的普遍发生的重要病害，一般可造成番茄减产 20%～30%，严重影响番茄的产量和品质[1]。目前生产上主要依靠化学药剂防治灰霉病，但一些常用杀菌剂如苯丙咪唑类多菌灵、二甲酰亚胺类菌核净、苯胺基嘧啶类嘧霉胺、氨基甲酸酯类乙霉威等已产生了严重的抗药性，同时由于产生交互抗药性，致使许多同类药剂也产生的抗药性，防治效果大大降低[2-3]。因此，急须开发新型杀菌剂来防治灰霉病及缓解抗药性。

磺酰胺类杀菌剂是 2000 年以来国外报道的一类新型杀菌剂。因其结构新颖，商品化的杀菌剂品种较少，而日益受到人们的重视，成为农药研发的新热点[4]。目前，国内外报道

* 基金项目：公益性行业（农业）科研专项经费资助（201303025）

** 作者简介：祁之秋，女，副教授，博士，主要从事农药毒理及抗药性研究；E-mail：syqizhiqiu@sina.com

的许多合成的磺酰胺类活性化合物基本上都含有环烷基，其抗菌谱较广，与已开发的杀菌剂品种明显不同[5]，但少有进一步研发。本课题组合成的新型环烷基磺酰胺类 *N*－（2，4，5－三氯苯基）－2－氧代环己烷基磺酰胺（简称 SYAUP108）经初步研究，发现其对灰葡萄孢菌有较强的离体和活体活性[6]，能破坏菌丝体的形态和结构[7]。本文将研究 SYAUP108 防治灰霉病的内吸输导性及持效期，旨在为该类化合物进一步开发提供理论依据。

1 材料与方法

1.1 供试材料

灰霉病菌 *Botrytis cinerea* Pers.，分离自番茄灰霉病病果。

2.5%的 SYAUP108 乳油的制备：将化合物 SYAUP108（结构式为 [structure: 2-oxocyclohexyl-SO_2NH-2,4,5-trichlorophenyl]）用甲醇溶解，再与农乳、二甲苯配制的乳化剂混匀，配制成 2.5% SYAUP108 乳油。

1.2 试验方法

1.2.1 保护作用测定

将 2.5% SYAUP108 乳油用去离子水分别配制成 80、40、20、10μg/mL 的药液。挑取长势及大小都一致的番茄叶片，均匀喷于番茄叶片上，直到药液在叶片上开始悬而未滴为宜。对照为喷施去离子水的处理。每处理 15 片叶，喷药后晾干。直径 5.00mm 的灰霉菌碟接种于人工处理伤口的番茄叶片上。23℃保湿培养 4d，十字交叉法测病斑直径，计算药剂防治效果。

1.2.2 治疗作用测定

人工处理番茄叶片产生伤口，于伤口处接直径 5.00mm 的灰霉菌碟。23℃保湿培养 24h，于叶片表面分别用 80、40、20、10、0μg/mL SYAUP108 药液均匀喷雾，每处理 15 片叶子。23℃保湿培养 3d，十字交叉法测病斑直径，计算药剂防治效果。

1.2.3 内吸性测定

分别用含 80、40、20、10、0μg/mL SYAUP108 药液浸泡的脱脂棉包裹叶柄基部，每处理 15 片叶。于距叶基部 4cm 处人工处理叶片产生伤口，于伤口处接直径 5.00mm 的灰霉菌碟。23℃保湿培养 3d，十字交叉法测病斑直径，计算药剂防治效果。

1.2.4 持效性测定

将 SYAUP108（浓度均为 80、40、20μg/mL）均匀喷洒于番茄叶片表面至流失，对照喷清水。每个浓度的番茄叶片于处理后的 3、7、14d 接灰霉菌菌碟。每处理 20 片叶，23℃保湿培养 3～6d。待对照发病充分后，调查病斑直径，计算相对防效。

1.3 数据统计分析

试验数据用 SPSS 19.0 软件进行显著差异（Duncan）统计分析。

2 结果与分析

2.1 SYAUP108 对番茄灰霉病的保护作用

由表 1 和图可知，随着药剂喷雾剂量的增大，SYAUP108 对番茄灰霉病的保护效果明显增强，当剂量为 80μg/mL 时，离体叶片色泽新鲜，几乎不发病，相对防效已达到 70.30% ± 0.87%。

表 1　SYAUP108 对番茄灰霉病的保护作用

Table 1　Protective efficacy of SYAUP108 against tomato grey mould

药剂浓度（μg/mL） concentration	病斑平均直径（cm） average diameters of lesions	相对防效（%） relative efficacy
0	2.12 b	
10	0.83 a	59.63 ±5.38 a
20	0.85 a	60.63 ±5.55 a
40	0.71 a	66.25 ±3.05 a
80	0.63 a	70.30 ±0.87 a

图　SYAUP108 对番茄灰霉病的保护作用

Fig　Protective efficacy of SYAUP108 against tomato grey mould

a. CK; b. 10μg/mL c. 20μg/mL; d. 40μg/mL; e. 80μg/mL

2.2　SYAUP108 对番茄灰霉病的治疗作用

由表 2 可以看出，SYAUP108 对番茄灰霉病具有一定的治疗活性，当处理剂量为 80μg/mL时，相对防效达 62.29% ±1.11%。

表 2　SYAUP108 的治疗作用

Table 2　Curative efficacy of SYAUP108 against tomato grey mould

药剂浓度（μg/mL） concentration	病斑平均直径（cm） average diameters of lesions	相对防效（%） relative efficacy
0	1.67 c	
10	1.12 b	32.80 ±5.82 b
20	1.03 b	38.00 ±3.96 b
40	0.98 b	41.14 ±3.50 b
80	0.63 a	62.29 ±1.11 a

2.3　SYAUP108 在番茄植株上的内吸输导性

叶柄处施药对叶片上的灰霉病具有一定的防治效果（表 3）。当药剂浓度为 40μg/mL 时，对灰霉病的防效达 29.34% ±4.61%。说明化合物 SYAUP18 可以通过番茄叶柄吸收，在导管或木质部中以短距离运输的方式向上输导。

表 3　SYAUP108 在番茄植株上的内吸输导性

Table 3　Systemic activity of SYAUP108 in tomato plants

药剂浓度（μg/mL）concentration	病斑平均直径（cm）average diameters of lesions	相对防效（%）relative efficacy
0	1. 84 b	
10	1. 48 ab	19. 86 ± 14. 94 b
20	1. 45 ab	21. 19 ± 5. 79 ab
40	1. 30 b	29. 34 ± 4. 61 a

2. 4　SYAUP108 对番茄灰霉病的持效期

随着 SYAUP108 处理浓度的降低、处理时间的延长，SYAUP108 对番茄灰霉病的防治效果明显下降（表4）。当 SYAUP108 浓度为 80μg/mL 时，第 3、7 和 14d 的相对防效分别为 65. 55%、62. 46%、29. 92%。说明 SYAUP108 化合物防治灰霉病的持效期可达 7d 仍能有效控制病害的发生。

表 4　SYAUP108 对番茄灰霉病的持效期（%）

药剂浓度（μg/mL）concentration	3d	7d	14d
0			
20	38. 28 ± 3. 271c	30. 81 ± 3. 762c	2. 64 ± 1. 312c
40	43. 69 ± 3. 972b	42. 96 ± 2. 124b	15. 01 ± 1. 723b
80	65. 55 ± 0. 981a	62. 46 ± 1. 703a	29. 92 ± 4. 522a

3　结论与讨论

杀菌剂的内吸输导性能直接关乎着其化学防治效果，因此要想达到好的防治效果，不仅要求药剂具有优良的杀菌活性，还需要在植物体内有良好的输导性能。本文研究结果表明，SYAUP108 具有一定的保护、治疗及弱内吸作用，持效期可达 7d。因此，SYAUP108 具有进一步开发成杀菌剂品种，应用于生产的价值。

生产上应用的磺酰胺类杀菌剂氰霜唑等主要用于防治卵菌病害。本实验室合成的环烷基磺酰胺类化合物 SYAUP108 却对灰霉病菌特效，其作用机理可能不同于氰霜唑。前期研究发现，SYAUP108 对灰霉病菌丝生长、孢子形成和萌发具有较强的抑制作用[8]，本研究进一步明确 SYAUP108 在活体植株上对灰霉病具有较强的防治效果，8μg/mL 时防治效果可达 60%，这是否表明 SYAUP108 除了对病原菌本身具有较强的活性，也能诱导寄主植物抗病性还需进一步研究。

参考文献

［1］　郑果，杜蕙. 几种新型药剂对番茄灰霉病的防治效果［J］. 中国蔬菜，2006（9）：22－23.

[2] 陈治芳，王文桥，韩秀英，等.灰霉病化学防治及抗药性研究进展［J］. 河北农业科学，2010，14（8）：19-23.

[3] 纪军建，张小风，王文桥，等.番茄灰霉病防治研究进展［J］. 中国农学通报，2012，28（31）：109-113.

[4] 陈美航.磺酰胺类化合物的研究进展［J］. 铜仁学院学报，2010，12（3）：116-122.

[5] Iraj R E，Charalabos C，Panagiotis Z，et al. Sulfonamide-1，2，4-triazole derivatives as antifungal and antibacterial agents：Synthesis，biological valuation，lipophilicity，and conformational studies［J］. Bio org Med Chem，2008，16：1 150-1 161.

[6] 李兴海，吴德财，祁之秋，等.2-氧代和2-羟基环烷基磺酰胺对14种病原真菌的杀菌活性［J］. 农药学学报，2011，13（4）：423-426.

[7] 祁之秋，孙青彬，李兴海，等.*N*-（2，4，5-三氯苯基）-2-氧代环己烷基磺酰胺对灰葡萄孢的抑制作用［J］. 农药学学报，2014，16（5）：523-528.

Fluorescence Microscopy to Track Phytopharmaceuticals in Microbes and Plants

Jean Marcseng[1,2], Viviane Calaora[2], Thierry Barchietto[2], Sergej Buchet[2 *]

(1. *Institut of Plant Science Paris-Saclay* (*IPS2*) *Université de Paris Sud*, *Bât/*630, *F*-91405 *Orsay France*; 2. *BIOtransfer Ltd*, 41 *rue Emile Zola*, *F*-93100 *Montreuil*, *France*)

In the past 20 years, functional microscopy has made significant progress in all areas of biology, especially in medical and brain research.

This approach is designed to reveal the morphology of cell constituents and their evolution in time, and to analyze metabolic modifications in response to environmental changes (xenobiotics, stress). Since 2000, BIOtransfer has developed a range of functional microscopy applications to study the mode of action of fungicides and herbicides in plant pathogens and weeds, respectively. We have adapted the technology to describe fungal cellular structures (cell wall, membranes, nuclei) bound by selective fluorescent molecules (fluorochromes). Fluorochromes may also serve as proxy to measure cell viability through the detection of cellular activity such as cytoplasmic, vacuolar, nuclear, -or mitochondrial activity. Our research team implemented this principle to demonstrate the potential of SDHI fungicides against fungal plant pathogen strains resistant to strobilurins and to assess the efficacy of single fungicide components in binary associations

Because of the poor penetration of flurochromes in plant tissues, this technology has been limited to surface investigation. BIOtransfer recently overcome this obstacle by enhancing the permeation of fluorochromes in plants. Applications involving *Plasmopara viticola* and *Phytophthora infestans* will be presented that demonstrate *in planta* the curative effects of specific treatments against these oomycetes.

The opportunities offered by fluorescence microscopy will be discussed, in comparison with other methods.

* Contact author: Jean-Marc Seng, jean-marc. seng@ u-psud. fr; Viviane Calaora, viviane. calaora@ biotransfer. fr

嘧菌酯对石榴干腐病菌的生物学活性研究*

杨 雪[1**]，张爱芳[1]，谷春艳[1]，陈 雨[1***]，徐义流[2***]
（1. 安徽省农业科学院植物保护与农产品质量安全研究所，农业部有害生物合肥科学观测实验站，农业部农产品质量安全风险评估实验室，合肥 230031；
2. 安徽省农业科学院，合肥 230031）

摘要：为探究石榴干腐病的病原菌种类及嘧菌酯对石榴干腐病的生物学活性，对石榴病果进行了病原菌分离纯化、分子鉴定和致病性测定，测定了嘧菌酯对其菌丝生长和孢子萌发的影响，并进行了连续两年的大田防治试验。结果表明，石榴干腐病的病原菌为石榴壳座月孢 *Pilidiella granati*，嘧菌酯在水杨肟酸（SHAM）的协同作用下，对石榴干腐病菌的菌丝生长和孢子萌发具有很强的抑制作用，其抑制菌丝生长和孢子萌发的平均 EC_{50} 值分别为 0.202 和 0.006 5μg/mL（含 100μg/mL SHAM）；大田防治试验结果表明，嘧菌酯对石榴干腐病具有良好的防治效果，其中 25% 嘧菌酯 SC 1000 在 2013 年和 2014 年对石榴干腐病的防效分别为 90.85% 和 81.91%，显著高于其他处理的防效，可作为防治石榴干腐病的候选药剂之一。

关键词：石榴干腐病；石榴壳座月孢；嘧菌酯；抑菌活性；田间防效

Biological Activity of Azoxystrobin Against *Pilidiella granati* Causing Pomegranate Dry Rot

Yang Xue[1], Zhang Aifang[1], Gu Chunyan[1], Chen Yu[1*], Xu Yiliu[2*]
(1. *Institute of Plant Protection and Agro－Products Safety*, *Anhui Academy of Agricultural Sciences*, *Scientific Observing and Experimental Station of Crop Pests in Hefei*, *Ministry of Agriculture*, *Laboratory of Quality & Safety Risk Assessment for Agro－Products* (*Hefei*), *Ministry of Agriculture*, *Hefei* 230031, *China*;
2. *Anhui Academy of Agricultural Sciences*, *Hefei* 230031, *China*)

Abstract: In order to investigate the pathogen causing pomegranate dry rot and the biological activity of funcide azoxystrobin against this pathogen, the isolated pathogen was purified and identified by molecular methods and pathogenicity test. The activity of azoxystrobin (amended with 100μg/mL SHAM) against mycelial growth and conidial germination was determined and the control efficacy of this fungicide was performed against pomegranate dry rot in two consecutive years (2013 and 2014). The results showed that the pathogen was *Pilidiella granati*, and azoxystrobin exhibited excellent control efficacy against this disease in which the treatment 25% azoxystrobin SC 1000 × provided 90.85% and 81.91% control efficacy in 2013 and 2014, respectively. Therefore, azoxystrobin could be a candidate

---

* 基金项目：安徽省果树产业体系（AHCYTX—14）；安徽省“115”产业创新团队

** 第一作者：杨雪，助理研究员，主要从事植物病理研究；E-mail：yangxue2121@163.com

*** 通讯作者：陈雨，副研究员，主要从事植物病害防治研究；E-mail：chenyu66891@sina.com
徐义流，研究员，主要从事果树研究；E-mail：yiliuxu@163.com

fungicide for the control of pomegranate dry rot.

Key words: Pomegranate dry rot; *Pilidiella granati*; Azoxystrobin; Fungicidal activity; Control efficacy

石榴干腐病是我国石榴生产上的一种重要病害，主要危害石榴的枝干，叶片和果实，是造成石榴减产，品质变劣的主要原因之一。该病害在希腊[1-3]，韩国[4]，土耳其[5]，伊朗[6]，西班牙[7]等国均有报道，在我国的陕西[8]、安徽[9]等省份也有报道。在管理粗放的果园，发病率可高达40%以上[8]，导致果实呈褐色腐烂。希腊学者于2011年与2012年调查了全国30多个石榴生产区，发现采摘前和采摘后石榴干腐病在果实上的发病率分别为50%和29%[3]。该病害在果实储藏期可以继续危害，是我国石榴生产、运输与销售过程中一个棘手的问题。

目前国内关于引起石榴干腐病的病原鉴定上还存在分歧，早期的研究认为该病害是由石榴鲜壳孢 *Zythia versoniana* 引起[10]，而有学者认为山东的石榴干腐病是由 *Dothiorella* 属真菌引起[11]。但据国外报道，希腊[1-3]，韩国[4]，土耳其[5]，伊朗[6]，西班牙[7]的石榴干腐病均由石榴壳座月孢 *Pilidiella granati*（= *Coniella granati*）引起。近年的研究表明，我国陕西省和安徽省的石榴干腐病确实由石榴壳座月孢 *Pilidiella granati* 引起[8-9]。

石榴干腐病目前在防治上还缺乏有效的药剂，对于选择何种药剂以及在哪个时期进行用药等目前都鲜有报道。嘧菌酯是一种以天然产物 strobilurins 为先导化合物而研制开发的甲氧基丙烯酸酯类杀菌剂[12]，该药剂也是近年来在全球使用广泛的一种杀菌剂，它通过阻止特定细胞色素，导致线粒体电子传递受到抑制。它具有杀菌活性高、杀菌谱广、内吸性强、对环境友好等特点，自问世以来，已在全世界60余种作物上进行了田间试验[13-14]，具有保护，铲除，渗透，内吸作用，能抑制孢子萌发和菌丝生长，几乎可以防治所有真菌病害，对多种作物和果树病害都有很好的防治效果。

本研究通过分离纯化石榴病果实上的干腐病菌，进行形态学和分子鉴定，明确了本地区石榴干腐病的病原菌种类。通过测定嘧菌酯对干腐病菌菌丝生长和孢子萌发的活性，并在石榴园里进行药效试验，以明确嘧菌酯对石榴干腐病的防治效果。而目前国内外对石榴干腐病菌的生物学特性及其发病机制研究甚少[3]，对石榴干腐病的发病机理及早期防治处理将是我们下一步的研究重点。

1 材料与方法

1.1 材料

1.1.1 培养基

水琼脂培养基（water agar，WA）：琼脂粉20g加水定容至1 000mL。

马铃薯葡萄糖琼脂培养基（potato dextrose agar，PDA）：马铃薯200g，葡萄糖20g，加水定容至1 000mL（固体培养基加琼脂粉20 g）。

1.1.2 试剂及药剂

25%嘧菌酯（azoxystrobin）悬浮剂（Suspension Concentrates，SC），由先正达中国有限公司提供；水杨肟酸（Salicylhydroxamic acid，SHAM），纯度为99%，由 Sigma - Aldrich 上海贸易有限公司提供；50% 多菌灵（carbendazim）可湿性粉剂（Water Powdery，WP），由四川国光农化股份有限公司提供。

1.1.3 仪器

恒温培养箱（LHP - 300H 型，上海三发科学仪器有限公司）；振荡培养箱（HZQ - F160

型，太仓市华美生化仪器厂）；电子显微镜（102M 型，麦克奥迪厦门销售有限公司）；湿热灭菌锅（GR60DF 型，厦门致微仪器有限公司）；PCR 仪（SEDI 型，WEALTEC 公司）。

1.1.4　数据处理系统

DPS 数据处理软件 v6.50 注册版。

1.1.5　供试病样

石榴干腐病果均采自安徽省蚌埠市禹会区、怀远县和淮南市、淮北市石榴果园，每个果园采集 5～10 个病样。将采集的病果和病枝用 75% 的酒精表面消毒后，取病健交界处的少许组织放在 WA 平板上，26℃培养至出现菌丝后，切取 0.5 cm×0.5 cm 大小的菌丝块放入 PDA 平板，26℃培养至其长出孢子后进行单孢纯化。

1.2　方法

1.2.1　菌株致病性测定

采用离体果实接种法进行致病性测定[8]。将分离纯化到的菌株接种的 PDA 平板上，在 26℃恒温培养至其长出孢子，用无菌水洗脱平板并制备孢子悬浮液（1×10^5 个孢子/mL）。用无菌水将健康石榴果实冲洗干净并用 75% 酒精表面消毒，晾干后将孢子液分别接种在健康和用消毒针刺伤的果实上，每果实接种 5 个点，以接种无菌的培养基块为对照，每个处理重复 10 次。将处理果实置于保湿容器中，在 26℃下 12h 光暗交替培养，定期观察并记录发病情况。

1.2.2　菌株种类形态学与分子鉴定

将分离纯化到的菌株接种的 PDA 平板上，26℃恒温培养箱内培养，每天观察菌落生长状况及形态特征。培养约 7d 待其产生孢子后，刮取孢子置于电子显微镜下观察孢子形态特征。

将 PDA 培养基平板上的供试病菌在 26℃培养 5d，从菌落边缘切取 10 块 2cm×2cm 菌落块，转至 PDA 液体培养基，26℃振荡培养 7d，过滤收集菌丝，经液氮冷冻研磨成粉，用 CTAB 法提取基因组 DNA。采用通用引物 ITS1 和 ITS4（ITS1：5′-TCCGTAGGTGAACCT-GCGG-3′；ITS4：5′-TCCTCCGCTTATTGATATGC-3′）对其 ITS 区域进行 PCR 扩增后，送样至上海生工进行测序，将测序结果在 NCBI 上进行比对以确定菌株种类。

1.2.3　嘧菌酯对菌丝生长的影响

将纯化后的菌株于 PDA 平板上 26℃培养 5d，在菌落边缘打取直径为 5mm 的菌碟，分别接种于含 0、0.1、0.5、1、5、10、50、100μg/mL 嘧菌酯（含或不含 100μg/mLSHAM）的 PDA 平板上，3 个重复，26℃培养至对照菌落布满培养皿 2/3 以上时，采用十字交叉法测定对照及处理的菌落直径，根据下列公式计算出不同浓度下的抑制率（%）。

$$\text{抑制率（\%）}=\frac{\text{空白对照菌落直径}-\text{药剂处理菌落直径}}{\text{空白对照菌落直径}}\times100$$

1.2.4　嘧菌酯抑制孢子萌发的活性及 MIC 值

从培养 7 d 的石榴干腐病菌 PDA 平板上收集分生孢子，用灭菌水配成 1×10^5 个孢子/mL 的孢子悬浮液。吸取 50μL 孢子悬浮液均匀涂布在含嘧菌酯浓度为 0、0.006 25、0.012 5、0.025、0.05、0.1μg/mL 的 WA 平板上（含或不含 100μg/mL SHAM），过夜培养后在显微镜下观察孢子的萌发情况并统计萌发率。

1.2.5　嘧菌酯对石榴干腐病的防治效果

于 2013 年和 2014 年在怀远县石榴园进行嘧菌酯防治石榴干腐病药效试验。每年施药 4

次，分别为5月20日（盛花期）、6月5日（盛花后期）、6月20日（幼果期）、7月5日（果实膨大期），于7月15日统计石榴干腐病的病果率，根据发病率计算防治效果。用药处理为：①25%嘧菌酯SC 1 000倍液；②25%嘧菌酯SC 1 500倍液；③25%嘧菌酯SC 2 000倍液；④50%多菌灵WP（对照药剂）2 000倍液；⑤清水对照。每处理喷施3棵石榴树，重复3次，随机区组设计。

1.3 数据分析

嘧菌酯对菌丝生长及孢子萌发的影响数据通过DPS6.50版本统计软件处理，求出毒力回归方程及有效抑制中浓度EC_{50}。

嘧菌酯对石榴干腐病的防治效果于最后一次施药后10d进行防效评估，石榴果实发病率（%）=（病果数量/调查果实总数量）×100进行评估。用Abbott公式计算：有效指数=（感染控制百分率－处理感染百分率）/感染控制百分率×100。

2 结果与分析

2.1 病菌致病性测定

致病性实验表明，分离纯化的病菌接种在健康和刺伤的石榴果实表皮上都能够致病，发病率为100%，并且症状与自然条件下侵染的症状相似。在刺伤的石榴表皮上接种，病斑的扩展速度较快，导致果实腐烂，同时在病斑上会长出白色菌丝，并产生黑色的分生孢子器；而在健康的石榴表皮上接种，致病过程较慢，只能在石榴表皮上形成褐色的小病斑。

2.2 菌株种类形态学与分子鉴定

从石榴果园里采集病枝和病果，分离纯化获得了石榴干腐病菌纯化菌株，在PDA上呈放射状生长，同心轮纹状，约7d后产生黑褐色分生孢子器（图1A），显微镜下呈球形，单生，器壁较薄，产孢区呈垫状隆起，器中央具一孔口，顶部不凸出。分生孢子呈纺锤形，淡褐色，表面光滑，直或微弯，无隔膜，10～15μm × 2.5～3.5μm（图1B）。该病菌的形态与石榴壳座月孢 *Pilidiella granati* 极为相似。

图1 病原菌的形态特征（A：菌丝；B：孢子）

Fig. 1 Morphological characteristers of the *Pilidiella granati*（A：mycelia；B：conidia）

采用ITS通用引物进行PCR扩增，得到一条大小约为610 bp的条带，送样至上海生工测序并用BLAST在GenBank中搜索比对，结果表明，分离物与石榴壳座月孢 *Pilidiella granati* ITS序列相似度达99.9%（GeneBank accession No. HQ166057）（图2），因此，初步判定分离纯化所得病原菌为石榴壳座月孢 *Pilidiella granati*。

从回接发病的果实上分离纯化病原菌并进行分子鉴定，结果与原接种菌株相同，因此可

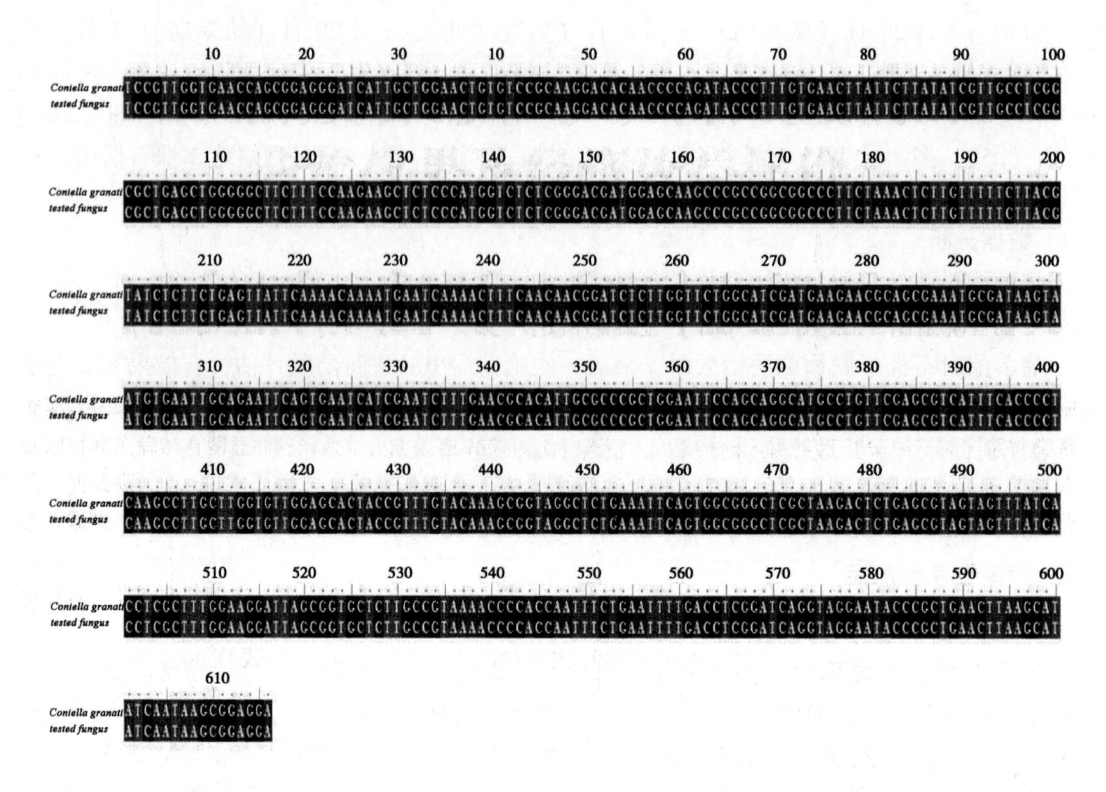

图 2　石榴干腐病原菌的分子鉴定

Fig. 2　Molecular identification of *Pilidiella granati* causing pomegranate dry rot

以确定导致石榴干腐病的病菌为石榴壳座月孢 *Pilidiella granati*。

2.3　嘧菌酯对菌丝生长的影响

嘧菌酯对石榴干腐病菌的菌丝生长有抑制作用（表 1），抑制菌丝生长的 EC_{50} 值为 1.041 ~ 18.868μg/mL，平均 EC_{50} 值为 7.260μg/mL。但在加入旁路氧化酶抑制剂 100μg/mL SHAM 时，嘧菌酯的抗菌活性大幅提高，抑制菌丝生长的 EC_{50} 值为 0.052 ~ 0.484μg/mL，平均 EC_{50} 值为 0.202μg/mL，使嘧菌酯的抑菌活性提高了约 35.94 倍，证明 SHAM 在嘧菌酯抑制石榴干腐病菌菌丝生长过程中有协同作用。

表 1　嘧菌酯对石榴干腐病菌菌丝生长的影响

Table 1　Activity of azoxystrobin against mycelial growth of *Pilidiella granati*

菌株名称 Strain name	菌株来源 Strain source	嘧菌酯 azoxystrobin			嘧菌酯 + 100μg/mL SHAM azoxystrobin + 100μg/mL SHAM		
		毒力回归方程 Regression equation	EC_{50} (μg/mL)	相关系数 (R^2)	毒力回归方程 Regression equation	EC_{50} (μg/mL)	相关系数 (r^2)
HY03	怀远县 Huaiyuan County	$y = 3.439 + 1.708x$	8.203	0.967	$y = 5.579 + 1.837x$	0.484	0.967
HY05	怀远县 Huaiyuan County	$y = 4.835 + 0.593x$	1.899	0.995	$y = 5.941 + 1.048x$	0.127	0.965
HY09	怀远县 Huaiyuan County	$y = 3.782 + 1.276x$	9.011	0.991	$y = 6.814 + 1.414x$	0.052	0.972

（续表）

菌株名称 Strain name	菌株来源 Strain source	嘧菌酯 azoxystrobin			嘧菌酯 +100μg/mL SHAM azoxystrobin +100μg/mL SHAM		
		毒力回归方程 Regression equation	EC_{50} (μg/mL)	相关系数 (R^2)	毒力回归方程 Regression equation	EC_{50} (μg/mL)	相关系数 (r^2)
YH02	禹会区 Yuhui District	$y=3.385+1.266x$	18.868	0.970	$y=6.985+1.356x$	0.188	0.963
YH03	禹会区 Yuhui District	$y=4.991+0.497x$	1.041	0.993	$y=6.482+1.648x$	0.126	0.943
HN01	淮南市 Huainan City	$y=3.459+1.850x$	6.808	0.994	$y=5.676+0.952x$	0.195	0.959
HN02	淮南市 Huainan City	$y=2.994+2.351x$	7.129	0.969	$y=5.607+0.821x$	0.182	0.928
HB01	淮北市 Huaibei City	$y=3.993+1.419x$	5.121	0.971	$y=5.371+0.638x$	0.262	0.983

2.4 嘧菌酯对孢子萌发的影响和抑制孢子萌发的 MIC 值

嘧菌酯对石榴干腐病菌的孢子萌发有抑制作用（表 2），抑制孢子萌发的 EC_{50} 值为 0.195～0.387μg/mL，平均 EC_{50} 值为 0.278μg/mL。但在加入旁路氧化酶抑制剂 100μg/mL SHAM 时，嘧菌酯的抗菌活性大幅提高，抑制孢子萌发的 EC_{50} 值约为 0.003～0.016μg/mL，平均 EC_{50} 值为 0.006 5μg/mL，使嘧菌酯的抑菌活性提高了约 42.77 倍。同时孢子萌发实验发现，当嘧菌酯的浓度为 0.04μg/mL（含 100μg/mL 水杨肟酸）时，所有的孢子都不能萌发，因此可以认为嘧菌酯抑制孢子萌发的 MIC 值为 0.04μg/mL。本实验结果表明，SHAM 在嘧菌酯抑制石榴干腐病菌孢子萌发过程中有协同作用，且分生孢子萌发比菌丝生长对嘧菌酯更加敏感。

表 2　嘧菌酯对石榴干腐病菌孢子萌发的影响

Table 2　Activity of azoxystrobin against the conidial germination of *Pilidiella granati*

菌株名称 Strain name	菌株来源 Strain source	嘧菌酯 azoxystrobin			嘧菌酯 +100μg/mL SHAM azoxystrobin +100μg/mL SHAM		
		毒力回归方程 Regression equation	EC_{50} (μg/mL)	相关系数 (R^2)	毒力回归方程 Regression equation	EC_{50} (μg/mL)	相关系数 (r^2)
HY03	怀远县 Huaiyuan County	$y=6.253+1.993x$	0.235	0.971	$y=5.742+0.699x$	0.007	0.981
HY05	怀远县 Huaiyuan County	$y=5.419+0.619x$	0.211	0.981	$y=5.903+0.056x$	0.005	0.991
HY09	怀远县 Huaiyuan County	$y=5.681+0.960x$	0.195	0.982	$y=5.885+0.900x$	0.004	0.980
YH02	禹会区 Yuhui District	$y=6.514+1.088x$	0.266	0.932	$y=5.940+0.911x$	0.003	0.997
YH03	禹会区 Yuhui District	$y=5.478+1.126x$	0.377	0.970	$y=5.598+0.932x$	0.005	0.992
HN01	淮南市 Huainan City	$y=5.507+0.942x$	0.289	0.958	$y=5.756+0.814x$	0.016	0.928

（续表）

菌株名称 Strain name	菌株来源 Strain source	嘧菌酯 azoxystrobin			嘧菌酯 + 100μg/mL SHAM azoxystrobin + 100μg/mL SHAM		
		毒力回归方程 Regression equation	EC_{50} (μg/mL)	相关系数 (R^2)	毒力回归方程 Regression equation	EC_{50} (μg/mL)	相关系数 (r^2)
HN02	淮南市 Huainan City	$y = 5.420 + 1.017x$	0.387	0.970	$y = 5.617 + 0.605x$	0.006	0.950
HB01	淮北市 Huaibei City	$y = 5.371 + 0.638x$	0.262	0.983	$y = 5.815 + 0.763x$	0.006	0.959

2.5 嘧菌酯对石榴干腐病的防治效果

在怀远县石榴园进行了连续2年的药剂防治石榴干腐病的试验，其中2014年的发病率要高于2013年。试验结果表明，25%嘧菌酯SC 1 000倍液在2013年和2014年对石榴干腐病的防效分别为90.85%和81.91%，显著高于25%嘧菌酯SC 1 500倍液、25%嘧菌酯SC 2 000倍液和50%多菌灵WP 2 000倍液的防效，显示了对该病害良好的防治效果。25%嘧菌酯SC 1 500倍液连续2年的防治效果也均超过70%，其中在2013年的防治效果为86.05%，而50%多菌灵WP 2 000倍液的防效分别为63.28%和54.91%，防治效果较低（表3）。

表3 嘧菌酯对石榴干腐病的防治效果（安徽怀远县，2013和2014）

Table 3 Control efficacy of azoxystrobin against Pomegranate dry rot (Huanyuan County, Anhui Province, 2013 & 2014)

试验地点 Experimental Site	药剂处理 Fungicide Treatment	剂量 Dosage（稀释倍数）	平均病果率 Average disease severity (%)		防治效果 Control efficacy (%)	
			2013	2014	2013	2014
怀远县 Huaiyuan County	25%嘧菌酯SC 25% azoxystrobin SC	1 000×	0.82±0.26	2.23±0.51	90.85±2.9a[1]	81.91±4.1a
	25%嘧菌酯SC 25% azoxystrobin SC	1 500×	1.25±0.4	3.65±0.58	86.05±4.5b	70.40±4.7b
	25%嘧菌酯SC 25% azoxystrobin SC	2 000×	2.25±0.49	4.42±0.69	74.89±5.5c	64.15±5.6c
	50%多菌灵WP 50% carbendazim WP	2 000×	3.29±0.55	5.56±0.8	63.28±6.1d	54.91±6.5d
	对照（CK）	—	8.96±0.69	12.33±0.72	—	—

注：同列数据后相同的字母表示没有显著差异（$P > 0.05$，Fisher's LSD）。Mean vales with the same letters within the same column were not significantly different ($P > 0.05$, Fisher's LSD)

3 讨论

目前，石榴干腐病是我国石榴上的最重要病害之一，对我国石榴生产造成了严重的威胁。该病害会导致新梢坏死，不能萌芽及产生新叶，感染的病果显现干腐症状，最终导致石榴的大幅度减产。早期国内关于石榴干腐病的研究表明石榴干腐病是由石榴鲜壳孢 *Zythia versoniana* 和 *Dothiorella* 属真菌引起[10-11]，而近年有学者研究发现，我国陕西省石榴干腐病是由石榴壳座月孢 *Pilidiella granati* 引起[8]。本研究分离纯化了安徽省石榴干腐病菌，并结合形态鉴定、分子鉴定以及致病性测定，也认为引起石榴干腐病的病原菌为石榴壳座月孢 *Pilidiella granati*，与国际上已经报道的石榴干腐病病原一致。

目前关于石榴干腐病药剂防治方面研究较少，现有的研究主要集中在使用传统药剂如百菌清、甲基托布津、多菌灵等药剂[10]，但目前这些药剂对石榴干腐病的防治效果较低，在雨水较多、石榴干腐病大发生的年份，这些药剂的防治效果均不理想。嘧菌酯是目前果树、蔬菜生产上最常用的甲氧基丙烯酸酯类杀菌剂之一，该药剂作用于真菌线粒体电子传递链的复合物Ⅲ，阻断电子传递，干扰能量合成，与现有的苯并咪唑类、三唑类、二甲酰亚胺类等杀菌剂无交互抗药性，同时该药剂对多种水果及蔬菜病害均有很好的防效[15]，而该药剂对石榴干腐病的防治效果未见报道。本研究通过菌丝生长速率法和孢子萌发测定法测定了嘧菌酯对石榴干腐病菌的抑菌活性，结果表明，该药剂对石榴干腐病菌具有较强的抑菌活性，为今后的抗药性监测奠定了良好的基础。同时生测实验表明，SHAM 在嘧菌酯抑制石榴干腐病菌过程中具有良好的协同作用，而 100μg/mL SHAM 本身并不能抑制石榴干腐病菌的菌丝生长和孢子萌发（数据未发表）。由于大多数病原真菌在受甲氧基丙烯酸酯类药剂处理后呼吸作用受抑制时，会启动旁路氧化途径（alternative respiration）获得能量，从而会表现出耐药性或抗药性[16-18]。SHAM 是一种旁路氧化酶抑制剂，而植物体内也存在抗氧素如黄酮类物质，这些物质与 SHAM 一样可以抑制病原真菌旁路氧化酶活性，从而阻止了旁路氧化途径[19-20]，因此在室内进行嘧菌酯的生测实验，加入 SHAM 更能准确反映此类药剂的抑菌活性。

本研究连续两年在安徽省怀远县石榴园进行嘧菌酯防治石榴干腐病试验，结果表明嘧菌酯对石榴干腐病具有良好的防治效果且显著高于其他处理的防效。同时从连续两年的防效试验可以看出，在石榴干腐病大发生、雨水较多的年份，应加大嘧菌酯的使用剂量，可以达到良好的防治效果。同时在药剂施用时，应尽量避开雨水天。由于嘧菌酯目前在我国果树和蔬菜上的使用量较大，具有较高的抗药性风险，因此在药剂使用后，应进行连年的抗药性监测，同时我们也在进行石榴干腐病菌对嘧菌酯的抗药性风险评估和抗药性分子机理研究，为该药剂的合理使用、延缓抗药性产生以及促进该病害的可持续控制奠定基础。

参考文献

[1] Tziros G T, Tzavella - Klonari K. Pomegranate fruit rot caused by *Coniella granati* confirmed in Greece [J]. Plant Pathology, 2008, 57: 783.

[2] Thomidis T, Exadaktylou E. First report of *Pilidiella granati* Saccardo causing crown rot of pomegranate in the prefecture of Xanthi, Greece [J]. Plant Disease, 2011, 95: 79.

[3] Thomidis T. Pathogenicity and characterization of *Pilidiella granati* causing pomegranate diseases in Greece [J]. European Journal of Plant Pathology, 2015, 141 (1): 45 - 50.

[4] Kwon J H, Park C S. Fruit rot of Pomegranate (*Punica granatum*) caused by *Coniella granati* in Korea [J]. Plant Pathology, 2002, 12: 45 - 50.

[5] Çeliker N M, Uysal A, Çetinel B, et al. Crown rot on pomegranate caused by *Coniella granati* in Turkey [J]. Australasian Plant Disease Note, 2012, 7: 161 - 162.

[6] Mirabolfathy M, Groenewald JZ, Crous PW. First report of *Pilidiella granati* causing dieback and fruit rot of pomegranate (*Punica granatum*) in Iran [J]. Plant Disease, 2012, 96: 461.

[7] Palou L, Guardado A, Montesinos - Herrero C. First report of *Penicillium* spp. and *Pilidiella granati* causing postharvest fruit rot of pomegranate in Spain [J]. New Disease Reports, 2010, 22: 21.

[8] Song X H, Sun D M, Wang M G, et al. Occurrence of the pomegranate fruit rot and identification of its pathogen [J]. Journal of Plant Protection, 2011, 38 (1): 93 -94.

[9] Chen Y, Shao D D, Zhang A F, et al. First report of a fruit rot and twig blight on pomegranate (Punica granatum) caused by *Pilidiella granati* in Anhui Province of China [J]. Plant Disease, 2014, 98: 695.

[10] Zhou Y S, Lu J, Zhu T G, et al. Bioecology and epidemics of *Zythia Versoniana* Sacc in pomegranate and its integrated control [J]. Journal of Southwest Agricultural University, 1999, 21 (6): 551 -555.

[11] Liu H S, Liang J, Zhao J P, et al. Study On the Etiology of Pomegranate Canker Diseases, Scientia silvae sinicae, 2007, 43 (4): 57 -58.

[12] Margot P, Huggenberger F, Amrein J, et al. A new broad - sprectrum strobilurin fungicide [C] // In: Brighton Crop protection Conference on Pests and Disease. Brighton, UK, 1998: 375 -382.

[13] Gisi U, Chin M K, Knapova G, et al. Recent development in elucidating modes of resistance to phenylamide, DMI and strobilurin fungicides [J]. Crop Protection, 2000, 19: 863 -872.

[14] Zhang G S. Current Status of Application, Development and Prospect of Strobin Fungicedes [J]. Pesticide Science and Administration, 2000, 24 (12): 30 -34.

[15] Dale S, Efficacy of ‘Amistar’ against Fruit and Vegetable Diseases in Asia [C] //Chemical control of plant disease in China. Bei Jing: China Agricultural Science and Technology Publishing House, 2002: 42 -49.

[16] Joseph - Home T, Hollomon D W. Functional diversity within the mitochondrial electron transport chain of plant pathogenic fungi [J]. Pest Management Science, 2000, 56: 24 -30.

[17] Köller W, Avila - Adame C, Olaya G, et al. Resistance to strobilurin fungicides [C] // Clark J. M., Yamaguchi I., Agrochemical resistance - extent, mechanism, and detection. Washington. DC: American Chemical Society, 2001: 215 -229.

[18] Joseph - Horne T, Hollomon D W, Wood P M. Fungal respiration: a fusion of standard and alternative components [J]. Biochimica et Biophysica Acta, 2001, 1504: 179 - 195.

[19] Yukioka H, Inagaki S, Tanaka R, et al. Transcriptional activation of alternative oxidase gene of the fungus *Magnaporthe grisea* by a respiratory - inhibiting fungicide and hydrogen peroxide [J]. Biochimica et Biophysica Acta, 1998, 1442: 161 - 169.

[20] Mizutani A, Miki N, Yakioka H, Tamura H, et al. A possible mechanism of control of rice blast disease by a novel alkoxyiminoacetamide fungicide, SSF126 [J]. Phyto Pathology, 1996, 86 (3): 295 -300.

6 种杀菌剂防治苹果褐斑病田间药效比较

时春喜，张　锐，胡文渊，杜学一
（西北农林科技大学植物保护学院，植保资源与病虫害治理
教育部重点实验室，杨凌　712100）

摘要：为筛选出防治苹果褐斑病的高效、低毒、安全的杀菌剂，通过进行田间药效试验对 6 中杀菌剂进行了药效比较。结果表明，25% 丙环唑水乳剂防治苹果褐斑病的防效为 81.49%；12.5% 氟环唑悬浮剂的防效为 76.64%；40% 肟菌酯悬浮剂的防效为 71.51%；50% 异菌脲可湿性粉剂的防效为 55.68%；80% 多菌灵可湿性粉剂的防效为 53.71%；80% 乙蒜素乳油的防效为 49.50%。在试验条件下 6 个试验药剂对苹果树无药害。防治苹果褐斑病应优先使用 25% 丙环唑水乳剂和 12.5% 氟环唑悬浮剂，为减缓抗药性的产生，应交替使用其他不同作用机制杀菌剂品种。

关键词：6 种杀菌剂；苹果褐斑病；药效

Comparison Field Efficiency with Six Fungicides Controlling Apple Brown Spot

Shi Chunxi, Zhang Rui, Hu Wenyuan, Du Xueyi
(*The Key Laboratory of Plant Protection Resources and Pest Management of Chinese Ministry of Education*, *College of Plant Protection*, *Northwest A& F University*, *Yangling Shaanxi* 712100, *China*)

Abstract: In order toscreenexcellent, low-toxicant and safety fungicides to control apple brown spot caused by *Marssonina mali* (*P. Henn.*) *Ito*, the control efficiency of six fungicides were compared byfield trials. The results showed that thefield efficiency of Propiconazole 25% EW was 81.49%, Epoxiconazole 12.5% SC was 76.64%, Trifloxystrobin 40% SC was 71.51%, Iprodione 50% WP was 55.68%, Carbendazim 80 WP was 53.71%, and Ethylicin 80% was 49.50%. All six fungicides were safe to apple plants. For preventing Apple brown spot, we commanded topreferentially use Propiconazole 25% EW and Epoxiconazole 12.5% SC suspension. And to alleviate the development of resistance, it is necessary to use different types of fungicides with other mechanisms alternatively.

Key words: six fungicides; apple brown spot; field efficiency

苹果是落叶果树中主要栽培树种之一，也是世界果树栽培面积较广、产量较多的树种之一[1]。苹果褐斑病［*Marssonina mali*（P. Henn.）Ito］是造成苹果早期落叶的主要病害之一。在陕西省各苹果产区均有不同程度的发生，主要危害苹果叶片，极少危害果实和枝干[2-3]。该病的发生与降雨和温度关系密切，冬季温暖潮湿，春雨早、雨量大，夏季阴雨连绵以及秋雨较多的年份，发病早且重；温度主要影响病害的潜育期，在较高温度下潜育期短（最短仅需 3 天），病害发展极为迅速，发生严重时，不仅造成果树大量落叶，而且影响果实的正常膨大和着色，致使果实品质下降，造成严重的经济损失[4]。为了满足果农在生产

中的需要，切实解决苹果褐斑病的防治问题，我们于2015年6—7月（苹果褐斑病高发期）在陕西省咸阳市乾县王村镇北索村进行了防治苹果褐斑病的田间药效试验。

目前，国内已登记的防治苹果褐斑病的杀菌剂有氟环唑、肟菌酯、异菌脲、多菌灵和丙环唑等[5]，但各种药剂防治苹果褐斑病的效果不同，本研究的目的是从以上药剂中筛选出对防治苹果褐斑病效果较好的药剂，从而给果农在防治苹果褐斑病方面提出建议，促进合理使用农药，减少给果农带来的损失。

1 材料与方法

1.1 试验处理

①12.5%氟环唑悬浮剂（市购）②40%肟菌酯悬浮剂（市购）③50%异菌脲可湿性粉剂（市购）④80%多菌灵可湿性粉剂（市购）⑤25%丙环唑水乳剂（市购）⑥80%乙蒜素乳油（市购）⑦清水处理。试验共7个处理，每个处理4次重复，共28个小区，每小区2株树，不同处理小区随机区组分布。

1.2 试验对象

苹果褐斑病

1.3 试验器材

新加坡利农HD-400型背负式手动喷雾器，量杯，吸管，玻璃棒等。

1.4 试验地情况

试验设在陕西省咸阳市乾县王村镇北索村苹果园内，面积2.5亩，供试品种为富士，24年生，株行距3.0m×3.5m，土壤类型为土娄土，pH值为中性，施农家肥2 000kg，复合肥300kg，浇水一次，管理水平一般，苹果褐斑病历年发生较重。

1.5 施药方法

采用全株茎叶均匀喷雾法，喷液时均匀喷湿叶片正反面，每株喷洒药液1.5L。苹果褐斑病防治自第1次施药时起至空白对照发病较重时止，第1次施药于2015年6月10日进行，第2次施药于2015年6月20日进行，整个试验共施2次药。

1.6 试验调查方法

施药当天调查各小区的病情基数，于每次施药前及末次药后10d各进行一次防效调查，共调查4次。

调查方法：每小区两株树均调查，每株树的东、南、西、北、中5个方位5点取样，每点选取2个当年生枝条调查全部叶片。分别记录各小区的发病状况：

叶片病情分级标准：

0级：无病斑；

1级：病斑面积占整个叶面积的5%以下；

3级：病斑面积占整个叶面积的6%~15%；

5级：病斑面积占整个叶面积的16%~25%；

7级：病斑面积占整个叶面积的26%~50%；

9级：病斑面积占整个叶面积的50%以上或落叶[6]。

$$\text{防治效果（\%）}=\left(1-\frac{CK_0\times PT_1}{CK_1\times PT_0}\right)$$

$$\text{病情指数}=\frac{\sum\text{（各级病叶数}\times\text{相对级数值）}}{\text{调查总叶数}\times 9}\times 100$$

式中：CK_0 为空白对照区施药前病情指数，CK_1 为空白对照区施药后病情指数；PT_0 为药剂处理区施药前病情指数，PT_1 为药剂处理区施药后病情指数。

2 结果与分析

2.1 防治效果

第一次施药后 10d 和第二次施药后 10d 的防治效果、生物学统计和方差分析见下表。由表 2 可以看出第一次施药后 10d 各药剂处理防治苹果褐斑病的效果分别为：12.5% 氟环唑悬浮剂 500 倍液处理的防效为 74.20%；40% 肟菌酯悬浮剂 5 500倍液处理的防效为 66.80%；50% 异菌脲可湿性粉剂 1 000倍液处理的防效为 58.21%；80% 多菌灵可湿性粉剂 1 000倍液处理的防效为 55.79%；25% 丙环唑水乳剂 1 000倍液处理的防效为 76.32%；80% 乙蒜素乳油 800 倍液处理的防效为 52.93%。由表 2 可以看出第二次施药后 10d 各药剂处理防治苹果褐斑病的效果分别为：12.5% 氟环唑悬浮剂 500 倍液处理的防效为 76.64%；40% 肟菌酯悬浮剂 5 500倍液处理的防效为 71.51%；50% 异菌脲可湿性粉剂 1 000倍液处理的防效为 55.68%；80% 多菌灵可湿性粉剂 1 000倍液处理的防效为 53.71%；25% 丙环唑水乳剂 1 000倍液处理的防效为 81.49%；80% 乙蒜素乳油 800 倍液处理的防效为 49.50%（表 1）。

结果表明：在以上药剂处理当中，25% 丙环唑水乳剂 1 000倍液处理和 12.5% 氟环唑悬浮剂 500 倍液处理对苹果褐斑病具有更好的防治效果；80% 乙蒜素乳油 800 倍液处理、80% 多菌灵可湿性粉剂 1 000倍液处理和 50% 异菌脲可湿性粉剂 1 000倍液处理对苹果褐斑病具有较差的防治效果。

表 不同药剂处理防治苹果褐斑病田间药效试验结果

Table Results of field test of 50% Tebuconazolenano-formulation against AppleLeaf Spots

处理 Treatment	稀释倍数 (times) dilution ratio	基数病指 Base index of disease	第一次施药后 10d The first 10 days after applying pesticide			第二次施药后 10d The second 10 days after applying pesticide		
			平均病指 Mean index of disease	平均防效 (%) Mean efficacy	差异显著性 Significance	平均病指 Mean index of disease	平均防效 (%) Mean efficacy	差异显著性 Significance
12.5% 氟环唑 SC	500	0.93	0.77	74.20	bB	1.67	76.64	bB
40% 肟菌酯 SC	5 500	0.89	0.94	66.80	cC	1.94	71.51	cC
50% 异菌脲 WP	1 000	0.89	1.19	58.21	dD	3.03	55.68	dD
80% 多菌灵 WP	1 000	0.86	1.22	55.79	eE	3.07	53.71	eE
25% 丙环唑 EW	1 000	0.88	0.66	76.32	aA	1.24	81.49	aA
80% 乙蒜素 EC	800	0.89	1.34	52.93	fF	3.45	49.50	fF
清水对照	CK	0.93	2.96			7.10		

注：不同大写字母表示显著水平 $\alpha<0.01$；不同小写字母表示显著水平 $\alpha<0.05$。

2.2 安全性

在本试验条件下，目测观察试验药剂对苹果生长的安全性，结果显示，6 种杀菌剂对苹果均无药害。

2.3 结果与分析

本试验中选用的 6 种杀菌剂对防治苹果褐斑病均有一定的效果。25% 丙环唑水乳剂 1 000倍液处理和 12.5% 氟环唑悬浮剂 500 倍液处理对于防治苹果褐斑病的效果比较好，在田间防治中应该优先使用；其他的试验药剂处理对于苹果褐斑病的防治效果较差，在田间防治中不提倡使用，避免耽误最佳的防治时期，给农户带来损失。

3 小结

在防治苹果褐斑病时可以选用丙环唑、氟环唑和肟菌酯等杀菌剂，优先使用丙环唑和氟环唑。同时在使用中为了延缓苹果褐斑病菌产生抗药性，可以通过交替使用其他不同作用机制杀菌剂品种，从而达到更为理想和安全的效果。

参考文献

[1] 吕佩珂，庞震，刘文珍，等. 中国果树病虫害原色图谱 [M]. 北京：华夏出版社，1993：4－5.

[2] 孙毅之，雷小英. 富士苹果褐斑病的防治效果调查 [J]. 山西果树，2008，122（2）：30－31.

[3] 刘川江，李海燕，赵守军. 山东半岛地区苹果褐斑病的发生与防治 [J]. 北方果树，2006，3：37.

[4] 冷鹏，陈力，白复芹，房孝叶. 喜瑞等 8 种不同杀菌剂防治苹果褐斑病药效试验 [J]. 烟台果树，2009，6：21－22.

[5] http：//www. chinapesticide. gov. cn/.

[6] 农业部农药检定所生测室. 农药田间药效试验准则（二）[M]. 北京：中国标准出版社，2004.

杀菌剂和硒的复配对梨炭疽菌的室内毒力测定

江　寒*，李　森，叶　磊，王文凤，檀根甲**
（安徽农业大学植物保护学院，合肥　230036）

摘要：硒是一种人体所需的微量元素，对人体有益，而亚硒酸钠不仅能被植物吸收，同时也有抑制梨炭疽菌的效果。本文主要研究亚硒酸钠对梨炭疽菌的菌丝生长的影响，结果表明亚硒酸钠对梨炭疽病原菌的菌丝生长有抑制作用，其 EC_{50} 达到 137.6037μg/mL。同时研究了几种杀菌剂对梨炭疽病原菌的菌丝生长影响，根据其结果筛选出 2 种对梨炭疽病菌效果较好的杀菌剂，戊唑醇和吡唑醚菌酯。将这两种杀菌剂和亚硒酸钠按不同比例复配，研究其对梨炭疽病原菌的菌丝生长的影响。结果表明：SW 混剂 B 效果最好，其共毒系数（CTC 值）能达到 117.34，具有增效作用；SB 混剂 C 和 SB 混剂 E 效果较好，其共毒系数（CTC 值）分别为 107.37 和 103.97，这两个混剂比例也具有一定的增效作用。
关键词：亚硒酸钠；梨炭疽；毒力测定；复配

梨炭疽病是梨树上的主要病害之一，其病原菌主要是胶孢炭疽菌（*Colletotrichum gloeosporioids*）[1]。梨炭疽病菌主要侵染梨树果实[2]；从花期开始侵染，伴随着果实的生长，在果实的中后期显症；能够潜伏侵染，经常在运输或者贮藏中发病，严重时甚至造成大批果实腐烂，病果率能高达40%，成为果实贮藏期的一种重要病害[3]。近年来，梨和苹果树之间的混栽，天气变暖，极端环境频繁，为梨炭疽病原菌提供了一个良好的环境，梨炭疽的发生愈发普遍，尤其是在贮藏期的发生。

目前对梨炭疽的控制主要采用的是清除田园、果实套袋、低温贮藏及药剂防治[4]。其中药剂防治主要采用的是吡唑醚菌酯、戊唑醇、苯醚甲环唑等化学药剂，而化学药剂的施用能带来环境问题[5]。微量元素硒与杀菌剂的配合施用能够很好的缓解这一问题。

硒（Se）是人和动物必需的微量元素，也对植物的生长和发育起到有益的作用，同时也能增强植物的抗逆能力[6]。有研究表明：适量的硒能够增强水稻的抗性；苹果和梨树轮纹、根腐等病害被硒处理后，也能明显降低发病率；硒对镰刀菌也有显著的抑制效果；适量的外源硒也能够明显的抑制油菜菌核病原菌的生长；另外，硒还明显降低了果实在驻藏期间的腐烂率[7-8]。但硒过量同样也会损害生物的生长，对人体的健康也不利。本论文研究微量的硒和杀菌剂协同对梨炭疽菌的抑制作用。

1　材料与方法

1.1　试验材料

1.1.1　供试菌株

梨炭疽病原菌来源于安徽农业大学植物保护学院植物病理研究室在安徽省砀山病梨果实

* 作者简介：江寒，男，硕士研究生，研究方向：植物病害生态与绿色防控；E-mail：jianghan@126.com
** 通讯作者：檀根甲，博士，教授，博士生导师，主要从事植病流行与绿色防控技术研究；E-mail：tgj63@163.com

上分离纯化获得。

1.1.2　供试培养基

马铃薯葡萄糖琼胶（PDA）培养基：马铃薯200g、葡萄糖20g、琼脂17g、水1 000mL。

1.1.3　供试药剂

亚硒酸钠（Na_2SeO_3）

80%多菌灵可湿性粉剂（宁国市朝农化工有限公司）

10%苯醚甲环唑水分散粒剂（先正达股份有限公司）

250g/L嘧菌酯悬浮剂（先正达股份有限公司）

250g/L吡唑醚菌酯乳油（巴斯夫欧洲公司）

戊唑醇原药（上海农乐生物制品股份有限公司）

1.2　试验方法

1.2.1　梨炭疽病原菌的活化

将实验室保存的梨炭疽病原菌（*Colletotrichum gloeosporioids*）接于PDA培养基上，放入28℃恒温培养箱中培养7d，备用。

1.2.2　亚硒酸钠对梨炭疽病原菌的菌丝生长的影响

采用生长速率法测定亚硒酸钠对梨炭疽病原菌的菌丝生长的影响。将亚硒酸钠配成有效成分为0、250、500、1 000、2 000和4 000μg/mL的浓度，每个浓度5mL。再将各浓度梯度的亚硒酸钠倒入冷却到45～55℃的PDA培养基上［V（药液）：V（培养基）=1∶9］，将固定量的含药培养基倒入培养皿中制成含药培养基平板，重复3次。制成的含药平板中的亚硒酸钠浓度变为：0、25、50、100、200和400μg/mL。

将在PDA培养了7d的梨炭疽病原菌用7mm的打孔器延菌落边缘打成菌碟。挑取一枚菌碟放入含药平板的中央，每个平板放入一枚菌碟，置于28℃恒温培养箱中培养，5d后用十字交叉法测量菌落直径。根据式（1）计算亚硒酸钠不同浓度的抑制率，按最小二乘法求回归式，计算EC_{50}。

$$抑制率（\%）=\frac{对照组菌落直径-处理组菌落直径}{对照组菌落直径-7}\times 100 \qquad (1)$$

1.2.3　杀菌剂对梨炭疽病原菌的菌丝生长的影响

杀菌剂对梨炭疽病菌的菌丝生长采用生长速率法。先对其做个预试验，将各杀菌剂浓度配成10 000、1 000、100、10μg/mL和0的浓度，再将杀菌剂与未冷却的PDA培养基1∶9的混合，倒入培养皿中，制成浓度为1 000、100、10、1μg/mL和0的含药平板。在平板中央接入菌碟，置28℃培养箱中培养，5d后用十字交叉法测量菌落直径。根据式（1）计算亚硒酸钠不同浓度的抑制率，按最小二乘法求回归方程，计算EC_{50}和EC_{90}。根据预实验的EC_{50}设置正式试验浓度，设置的浓度应涉及EC_{50}和EC_{90}所对应的浓度。浓度确定后，按预实验步骤重新试验，计算抑制率，回归方程、r值和EC_{50}等。

1.2.4　亚硒酸钠与杀菌剂协同使用对梨炭疽病原菌的菌丝生长的抑制作用

选取1.2.3结果中效果较好的杀菌剂与亚硒酸钠联用，先设置亚硒酸钠和杀菌剂的体积比为A∶B，C∶D，E∶F，G∶H，I∶J；再用无菌水将亚硒酸钠和杀菌剂配制成浓度为10μg/mL，按照设置的比例吸取相对应的药液。每个比例混合液依次稀释到1、0.1、0.01μg/mL和0的浓度梯度，将每个比例的混合液的不同浓度依次倒入PDA平板上［V（混合液）：V（培养基）=1∶9］，配成含药平板；每个浓度处理重复三次。将7mm的菌碟接于平板中央，放入25℃培养箱中培养，5d后用十字交叉法测量菌落直径。分别按照式（1）计算抑制率，EC_{50}。根据孙云沛法计算出药剂不用比例混合使用的共毒系数。

2 结果与分析

2.1 亚硒酸钠对菌丝生长的影响

结果表明（表1），不同浓度的亚硒酸钠的菌落直径均比对照组小（图），说明亚硒酸钠对梨炭疽病原菌的菌丝生长有抑制效果，且亚硒酸钠对梨炭疽病原菌的菌丝生长的抑制作用随着浓度的增加而增加。

表1 亚硒酸钠对梨炭疽病原菌的菌丝生长的影响

浓度（μg/mL）	5d的菌落直径（mm）						5d的菌落平均直径（mm）	生长抑制率（%）
400	22	20	25	25	22	24	23	70.09
200	30	30	31	30	30	30	30.17	57.48
100	36	36	36	35	36	36	35.83	47.51
50	47	43	47	48	45	45	45.83	29.91
25	53	52	55	55	54	53	53.67	16.13
ck	60	62	61	63	65	64	62.83	

图 亚硒酸钠对梨炭疽病原菌的菌丝生长的影响

注：上：从左到右亚硒酸钠的浓度依次为：400，200，100μg/mL；下：从左到右亚硒酸钠的浓度依次为：50，25μg/mL，CK。

2.2 杀菌剂对梨炭疽病原菌菌丝生长的影响

通过DPS软件处理，分析出杀菌剂的毒力回归方程、抑制中浓度EC_{50}值、相关系数r

值。结果（表2）表明，不用的杀菌剂对梨炭疽病原菌的菌丝抑制有不同的效果，其中对梨炭疽病原菌菌丝生长抑制效果最好的是吡唑醚菌酯和苯醚甲环唑，吡唑醚菌酯的 EC_{50} 能达到0.574 6μg/mL，苯醚甲环唑的 EC_{50} 为0.676 0μg/mL。抑制效果较好的是戊唑醇，EC_{50} 是2.200 6μg/mL。效果最差的是多菌灵和申嗪霉素，EC_{50} 分别高达到266.106 7μg/mL和213.823 9μg/mL。亚硒酸钠的 EC_{50} 为137.603 7μg/mL，有一定的抑制效果。

表2 亚硒酸钠及杀菌剂对梨炭疽病原菌的菌丝生长的抑制作用

杀菌剂	毒力回归方程	相关系数 r	EC_{50}（μg/mL）
吡唑醚菌酯	$y=5.238\ 2+0.889\ 9x$	0.979 5	0.539 7
苯醚甲环唑	$y=5.106\ 3+0.625\ 0x$	0.986 0	0.676 0
多菌灵	$y=3.701\ 7+0.535\ 4x$	0.988 7	266.106 7
申嗪霉素	$y=3.784+0.521\ 9x$	0.961 5	213.823 9
戊唑醇	$y=4.512\ 5+1.423\ 1x$	0.975 0	2.200 6
亚硒酸钠	$y=2.328\ 0+1.249\ 1x$	0.993 1	137.603 7

2.3 亚硒酸钠和杀菌剂协同使用对梨炭疽病原菌的菌丝生长的影响

通过DPS软件处理，分析出毒力回归方程、抑制中浓度 EC_{50} 值、相关系数r值，95%置信限。再根据孙云沛法，计算出共毒系数（CTC）值。

根据杀菌剂对梨炭疽病原菌菌丝生长的抑制结果（表2），筛选出戊唑醇、吡唑醚菌酯和这两个杀菌单剂和亚硒酸钠混合配用，来协同控制梨炭疽病原菌的菌丝的生长。

亚硒酸钠和戊唑醇按照1:4，1:1，4:1的比例进行混合使用。混剂结果显示（表3）：SW混剂C对梨炭疽病原菌的菌丝生长效果最好，但共毒系数不高，作为混剂使用后没有明显的增效作用，甚至两者按这种比例混用有拮抗作用；SW混剂A对梨炭疽病原菌的菌丝生长效果最差，但共毒系数比混剂C高，CTC为79.32，增效不明显，表现出的是一种相加效应。SW混剂B对梨炭疽病原菌的菌丝生长效果较好，且共毒系数较高，其CTC为117.34，表现出一种增效作用，SW混剂B协同抑制梨炭疽病原菌的菌丝生长的作用最强。

表3 亚硒酸钠和戊唑醇混合使用对梨炭疽病原菌菌丝生长影响

处理液	毒力回归方程	相关系数 r	EC_{50}（μg/mL）（95%置信限）	EC_{90}（μg/mL）（95%置信限）	CTC
亚硒酸钠	$y=2.328\ 0+1.249\ 1x$	0.993 1	137.603 7（119.84~158.00）	1 472.260 9（100 3.06~216 0.93）	—
戊唑醇	$y=4.529\ 0+1.433\ 2x$	0.975 2	2.131 1（1.45~3.12）	16.704 1（7.55~36.94）	—
SW混剂A	$y=4.103\ 6+0.502\ 7x$	0.887 0	60.694 1（0.89~4 139.20）	21 497.517 0（5.68~81 325 619.80）	79.32

（续表）

处理液	毒力回归方程	相关系数 r	EC_{50}（μg/mL）（95%置信限）	EC_{90}（μg/mL）（95%置信限）	CTC
SW 混剂 B	$y=4.4361+0.9941x$	0.991 9	3.691 8（1.96～6.94）	71.856 2（24.79～208.32）	117.34
SW 混剂 C	$y=4.2876+1.3233x$	0.973 0	3.454 3（1.09～10.95）	32.120 1（5.65～182.54）	17.04

亚硒酸钠和吡唑嘧菌酯按照1∶4，2∶3，1∶1，3∶2，4∶1的比例进行混剂使用。混剂结果显示（表4）：对梨炭疽病原菌菌丝抑制效果最好的是SB混剂B和混剂C，其EC_{50}分别是1.04和1.0659μg/mL，但从EC_{90}看，SB混剂C明显优于混剂B。而且SB混剂B共毒系数（CTC）值低于混剂C的共毒系数，两者共毒系数均表现出一定的相加作用，SB混剂C同时还有一定的增效作用。SB混剂A和混剂D抑制效果相对一般，其共毒系数也不高，CTC值分别为64.11和47.64，相加作用菌不明显。SB混剂E对梨炭疽病原菌的菌丝抑制的EC_{50}一般，但EC_{90}较好，EC_{90}为251.9875μg/mL。其共毒系数也能表现一定的增效作用，其CTC值为103.97。

表4　亚硒酸钠和吡唑醚菌酯混合使用对梨炭疽病菌菌丝生长的影响

处理液	毒力回归方程	相关系数 r	EC_{50}（μg/mL）（95%置信限）	EC_{90}（μg/mL）（95%置信限）	CTC
亚硒酸钠	$y=2.3280+1.2491x$	0.993 1	137.603 7（119.84～158.00）	1 472.260 9（1 003.06～2 160.93）	—
吡唑醚菌酯	$y=5.2384+0.8899x$	0.979 5	0.539 7（0.320～0.91）	14.866 8（4.54～48.66）	—
SB 混剂 A	$y=4.9793+0.4238x$	0.974 7	1.119 1（0.449 7～2.784 9）	1 182.124 6（76.31～18 313.14）	64.11
SB 混剂 B	$y=4.9943+0.3336x$	0.907 8	1.04（0.17～6.40）	7 219.936 2（9.48～96 655.57）	91.83
SB 混剂 C	$y=4.9896+0.3747x$	0.943 6	1.065 9（0.27～4.26）	2 805.945 5（28.36～77 667.85）	107.37
SB 混剂 D	$y=4.7325+0.5613x$	0.963 9	2.996 4（0.81～11.10）	574.988 1（27.53～12 007.83）	47.64
SB 混剂 E	$y=4.7171+0.6515x$	0.921 2	2.717 9（0.38～19.38）	251.987 5（3.81～16 675.46）	103.97

3　结论与讨论

结果表明，亚硒酸钠对梨炭疽病原菌的菌丝生长有抑制作用，但对梨炭疽抑制时亚硒酸钠的浓度较高，考虑到硒是植物中的一种微量元素，也是人体的一种微量元素，硒含量不宜过高。如果单纯的把硒作为一种药剂来防治梨炭疽病原菌，可能会导致梨吸收硒含量过高，产生毒害，人食用也可能会有对健康产生问题。如要解决这一问题，本文采用亚硒酸钠和杀菌剂协同使用来控制梨炭疽病原菌，这样既可以控制病害；而且硒含量降低了，植物可以吸收安全状态下的硒，人类食用也处于安全状态；同时，硒含量降低了，对植株的生长有利，对人体健康也有帮助。

在杀菌剂对梨炭疽病菌的室内毒力测定结果中，吡唑醚菌酯、苯醚甲环唑和戊唑醇对梨炭疽病原菌的菌丝抑制效果好，EC_{50}和EC_{90}都较低，其中吡唑醚菌酯效果最好。本文就效果最好三个杀菌剂中，选择戊唑醇和吡唑醚菌酯用作和亚硒酸钠进行协同配合来控制梨炭疽病菌。根据杀菌剂和亚硒酸钠的混剂可以看出，亚硒酸钠和戊唑醇按照SW混剂B的配比时增效最为明显，防治效果也明显，其EC_{50}为3.454 3μg/mL，按其比例，亚硒酸钠的浓度在很低的范围，符合使用安全标准。亚硒酸钠和吡唑醚菌酯按照SB混剂C的配比时增效较明显，防治效果明显，其EC_{50}为1.065 9μg/mL，亚硒酸钠浓度较SW混剂B更低，更加符合使用安全标准，但其增效不如SW混剂B。本文筛选出的混剂配比为防治梨炭疽病提供了一个理论基础。

本文只研究了室内硒和杀菌剂的混用对梨炭疽病菌菌丝的影响，对孢子萌发的影响尚未研究，这需要后者进行下一步试验。另外，本文只研究了室内的毒力测定，并未对活体防效及田间防效做出研究，筛选出的混剂配比是否在生产应用中有效，还需要进一步考察。硒对病害的防治属于一个全新的领域，其研究前景非常可观。

参考文献

[1] 吴良庆，朱立武，衡伟，等. 砀山梨炭疽病病原鉴定及其抑菌药剂筛选［J］. 中国农业科学，2010，43（18）：3 750－3 758.

[2] 刘油洲，陈志谊，刘永锋，等. 枯草芽孢杆菌 *sf*628 对梨炭疽的控制作用［J］. 植物保护学报，2012，39（6）：492－496.

[3] 沈静霆. 砀山酥梨炭疽病菌生物学特性及有效药剂筛选［D］. 安徽：安徽农业大学，2012.

[4] 高正辉，张金云，伊兴凯，等. 砀山酥梨炭疽病发生特征与防治技术［J］. 安徽农业科学，2010，38（19）：10 445－10 446.

[5] 周增强，侯珲，王丽. 九种杀菌剂对梨炭疽病菌的抑制效果［J］. 北方园艺，2012（10）：157－158.

[6] 贾玮，吴隽，屈婵娟，等. 硒增强植物抗逆能力及其机理研究进展［J］. 中国农业通报，2015，31（14）：171－176.

[7] 吴之琳，尹雪斌，袁林喜，等. 硒在植物抗逆境胁迫耐受中的作用［J］. 粮食食品与经济，2014，39（3）：26－31.

[8] 贾芬，朱婷，赵小虎，等. 硒对油菜菌核病的控制作用［J］. 中国农学通报，2015，31（25）：176－181.

[9] 檀根甲，李增智，薛莲，等. 枯草芽孢杆菌和丙环唑对采后苹果炭疽病的防治效果［J］. 激光生物学报，2008，17（2）：245－249.

[10] Seppanen M，Turakainen M，Hartikainen H. Selenium effects on oxidative stress in potato［J］. Plant Science，2003，165（2）：311－319.

微生物源"吩嗪-1-甲酰胺"对赤霉病菌作用机制研究

杨　楠，王　静，徐孙德，郑诗昱，陈　云，马忠华
（浙江大学生物技术研究所，杭州　310058）

摘要： 假单胞菌PC60是本实验室从小麦根围分离得到的对小麦赤霉病菌有显著抑制作用的生防细菌。基因组测序结果显示，该菌株基因组大小为6.82M，共编码6 162个蛋白，GC含量62.8%，系统进化树结果表明，PC60为绿针假单胞菌。研究通过次生代谢基因簇预测、突变体构建并结合活性物质分离、鉴定，明确了PC60抑制赤霉病菌的活性化合物为吩嗪－1－甲酰胺（phenazine-1-carboxamide，PCN）。该菌株在金氏培养液中振荡培养72h后，菌液经乙酸乙酯萃取和纯化，PCN的产量能达到1g/L，是已报道PCN产量最高的微生物菌株。PCN对酵母、小麦赤霉病菌、稻瘟病菌、小麦纹枯病、灰霉病等植物病原真菌都具有显著的抑制活性；且与常用杀菌剂无交互抗性，不易产生抗药性，是有开发潜力的新型杀菌剂。为探索PCN的抑菌机制，研究采用HIP-HOP筛选酵母突变体库方法，获得对PCN敏感性有显著差异的酵母突变体共21株。对21个酵母突变体所对应基因的功能分析，推测PCN可通过干扰酵母的氧化胁迫调节、细胞周期、蛋白质合成等途径抑制酵母的生长。有趣的是，在赤霉病菌中利用同源重组法敲除了21个酵母突变体的同源基因，并测定所有赤霉突变体对PCN的EC_{50}，结果发现：与野生型菌株PH-1相比，所有突变体对PCN的敏感性均无显著差异，表明PCN抑制丝状真菌和酵母的作用机制不同。研究将进一步以赤霉病菌为靶标生物，解析PCN抑制丝状真菌的作用机制，为后续创制以PCN为主效成分的微生物杀菌剂奠定基础。

The Microtubule End-Binding Protein FgEB1 Regulates Polar Growth and Fungicide Sensitivity Via Different Interactors in *Fusarium graminearum*

Liu Zunyong[1], Wu Sisi[1], Chen Yun[1], Han Xinyue[1], Gu Qin[2], Yin Yanni[1*], Ma Zhonghua[1*]

(1. *Institute of Biotechnology*, *Zhejiang University*, 866 *Yuhangtang Road*, *Hangzhou* 310058, *China*; 2. *Department of Plant Pathology*, *College of Plant Protection*, *Nanjing Agricultural University*, *and Key Laboratory of Integrated Management of Crop Diseases and Pests*, *Ministry of Education*, *Nanjing* 210095, *China*)

Abstract: In yeasts, the end-binding protein 1 (EB1) homologs regulate microtubule dynamics, cell polarization, and chromosome stability. However, the functions of EB1 orthologs in plant pathogenic fungi have not been characterized yet. Here, we observed that the *FgEB*1 deletion mutant (ΔFgEB1) of *Fusarium graminearum* exhibits twisted hyphae, increased hyphal branching and curved conidia, indicating that FgEB1 is involved in the regulation of cellular polarity. Microscopic examination further showed that the microtubules of ΔFgEB1 exhibited less organized in comparison with those of the wild type. In addition, the lack of *FgEB*1 also altered the distribution of polarity-related class I myosin via the interaction with the actin. On the other hand, we identified four core septins as FgEB1-interacting proteins and found that FgEB1-septin complex regulates conidial polar growth in the opposite orientation. Interestingly, FgEB1 and FgKar9 constitute another complex that modulates the response to carbendazim, a microtubule-damaging agent specifically. In addition, the deletion of *FgEB*1 led to dramatically decreased deoxynivalenol (DON) biosynthesis. Taken together, results of this study indicate that FgEB1 regulates cellular polarity, fungicide sensitivity and DON biosynthesis via different interactors in *F. graminarum*, which provides a novel insight into understanding of the biological functions of EB1 in filamentous fungi.

* Corresponding author: Yin Yanni; E-mail: ynyin@zju.edu.cn; Ma Zhonghua; E-mail: zhma@zju.edu.cn

Recent Progress of Research on SDHI Fungicide Resistance

Hideo Ishii

(*Kibi International University*, *Sareo* 370 - 1, *Shichi*, *Minami - awaji*, *Hyogo*, 656 - 0484 *Japan*)

Abstract: In recent years, a new generation of succinate dehydrogenase inhibitors (SDHIs, mitochondrial complex II inhibitors) such as boscalid and penthiopyrad was intensively developed. Other fungicides which possess the same mode of action have also been commercialized and at present twenty SDHI fungicides are known in total. The risk of resistance development to SDHI fungicides is regarded medium to high, and SDHI fungicide resistance has already been reported in many pathogen - crop combinations. However, the pattern of cross - resistance is unique among SDHI fungicides. The lack of cross - resistance to fluopyram was found in highly boscalid - resistant isolates of some fungi including *Corynespora cassiicola*, *Podoshaera xanthii*, *Botrytis cinerea* and others. This interesting phenomenon in boscalid - resistant isolates was correlated with the H 277 Y/R mutations of *sdhB* genes, encoding a subunit of fungicide - targeted succinate dehydrogenase, suggesting differential binding of individual fungicides to the target site of protein molecule. More recently, a novel SDHI fungicide benzovindiflupyr has been shown to be effective against *Colletotrichum* species, inherently insensitive (naturally resistant) to boscalid, fluxapyroxad, and fluopyram. The *sdhB*, *sdhC* and *sdhD* genes were partially sequenced, but, despite high polymorphisms, no apparent resistance mutations were found in *Colletotrichum* species. Other mechanism (s) than fungicide target - site modification might be responsible for differential sensitivity of *Colletotrichum* species to SDHI fungicides. In this paper, recent development of SDHI fungicides and the progress of research on resistance to this class of fungicides will be discussed.

Progress in Studies on Resistance of Carboxylic Acid Amides in Oomycetes

Gerd Stammler *

(*BASF SE*, *Agricultural Center*, *Fungicide Resistance Management*, *Speyerer Strasse* 2, 67117 *Limburgerhof*, *Germany*)

Abstract: Carboxylic acid amides (CAA) are active fungicides against different genera of Oomycetes such as *Bremia*, *Peronospora*, *Peronophythora*, *Phytophthora*, *Plasmopara*, *Pseudoperonospora* and others, but not *Pythium* species. CAAs are also not active on other fungi outside the Oomycetes. CAAs have excellent preventive, curative and also antisporulant effects and have, depending on the compound, more or less pronounced systemic properties.

The mode of action CAA was not fully elucidated for several years after market introduction of the first CAAs. Morphological studies from Albert et al. with *Plasmopara viticola* and Kuhn et al. with *Phytophthora* spp. indicated that the CAA dimethomorph interferes with processes involved in fungal cell wall biogenesis: inhibition of mycelial growth, germination of zoospores and sporangia and germ tube elongation, while other phases of the life cycle such as zoospore release and zoospore motility are not affected. Later studies with the CAA mandipropamid proposed an involvement in phospholipid biosynthesis, with an inhibition of phosphatidyl biosynthesis. However, subsequent studies with CAA resistant lab mutants of *Phytophthora capsici* did not show any mutations in the two potential target genes . Finally, Blum et al. identified cellulose synthesis and the enzyme cesA3 (cellulose synthase) as the target of CAAs by a combination of phenotypic observations, incorporation assays, mutation of *Phytophthora infestans*, candidate gene sequencing and complementation in *P. infestans* transformants. Further studies with CAA resistant field isolates of *P. viticola* and *Pseudoperonospora cubensis* confirmed that the target of CAAs is cesA3 .

After the elucidation of the mode of action, the CAA fungicides were regrouped by the Fungicide Resistance Action Committee (FRAC) in the FRAC Group H (cell wall biosynthesis), here in the group H5, which means inhibition of cellulose synthase and the FRAC group number # 40, which are the carboxylic acid amides (CAAs). The CAA fungicide group includes three subclasses, which are the cinnamic acid amides (dimethomorph, flumorph and pyrimorph), valinamide carbamates (benthiavalicarb, iprovalicarb, valifenalate), and mandelic acid amides (mandipropamid).

Four cesA3 genes have been identified in Oomycetes and the mutations responsible for CAA resistance have so far been only detected in the cesA3 gene. The main mutation site, which confers CAA resistance is the G1105. Mutations leading to exchanges G1105S (and to a lesser extent to G1105V) were found in *P. viticola* and mutations leading to G1105V or G1105W in *P. cubensis* . CAA – resistant lab mutants of *P. capsici* and *Phytophthora melonis* showed different mutations in the cesA3 gene leading to the exchange Q1077K and V1109L, respectively .

Inheritance studies showed that sexual crosses between CAA sensitive and resistant isolates of *P. viticola* lead to a co – segregation of resistance to CAA fungicides, but not to other modes of action. Further, the inheritance studies showed that the gene for resistance to CAA fungicides is inherited in a recessive manner. Therefore, the entire F1 generation of crosses between sensitive and CAA resistant isolates was sensitive, and only in the F2 progeny did CAA resistance reappear in some isolates (FRAC 2016).

* Corresponding author: Gerd Stammler; E-mail: gerd. stammler@ basf. com

CAA resistance was detected in *P. viticola* soon after market launch of dimethomorph in Southwest France and appeared over the years in the most grapevine growing regions in Europe, where a high disease pressure of *P. viticola* is present. In recent years, the frequency of CAA resistance in Europe remains heterogeneous, depending on the vine growing region, with areas of no, low, moderate and high resistance frequencies. Monitoring data are exchanged in the CAA FRAC Working Group on an annual basis and the data are provided per region in the FRAC meeting minutes (FRAC 2016). The activity of CAAs is not completely lost in case of resistance development. While resistant strains are not controlled by CAAs when applied curatively, studies with dimethomorph showed that there is still a preventive activity, which contributes to disease control.

CAA resistance also been detected in field isolates of *P. cubensis* and occurrence of CAA resistance has described by Blum et al. and FRAC for some samples in US, Israel, China and Spain (FRAC 2016).

As a consequence of resistance development in *P. viticola* and *P. cubensis*, CAAs are recommended in a preventive manner and only in mixture with other effective partners with a maximum of 4 applications (not more than 50% of the total number of applications). In areas with high frequency of CAA resistance, this number is limited to 3 applications per season.

No CAA resistance has been detected in field isolates of *P. infestans* (FRAC 2016), *Peronophythora litchii*, *Bremia lactucae* (FRAC 2016) or *Peronospora destructor* (BASF monitoring data from 2016). For these and other Oomycetes, FRAC recommends to use CAAs in a preventive manner and that CAA containing sprays should not exceed 50% of the total number of applications.

Identification of A Novel Phenamacril-resistance Related Gene by cDNA-RAPD Method in *Fusarium asiaticum*

Ren Weichao, Zhao Hu, Shao Wenyong, Ma Weiwei,
Wang Jianxin, Zhou Mingguo, Chen Changjun *
(*College of Plant Protection*, *Nanjing Agricultural University*, *Nanjing* 210095, *China*)

Abstract: *Fusarium asiaticum*, a dominant pathogen of Fusarium head blight (FHB) in East Asia, causes huge economic losses. Phenamacril, a novel cyanoacrylate fungicide, has been increasingly applying to control FHB in China, especially where resistance of *F. asiaticum* against carbendazim was severe. It is important to clarify the resistance related mechanisms of *F. asiaticum* to phenamacril so as to avoid control failures and to sustain the usefulness of the new product. In this study, a novel phenamacril-resistance related gene *Famfs*1 was obtained by employing cDNA random amplified polymorphic DNA (cDNA-RAPD) technique and was validated by genetic and biochemical assays. Compared with the corresponding progenitors, deletion of *Famfs*1 in the phenamacril-sensitive or -highly resistant strain caused significant decrease in effective concentrations inhibiting radial growth by 50% (EC_{50} value). Additionally, the biological fitness parameters (including mycelial growth under different stresses, conidiation, perithecia formation and virulence) of the deletion mutants attenuated significantly. *Famfs*1 was not only involved in the resistance of *F. asiaticum* to phenamacril, but also played important roles in adaptation of *F. asiaticum* to environment. Moreover, our data appear that cDNA-RAPD method can be a candidate technique to clone resistance related genes in fungi.

Key words: *Fusarium asiaticum*; Phenamacril; Resistance; cDNA-RAPD

* Corresponding author: Chen Changjun; E-mail: changjun-chen@ njau. edu. cn

Current status on SDHI Sensitivity of Cereal Pathogens

Gerd Stammler *

(*BASF SE*, *Agricultural Center*, *Fungicide Resistance Management*, *Speyerer Strasse* 2, 67117 *Limburgerhof*, *Germany*)

Abstract: Succinate dehydrogenase inhibitors (SDHIs) are very effective fungicides for the control of many important plant pathogens in cereal crops. Besides net blotch caused by *Pyrenophora teres*, scald caused by *Rhynchosporium secalis* and powdery mildew (*Blumeria graminis* f. sp. *hordei*) in barley, leaf blotch in wheat caused by *Zymoseptoria tritici*, brown rust caused by *Puccinia triticina* and powdery mildew (*Blumeria graminis* f. sp. *tritici*) are destructive cereal diseases. In the background of (partial) resistance to strobilurins and the adaptation to azoles of some pathogens, the use of SDHIs in spray programmes is becoming more relevant.

First isolates of *P. teres* and *Z. tritici* with a reduced sensitivity to SDHIs were found in 2012 in Europe. In both cases, a target site mutation in SDH genes leading to an amino acid exchange in the SDH complex, B – H277Y in *P. teres* and C – T79N in *Z. tritici*, were detected. In case of net blotch, an increase of less sensitive isolates was detected in the following years mainly in Northern France and Germany, which are intensive barley growing regions. In the following years, SDHI adaptation in *P. teres* was caused by amino acid exchanges B – H277Y, C – N75S, C – G79R, C – H134R, C – S135R, D – D124N/E, D – H134R, D – D145G and D – E178K in the SDH complex resulting in different resistance levels. The majority of less sensitive isolates contained the C – G79R substitution and showed a relatively high efficacy loss in microtiter tests with resistance factors of ~100 for all SDHIs currently registered in barley (Rehfus et al. 2016).

In *Z. tritici*, further SDHI adapted isolates were sporadically found in Europe in 2013, 2014 and 2015 carrying B – N225T, B – T268I, C – W80S, C – N86S and C – H152R (FRAC 2016). However, in contrast to *P. teres*, isolates of *Z. tritici* with a reduced SDHI sensitivity did not propagate and spread in Europe, so far. Monitoring studies from 2016 are currently under further investigation.

Extensive monitoring programmes were also carried out for *R. secalis*, *B. graminis* f. sp. *tritici* and *Puccinia triticina*. All isolates tested were fully sensitive to SDHIs.

In contrast to the rapid development and spread of QoI resistant in *Z. tritici* isolates in European countries since 2003, the evolving situation of SDHI resistance in *P. teres* and *Z. tritici* is more complex and seems highly dynamic. To maintain SDHIs as effective tools for the control of cereal pathogens, resistance management strategies as recommended by FRAC should be strictly followed.

* Corresponding author: Gerd Stammler; E-mail: gerd.stammler@basf.com

禾谷镰孢菌 Hsp70 蛋白复合体 FgSsb-FgZuo1-FgSsz 调控 β_2 -tubulin 稳定性和液胞完整性

黄蒙蒙，王芝慧，刘尊勇，尹燕妮，马忠华

（浙江大学生物技术研究所，杭州 310025）

摘要：多菌灵是生产上防治小麦赤霉病的常用杀菌剂。本文通过测定 600 多个禾谷镰孢菌基因缺失突变体多菌灵敏感性，发现 70ku 热激蛋白（70ku heat shock proteins，Hsp70s）FgSsb 缺失突变体对多菌灵敏感性显著增加。进一步分析发现，ΔFgSsb 突变体中 β_2-tubulin 稳定性明显降低。为深入研究 FgSsb 对 β2-tubulin 稳定性的调控作用，通过亲和捕获（Affinity capture）-MS 方法鉴定出 FgSsb 互作蛋白 FgSsz 和 FgZuo1；免疫共沉淀（Co-immunoprecipitation，Co-IP）试验证明了这 3 个蛋白在病菌体内形成一个复合体，该复合体能直接作用于 β_2-tubulin。基因敲除试验发现，ΔFgSsz 和 ΔFgSuo1 对多菌灵敏感性也显著增加，并且这两个突变体中 β_2-tubulin 的稳定性也显著下降。Hsp70s 作为细胞内主要的一类分子伴侣，参与细胞内蛋白的折叠、降解和运输等过程。本研究结果表明，FgSsb-FgZuo1-FgSsz 复合体通过调控 β_2-tubulin 的稳定性，进而影响病菌对多菌灵的敏感性。此外，研究还发现 FgSsb 能够与囊泡形成蛋白互作。CMAC 染色和透射电镜发现，FgSsb/FgZuo1/FgSsz 基因突变体中液泡完整性受损、细胞自噬提前、DON 合成能力及致病力均显著下降，表明该复合体作为分子伴侣调控病菌真菌多个重要的生命活动。

禾谷镰孢菌中 ABC 转运蛋白组的功能分析

尹燕妮，陈　骧，韩馨月，王敏慧，马忠华
（浙江大学生物技术研究所，杭州　310058）

摘要：禾谷镰孢菌（*Fusarium graminearum*）基因组中共有 62 个 ABC（ATP-binding cassette）转运蛋白。通过基因敲除的方法，获得 60 个 ABC 转运蛋白基因的敲除突变体；另外 2 个基因 FgABC-01 和 FgABC-11 是病菌的持家基因，尚未获得敲除突变体。在 60 个基因突变体中，ΔFgABC-11、ΔFgABC-16、ΔFgABC-32 和 ΔFgABC-25 表现严重的菌丝生长缺陷。由于 ΔFgABC-11 在 MM 培养基上不能生长，本课题重点研究了 59 个 ABC 基因的生物学功能。对 59 个突变体进行产孢量、孢子形态、孢子萌发、致病性、DON 和及 23 种环境胁迫的测定发现：3 个生长减慢的突变体 ΔFgABC-16、ΔFgABC-32 和 ΔFgABC-25 均出现产孢、致病和 DON 合成的严重缺陷，其中 ΔFgABC-25 对麦穗、小麦叶片和玉米须完全丧失致病力；进一步试验发现该突变体不能穿透小麦叶片和玻璃纸，表明该 ABC 转运蛋白在调控病菌侵入中起重要作用。对 10 种药剂敏感性测定发现，FgABC-60、FgABC-30 和 FgABC-12 的单敲突变体对三唑类药剂表现敏感，这 3 个基因中任意 2 个基因双敲突变体多三唑类药剂的敏感性进一步增加，表明 3 个转运 ABC 蛋白均参与三唑类药剂的运输过程。此外，FgABC-30 和 FgABC-27 分别参与病菌对多菌灵和咯菌腈的敏感性；FgABC-33 和 FgABC-30 参与调控病菌对天然产物 PCN（phenazine-1-carboxamide）的敏感性。这些结果表明，FgABC-30 能够运输多种结构不同的有毒物质，而其他几个 ABC 转运蛋白特异性运输一类（种）物质。逆境胁迫试验发现，ΔFgABC-25 对氧化胁迫的敏感性显著增加，该突变体中 ROS 含量也显著增加，且突变体中参与 ROS 降解的 5 个酶活基因表达量显著下降；ΔFg02025 对细胞壁胁迫的敏感性显著增加，透射电镜发现该突变体菌丝中出现“intrahyphal hyphae”（内生菌丝的菌丝）现象；进一步解析了该 ABC 转运蛋白参与多种生命活动的分子机制。本研究结果有助于系统理解丝状真菌中 ABC 转运蛋白的生物学功能，为研发 ABC 转运蛋白抑制剂提供科学依据。

禾谷镰刀菌响应三唑类杀菌剂的转录因子鉴定及功能分析

刘尊勇，尹燕妮，马忠华
（浙江大学生物技术研究所，杭州　310058）

摘要： 甾醇脱甲基抑制剂类（DMI）药剂对小麦赤霉病有良好的防治效果。笔者前期研究发现，禾谷镰刀菌中存在三个编码 DMI 类药剂靶标的基因（命名为 FgCYP51A/B/C，DMI 类药剂能诱导 FgCYP51A 和 B 的高水平表达。为此，本文进一步解析 DMI 药剂诱导靶标基因高表达的分子机制。以 FgCYP51A 基因的启动子区为诱饵，筛选了禾谷镰刀菌的 cDNA 文库，发现了 3 个转录因子可以与 FgCYP51A 基因的启动子区；对这 3 个基因分别进行了敲出后检测了这 3 个转录因子突变体对药剂的敏感性发现，转录因子 FgTeb 突变体对 DMI 类药剂敏感性显著增加，且突变体内麦角甾醇含量降低。在 FgTeb 突变体中，DMI 药剂不能诱导 FgCYP51A、FgCYP51B 基因的高水平表达。利用 ChIP-qPCR 技术，我们发现 FgTeb 可以结合 FgCYP51A/B 基因的启动子区，但在不同基因启动子区域的富集程度不同；启动子互换试验表明，FgCYP51A/B 基因具有相互替代的功能。利用 ChIP-seq 结合表达谱的数据，分析得到了一个长度为 16bp 顺式作用元件；利用酵母双杂交试验表明，FgTeb 可以直接与该顺式元件结合。深度分析 ChIP-seq 和表达谱的数据发现，FgTeb 可以结合 121 个基因的启动子区，其中包括麦角甾醇合成途径中 19 个相关基因、氨基酸合成途径相关基因。药剂敏感性试验还发现，FgTeb 突变体对 3-AT 和 4-NQ 的敏感性显著增加，这表明 FgTeb 调控病菌对多种逆境反应。致病性试验表明，FgTeb 突变体在麦穗上致病力显著降低，致病力下降是突变体对植保素敏感性增加所致。此外，还发现 FgTeb 与转运蛋白 impotin1 直接互作，表明 FgTeb 是在 importin1 作用下进入细胞核，调控多个基因的表达。本文结果表明，FgTeb 是一个新的 SREBP 类转录调控因子，它能与不同蛋白互作调控病菌对多种胁迫反应及病菌致病性。

山西省苹果黑腐皮病菌 *Valsa mali* 对 3 种药剂的敏感性及交互抗性研究*

周建波[1,2]**，任　璐[2,3]，殷　辉[1,2]，秦　楠[1,2]，张志斌[1,2]，赵晓军[1,2]***

（1. 山西省农业科学院植物保护研究所，太原　030031；2. 农业有害生物综合治理山西省重点实验室，太原　030031；3. 山西农业大学，太谷　030801）

摘要： 苹果树腐烂病是一种为害严重且难以防治的苹果树病害，主要为害主枝干，也可感染幼树、幼苗和果实，造成树势衰弱和产量下降，甚至毁园。山西省苹果树腐烂病病株率为50%以上，主产区运城果园为害率达85%以上，已严重威胁着当地苹果产业的发展。

为明确山西省苹果树腐烂病主要致病菌 *Valsa mali* Mayabe et Yamada 对甲基硫菌灵、苯醚甲环唑和嘧菌酯 3 种常用杀菌剂的抗性水平，采用菌落生长速率法测定了山西省 8 个苹果产区分离到的 53 个 *V. mali* 菌株对 3 种杀菌剂的敏感性。测定结果表明，3 种杀菌剂的 EC_{50} 范围依次为 0.013～1.027μg/mL（均值为 0.516μg/mL）、0.042～41.372μg/mL（均值为 7.509μg/mL）和 0.003～0.309μg/mL（均值为 0.035μg/mL）。Kolmogrov-Smirnov 频次正态分布和 Euclidean 距离法聚类分析结果表明：苹果树腐烂病菌群体中均出现了对 3 种药剂敏感性降低的亚群体；山西省苹果树腐烂病菌株对药剂的敏感性无地域性差异；3 种药剂两两间无交互抗性。

山西省苹果树整个生长季中 3 种供试药剂常被大量使用以综合防治多种病害，当腐烂病与其他病害混合发生时，频繁用药可增加腐烂病菌抗药性产生的概率，嘧菌酯抗性严重可能与其大剂量高频率使用相关。为延缓和降低苹果树腐烂病菌抗药性的产生，在加强腐烂病菌抗药性监测的同时，建议在腐烂病和其他病害混发果园尽量减少以上 3 种药剂的使用频率。

---

* 基金项目：山西省重点研发计划项目（201603D221013-3）；山西省青年科技研究基金（201601D202073）；山西省农科院博士基金项目（YBSJJ1408）；山西省农科院重点项目（YZD1502）；山西省农业科学院科技自主创新能力提升工程项目（2016ZZCX-15）

** 作者简介：周建波，男，助理研究员，从事果蔬病害综合治理及病原菌抗药性研究；E-mail：zhoujianbo6067@163.com

*** 通讯作者：赵晓军，副研究员，从事果蔬病害综合治理及病原菌抗药性研究；E-mail：zhaoxiaojun0218@163.com

山西省辣椒炭疽病菌 *Colletotrichum gloeosporioides* 对啶氧菌酯的敏感性基线及生物学特性研究*

任　璐[1,2]**，周建波[1,3]，殷　辉[1,3]，曹俊宇[2]，赵晓军[1,3]***
（1. 农业有害生物综合治理山西省重点实验室，太原　030006；2. 山西农业大学农学院，太谷　030801；3. 山西省农业科学院植物保护研究所，太原　030006）

摘要：为评估辣椒炭疽病菌对啶氧菌酯的抗性风险，本研究建立了辣椒炭疽病菌对啶氧菌酯的敏感性基线，在此基础上室内诱导获得了辣椒炭疽病菌抗啶氧菌酯突变体，并对突变体生物学特性进行了研究。结果表明，在山西省晋中市 3 个未使用过啶氧菌酯及同类药剂的地区采集并分离到 45 株辣椒炭疽病菌菌株（*Colletotrichum gloeosporioides*）。采用菌丝生长速率法测定其对啶氧菌酯的敏感性，结果显示，EC_{50}值介于 1.404 1 ~ 16.649 7μg/mL，平均 EC_{50}值为（6.783 1 ± 3.499 4）μg/mL。在水杨肟酸（SHAM）的处理浓度为 150μg/mL 时，EC_{50+S}介于 0.022 1 ~ 0.275 2μg/mL，平均值为（0.109 0 ± 0.058）μg/mL，EC_{50}值呈连续性单峰曲线，且敏感性频率分布呈近似正态分布，因此 EC_{50}平均值可作为辣椒炭疽病菌对啶氧菌酯的敏感基线。旁路氧化所表现出的增效值 F 最低为 6.039 1，最高为 301.441 2，平均为 78.026 2。通过室内抗性诱导，共获得 8 株抗性突变体，其中低抗菌株 6 株，中抗菌株 2 株。抗性突变体在无药培养基无性培养 10 代，抗性不丧失，可稳定遗传；抗性突变体生物学特性测定结果表明，突变体产孢量、菌丝生长速率、致病力与敏感菌株相比无显著差异；在不同营养条件、pH 值、温度条件下，抗感菌株均以淀粉作为碳源、硝酸钾作为氮源时利用率最高；均在中性偏酸条件下生长较旺盛；菌丝的最适生长温度均为 25℃。研究结果表明，辣椒炭疽病菌抗性突变体具有较高的适合度。因此，辣椒炭疽病菌对啶氧菌酯具有较高的抗性风险，研究结论为指导生产用药，为延缓抗药性的发展提供理论依据。

关键词：*Colletotrichum gloeosporioides*；啶氧菌酯；敏感基线；生物学特性

---

* 基金项目：山西省科技基础条件平台建设项目（2014091003 - 0102）；山西省农科院育种工程项目（16yzgc120）；山西省农科院重点攻关项目（YGG1603）；农业有害生物综合治理山西省重点实验室开放课题（YHSW2015002）

** 第一作者：任璐，女，山西太原人，副教授，主要从事杀菌剂毒理与病原菌抗药性研究；E-mail：renlubaby@163.com

*** 通讯作者：赵晓军，副研究员，主要从事果蔬病害综合治理与病原菌抗药性研究；E-mail：zhaoxiaojun0218@163.com

黄瓜多主棒孢菌对 3 种 DMIs 类杀菌剂的敏感性检测

高　苇*，王　勇，张春祥
（天津市植物保护研究所，天津　300381）

摘要： 多主棒孢菌（*Corynespora cassiicola*）是一种重要的植物病原真菌，可侵染蔬菜和园艺观赏作物等。由多主棒孢菌引起的黄瓜棒孢叶斑病是我国近几年来新流行的一种病害，对我国黄瓜生产造成了严重的威胁。化学防治是防治黄瓜棒孢叶斑病最常用且最有效的措施。苯并咪唑类杀菌剂如多菌灵、甲基托布津，二甲酰亚胺类的异菌脲，甲氧基丙烯酸酯类（QoI 类）的嘧菌酯，苯吡咯类的咯菌腈和烟酰胺类的啶酰菌胺等都是常用于黄瓜棒孢叶斑病防治的杀菌剂。然而，随着杀菌剂使用时间的延长及剂量的增加，多主棒孢菌已有对这些药剂产生了抗性群体，尤其是对苯并咪唑类、甲氧基丙烯酸酯类及 QoI 类杀菌剂已产生了明显的抗药性的报道。DMIs 类杀菌剂是 20 世纪 70—80 年代开发的杀菌剂，具有杀菌谱广、活性高、作用位点单一及种类多等特点，对许多蔬菜病原真菌具有较强的抑制作用，在田间黄瓜棒孢叶斑病的防治中发挥了重要作用。但目前关于黄瓜多主棒孢菌对 DMIs 杀菌剂的敏感性研究尚未见报道。

本研究以咪鲜胺、苯醚甲环唑和戊唑醇 3 种 DMIs 药剂作为供试药剂，以黄瓜多主棒孢菌作为试验靶标菌株，系统地检测天津市多主棒孢菌对该 3 种药剂的敏感性。2013—2015 年在天津地区黄瓜栽培地分离获得 16 株黄瓜多主棒孢菌，通过菌丝生长速率法测定该 16 株菌株对咪鲜胺、苯醚甲环唑和戊唑醇 3 种药剂的敏感性。试验结果表明：16 株菌株对 3 种药剂的敏感性存在着显著的差异，其中对咪鲜胺的 EC_{50} 值 0.001 6 ~ 2.515 9μg/mL，对戊唑醇的 EC_{50} 值 0.144 4 ~ 8.641 3μg/mL，对苯醚甲环唑的 EC_{50} 值 0.071 4 ~ 23.464 6μg/mL。通过最小抑制浓度法，以 MIC 值 5μg/mL 为标准，划分黄瓜多主棒孢对 DMIs 杀菌剂抗性和敏感性，发现多主棒孢对咪鲜胺、戊唑醇和苯醚甲环唑已发现产生不同程度抗性的菌株。因此，认为 DMIs 类杀菌剂对黄瓜多主棒孢菌具有较强的抑制作用，不同菌株对同一种药剂的敏感性存在着一定的差异，已检测到对其有不同程度抗药性菌株的存在。因此，生产上应合理用药，控制药剂使用浓度，通过多种杀菌剂轮换和混配处理以保证对该病害的可持续控制。

* 作者简介：高苇，女，助理研究员，主要从事植物病原真菌的检测和防治方面的研究；E-mail：gaowei5277@163.com

桃褐腐病菌中 Mona 遗传元件引起 DMI 杀菌剂抗性的功能研究及检测方法*

陈淑宁[1**]，袁楠楠[1]，罗朝喜[1,2***]
（1. 华中农业大学植物科技学院，武汉 430070；
2. 湖北省作物病害监测和安全控制重点实验室，武汉 430070）

摘要：在桃褐腐病菌（*Monilinia* spp.）中，DMI 杀菌剂抗药性产生的机理已经被证明可由靶标基因 *MfCYP*51 的过量表达所造成。而 *MfCYP*51 的过量表达可能和其启动子区域的一个 65bp 的 Mona 片段的插入相关。本实验中，构建了 Mona 驱动报告基因 *NPT*2 的表达载体 pMona－NPT，并转化野生型桃褐腐菌株获得 pMona－NPT 转化子。通过检测 NPT2 基因功能的启动，验证了 Mona 的启动子活性。

为了进一步确定 Mona 的启动子活性区域，构建了 6 个依次删减 Mona 启动子区域的 *NPT*2 报告基因载体，转化野生型桃褐腐菌株，检测不同长度启动子活性，最终将启动子活性区域确定到其中的 20bp 片段。

为了进一步证明 Mona 的插入和 DMI 抗药性的产生有直接关系，本研究还对 DMI 抗药性菌株 Bmpc7 进行了 Mona 敲除实验，同时在 DMI 敏感菌株 HG3 的 *MfCYP*51 基因上游人工插入了 Mona 片段。获得的敲除子在 EC_{50} 数值和 *MfCYP*51 基因表达量上都低于亲本菌株。而插入子的 EC_{50} 和 *MfCYP*51 的表达量都高于亲本菌株。这些结果表明 Mona 元件通过上调其下游的 *MfCYP*51 基因的表达量，引起了 DMI 抗药性的产生。Mona 敲除子和插入子都未表现出适合度下降的现象，具体而言，转化子和亲本表现出相同的菌丝生长速率，产孢量，以及在果实上的侵染力，说明 Mona 元件不影响桃褐腐菌株基本的生命活性。

同时，为了在田间快速检测桃褐腐 DMI 抗药性菌株，本研究借助环介导等温扩增技术（LAMP），建立了一种灵敏、特异、快速的分子生物学检测方法。设计了 4 条针对 Mona 及其上下游片段 6 个位点的特异性引物，建立了一套从采样到检测仅需不到 2h 的快速检测技术。结果显示，用建立的 LAMP 方法对抗性菌株扩增产物的电泳呈特征性梯状条带，而敏感菌株则无扩增产物。建立的 LAMP 方法对 DNA 模板的最低检出量为 10fg/μL，而 PCR 检测方法的最低检出量为 10^3fg/μL。对 42 份样本的检测结果显示，抗性菌株的检出率为 100%。本试验为田间快速检测 DMI 抗性桃褐腐菌提供了新的方法。

关键词：桃褐腐；抗药性；DMI 杀菌剂；LAMP

---

* 基金项目：国家自然科学基金（31371896）；公益性行业（农业）科研专项（201303025）资助

** 第一作者：陈淑宁，女，博士研究生

*** 通讯作者：罗朝喜，教授，主要从事病原真菌抗药性分子机理及稻曲病与水稻互作研究；E-mail：cxluo@mail. hazu. edu. cn

桃褐腐病菌 *Monilia mumecola* 种群对常用杀菌剂的敏感性及遗传多样性分析*

都胜芳[1**]，罗朝喜[1,2***]

（1. 华中农业大学植物科技学院，武汉 430070；2. 湖北省作物病害监测和安全控制重点实验室，武汉 430070）

摘要：本研究以93株桃褐腐病菌（*Monilia mumecola*）为供试菌株，将 *M. mumecola* 菌群对常用杀菌剂的敏感性及遗传多样性进行了分析，主要结果如下：

采用菌丝生长速率法测定了93个 *M. mumecola* 菌株对常用杀菌剂的敏感性。结果显示，*M. mumecola* 菌株对多菌灵、戊唑醇、嘧菌酯、啶酰菌胺4种杀菌剂的平均 EC_{50} 值分别为0. 103μg/mL、0. 034μg/mL、0. 325μg/mL、0. 419μg/mL，且敏感性分布均呈连续的单峰曲线，接近正态分布，没有出现抗药性菌株。比较 *M. mumecola* 菌株对同一杀菌剂的 EC_{50}，结果显示不同地区菌株的平均 EC_{50} 值均不存在显著差异，表明不同采集地的 *M. mumecola* 菌株对相应杀菌剂的敏感性没有差异。

采用SNP分子标记对93株 *M. mumecola* 菌株进行遗传多样性分析。结果显示，重庆与襄樊菌株之间遗传差异性较小，武汉及鄂州菌株间几乎没有差异，但重庆、襄樊与武汉、鄂州菌株间的差异十分显著，表明种群的差异性受区域的影响。同一地区采自杏树与桃树上的菌株几乎完全聚在一起，表明同一地区不同寄主来源的菌株间遗传差异不明显，预示着 *M. mumecola* 种群在相同地方可以侵染不同的寄主植物。

关键词：桃褐腐病；杀菌剂；敏感性测定；SNP；遗传多样性

---

* 基金项目：公益性行业（农业）科研专项（201303025）

** 第一作者：都胜芳，女，硕士研究生

*** 通讯作者：罗朝喜，教授，主要从事病原真菌抗药性分子机理及稻曲病与水稻互作研究；E-mail：cxluo@mail. hazu. edu. cn

辣椒疫霉胞外囊泡形态特点及蛋白组分析*

方　媛**，王治文，彭　钦，刘西莉***
（中国农业大学植物病理学系，北京　100193）

摘要：辣椒疫霉（*Phytophthora capsici*）是一类重要的病原卵菌，其寄主范围广泛，可侵染70多种植物，给农作物生产造成巨大危害。目前针对卵菌病害防治的高效化学防治药剂种类较少，主要以苯甲酰胺类、甲氧基丙烯酸酯类和羧酸酰胺类等内吸性杀菌剂为主。尽管这些内吸性杀菌剂防效显著，但由于其作用位点单一，并在植物病害防治中被广泛频繁使用，已经造成病原菌对杀菌剂敏感性的下降，抗药性问题日趋严重，给农业生产带来了潜在的巨大风险。因此，发掘新的药剂分子靶标对于新型杀菌剂的创制具有重要的意义。

胞外囊泡是一类包裹着细胞质基质及多种生物大分子的具有膜结构的囊泡，它们由细胞分泌到胞外空间，并在细胞间的信息交流、物质传递及致病过程中发挥着重要作用。胞外囊泡可以对包括蛋白质、糖类、脂质等多种生物大分子及mRNA，miRNA等多种小RNA物质进行运输。推测辣椒疫霉胞外囊泡在病原菌致病过程、病原菌与植物互作中可能发挥着重要的作用，但目前尚未见有关辣椒疫霉胞外囊泡的形态特征及内含物质的相关研究报道，明确辣椒疫霉胞外囊泡内致病相关的重要蛋白质及小RNA种类和功能，将有助于揭示植物病原卵菌与植物互作机制，并为新型药剂靶标设计提供新的思路。

本研究通过超速离心技术对辣椒疫霉侵染叶片过程中的胞外囊泡进行了提取，并通过扫描电镜对胞外囊泡的形态及粒径分布进行了观察和统计。从辣椒疫霉中提取的胞外囊泡形态近球形，大小均一，粒径在40～110nm，其中90%以上的胞外囊泡大小在60～90nm。通过LC－MS/MS对胞外囊泡的蛋白组进行了分析，并根据GO注释分析结果，发现辣椒疫霉的胞外囊泡中包裹的大分子蛋白参与了多个生物反应过程，包括应激反应、翻译、磷酸代谢、氧化还原、糖代谢、蛋白质代谢、核酸代谢、信号转导和物质运输等，其中在这些已注释的蛋白中还包括2个效应子及2个激发子蛋白。

关键词：辣椒疫霉；胞外囊泡；形态特征；蛋白组；效应子蛋白

---

* 基金项目：北京市自然科学基金（编号：6162018）

** 第一作者：方媛，女，在读博士研究生；E-mail：fangyuan7852@163.com

*** 通讯作者：刘西莉，女，教授，主要从事杀菌剂药理学及病原菌抗药性研究；E-mail：seedling@cau.edu.cn

新型杀菌剂 R031－1 的抑菌谱测定及其对辣椒疫霉不同发育阶段的影响*

林　东**，黄中乔，刘　莹，薛昭霖，刘西莉***
（中国农业大学植物病理学系，北京　100193）

摘要：R031－1 是由北京迪尔乐农业高新技术研发中心创制的一种结构新颖的化合物，目前已申请发明专利（CN201510926531.8），其作用机制尚不明确。本研究测定了 R031－1 的抑菌谱，明确了其对辣椒疫霉不同发育阶段的影响，为该化合物的商品化登记和作用机制的研究提供了参考。

本研究采用微孔板法及菌丝生长速率法测定了 R031－1 对 8 种植物病原卵菌和 6 种植物病原真菌的抑菌活性。结果表明，R031－1 对供试的荔枝霜疫霉（*Peronophythora litchi*）、致病疫霉（*Phytophthora infestans*）、辣椒疫霉（*Phytophthora capsici*）、大豆疫霉（*Phytophthora sojae*）、烟草疫霉（*Phytophthora nicotianae*）和终极腐霉（*Pythium ultimum*）具有较好的抑菌活性（EC_{50}为 0.023～0.59μg/mL），而对瓜果腐霉（*Pythium aphanidermatum*）、德里腐霉（*Pythium deliense*）及供试的 6 种植物病原真菌，包括桃褐腐病菌（*Monilinia fructicola*）、稻瘟病菌（*Magnaporthe oryzae*）、灰霉病菌（*Botrytis cinerea*）、水稻纹枯病菌（*Rhizoctonia solani*）、辣椒炭疽病菌（*Colletotrichum truncatum*）和禾谷镰刀病菌（*Fusarium graminearum*）的菌丝生长无明显抑制作用。采用活体叶盘法测定了 R031－1 对黄瓜霜霉病菌（*Pseudoperonospora cubensis*）的抑菌活性。结果表明，R031－1 对黄瓜霜霉病菌的 EC_{50}为 0.068μg/mL，对照药剂烯酰吗啉的 EC_{50}为 0.158μg/mL。

研究了 R031－1 对两株辣椒疫霉不同发育阶段的影响。结果表明，R031－1 对供试辣椒疫霉菌丝扩展（EC_{50}为 0.17～0.18μg/mL）和孢子囊产生（EC_{50}为 0.003 6～0.006 4μg/mL）均具有显著的抑制活性，对孢子囊和休止孢萌发的抑制作用较弱（EC_{50}为 9.8～21μg/mL），对游动孢子释放无明显影响。

上述研究结果为 R031－1 的商品化应用以及田间相关靶标病害的科学防治提供了理论依据，同时也为进一步开展该新型化合物的作用机制研究提供了参考。

* 基金项目：公益性行业（农业）科研专项（编号：201303023）
** 第一作者：林东，男，在读博士研究生；E-mail：lindong2012@cau.edu.cn
*** 通讯作者：刘西莉，女，教授，主要从事杀菌剂药理学及病原菌抗药性研究；E-mail：seedling@cau.edu.cn

辣椒疫霉对氟噻唑吡乙酮的抗性分子机制[*]

苗建强[**]，迟源东，董　雪，刘西莉[***]
（中国农业大学植物病理学系，北京　100193）

摘要：氟噻唑吡乙酮是美国杜邦公司 2007 年研发的具有全新作用机制的卵菌抑制剂，于 2015 年 12 月在我国获得登记，用于辣椒疫霉、致病疫霉和黄瓜霜霉等植物病原卵菌的防治。杜邦公司前期通过亲和色谱等研究表明，其靶标蛋白为氧化固醇结合相关蛋白（oxysterol binding – related protein，ORP）。ORPs 家族是真核生物中一个保守的大家族，它广泛存在于真菌、植物以及动物的细胞质中，至今为止，关于 ORPs 家族基因的报道主要集中在酵母和哺乳动物上，国际上有关植物病原菌的 ORPs 蛋白研究鲜有报道。本研究前期通过药剂驯化的方法，以两株敏感野生型菌株 LP3 及 HNJZ10 为亲本，共筛选获得了 12 株抗性稳定且抗性倍数均大于 300 倍的抗药性突变体，研究了突变体的生存适合度并结合辣椒疫霉的特性，初步推测辣椒疫霉对氟噻唑吡乙酮存在中等抗性风险。本研究拟以前期获得的抗药性突变体为材料，分析并验证辣椒疫霉对氟噻唑吡乙酮的抗性分子机制，为未来田间抗性监测及反抗性药剂的设计提供科学依据。

结合 JGI 已公布的辣椒疫霉基因组及美国杜邦专利（WO2013009971）中的靶标蛋白序列信息，确定氟噻唑吡乙酮的靶标蛋白 ID 为 564296（PHYCAscaffold_ 14：545241 – 548188），由于目前该蛋白在辣椒疫霉等植物病原卵菌中的具体功能尚属未知，因此我们将其命名为 PcORP1，该蛋白编码基因全长 2 948bp，包含一个 74bp 的内含子，共编码 957 个氨基酸。通过聚类分析人类、植物、真菌及卵菌中的 ORPs，结果显示植物病原卵菌中的 ORPs 与其他生物体具有较低的同源性，分析这可能是氟噻唑吡乙酮主要为植物病原卵菌特异性抑制剂的原因。

由于前期生存适合度研究结果显示 LP3 系列突变体具有优异的田间生存适合度，因此本研究主要以 LP3 系列突变体为材料，进行了抗药性分子机制的研究。首先，利用 Q – PCR 分析了药剂处理前后病原菌中靶标蛋白编码基因的表达量，结果显示 LP3 – F 及 LP3 – H 与亲本菌株 LP3 没有显著性差异，而其他 4 株突变体的表达量均显著低于亲本菌株，各菌株用药前后，*PcORP*1 的表达量均没有显著性的变化，因此，认为抗药性机制不是由靶标蛋白的过量表达所致。进一步扩增亲本菌株及其突变体的靶标蛋白基因全长，进行测序比对分析。结果表明，突变体 769 位密码子 GGG 杂合突变为 TGG，导致了 769 位甘氨酸突变为色氨酸。通过遗传转化的方法，将具有 769 位密码子突变位点的 *PcORP*1 转化到野生敏感菌株 BYA5 中，共获得了 5 株转化子，且对氟噻唑吡乙酮均表现出抗性，证实了靶标蛋白点突变 G769W 导致了辣椒疫霉对氟噻唑吡乙酮的高水平抗性。

* 基金项目：国家自然科学基金（31471791）
** 第一作者：苗建强，在读博士研究生；E-mail：mjq2014@ cau. edu. cn
*** 通讯作者：刘西莉，女，教授，主要从事杀菌剂药理学及病原物抗药性研究；E-mail：seedling@ cau. edu. cn

噻唑菌胺对辣椒疫霉不同发育阶段的影响*

彭　钦**，方　媛，王治文，刘西莉***
（中国农业大学植物病理学系，北京　100193）

摘要： 噻唑菌胺（ethaboxam）是韩国 LG 生命科学公司（原 LG 化学有限公司）开发并于 1998 年首先在韩国获准登记的噻唑酰胺类杀菌剂，作用机制尚不明确。其防治对象主要是致病疫霉（*Phytophthora infestans*）、葡萄霜霉（*Plasmopara viticola*）、辣椒疫霉（*Phytophthora capsici*）和古巴假霜霉（*Pseudoperonospora cubensis*）等植物病原卵菌。本研究测定了噻唑菌胺对辣椒疫霉不同发育阶段的影响，建立了辣椒疫霉对噻唑菌胺的敏感基线，为生产中选择合适的施药时期及进行抗性监测提供了参考。

结果表明，噻唑菌胺对两株供试辣椒疫霉的菌丝扩展（EC_{50}为 0.287 ~ 0.337μg/mL）、孢子囊产生（EC_{50}为 0.014 ~ 0.018μg/mL）和休止孢萌发（EC_{50}为 0.045 ~ 0.092μg/mL）均具有良好的抑菌活性，而对游动孢子释放和游动孢子游动没有明显的抑制作用。

采用菌丝生长速率法，测定了采自全国不同省区的 118 株辣椒疫霉对噻唑菌胺的敏感性。结果表明，供试辣椒疫霉对噻唑菌胺的 EC_{50}介于 0.015 ~ 1.353μg/mL，平均 EC_{50}值为（0.367 ±0.219）μg/mL。研究显示，该方法测得的辣椒疫霉对噻唑菌胺的敏感性频率均呈连续分布，未出现敏感性下降的亚群体。因此，可将其作为辣椒疫霉对噻唑菌胺的敏感性基线，该结果为田间抗药性监测提供了参考和依据。

---

* 基金项目：公益性行业（农业）科研专项（编号：201303023）

** 第一作者：彭钦，在读博士研究生；E-mail：pengqin1991@126.com

*** 通讯作者：刘西莉，女，教授，主要从事杀菌剂药理学及病原物抗药性研究；E-mail：seedling@cau.edu.cn

氟吡菌胺对辣椒疫霉不同发育阶段的影响*

薛昭霖[1,2]**，吴　杰[1]，林　东[1]，刘西莉***
（1. 中国农业大学植物病理学系，北京　100193；
2. 河南农业大学植物保护学院，郑州　450002）

摘要：辣椒疫霉（*Phytophthora capsici*）引起的植物病害是世界各地以及我国蔬菜作物生产上的毁灭性病害，造成了严重的经济损失。因为该病原菌寄主范围广，在土壤中可长期存活，使得疫病的防治愈发困难。化学防治是目前防治辣椒疫霉最为重要和有效的措施，在病害防治中发挥着不可取代的作用。氟吡菌胺（fluopicolide）是德国拜耳公司研发的一种新型吡啶酰胺类内吸性杀菌剂，在生产应用中与霜霉威复配（商品名：银法利）用于植物卵菌病害的防治。本研究明确了氟吡菌胺对辣椒疫霉不同发育阶段的影响，为该药剂的科学使用提供了理论依据。

本研究测定了氟吡菌胺对辣椒疫霉的菌丝生长、孢子囊形成、游动孢子释放和游动以及休止孢萌发等不同发育阶段的抑菌作用。结果表明，氟吡菌胺对辣椒疫霉的菌丝生长、孢子囊形成、游动孢子释放以及休止孢萌发等各个阶段均具有良好的抑制活性，其 EC_{50} 在 0.09～1.58μg/mL，EC_{90} 在 0.40～4.27μg/mL。其中，对菌丝生长、孢子囊形成以及游动孢子释放阶段的 EC_{50} 均小于 1.00μg/mL，而对休止孢萌发阶段的 EC_{50} 大于 1.00μg/mL。对照药剂烯酰吗啉对供试辣椒疫霉的菌丝生长、孢子囊形成以及休止孢萌发阶段的 EC_{50} 在 0.03～0.66μg/mL，EC_{90} 在 0.14～11.63μg/mL，而在游动孢子释放阶段，其 EC_{50} 和 EC_{90} 均大于 100μg/mL。另外，氟吡菌胺对辣椒疫霉游动孢子游动阶段的抑制活性高于对照药剂烯酰吗啉和嘧菌酯，研究发现在 0.01μg/mL 氟吡菌胺处理下，辣椒疫霉游动孢子游动缓慢；在 0.1、1、10μg/mL 氟吡菌胺处理下，游动孢子休止甚至裂解，且随着浓度升高，游动孢子裂解的数量增多。推测该裂解现象可能与氟吡菌胺的疑似作用靶标类血影蛋白被抑制有关，已有文献报道类血影蛋白可能在维持细胞膜稳定性上发挥着重要的作用，具体机制需要进一步进行探究。

---

* 基金项目：公益性行业（农业）科研专项（编号：201303023）

** 第一作者：薛昭霖，女，在读硕士生；E-mail：xuezhaolin1215@163.com；吴杰，男，在读博士生；E-mail：wujiecarlos@163.com

*** 通讯作者：刘西莉，女，教授，主要从事杀菌剂药理学及病原菌抗药性研究；E-mail：seedling@cau.edu.cn

辣椒平头炭疽病菌（*Colletotrichum truncatum*）对5种常用DMIs杀菌剂的敏感性分化及其对咪鲜胺、氟环唑和苯醚甲环唑的抗性分析*

张　灿**，王为镇，刁永朝，刘　利，刘西莉***
（中国农业大学植物病理学系，北京　100193）

摘要： 炭疽病是辣椒生产上的重要病害，在我国各辣椒种植区均有发生。平头炭疽病菌（*Colletotrichum truncatum*）是导致我国很多辣椒产区炭疽病发生和流行的优势病原菌之一，给辣椒生产带来严重损失。此外，平头炭疽病菌还可以引起人类眼部疾病如真菌性角膜炎和眼内炎等。目前在农业生产和临床实践中，DMIs杀菌剂被广泛应用于炭疽病菌的防治，然而有关平头炭疽病菌对不同DMIs杀菌剂的敏感性以及抗性机制尚未见有报道。

本研究选用咪鲜胺、氟环唑、苯醚甲环唑、戊唑醇和腈菌唑5种生产中常用的DMIs药剂作为供试药剂，以平头炭疽病菌为靶标菌，系统检测了我国多省市的平头炭疽病菌对5种供试药剂的敏感性。结果表明，平头炭疽病菌对咪鲜胺、氟环唑和苯醚甲环唑均表现为敏感，EC_{50}介于0.05～1.96μg/mL；而其对戊唑醇和腈菌唑的敏感性较低，EC_{50}分别达到40和100μg/mL以上，表明平头炭疽病菌对这5种常用的唑类药剂存在显著的敏感性分化。分别建立了平头炭疽病菌对咪鲜胺、氟环唑和苯醚甲环唑的敏感基线，112株平头炭疽菌株对3种供试药剂的敏感性频率分布均呈单峰曲线，表明并未在田间检测到敏感性下降的平头炭疽亚群体。

通过药剂驯化的方法筛选获得了11株平头炭疽病菌对咪鲜胺、氟环唑和苯醚甲环唑的抗性突变体，并评估了平头炭疽对3种供试药剂的抗性风险，同时对其抗性分子机制进行了研究。综合分析了突变体的生存适合度及突变频率，明确了平头炭疽病菌对唑类药剂具有低等抗性风险。进一步检测了3种供试药剂与田间及临床中用于防治炭疽病菌的其他药剂之间的交互抗药性，结果表明，供试3种药剂与嘧菌酯、多菌灵和代森锰锌之间无交互抗药性，而与医用唑类药剂氟康唑和酮康唑存在正交互抗药性。抗性分子机制研究表明，获得的11株突变体对3种供试药剂的抗性不是由CtCYP51蛋白上的氨基酸点突变造成，而是与*CtCYP51*转录水平上的过量表达有关。

综合以上结果，分析认为我国大部分地区的平头炭疽病菌对DMIs杀菌剂尚未产生抗药性，且其对该类药剂的抗性风险较低，DMIs杀菌剂仍可以作为生产中防治辣椒炭疽病的重要药剂。然而平头炭疽病菌对不同DMIs杀菌剂的敏感性存在着显著的分化，因此在田间生产中，如果某地区的优势种群为平头炭疽，则应避免使用戊唑醇和腈菌唑这两种药剂进行病

* 基金项目：公益性行业（农业）科研专项（编号：201303023）
** 第一作者：张灿，女，在读博士研究生；E-mail：czhang@cau.edu.cn
*** 通讯作者：刘西莉，女，教授，主要从事杀菌剂药理学及病原物抗药性研究；E-mail：seedling@cau.edu.cn

害防治。此外，供试的3种DMIs杀菌剂与医用药剂氟康唑和酮康唑存在正交互抗药性，表明农用DMIs杀菌剂的大量使用可能造成田间平头炭疽抗性突变体的产生，并进一步导致DMIs杀菌剂在生产实际和临床实践中的治疗效果下降。因此在田间必须科学合理使用DMIs杀菌剂，如将其与不同作用机制的药剂复配用来防治炭疽病等，以延缓抗药性的产生和发展。

湖北省番茄和草莓保护地灰霉病菌的抗药性研究*

范　飞[1**]，李　娜[1]，李国庆[1,2]，罗朝喜[1,2***]
（1. 华中农业大学植物科技学院，武汉　430070；2. 湖北省作物病害监测和安全控制重点实验室，武汉　430070）

摘要：近些年，湖北省番茄和草莓的种植规模不断扩大，而许多病害也随之日益严重。灰霉病作为番茄和草莓上的常见病害之一，给其生产带来了严重的经济损失。生产上防治灰霉病常以化学防治为主，但由于杀菌剂的长期大量使用，抗药性问题逐渐突出，已成为病害防治中新的挑战。在2012年和2013年的初夏，从湖北省8个地级市的番茄保护地和10个地级市的草莓保护地中分别采集分离了221个和240个灰霉单孢菌株，并检测了其对多菌灵、乙霉威、啶酰菌胺、咯菌腈和嘧菌环胺的敏感性。结果显示，多菌灵、乙霉威和嘧菌环胺的抗性在湖北省番茄和草莓保护地分布广泛，但只在草莓保护地检测到了啶酰菌胺的抗性。在番茄和草莓保护地，均未发现咯菌腈的抗性。此次关于田间灰霉菌株对啶酰菌胺抗性的报道在国内尚属首次。将从番茄保护地和草莓保护地各选取的3个多菌灵抗性菌株，6个多菌灵、乙霉威和嘧菌环胺的三抗菌株，以及从草莓保护地选取的两个啶酰菌胺抗性菌株在不含药的培养基上继代培养10代后，这些菌株仍然保持了对杀菌剂的较高抗性，表明这些抗性是稳定的。此外，还测定了以上菌株的环境适合度参数，包括菌丝生长速率、对NaCl的渗透压敏感性、致病力和离体、活体产孢量。结果表明，从番茄保护地和草莓保护地选取的多菌灵抗性菌株和敏感菌株在各参数之间无显著差异。从番茄保护地选取的多菌灵、乙霉威和嘧菌环胺的三抗菌株的菌丝生长速率和致病力显著小于敏感菌株，但从草莓保护地选取的三抗菌株则不存在此类现象，这可能与不同的寄主有关。从草莓保护地选取的啶酰菌胺抗性菌株的致病力也显著小于敏感菌株，表明啶酰菌胺抗性菌株侵染寄主的能力有所下降。为了研究抗性菌株的抗性机理，克隆并测序了23个多菌灵抗性菌株和10个多菌灵和乙霉威抗性菌株的β微管蛋白基因片段，发现23个多菌灵抗性菌株均含有E198V或E198A突变，而10个抗多菌灵和乙霉威的抗性菌株均含有E198K突变。此外，克隆并测序了2个啶酰菌胺抗性菌株的琥珀酸脱氢酶B亚基（SdhB）的基因片段，发现2个啶酰菌胺抗性菌株均含有H172R突变。此次研究结果对制定湖北省番茄和草莓保护地的灰霉病防治策略具有重要的指导意义。

关键词：灰霉病；番茄保护地；草莓保护地；杀菌剂；抗药性

---

* 基金项目：公益性行业（农业）科研专项经费（201303025）
** 第一作者：范飞，男，博士研究生
*** 通讯作者：罗朝喜，教授，主要从事病原真菌抗药性分子机理及稻曲病与水稻互作研究；E-mail：cxluo@mail. hazu. edu. cn

天津地区黄瓜霜霉病菌对多种杀菌剂的抗药性检测

王　勇，张春祥，高　苇，王万立
（天津市植物保护研究所，天津　300381）

摘要：由古巴假霜霉菌（*Pseudoperonospora cubensis*）引起的黄瓜霜霉病是一种严重影响黄瓜生产的病害，当条件适合病情发展时，从发现中心病株开始一周左右，便可造成全棚黄瓜发病。该病原菌具有潜育期短、传播速度快、对寄主破坏性强等特点，生产上最有效的控制办法仍以化学防治为主。目前防治黄瓜霜霉病的药剂种类虽然较多，但因长期连续使用，病菌对药剂均已产生不同程度的抗药性，其防治效果均不够理想。为了摸清天津地区黄瓜霜霉病田间抗药性发生情况，指导生产上药剂的科学使用。我们针对目前生产中常用药剂：吗啉类杀菌剂烯酰吗啉、甲氧基丙烯酸酯类杀菌剂嘧菌酯、甲氧基氨基丙酸酯类杀菌剂甲霜灵、二硫代氨基甲酸类杀菌剂代森锰锌、氨基甲酸酯类杀菌剂霜霉威盐酸盐、四氯间苯二腈类杀菌剂百菌清。自 2016 年 4 月至 5 月期间，于天津宝坻、宁河、西青、静海、武清等蔬菜主产区采集黄瓜霜霉病样，采用叶盘漂浮法进行了天津地区田间黄瓜霜霉病菌对多种杀菌剂的抗药性检测。由试验结果显示，发现天津地区采集古巴假霜霉菌对吗啉类杀菌剂烯酰吗啉、甲氧基氨基丙酸酯类杀菌剂甲霜灵、二硫代氨基甲酸类杀菌剂代森锰锌的药剂敏感性较低，大部分菌株的 EC_{50}值为 2. 569 4 ~ 145. 908 7μg/mL；对氨基甲酸酯类杀菌剂霜霉威盐酸盐、四氯间苯二腈类杀菌剂百菌清的药剂敏感性稍高，EC_{50} 值为 0. 310 3 ~ 8. 990 4μg/mL；对甲氧基丙烯酸酯类杀菌剂嘧菌酯的敏感性较高，EC_{50}值为 0. 016 8 ~ 0. 452 6μg/mL。

将本试验中采集古巴假霜霉菌进行抗药水平统计，其中嘧菌酯的敏感基线采用韩秀英等测得的 2.317×10^{-4}μg/mL，甲霜灵的敏感基线采用王文桥等测得的 0. 047μg/mL。经统计分析，天津地区黄瓜霜霉病菌对嘧菌酯存在抗药性，抗性在 72. 51 ~ 1 953. 39倍，其中以武清、宁河和宝坻地区抗药性较高，静海和西青区次之。同时，天津地区古巴假霜霉菌对甲霜灵的抗药性也普遍较高，抗性在 54. 67 ~ 665. 51 倍，除静海抗药性较低外，宁河、武清、宝坻、西青等地区均存在较高的抗性。

据此，黄瓜霜霉病菌抗药性已成为天津地区黄瓜生产上的一个重要问题，开展黄瓜霜霉病菌对化学药剂的抗药性治理势在必行。建议加强田间抗药性监测，科学使用农药，减少甚至停用已产生抗药性的同类型药剂，寻求新的替代类型，并合理进行杀菌剂的轮换和混配。

辣椒疫霉对氟吡菌胺的室内抗性风险评估*

吴　杰[1,2]**，薛昭霖[1]，林　东[1]，刘西莉[1]***

（1. 中国农业大学植物病理学系，北京　100193；

2. 河北省农林科学院植物保护研究所，保定　071000）

摘要：辣椒疫霉（*Phytophthora capsici* Leonian）是具有重大危害性、全球广泛分布的植物病原卵菌，其寄主范围广泛，在植物各个生长阶段均可侵染致病，给农业生产造成巨大损失。目前生产上化学防治仍然是防治由辣椒疫霉所引起的植物病害的主要措施。氟吡菌胺（fluopicolide）是拜耳作物科学公司开发的一种酰胺类杀菌剂，对植物病原卵菌表现出优异的生物活性，本研究室通过田间抗药性检测，未获得辣椒疫霉对氟吡菌胺的田间抗性菌株，进而采用室内药剂驯化的方法筛选获得了抗氟吡菌胺的辣椒疫霉突变体，并开展了辣椒疫霉对氟吡菌胺的抗性风险评估，为指导药剂科学使用、避免和延缓抗药性发生发展提供理论依据。

本研究以 6 株辣椒疫霉敏感菌株为亲本，进行了突变体筛选，共获得 44 株抗氟吡菌胺的突变体。选取 23 株抗药性突变体在无药培养基平板上连续转接培养 10 代后，发现其中 13 株突变体的抗性水平有所上升，10 株略有下降，综合分析表明，突变体对氟吡菌胺的抗性可稳定遗传。

比较敏感亲本菌株和抗性突变的生物学性状，结果表明，以 BYA5、Pc1723、JA8、12－11 为亲本筛选获得的突变体离体生长速率明显大于亲本或与亲本菌株相当，孢子囊产生量与亲本相比无明显差异；以 A1、LP3 为亲本筛选获得的的突变体，其离体生长速率及孢子囊产生数量均明显低于亲本；所有突变体的活体致病力与亲本菌株相比无显著差异。

采用秩相关分析方法分别研究了氟吡菌胺与百菌清、氰霜唑、氟噻唑吡乙酮、霜脲氰、嘧菌酯、烯酰吗啉、氟啶胺、苯酰菌胺、甲霜灵和氟醚菌酰胺等 10 种药剂间的交互抗药性。结果显示，氟吡菌胺与氟醚菌酰胺之间存正交互抗药性，与其他 9 种供试药剂无交互抗药性。

综合上述研究表明，大部分抗氟吡菌胺的辣椒疫霉突变体具有较高的生存适合度，在药剂选择压下将有利于抗性群体的发展，表明辣椒疫霉对氟吡菌胺存在一定的抗药性风险；交互抗药性结果及药剂化学结构分析表明，氟吡菌胺与氟醚菌酰胺可能具有相同的作用机制，在生产中应将氟吡菌胺与其他无交互抗药性的药剂轮换或混合使用，以避免和延缓抗药性的产生。

关键词：辣椒疫霉；氟吡菌胺；抗性风险；交互抗药性

---

* 基金项目：公益性行业（农业）科研专项（编号：201303023）

** 第一作者：吴杰，在读博士研究生；E-mail：wujiecarlos@163.com

*** 通讯作者：刘西莉，教授，主要从事杀菌剂药理学及病原菌抗药性研究；E-mail：seedling@cau.edu.cn

抑制树木腐烂病菌金黄壳囊孢的杀菌剂筛选研究*

刘础荣[1]**，罗来鑫[1]，李志军[2]，李健强***

（1. 中国农业大学植物病理学系，种子病害检验与防控北京市重点实验室，北京 100193；
2. 塔里木大学植物科技学院，阿拉尔 843300）

摘要：金黄壳囊孢（*Cytospora chrysosperma*）是一种世界性分布、寄主范围广泛的病原菌，在我国主要分部于东北、西北、华北等地区。该菌可引起杨树、核桃等多种经济林树木腐烂病，损失严重。目前，树木腐烂病主要以机械刮除病斑部位并涂抹化学药剂进行防治，筛选安全、高效的药剂在生产中具有重要意义。本研究以分离自新疆阿克苏地区温宿县核桃树腐烂病病枝上的金黄壳囊孢（*C. chrysosperma*）为靶标菌，以抑制病菌的呼吸作用、信号转导、细胞骨架及运动蛋白、细胞膜甾醇合成和氨基酸及蛋白质合成的5种不同作用机制的9种杀菌剂（选择其原药，分别为98.6%吡唑醚菌酯、98.6%嘧菌酯、84.0%氰霜唑、99.5%咯菌腈、98.1%甲基硫菌灵、95.8%己唑醇、96.0%苯醚甲环唑、95.0%腐霉利和98.4%嘧菌环胺）作为供试药剂，采用农业行业标准NY/T1156.2—2006《农药室内生物测定试验测定准则杀菌剂第2部分：抑制病原真菌菌丝生长试验平皿法》，测定供试药剂对靶标菌的抑制作用，比较分析不同药剂的毒力差异。结果表明，苯醚甲环唑和己唑醇对靶标菌具有极高的抑菌活性，其EC_{50}值分别为0.003 9、0.008 3μg/mL；吡唑醚菌酯、咯菌腈、嘧菌酯、腐霉利、甲基硫菌灵和嘧菌环胺亦具有较强的抑菌活性，其EC_{50}值分别为0.014 8、0.017 2、0.056 0、0.293 2、2.782 5和2.791 4μg/mL；而氰霜唑抑菌效果较差其EC_{50}值为11.670 7 μg/mL。本研究使用苯醚甲环唑、吡唑醚菌酯等原药测定获得的抑菌活性明显强高于使用25%苯醚甲环唑微乳剂和25%吡唑醚菌酯乳油等制剂的结果，更好的排除了制剂中非活性成分对抑菌作用测定的干扰，方法规范科学；首次测定了咯菌腈对靶标菌的抑菌效果，为后续田间试验和药剂研发提供了基础支持。

关键词：金黄壳囊孢（*C. chrysosperma*）；杀菌剂；毒力测定

---

* 基金项目：国家科技支撑计划专题（2014BAC14B04－3）、自治区科技支疆项目（201491150）

** 作者简介：刘础荣，硕士研究生；E-mail：x8281288@163.com

*** 通讯作者：李健强，教授，主要研究方向为种子病理与杀菌剂药理学；E-mail：lijq231@cau.edu.cn

双苯菌胺对灰葡萄孢菌代谢组的影响分析

代　探*，胡志宏，刘盼晴，梁　莉，刘鹏飞**
（中国农业大学植物保护学院植物病理学系，北京　100193）

摘要：双苯菌胺（SYP－14288）是我国沈阳化工研究院自主研发的一个新型杀菌剂，具有高效、广谱的抑菌活性。其结构与氟啶胺（fluazinam）类似，有报道称氟啶胺作用机制与氧化磷酸化解偶联、ATP 合成或呼吸电子传递链复合物Ⅱ的作用相关，而双苯菌胺作用机制目前尚不清楚。本研究基于 GC－MS 分析平台，比对分析了双苯菌胺或氟啶胺 EC_{50}浓度下处理与未用药处理的 *B. cinerea* 菌丝代谢组的差异，根据标准物质的保留时间和质谱特征对色谱峰进行定性，并结合 NIST 谱库检索结果对其中匹配度大于 80% 的代谢物进行了指认。与对照相比双苯菌胺引起了 *B. cinerea* 的 36 种代谢物上调，包括 13 种有机酸类代谢物如 2－吡咯烷酮－5－羧酸、半乳糖醛酸、4－氨基丁酸、2－吡咯烷酮－5－羧酸、粘酸、D－葡萄糖酸等，1 种氨基酸：L－天冬氨酸，4 种糖类代谢物：D－半乳糖、来苏糖、D－果糖、D－木糖，4 种醇类代谢物：异喹醇、肌醇、木糖醇、丁二醇，还有 14 个其他类代谢物发生了不同程度的上调；同时，有 11 种代谢物发生下调，包括 2 种醇类代谢物：1，4－二噁烷－2，3－二醇、半乳糖醇，2 种有机酸：丁酸和丙二酸类代谢物，1 种氨基酸：丙氨酸，1 种糖类代谢物：景天庚酮糖，以及 5 种其他类代谢物。双苯菌胺作用下灰葡萄孢菌发生变化的这些代谢物中，有 32 种上调的代谢物和 9 种下调的代谢物在氟啶胺的作用下亦发生了一致的变化，表明两个药剂有着相似的作用机制，对双苯菌胺与氟啶胺作用下灰葡萄孢代谢组数据的深入分析，可为双苯菌胺靶标代谢途径及作用机制的解析提供参考。

关键词：灰葡萄孢菌；代谢组；气相色谱－质谱；双苯菌胺；作用机制

---

* 第一作者：代探，硕士研究生，研究方向为植物病理学；E-mail：daitan@ cau. edu. cn
** 通讯作者：刘鹏飞，副教授，博士，研究方向为植物病理学；E-mail：pengfeiliu@ cau. edu. cn

13种杀菌剂对玉米大斑病菌和弯孢霉叶斑病菌的毒力测定*

甘　林[1,2]**，代玉立[1,2]，杨秀娟[1,2]，杜宜新[1,2]，阮宏椿[1,2]，石妞妞[1,2]，陈福如[1,2]***
（1. 福建省农业科学院植物保护研究所，福州　350013；
2. 福建省作物有害生物监测与治理重点实验室，福州　350003）

摘要： 玉米叶斑病是玉米生产上一类主要的病害。为了筛选出高效、低毒的杀菌剂，本文采用生长速率法和孢子萌发法比较了13种杀菌剂对玉米大斑病菌和弯孢霉叶斑病菌的毒力作用。结果表明，不同杀菌剂对2种病菌菌丝生长和孢子萌发的抑制活性之间存在显著差异。其中氟啶胺、苯醚甲环唑、烯唑醇、丙环唑、腈苯唑、咪鲜胺锰盐等对病菌菌丝生长具有较强的抑制效果，其 EC_{50} 值均低于1μg/mL，而代森锰锌的抑制作用较差，对2种病菌的 EC_{50} 值分别为5.374 5、6.465 0μg/mL；6种不同作用机理的杀菌剂中，氟啶胺和异菌脲对大斑病菌孢子萌发抑制效果较好，EC_{50} 值分别为0.314 7、1.515 6μg/mL，而丙环唑和吡唑醚菌酯则对弯孢霉叶斑病菌孢子萌发抑制作用较佳，EC_{50} 值分别为5.049 0、18.843 9μg/mL。本研究结果为有效控制玉米叶斑病提供了试验依据。

关键词： 玉米；大斑病菌；弯孢霉叶斑病菌；杀菌剂；毒力测定

* 基金项目：福建省属公益类科研院所专项（2014R1024－5），福建省农业科学院博士启动基金（2015BS－4）
** 作者简介：甘林，男，硕士，助理研究员，主要从事植物病害防治；E-mail：millergan@yeah.net
*** 通讯作者：陈福如，研究员；E-mail：chenfuruzb@163.com

福建省玉米小斑病菌对戊唑醇、吡唑醚菌酯和硝苯菌酯的敏感性*

杜宜新**，阮宏椿，石妞妞，甘　林，杨秀娟，陈福如***
（福建省农业科学院植物保护研究所，福州　350013）

摘要：为明确福建省玉米小斑病菌对戊唑醇、吡唑醚菌酯和硝苯菌酯的敏感性，采用菌丝生长速率法测定了戊唑醇、吡唑醚菌酯和硝苯菌酯对供试的214株玉米小斑病菌的抑制作用。结果表明供试菌株对戊唑醇、吡唑醚菌酯和硝苯菌酯的 EC_{50} 值分别为0.024 9～21.582 3μg/mL、0.032 1～0.724 9μg/mL和0.146 3～3.412 7μg/mL，敏感性频率分布显示福建省玉米小斑病菌对戊唑醇出现了敏感性下降的亚群体，供试菌株对吡唑醚菌酯和硝苯菌酯均未出现敏感性下降的亚群体。可分别将供试菌株对唑醚菌酯和硝苯菌酯 EC_{50} 平均值0.266 2μg/mL和1.340 6μg/mL作为其敏感基线用于田间菌株的抗药性检测；供试菌株对戊唑醇、吡唑醚菌酯和硝苯菌酯的敏感性相互之间均无显著相关性。

关键词：玉米小斑病菌；杀菌剂；抑制作用；敏感基线

---

* 基金项目：福建省种业工程项目（FJZZZY－1526）；福建省省属公益类科研院所专项（2014R1024－5）
** 作者简介：杜宜新，男，副研究员，主要从事植物真菌病害及其防治研究；E-mail：yixindu@ 163. com
*** 通讯作者：陈福如，男，研究员，主要从事植物真菌病害研究；E-mail：chenfuruzb@163. com

烟草黑胫病菌对烯酰吗啉的抗性机制研究*

牟文君**，胡利伟，张艳玲，尹启生，宋纪真
（中国烟草总公司郑州烟草研究院，郑州　450001）

摘要：烟草黑胫病（Tobacco Black Shank）是烟草生产上的重要土传病害，烯酰吗啉是用于该病害防治的羧酸酰胺类（CAA 类）杀菌剂，目前已有报道检测到对烯酰吗啉产生抗性的烟草黑胫病菌菌株。CAA 类杀菌剂可能作用于病原菌的纤维素合酶 CesA3，影响细胞壁纤维素的合成或者运输，并且在对 CAA 类杀菌剂产生抗性的不同靶标菌中，发现其纤维素合酶 CesA3 上均发生了点突变。

本研究采用室内药剂驯化法，获得了系列抗性突变体，其中，4 株为抗性倍数在 2 ~ 5 倍的低中抗突变体，4 株为抗性倍数大于 250 倍的高抗突变体。通过同源序列比对的方法，扩增获得了烟草黑胫病菌的 4 个纤维素合酶基因，其中，*CesA*1，*CesA*2，*CesA*3，*CesA*4 基因的 DNA 序列分别为 3 051 bp、3 084 bp、3 528 bp 和 3 135 bp，分别编码氨基酸 1 016个、977 个、1 142个和 1 019个。将敏感与抗性菌株的纤维素合酶基因进行序列比对，结果表明，高抗菌株仅在 *CesA*3 基因上发现点突变，其中，3 株菌株纤维素合酶 CesA3 检测到 V1109L 突变，1 株菌株的 CesA3 检测到 Q1077H 突变，说明烟草黑胫病菌同时存在两种突变位点；4 株低中抗菌株在 *CesA*1，*CesA*2，*CesA*3，*CesA*4 基因上均未发现点突变，说明其可能存在其他抗性机制。

* 基金项目：郑州烟草研究院院长科技发展基金项目（122015CA0190）

** 第一作者：牟文君，博士，工程师，主要从事烟草病害防治及病原菌抗药性研究；E-mail：muwenjun@126.com

抗腐霉利灰葡萄孢菌代谢组分析

陈　晨，王小雨，刘子淇，胡志宏，刘鹏飞
（中国农业大学植物保护学院，北京　100193）

摘要： 腐霉利是田间灰霉病害防治的重要农药品种，由于长期频繁用药，造成一些地区抗药性问题频发。根据已有研究报道，灰葡萄孢菌对腐霉利作用机制可能与病原菌体内的渗透信号传导及甘油累积有关，但是抗药性机制目前尚不清楚。比对分析田间抗感性菌株的代谢组特征，寻找抗性菌株特有代谢标志物，可为腐霉利抗性机制分析提供有益参考。

本研究中采用 GC/MS 代谢组分析平台，对采自福建、辽宁和上海等地的 3 株高抗菌株（≥10μg/mL 带药平板上可生长）、5 株中抗菌株（≥5μg/mL 且 <10μg/mL 带药平板上可生长）和 1 株敏感菌株（ >5μg/mL 带药平板上不可生长），分别在 EC_{50}（0. 2μg/mL）的腐霉利处理和未用药处理下，进行了菌丝代谢组的检测。

基于代谢组的主成分分析表明，采自不同地区的高抗菌株包括用药和未用药的处理均聚为一组，中抗菌株单独聚为一组，表明菌株的抗性水平与其代谢组有相关性。通过对腐霉利田间抗性和敏感灰霉菌代谢组的比对分析，寻找抗性菌株代谢特征标志物。与敏感菌株相比，各中抗菌株中葡萄糖醇、琥珀酸、丙酸等代谢物含量显著上调，各高抗菌株中丝氨酸、丁酸、草酸含量显著上调。这些代谢物可能与菌株田间抗性的产生密切相关，是腐霉利抗性菌株鉴定的潜在的生物标志物。进一步的数据分析将为代谢水平上揭示灰霉病菌对腐霉利的抗性机制作用提供参考。

基于灰葡萄孢代谢组的杀菌剂作用机制区组研究*

胡志宏**，刘鹏飞***，刘西莉
（中国农业大学植物保护学院，北京 100193）

摘要：近年来，代谢组学的发展为农药作用机制研究提供了强大的分析工具。为了建立杀菌剂作用机制快速分类识别方法，寻找不同机制杀菌剂对灰葡萄孢（*Botrytis cinerea*）作用的生物标志物，本文对 *B. cinerea* 代谢组及其在6类杀菌剂作用机制区组中的应用进行了系统研究。主要获得了以下几方面的研究结果：基于GC－MS分析平台建立了 *B. cinerea* 菌丝代谢组提取检测方法，代谢组提取溶剂为甲醇/水（v/v，80/20），提取剂量为每0.1g菌丝4mL溶剂，两步衍生化：30℃下甲氧基胺盐酸盐衍生化2h，37℃下BSTFA硅烷化6h。检测获得的 *B. cinerea* 代谢组，包含了245种代谢物，主要由氨基酸类、有机酸类、醇类、糖类等组成。其中，与NIST 2005匹配度大于80%的代谢物有56种。基于浓度为EC_{50}的杀菌剂作用下GC－MS检测到的 *B. cinerea* 代谢指纹，采用IBM SPSS Statistics 21软件对供试的14种杀菌剂进行了层次聚类分析，其中呼吸作用抑制剂嘧菌酯和醚菌酯聚为一类；蛋氨酸合成抑制剂嘧霉胺和嘧菌环胺聚为一类；信号传导抑制剂腐霉利、异菌脲和咯菌腈聚为一类；甾醇合成抑制剂苯醚甲环唑和四氟醚唑聚为一类；多作用位点抑制剂福美双和百菌清聚为一类；聚类结果与FRAC作用机制类别一致，表明通过代谢指纹的系统聚类可以实现对杀菌剂作用机制的区组。β微管蛋白合成抑制剂多菌灵和甲基硫菌灵处理的菌株代谢组存在显著差异，聚类的距离较远，可能与甲基硫菌灵未经寄主活化有关。本研究中氟啶胺与嘧菌酯和醚菌酯距离较远，表明虽然同为呼吸作用抑制剂，但解偶联剂与复合物Ⅲ抑制剂在代谢组上存在显著差异。分析了多种作用机制杀菌剂在EC_{50}浓度下对 *B. cinerea* 代谢组的影响，并获得了各类作用机制杀菌剂处理下病原菌代谢组中特征性调整的代谢物，这些代谢物为该类作用机制杀菌剂的潜在生物标志物，研究结果将为新杀菌剂作用机制的快速预测识别提供有效方法。

---

* 基金项目：保护地果蔬灰霉病绿色防控技术研究与示范（21173088）
** 第一作者：胡志宏，女，在读博士研究生；E-mail：zhihonghu186@126.com
*** 通讯作者：刘鹏飞，女，副教授，主要从事植物病理学与杀菌剂药理学研究；E-mail：pengfeiliu@cau.edu.cn